CLASSICAL AND QUANTUM MECHANICS

NONLINEAR QUANTUM MECHANICS AND ITS APPLICATIONS

CLASSICAL AND QUANTUM MECHANICS

Additional books in this series can be found on Nova's website under the Series tab.

Additional e-books in this series can be found on Nova's website under the e-book tab.

CLASSICAL AND QUANTUM MECHANICS

NONLINEAR QUANTUM MECHANICS AND ITS APPLICATIONS

PANG XIAO FENG

nova publishers
New York

Library of Congress Cataloging-in-Publication Data

Nonlinear quantum mechanics and its applications / edited by Pang Xiao-Feng (Institute of Life Science and Technology, University of Electronic Science and Technology of China, Sichuan, China) .
pages cm -- (Classical and quantum mechanics)
Includes index.
ISBN 978-1-61122-011-7 (hardcover)
1. Nonlinear theories. 2. Quantum theory. I. Pang, Xiao-Feng, 1945- editor. II. Series: Classical and quantum mechanics series.
QC20.7.N6N6625 2015
530.12--dc23
2014041534

Published by Nova Science Publishers, Inc. † New York

CONTENTS

Preface		**ix**
Chapter 1	**The Wave-Corpuscle Duality of Microscopic Particles and Difficulties of Quantum Mechanics**	**1**
	1.1. The Experimental Evidences of Wave-Corpuscle Duality of Microscopic Particles	1
	1.2. The Difficulties of Quantum Mechanics and Its Limitations	10
	1.3. The Improvements of Description in Quantum Mechanics for Eliminating Its Difficulties	14
	1.4. The Disputations on The Difficulties in Quantum Mechanics	17
	1.5. The Reasons of Difficulty of Linear Quantum Mechanics and Its Direction Eliminating These Difficulties	22
	References	24
Chapter 2	**Proposal and Build of Different Nonlinear Quantum Mechanical Models and Theories**	**29**
	2.1. De Broglie's Idea 25for the Nonlinear Quantum Mechanical Theory	29
	2.2. Weinberg's Model for Nonlinear Quantum Mechanics and Some Comments	31
	2.3. Simple Descriptions of Various Different Nonlinear Quantum Mechanical Ideas and Suggestions	55
	2.4. Simple Survey of Other Ideas of Nonlinear Quantum Mechanics	65
	2.5. Comment on the Correctness of the Various Nonlinear Quantum Mechanical Ideas and Models	72
	2.6. Pang's Nonlinear Quantum Mechanic Theory	73
	References	83
Chapter 3	**The Wave-Corpuscle Duality of Microscopic Particles in Nonlinear Quantum Mechanics and Its Experimental Evidences**	**89**
	3.1. The Wave-Corpuscle Duality of Microscopic Particles Confirmed by the DynamicEquation without Externally Applied Potential	89

3.2. The Wave-Corpuscle Duality of the Solutions of the Dynamic Equation with Different Potentials 93
3.3. The Localized Property of Microscopic Particles in Nonlinear Quantum Mechanics 97
3.4. The Stability of Microscopic Particles with the Wave-Corpuscle Duality 99
3.5. The Corpuscle and Classical Characteristics 105
of Motion of Microscopic Particles 105
3.6. The Conservation Laws of Motion of Microscopic Particles 115
3.7. The Properties of Motion and Collision of Microscopic Particles with Attractive Nonlinear Interactions 121
3.8. The Features of Collisions of Microscopic Particles Under Action ofRepulsiveNonlinear Interaction 128
3.9. The Properties of Propagation of Microscopic Particles at Interfaces of Media 136
3.10. The Properties of QuantumFluctuationin Nonlinear Systems 139
3.11. The Experimental Confirmation of Wave-Corpuscle Duality and Soliton Motion of Microscopic Particles in Nonlinear Quantum Mechanics 147
References 160

Chapter 4 The Differences of Nonlinear Quantum Mechanics from Linear Quantum Mechanics As Well As the Effects and Mechanism of Generation of Nonlinear Interaction 165
4.1. The Variations of Physical Significances and Natures of the Wave Functions of Microscopic Particle 165
4.2. The Changes of Meaning and Property of Mechanical Quantities for the Microscopic Particles 173
4.3. The Changes of Uncertainty Relation for Microscopic Particles in Two Theories 176
4.4. The Variations of Natures of Microscopic Particles Arising from the Nonlinear Interaction 184
4.5. The Alternations of Eigenvalue Problems in Two Theories 187
4.6. The Eigenvalue Problems ofNonlinear SchrödingerEquation 193
4.7. The Variations of Method Solving theDynamic Equations in Two Theories 199
4.8. Distinctions of Feature of Microscopic Particles 210
in Two Theories at MinimumNonlinear Interaction 210
4.9. The Reasons of the Differences between the Two Theories As Well As the Physical Effects of the Nonlinear Interaction 212
4.10. The Real Existences of the Nonlinear Interaction in Physical Systems and Its Different Mechanism of Generation 214
References 229

Chapter 5 **The Nonlinear Quantum Properties of the Exciton and Its Experimental Confirmation in Molecular Crystals** **233**

5.1. The Nonlinear Quantum Properties of the Excitons in Molecular Crystal and Its Self-Trapping Mechanism of the Nonlinear Interaction 234

5.2. The Behaviors of the Nonlinear Excitation of the Exciton in the Case of Anharmonic Vibration of Lattices in Acetanilide 245

5.3. The Raman Scattering Arising from the Motions of the Exciton-Solitons in Molecular Crystal 247

5.4. The Characters of Infrared Absorption of Acetanilide in the Nonlinear Quantum Excitation Case in Acetanilide 255

5.5. Properties of Nonlinear Energy-Spectrum in Molecular Crystals 262

5.6. The Features of Specific Heat of Acetanilide 266

5.7. Experimental Verifications of Nonlinear Quantum Mechanical Property of the Soliton in Acetanilide 269

References 280

Chapter 6 **The Nonlinear Quantum Mechanical Features of Boson-Einstein Condensed States** **283**

6.1. Boson-Einstein Condensation Phenomenon of Microscopic Particles 283

6.2. The Theoretical Model Describing the Boson-Einstein Condensation 285

6.3. Argument of Condensation Feature of the Particles in BEC System 288

6.4. The Essences of Boson-Einstein Condensed State of Microscopic Particles 294

6.5. Experimental Observation of Property of the Boson-Einstein Condensation 300

6.6. The Results of Gross–Pitaevskii Theory of Boson-Einstein Condensation 304

6.7. The Nonlinear Quantum-Mechanical Features of BEC States 308

6.8. Different Kind of Boson-Einstein Condensation and Their Properties 313

References 325

Chapter 7 **Nonlinear Quantum Dynamic Properties of Electrons and Polarons in Atoms and Condensed-State Systems** **329**

7.1. The Nonlinear Quantum Properties of the Electron in Atomic Systems 329

7.2. Nonlinear Quantum Dynamic Properties of Electron in Coupled Electron-Phonon Systems 344

7.3. New State-Ansatz for Microscopic Particles and Corresponding Uncertainty Relationship in Strong Electron-Phonon Coupling Systems 350

7.4. Properties of Nonadiabatic Quantum Fluctuations for Phonons and Electrons in the Strongly Coupled Electron-Phonon Systems 359
7.5. Nonlinear Quantum-Mechanical Properties of Electrons in Superconductors 366
7.6. Nonlinear Quantum-Mechanical Properties of Polarons in Ionic Crystals 382
7.7. The Nonlinear Localization of Small Polarons and Its Properties in Semiconductors 392
7.8. The Theory of Motion of the Polarons and Its Properties in One-Dimensional Molecular Crystals 398
References 401

Chapter 8 Nonlinear Quantum-Mechanical Properties of the Magnons in the Spin-Spin Interaction Systems 407
8.1. The Spin Wave and Its Features in the Spin-Spin Interaction Systems 407
8.2. The Wave Features of the Magnons in the Ferromagnetic Systems 413
8.3. The Wave Features of the Magnons in the Antiferromagnetic Systems 418
8.4. Nonlinear Quantum-Mechanical Properties of Magnons in Ferromagnetic Systems 422
8.5. Nonlinear Quantum Properties of Solitary Magnons in Different Ferromagnetic Chains 429
8.6. The Nonlinear Features of Magnons in Antiferromagnetic Systems 436
References 453

Index 455

PREFACE

This book describesa completenonlinear quantum mechanics, in whichwe fundamentally and, out ofnecessity, built this theory,whichbasicallyconsists of theoretical principles, wave-corpuscle duality of microscopic particles and its experimental evidences, the mechanisms of generation of nonlinear interactions and their effects, the methods solving nonlinear quantum mechanical problems, its distinctions with linear quantum mechanics and early nonlinear quantum mechanical ideas and models, the completeness, correctness and universality of this new theory as well as its applications in different systems containing polymers, physical and biological systems. We have studied and exhibitedinteresting results of these systems and a large number of novel properties of microscopic particles, including the electron, proton, phonon, photon, exciton, polaron, magnon and Boson, involving their localizations and classical features, which were statedin detail. It is intended for researchers, teachers, graduate students, and upper level undergraduate students.

THE BOOK

Nonlinear quantum mechanics (NLQM) is a new theory, which is used to research the properties and movement rulesof microscopic particles (MIPs) in nonlinear systems.Therefore, it isthefoundationof modern science. Obviously, it should be based on linear quantum mechanics, which was established by Bohr, Heisenberg, Schrödinger, and many others, because the latter cannot give the wave-corpuscle duality of microscopic particles, which were confirmed by many experiments. The topic of wave-corpuscle duality of microscopic particles and linear quantum mechanical difficulties, problems of variously different nonlinear quantum mechanical ideas, models and theories, the wave-corpuscle duality of microscopic particles in new nonlinear quantum mechanics and its experimental evidences, differences of nonlinear quantum mechanics with linear quantum mechanics and the mechanism of generation of nonlinear interaction as well as the nonlinear quantum mechanical properties of the exciton in molecular crystals, Boson particles in Boson-Einstein condensed states, electrons and polarons in atoms and electron-phonon systems as well as magnons in spin-spin interaction systems, and their experimental confirmations, are covered in 8 chapters in this book. An in-depth discussion on the wave-corpuscle duality, localization of microscopic particles, different mechanisms of generation of different nonlinear interactions, and much experimental evidence regarding wave-corpuscle of microscopic particles innonlinear systems,are described in detail in this book.

The book is organized as follows. We start witha review on the experimental data of wave-corpuscle duality of microscopic particles and difficulties of linear quantum mechanicsin Chapter 1, which clearly indicated the necessity to establish nonlinear quantum mechanics. In Chapter 2, about 30 different nonlinear quantum mechanical ideas, models, and theories, including Weinberg's model and their properties, are briefly stated and reviewed. This also exhibited the necessity of building new nonlinear quantum mechanics because there are a lot of questions. In this case, the new nonlinear quantum mechanics proposed by Pang including the theoretical, mathematical, and experimental foundations and its characteristics are depicted in Section 2.6. In Chapter 3, we examinein detail the wave-corpuscle duality of microscopic particles and its experimental confirmations, includingthe wave-corpuscle duality of the solutions of different nonlinear Schrödinger equations, their dynamic features obtained by both nonlinear Schrödinger equations and the classical Euler-Lagrange and Hamilton equations and Newton's motion equation, the determinations of the mass, momentum and energyof the particles and corresponding conservation laws of mass, momentum and energy, which display the corpuscle feature of the particles, are depicted completely. Except for the corpuscle feature, they have also the wave feature because the solutions of the nonlinear Schrödinger equations are composed of a bell-type soliton and a carrier wave.There is a superposition of waves in their collisions and the reflection and transmission occur in the scatteringon the interface between two mediums, but their wave feature differs not only by the KdV solitary wave but the linear wave. Finally, we used the experimental results of photon soliton in optic fiber and water soliton in water, which are described by the same dynamic equation, as well as the wave-corpuscle duality of microscopic particles, depicted in Chapter 1, to confirm the validity of the wave-corpuscle duality of microscopic particles depicted by the new NLQM. These problems are also stated in Chapter 3. Therefore, the properties of the microscopic particle are completely different from those in linear quantum mechanics. Their distinctions are stated in detail and systematically in Chapter 4, which describes the significance and properties of wave function and mechanical quantities, the solutions of dynamic equations, the uncertainty relation of mechanic qualities and methods solving dynamic equations, and so on.The reasons causing these distinctions are also given.They are due to the nonlinear interactions in the dynamic equations, where these nonlinear interactions are produced from five mechanisms of self-interaction, self-condensation, self-focus, self–trapping, and self-localization in different systems. The latter are due to the interactions, among the moved particles or between moved particles, and the moved background fields such as lattice in solids and nucleons in atoms, or the special structural features of the systems, such as anisotropy, double-well, or many-well potentials and nonlinear features of the structure. These results sufficiently display the truth and correctness of the new NLQM and approximation of linear quantum mechanics(LQM). In Chapter 5-8, we depicted in detail the applications of a new nonlinear theory and gave further the nonlinear quantum mechanical properties of the proton, electron, phonon, photon, exciton, polaron, magnon, and Boson in the molecular crystals, atoms, Boson-Einstein condensed systems, electron-phonon coupling systems, superconductor, ion-crystal, semiconductor, ferromagnetic and antiferromagnetic systems, using the new NLQM theory and methods, respectively. A large number of interesting results are obtained. These investigations exhibited the correctness and practicability of the new nonlinear theory.

This book is a summary of results of my own research, as the author, on problems of nonlinear quantum mechanics over 30 years. A great number of articles,published in journals at home and abroad, as well as lecture notes for my1980s graduate course, "Problems in nonlinear quantum theory", and my book entitled "Theory of Nonlinear Quantum Mechanics" (in Chinese) in 1994 and "nonlinear quantum mechanics"(in Chinese) in 2009. However, the contents were completely rewritten in this English edition. In the process we also incorporated some results related to the nonlinear Schrödinger equations and the early nonlinear quantum mechanics by other scientists and researchers.

THE OVERVIEW

Looking back at the modern history of physics, we understand that quantum mechanics established by Bohr,Heidelberg, Schrödinger, etc., in the 1900's, is the foundation of modern science. It had great successesin solving some physical problems, such as the light spectra of hydrogen and hydrogen-like atoms, the Lamb shift in these atoms, and so on. However, it was known that LQM is a linear theory and has plenty of difficulties. A basic difficulty is that it gave only the wave feature of microscopic particles, which cannot describe their wave-corpuscle duality. However, they were confirmed by many experiments, thus some contradictions between the theory and results occur in LQM. In this case, different opinions on how to resolve such issues and further develop quantum mechanics lead to intense arguments and debates, which lasted more than a century. The long-time controversy between Bohr and Einstein showed that these problems couldn't be solved within the framework of LQM. Therefore, it was necessary to develop the quantum mechanics.

However, how does the quantum mechanics develop? This is a controversial problem. Plenty of researchers thought that the direction of development was to establish nonlinear quantum mechanics. Thus, a large number of nonlinear quantum mechanical ideas, suggestions, and about 30 models including Weinberg's model, were proposed from 1927 to 2008. Clearly, the criterions judged the correctness of the NLQM models as to whether or not they could give the wave-corpuscle duality of microscopic particles. In accordance with this criterion, these NLQM models were not successful. Most of the NLQM and their properties were reviewed and commented briefly upon in Chapter 2 in this book. This implies that it was necessary to establish a correct NLQM, which we chose to be our task and responsibility.

We thought that the correct NLQM should be established based on LQM, but its basic hypotheses of independences of Hamiltonian of the systems on the wave functions of microscopic particles and the linearity must be broken down for establishment of new NLQM.

Based on the experimental properties of macroscopic quantum effects, which is a nonlinear quantum effect, and the modern theories of superconductivity, and superfluids, and their experimental properties as well as soliton theory, we established the fundamental principles and theory of new NLQM in 1985-1993, described in Section 2.6. We also studied the properties of microscopic particles and all related problems of the new nonlinear theory by using these fundamental principles and theory, in which the wave-corpuscle duality of microscopic particles, the validity, self-consistence, practicability and universality of the nonlinear theory were obtained. Its applications in variously different systems have already acquired great success.The nonlinear quantum mechanical features of the particles were further confirmed by many experiments. We estimated that about 100 papers on this topic were published in journals, at home and abroad. These results were described systematically and completely in this book.

The author would like to acknowledge the National Basic Research Program of China("973" program) for their financial support (GrantNo. 2011CB503701).

Pang Xiao-Feng
July 31,2014
University of Electronic Science and Technology of China, Chengdu
610054, Sichuan, China
Email: pangxf2006@aliyan.com

Chapter 1

THE WAVE-CORPUSCLE DUALITY OF MICROSCOPIC PARTICLES AND DIFFICULTIES OF QUANTUM MECHANICS

From the diffraction of electronsin nickel by Davisson and Gerner, diffractions of electrons and neutrons in multi-crystals by Thomson et al., the diffraction of the He atom beam on LiF, and Young's slit interference experimental results of the proton, electron, and photon, as well as Planck's radiationof blackbody, the photon-electric effect, and Compton scattering effect of the photon and electron, etc., we affirmed that microscopic particles have both a wave and corpuscle features. However, linear quantum mechanics can give only the wave feature, but not the corpuscle feature of microscopic particle, no matter how the potential is changed. This is its intrinsic difficulty, which cannot be eliminated in its framework, although these problems were studied and disputed for more a century. This means that quantum mechanics must be further developed.

1.1. THEEXPERIMENTAL EVIDENCES OF WAVE-CORPUSCLE DUALITY OF MICROSCOPIC PARTICLES

It is well known that microscopic particles have a wave-corpuscle duality, which was confirmed by many experiments, such as the diffractions of electrons on surfaces of Ni-mono-crystals by Davisson and Germer, in 1927, and the electron beams with de Broglie wavelength of 0.00645nm, which penetrated over the multi-crystal powders and Ag-sheet metal by Thomson and Taltakovsky, as well as the diffractions and inflections of the beams of neutrons and He and H_2 atoms, which penetrated or passed LiF crystals at $T=295^0C$ by Stern, Gerlach and Easterman [1-18], and Young's single and double-slit interference of the wave character of photons by Tuan and Gerjuoy et al., [19-20], respectively. These experiments and their results were described in detail books of quantum mechanics. Based on these experimental results, Bohr, Born, Schrödinger and Heisenberg, etc., established quantum mechanicsin the early 1900s [21-32],which provided the foundation of modern science. These experimental results and their analyses are described briefly as follows.

In Davisson and Gerner's experiment, they measured the diffraction of the electrons on nickel crystal[33].Their results are shown in Figure 1.1; they injected the electronic beam, which was launched by an electron gun, to the surface of this crystal along its vertical direction. In this experiment, they observed the reflect and phenomenon of the electrons from the surface of the crystal. If ϑ is used to denote the reflected angle, the wavelength of the electron is represented by λ, which is related to the electric-potential V accelerating the electronic beam through the relationship of $\lambda = \sqrt{(150/V)}\ \overset{0}{A}$, then this phenomenon can be described by formula: $n\lambda = d\sin\vartheta$, where n is the order number of maximum diffraction, d is a constant of the surface grid of the crystal. This relation is similar to the diffraction effect of an X ray. Therefore, it shows that the electrons have a wave feature. From this experiment Davisson and Gerner obtained the De Broglie relationship of $\vec{P} = (h/\lambda)\vec{n}$. Thus, this experiment indicated the wave-corpuscle duality of the electrons.

Thomson [34-36] also observed the diffraction of the electrons on the multi-crystal, in which the electronic beam was penetrated over the multi-crystal film with the thickness of 10^{-5}cm. This experimental theorem is shown in Figure 1.2. They also obtained $n\lambda = 2d\sin\vartheta'$ from this experiment, where n is also the order number of maximum diffraction, d is the space constant of the surface grid of the crystal, and ϑ' is the angle between the electronic beam and grid plane of the multi-crystal. At the same time, they observed the diffraction rings shown in Figure 1.3, which are similar with the diffraction of X -ray, where the electronic beam was accelerated by the electric-potential of $36\text{x}10^3$eV, which corresponds to the De Broglie wavelengths of 0.00645nm.

Stern and Fasterman also inspected the reflect and phenomena of the He atom and the H_2 molecule on the surface of the LiF crystal [11,18], in which the energies and corresponding De Broglie wavelength of these corpuscles were changed gradually by modulating the electric-potential.A sensitive barometer recorded the scattering strengths of these corpuscles on the crystal.

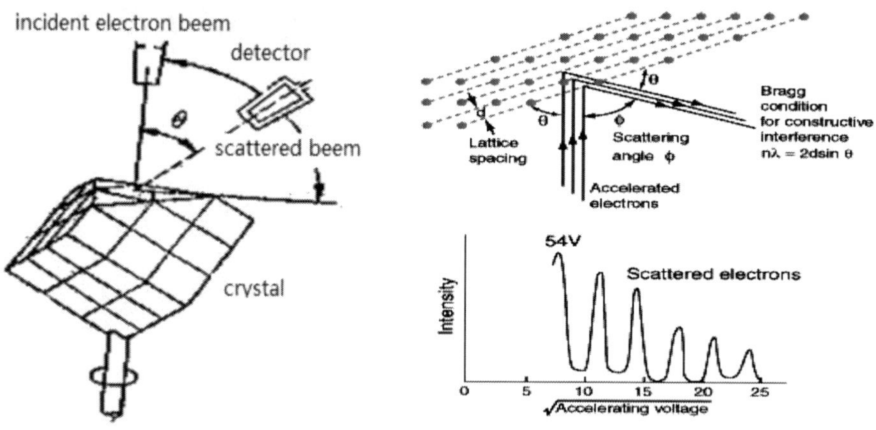

(a) (b)

Figure 1.1.The diffraction of electrons in nickel, (a) is experimental device, (b) is the result[18, 33].

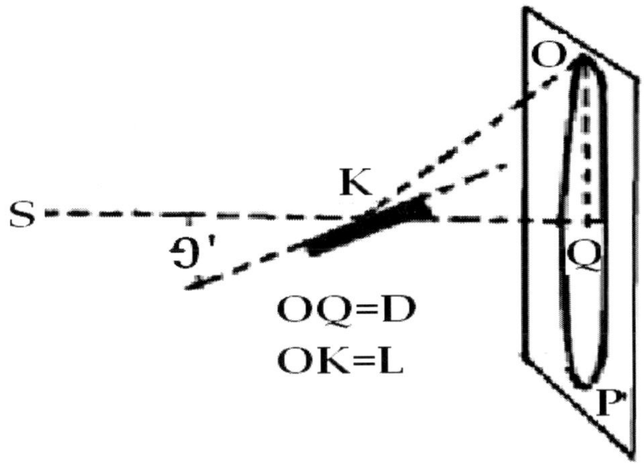

Figure 1.2. Diffraction of electrons in multi-crystal [18,34-36].

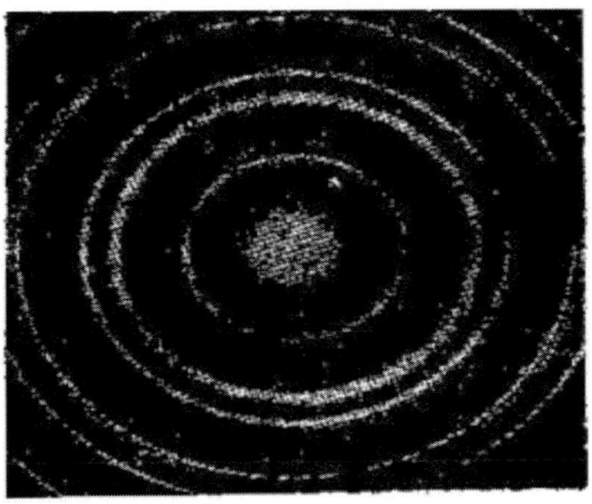

Figure 1.3. The image of diffraction of electronsin multi-crystal [11]

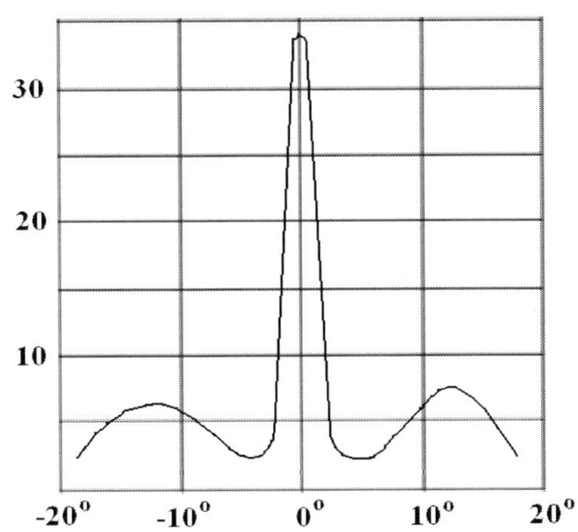

Figure 1.4. The distribution of reflected strength of He atom beam on LiF at $T=295^0C$ [Ref.11].

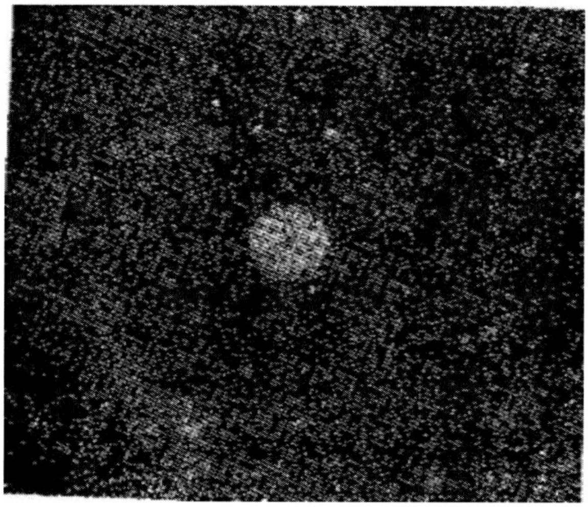

Figure 1.5. The diffraction of neutrons (Laue diffraction) at NaCl crystal [Ref.11].

The result of distribution of reflected strength of the He atom beam on LiF crystal at $T=295^0C$ was shown in Figure 1.4, where the result at the reflective angle 0^0 corresponds to the normal reflectance of the He atom. Its strength is maximal, but two small peaks at the reflective angles of 10^0 and -10^0 occur also due to the diffraction of the He atom. The analogous results for H_2 molecule were also attributed to them.

On the other hand, the diffraction of the neutrons has been observed also. Owing to the fact that the mass of the neutron is $m_0 =1.66x10^{-24}g$, then the neutrons with energy of 0.01 eV can generate a stronger diffraction resembling that of the X-ray mentioned above. The diffraction effect of neutrons (Laue diffraction) on the NaCl crystal is shown in Figure 1.5 [11,18].

It is known that Young's slit interference experimental results of proton, electron and photon and their double-center diffraction images can indicate the wave feature of these microscopic particles. In fact, Tuan and Gerjuoy in 1960 [37] discussed and observed an atomic version of Young's double slit experiment for matter waves in ion-atom collisions. They also debated the capture processes in collisions of protons with H_2 and predicted that the diffraction of the protons from the two atomic centers of the molecule can lead to interference effects, in which the captured cross-sections are related to the molecular orientation, which may be due to the interference effects [38–41].Schmidtet al., [42] observed the interference effect caused by the pronounced matter wave in multiple differential momentum spectra of recoil ions produced in dissociative capture in $H_2^+ + He$ collisions. However, the interference effects in ionization processes are quite difficult to observe [43] because the final state of the collision involves three unbound particles, opposed to only two in a capture process, where the phase angle between the amplitudes associated to the two molecular centers is accessible through both the momentum information of the collision and the molecular orientation. Data for ionization, in which the phase angle was determined, are only available for electron impact [44]. Forion impact, interference in the projectile diffraction was observed for fixed projectile energy losses in the scatterin gangle dependence [45, 46]. The interference structures were also discussed in double differential energy distributions of the ejected electrons in fixed emission angles [47–53]. Two-center interference should also occur in the ion impact induced excitation of a molecule. Zhang et al., [54] recently studied the diffraction process, using an indirect approach, by investigating simultaneous excitation of H_2^+ projectile ions, resulting in dissociation, and target ionization, in collision with He, i.e. $H_2^+ + He \to H^+ + H + He + e$, in a kinematically complete experiment providing the fullphase information. They observed the interference structures due to scattering of helium matter waves from the two atomic centers in the H_2^+ ions. Two different interaction channels were identified, the first one is based on the electron-electron interaction,the second one includes two independent electron-nuclear interactions, the molecule in the two-electron process may be thought of as a double slit in two different ways, which are reflected by different interference patterns, depending on which of these two channels is selected. In the first case, the double slit can be denoted by the two-center electronic wave function and in the second case by the additional two nuclei of the molecule.The experiment was finished at the van de Graaff accelerator of the Max Planck Institute for Nuclear Physicsin Heidelberg.

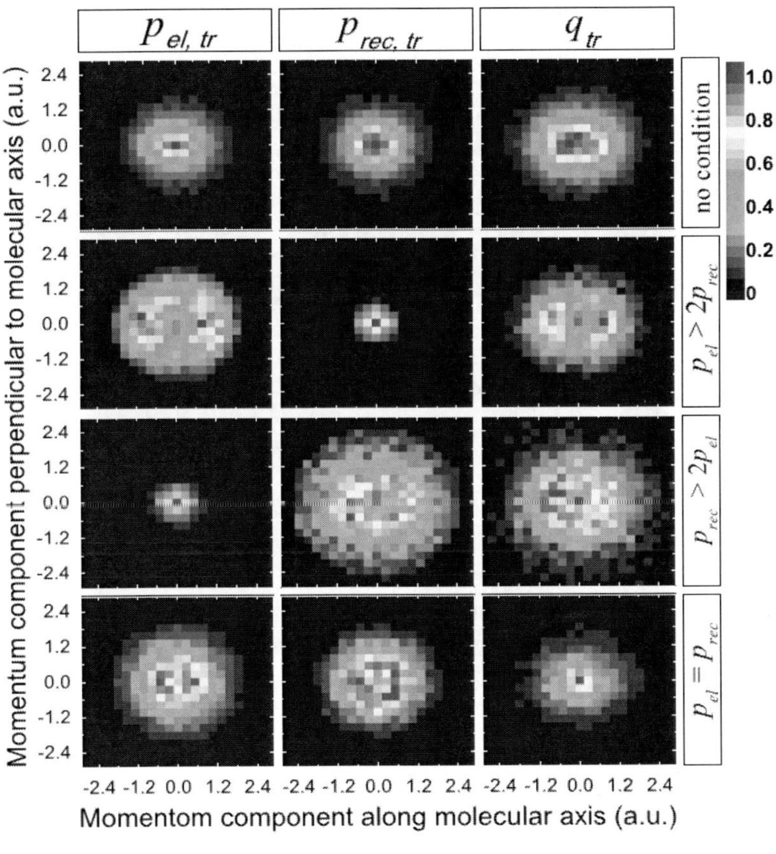

Figure 1.6. (color online). The two-dimensional momentum distributions in the x-y plane of the molecular frame [Ref.54].

A beam of 1 MeV H_2^+ molecular ions was intersected with a supersonic helium gas jet with a temperature of less than 2 K. The momenta of the recoiling heliumions and the ejected low-energy electrons were measured with a conventional reaction microscope[55]. From this experiment, they obtained a dependence of the differential cross-sections on the orientation of molecular ion. Compared to earlier studies onYoung-type interference in atomic collisions, the present collision system features more degrees of freedom, because it represents an effective five-body system, consisting of the two active electrons, the two protons in the molecule, and the Hecore. In the perturbative description, different first- and higher-order terms contribute to the cross section, a correlated first-order mechanism, and an independent higher-order process. This results in a blurring of oscillating patterns in the integrated cross sections. However, in spite of this rather complex collision system, the two contributing mechanisms can be separated by choosing appropriate kinematic conditions revealing clear interference structures. The observed features agree with of two center interference effects in the first and second order perturbation theory. Their experimental results were shown in Figures 1.6-1.7, respectively.

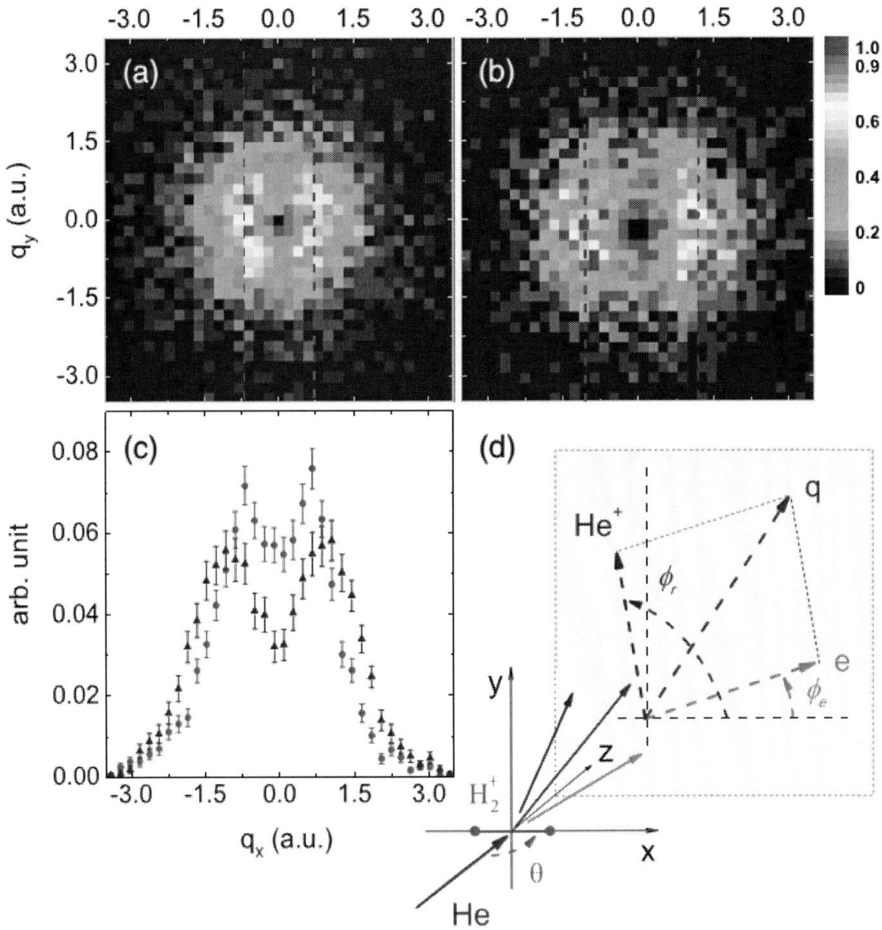

Figure 1.7. (color online). The two-dimensional momentum transfer distributions in the transverse plane under different conditionsof internuclear distances [Ref.54].

The two-dimensional momentum distributions in the x-yplane of the molecular frameis shown in Figure 1.6, where the columns from left toright are electron momentum ($P_{el;tr}$), recoil-ion momentum ($P_{rec;tr}$),and momentum transferred from the molecular ion to the target atom (q_{tr}), respectively. Also, no indication of interferences was obtained when no additional condition was applied. When $P_{el} > 2P_{rec}$, the interferences in both the electron and the momentum transfer spectra were obtained, whereas for $P_{rec} > 2P_{el}$, the interferences in both the recoil and the momentum transfer spectra were observed. When $P_{el} = P_{rec}$ (i.e., $0.9P_{rec} \leq P_{el} \leq 1.1P_{rec}$), no clear indication for the interference effect can be found.

Figure.1.7 showed the two-dimensional momentum transfer distributions in the transverse plane under different conditions of internuclear distances, where the molecular axis is perpendicular to the beam direction: (a) KER<2.5eV; (b) KER>8eV; (c) their q_x distributions: filled circle for (a), and filled upward triangle for (b). (d) Sketch of the molecular frame and corresponding definitions [54]. These results exhibit clearly the wave feature of these particles.

The above results shown in Figures 1.1-1.7 indicated clearly that the microscope particles including the electron, photon and neutron all have a wave feature, no matter what the nature of these particles. This means that wave feature is a basic nature of microscopic particles. However, the microscopic particles have also a corpuscle features. This statement can be confirmed by the following experimental results.

(1) The microscopic particlesall have certain sizes and masses, for example, the masses of electron, proton, and neutrons are $9.10953\text{x}10^{-31}$kg, $1.67265\text{x}10^{-27}$kg, and $1.67495\text{x}10^{-27}$kg, respectively, the electron's classical radius is 2.817938Fm.

(2) Planck's radiation theory of black-body affirmed the corpuscle feature of microscopic particles [1-18]. So-called black-body is that the body can absorb completely all incident lights without any reflection. Planck proposed the concept of light quantum (or photon quantum) from this experiment and thought that the black-body is composed of a large number of resonant-oscillators with certain charges and quantized energies of $\varepsilon_0, 2\varepsilon_0, 3\varepsilon_0......n\varepsilon_0...$. Based on these hypothesizes he derived the formula of distribution of the energy or photon quanta radiated by the black-body, which is represented by

$$\rho_v dv = (8\pi h v^2 / c^2)\{\exp[(hv / KT) - 1]\}^{-1} dv \qquad (1.1)$$

from v to $v + dv$, where $\varepsilon_0 = hv$ is a minimal energy, which is completely consistent with the experimental results. This verified that the photon quanta in the radiation of the black-body has acorpuscle feature, its energy is quantized, i.e., $E = nhv$, n is a integer.

In accordance with Planck's theory, the specific heat of solid can be found, which we regarded as that the atoms in the solid are some resonant oscillators (or phonons). Then the average energy of the atom can be denoted by $E = hv\{\exp[(hv / KT) - 1]\}^{-1}$. Thus, the specific heat can be expressed by $C = dE / dT = (hv)^2[\exp[(hv / KT)]\{\exp[(hv / KT) - 1]\}^{-1}$. Hence, he obtained that the specific heat of the solid is k at high temperate $(hv / KT << 1)$ and zero at low temperate $(hv / KT >> 1)$, respectively. This result approaches the experimental value. This means that the oscillators or phonons in the solids also have a corpuscle feature.

(3)Einstien explained the photon-electric effect occurred in metal using Planck's concept of photon quantum [1-18]. In this phenomenon, Einstein thought that the photons with the energy of $nh\nu$ are absorbed by the electrons in the metal, in which a part of energy absorbed by the escaped electrons from the photons is expended to overcome the attraction force of the surface of metal on it, but the residual energies are transformed as its kinetic energy of motion. Thus the change of the escaped electrons can be denoted by $(1/2)mv^2 = h\nu - W_0$ in this process, where m and v are the mass and moved the velocity of the electron, respectively, W_0 is the function of the electron, which is used to escape from the metal. The above result manifested that the velocity or energy of escaped electrons depends only on the frequency of the photons absorbed, instead of the intensity of the light. This is concordant with the experimental data. Therefore, this experimental result affirmed again the corpuscle features of both the electron and photon.

On the other hand, from the relationship of the energy-momentum in the relativity theory, $E^2 = m^2c^2 + c^2P^2$, we can obtain the relationship of the energy-momentum of the photon with light velocity c is $E = cP$. Thus, we can gain the Broglie relationship:

$$E = h\nu = \hbar\omega, \vec{P} = \frac{h\nu}{c}\vec{n} = \frac{h}{\lambda}\vec{n} = \hbar\vec{k} \tag{1.2}$$

This representation indicated clearly that the photons possess both corpuscle features, which is embodied by E and $\vec{p}$, and the wave feature, which is embodied by $\vec{k}$ and λ. In fact, the Broglie relationship is appropriate for all microscopic particles.

(4). The Compton scattering effect of the photon and electron, which was observed by Compton and Wu You-xun [1-18], respectively. In this experiment, the X-ray with high frequency is scattered by the electrons in matter, where the wavelength of the X -ray increases with increasing the scattering angle. The phenomenon is, in essence, due to the collisions of the photon with the electron as is shown in Figure 1.6. Based on this idea the conservation relations of the energy and momentum in the collision process can be discovered, they are expressed by

$$\hbar\omega = \hbar\omega' + mc^2[\frac{1}{\sqrt{1-\beta^2}} - 1], \tag{1.3}$$

$$\hbar\omega / c = \hbar\omega' con\vartheta / c + con\vartheta'[\frac{mv}{\sqrt{1-\beta^2}}], \tag{1.4}$$

$$0 = \hbar\omega' \sin\vartheta / c - \sin\vartheta'[\frac{mv}{\sqrt{1-\beta^2}}], \tag{1.5}$$

respectively, where $\beta = v/c$, Eq.(1.3) marks the energy conservation, Eqs.(1.4) and (1.5) represent the momentum conservations along the OA direction and the vertical direction of OA in Figure 1.6, respectively, where ϑ and ϑ' are the angle between OB and OA as well as

OC and OA, respectively. From Eqs.(1.3)-(1.5) we can obtain the following relationships of change of frequency or wavelength of the photon in this scattering process.

$$\omega-\omega'=\omega\omega'\frac{2\hbar}{mc^2}\sin^2(\vartheta'/2) \quad \text{or} \quad \Delta\lambda=\lambda'-\lambda=\frac{4\pi\hbar}{mc}\sin^2(\vartheta'/2) \tag{1.6}$$

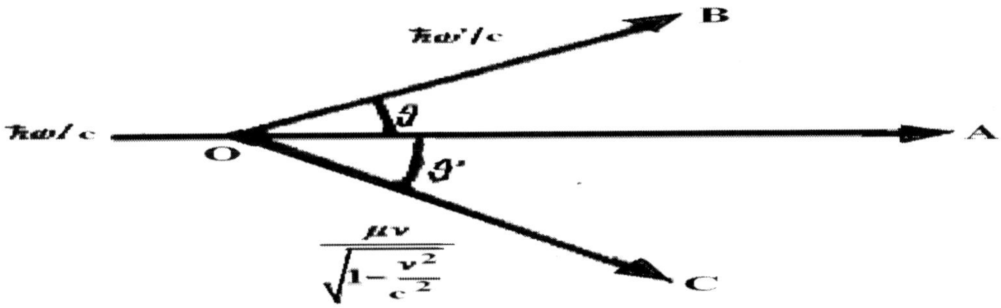

Figure 1.6.The scheme of the Compton scattering effect of the photon on the electron [Ref.18].

(5).The quantum theory of the light spectrum of hydrogen atom, which was obtained by Bohr [21-23]. In this phenomenon, Bohr assumed the electrons with quantum feature move along the orbit's circumference around the nucleon. In this case the quantization condition of the circle orbits is denoted by $\oint pdq = nh$, or $\int_0^{2\pi} p_\theta d\theta = nh$, where n is the quantum number, and p and q or p_θ and θ are the generalized momentum and coordinate, respectively. The energy of the electron in motion of circumference of the orbitscan be represented by

$$E=T+U=e^2/2a-e^2/a=-e^2/2a=-me^4/2n^2\hbar^2,(n=1,2,3,....)$$

Therefore, the energy of hydrogen atom is quantized. Bohr thought further that the light spectrum of the hydrogen atom is due only to the transitions of the electron between the above energy levels, and then the frequencies of photons radiated in the hydrogen atom can be expressed by

$$\nu=\frac{2m\pi^2e^4}{h^3}\left(\frac{1}{m^2}-\frac{1}{n^2}\right) \tag{1.7}$$

Utilizing Eq.(1.7) the light spectrum of the hydrogen atom can be completely explained, thus we confirmed again the corpuscle feature of the electron.

(6). In a modern technique, the different accelerators, such as the electron accelerator and proton accelerator, were widely used in industry, medicine, and science research. Very clearly, these accelerators accelerated these particles, instead of the waves. Then, we should affirm the corpuscle feature of the microscopic particles from these experiments.

The above results studied show clearly that all microscopic particles, including electron, proton, neutron, photon, hydrogen atom and He molecule,all possessboth corpuscle and wave features,therefore they have wave-corpuscle duality[1-18]. Just so, the natures of microscopic particles were represented in the well-known De Broglie relationship of $\oint pdq = nh$, or $\int_0^{2\pi} p_\theta d\theta = nh$ [10-18]. Thus we have to represent the states and properties of microscopic particles by a wave function, $\psi(\vec{r},t)$ or $\phi(\vec{r},t)$, where $\vec{r}$ denoted the positionof the particle at time t. This means that $\phi(\vec{r},t)$ embodies both its wave feature and corpuscle feature. Therefore, we can affirm and conform that the microscopic particles should have a wave-corpuscle duality. Then we must establish a correct theory, which can give and delineate the wave-corpuscle duality of microscopic particles, or else, this theory is not correct, or it must be improved further. The quantum mechanics established by Bohr [21-23] is like this. The reasons and results are described in the following sections.

1.2. The Difficulties of Quantum Mechanics and Its Limitations

Quantum mechanics cannot give the above wave-corpuscle duality of microscopic particles at all. Why? The reasons are described in detail as follows.

At the end of the 19 th century, it was known that classical mechanics encountered many difficultiesin studying the motions of microscopic particles with very small masses(10^{-25}-10^{-31}kg) and high velocities, which constitute macroscopic matters, or speaking, it cannot also explain the physical phenomena caused by the motions of the microscopic particles. Thus, Bohr [21-23], de Broglie [32] and others [24-31] proposed boldly that the states of the microscopic particles should be represented by a wave function. On the basis of this revolutionary idea, and some fundamental hypotheses, Bohr, Broglie, Schrödinger, Heisenberg, Born, Born, and Dirac et al. [21-32], established quantum mechanics to describe the states and properties of microscopic particles. This theory promoted the developments of modern science including physics, chemistry, and biology. In such case, the quantum mechanics (here called linear quantum mechanism)becomes the foundation and pillar of modern science. In this theory,the states and properties of microscopic particles are described by the Schrödinger equation (here called linear Schrödinger equation) for the wave function, $\phi(\vec{r},t)$, which is used to express the states of microscopic particles, this equation is the form of

$$i\hbar\frac{\partial\phi(\vec{r},t)}{\partial t} = -\frac{\hbar^2}{2m}\nabla^2\phi(\vec{r},t) + V(\vec{r},t)\phi(\vec{r},t) \tag{1.8}$$

where $\hbar^2\nabla^2/2m$ is the kinetic energy operator, $V(\vec{r},t)$ is the externally applied potential operator, m is the mass of the particles, $\vec{r}$ is the coordinate or position of the particle, and t is the time.This theory tells us that once the externally applied potential field and initial states of the microscopic particles are first known or given, then the states and properties of the microscopic particles at any time later and any position can be determined by the linear

Schrödinger equation (1.8). This is a great creation; it promotes the development of science. From this theory we can very easily find and obtain the movement rules of microscopic particles at any times and positions using Eq.(1.8).

For example, at $V(\vec{r},t)$ =0, the solution of Eq.(1.8) is a plane wave, which is represented by

$$\phi(\vec{r},t) = A'\exp[i(\vec{k}\cdot\vec{r}-\omega t)] \tag{1.9}$$

where k, ω, A' and are the wave vector, frequency, and amplitude of the wave, respectively.The particle has an eigen energy, which is as follows

$$E=\frac{p^2}{2m}=\frac{1}{2m}(p_x^2+p_y^2+p_z^2),(-\infty<p_x,p_y,p_y<\infty) \tag{1.10}$$

This is a continuous spectrum of the energy of the particle. The wave function in Eq.(1.9) states that the probability of the particle to appear at any point in space is the same. This implies that the microscopic particle described by Eq.(1.9) propagates freely in a wave and distributes uniformly in total space, then the microscopic particle cannot be localized and has nothing about the corpuscle feature.

If a free particle can be confined in a small finite space, such as, a rectangular box of dimensions a, b and c, the solution of Eq. (1.9) is standing waves as follows:

$$\phi(x,y,z,t) = A\sin\left(\frac{n_1\pi x}{a}\right)\sin\left(\frac{n_2\pi y}{b}\right)\sin\left(\frac{n_3\pi z}{c}\right)e^{-iEt/\hbar}$$

where n_1, n_2 and n_3 are three integers.In this case, the particle is still not localized, it appears also at each point in the box with a determinant probability.In this case, the eigen energy of the particle is quantized as follows:

$$E=\frac{\pi^2\hbar^2}{2m}\left(\frac{n_1^2}{a^2}+\frac{n_2^2}{b^2}+\frac{n_3^2}{c^2}\right) \tag{1.11}$$

The corresponding momentum is also quantized. This means that the wave feature of the microscopic particle has been changed because of the variation of its own boundary condition.

If the potential field is further varied, for example, the microscopic particle is subject to a conservative time-independent field, $V(\vec{r},t)=V(\vec{r})\neq 0$, then the microscopic particle satisfies the time-independent linear Schrödinger equation:

$$-\frac{\hbar^2}{2m}\nabla^2\phi'+V\left(\vec{r}\right)\phi'=E\phi' \tag{1.12}$$

where $\phi\left(\vec{r},t\right)=\phi'\left(\vec{r}\right)e^{-iEt/\hbar}$.

When $V=\vec{F}\cdot\vec{r}$, here $\vec{F}$ is a constant field force, such as a one dimensional uniform electric field E', then $V(x)=-eE'x$, thus its solution is

$$\phi'=A\sqrt{\xi}H^{(1)}_{1/2}\left(\frac{2}{3}\overline{\xi}^{3/2}\right),\left(\overline{\xi}=\frac{x}{l}+\overline{\lambda}\right) \tag{1.13}$$

where$H^{(1)}(x)$ is the first kind of Hankel function, A is a normalized constant, l is the characteristic length, and is a dimensionless quantity. The solution remains a dispersed wave. When $\xi\to\infty$, it approaches

$$\phi'(\xi)=A'\xi^{-1/4}e^{-2\xi^{3/2}/3}, \tag{1.14}$$

which is a damped wave.

If $V(x) = ax^2$, the Eigen wave function and eigen energy are

$$\phi'(x)=N_n e^{-a^2x^2/2}H_n(\alpha x) \tag{1.15}$$

and

$$E_n=(n+\frac{1}{2})\hbar\omega,\ (n=0,1,2,\ldots) \tag{1.16}$$

respectively, here $H_n(\alpha x)$ is the Hermite polynomial. The solution obviously has a decaying feature.

For the hydrogen atom, the potential is the Coulomb interaction between the electron and nucleon, which is denoted in $V\left(\vec{r}\right)=-e^2/\vec{r}$, which $\vec{r}$ is the position of the electron related to the origin of the coordinate system or the nucleon. Thus, the time-independent Schrödinger equation in Eq.(1.12) or eigen equation of the electron in the hydrogen atom is represented by

$$-\frac{\hbar^2}{2m}\nabla^2\phi'+\frac{e^2}{\vec{r}}\phi'=E\phi' \tag{1.17}$$

Clearly, its solution is denoted in a product of associated Legendre polynomial $R_{nl}(r)$and spherical function $Y(\theta,\vartheta)$ [21-23], i.e.,

$$\phi'(r,\theta,\vartheta) = R(r)Y_{m'l}(\theta,\vartheta) = N_{lm'}R_{nl}(r)P_l^{m'}(\cos\theta)e^{im'\vartheta} \quad (1.18)$$

where $Y_{m'l}(\theta,\vartheta) = P_l^{m'}(\cos\theta)e^{-im'\vartheta}$, $P_l^{m'}(\cos\theta)$ is the associated Laguedre polynomial, $N_{lm'}$ is the normalized coefficient, θ and ϑ are the angles in a spherical coordinate, n, l and m' are main, trajectory and magnetic quantum numbers of the electron, respectively. The eigen energy E of the electron is determined by the boundary condition of the wave function, $\phi'(r,\theta,\vartheta)$,(i.e., we demand that $R(r)$ is finite at $r \to 0$, thus the series solution,$R(r)$, becomes as a polynomial $R_{nl}(r)$, in this case the eenergy of the electron is denoted by

$$E= - me^4/2\hbar^2 n^2,(n=1,2,3......) \quad (1.19)$$

Therefore, the electron in the hydrogen atom is distributed in accordance with the energy levels.

Obviously, the solution in Eq.(1.18) is a polynomial.From the images of fore-terms in the polynomials,$R_{ml}(r)$ and $P_l^{m'}(\cos\theta)$, we see that $R_{ml}(r)$ containing exp $[-r/2a_0]$ is only a decaying function, where $a_0 = \hbar^2/me^2$is the Bohr radius, with some sine and cosine functions. Then the electron in the hydrogen atom described by Eq.(1.17) is still a wave having a dispersive and decaying feature [18,57-58].Therefore, the electron cannot be localized and has no determinant position.

If the potential fields are successively varied, we can also find that the wave nature of the solutions in Eq.(1.8) does not change no matter what the forms of interaction potential. This shows clearly that the wave nature of the microscopic particles is intrinsic in quantum mechanics.Therefore, we affirm from the above results that quantum mechanics can only give the wave feature of microscopic particles, but cannot describe their corpuscle feature at all [57-58]. However, the electron and other particles really have a wave-corpuscle duality as mentioned in Section 1.1. This is just a basic difficulty of linear quantum mechanics.

Otherwise, if the above results in the hydrogen atom are carefully analyzed then we obtained a contradictory conclusion of the researched results with first assumptions. Why?It is well known that the Coulomb potential represents the interaction between two point charges with determinant positions in macroscopic world, which in this instance is $\vec{r}$. The Coulomb potential, $V= -e^2/\vec{r}$, is now used in the dynamic equation(1.17) of the electron in the hydrogen atom .We have already stated that the electron and nucleon are all particles with corpuscle features, they all have determinant positions, their relative position is denoted by $\vec{r}$ in dynamic equation (1.17) and the wave function $\phi(\vec{r})$ in Eq.(1.18). However, we finally obtained the solution in Eq.(1.18) from the dynamic equation(1.17), which manifested that the electron possesses only a wave feature, instead of a particle with corpuscle features in the hydrogen atom. Thus, we affirmed that the first assumption and the final result are contradictory to each other in the hydrogen atom. In fact, the contradiction between the first assumption and the final resultalways occurs in all physical systems, only if the linear quantum mechanics are used to study the properties and natures of the microscopic particles in them. This means that the contradiction between the first assumption and final results is intrinsic for linear quantum mechanics, it cannot be overcome and changed in the linear quantum mechanical framework. Obviously, this contradiction is due to the linear Schrödinger equation (1.17) because it only has a wave solution, it is not the solution with wave-corpuscle duality. This indicates clearly that linear quantum mechanics cannot be used to correctly describe the nature of microscopic particles, such as the electron. This exhibits again the difficulty of linear quantum mechanics. Therefore, it is necessary to improve and develop the theory of linear quantum mechanics.

On the other hand, it is also quite difficult to correctly explain the light spectrum of hydrogen ation by using Eq. (1.19), although Eq.(1.7) can be obtained from it. This is because the electron described by Eq.(1.19) is now a wave, and so it cannot carry out the transition from one energy level to another in the hydrogen atom.Thus, the solution in Eq.(1.18) cannot be used to elucidate the luminescence phenomenon of the hydrogen atom. This is another difficulty of linear quantum mechanics.

However, the above difficulties and contradictions can be solved in nonlinear quantum mechanics, which will be described in chapters 3-4.

1.3. The Improvements of Description in Quantum Mechanics for Eliminating Its Difficulties

1.3.1. Born's Statistic Explanation of Wave Function Cannot Affirm the Wave-Corpuscle Duality of Microscopic Particles

In order to make up for the above difficulties of linear quantum mechanics, plenty of men and women improved the described manners ofmicroscopic particles and expressed again the significances of the wave functions. For example, Born proposed the statistic explanation of wave function of the particles, namely he used the square of the absolute value of wave function to represent the probability of the microscopic particle occurred at any point in the space, which tries to express the corpuscle feature of the microscopic particles and make up further for the above difficulties of linear quantum mechanics. Because this hypothes is relates to the state of the particle, some men thought erroneously that Born's statistic explanation [27-28] exhibited the corpuscle feature of the particle. Is it real? In fact, this is impossible because the probability of the microscopic particle at any point in the space is not zero.For the wave function in Eq.(1.9), the probability of the particle denoted by it to appear at any point in space is the same, which means that the microscopic particle still moves freely and distributesin total space. This is still the feature of wave, instead of corpuscle, because the microscopic particlehas not been localized in Born'srepresentation.For the wave functions in Eqs.(1.13)-(1.16) and Eq.(1.18), the particles represented by them still have similar features. Therefore, Born's statistic explanation of wave function [27-28] cannot change the wave feature of microscopic particles described by linear quantum mechanics because the particle is in essence a wave in this theory, which cannot be varied by its change of description or representation in linear quantum mechanics.

1.3.2. The Wave Packet Assumption Cannot Be Used to Embody and Exhibit the Corpuscle Feature of Microscopic Particles

In investigations of linear quantum mechanics, a lot of men suggest using the wave-packet for representing the localized or corpuscle feature of the microscopic particle. However, this idea or suggestion still cannot change the difficulty of linear quantum mechanics due to the properties of the wave-packet.

It is known that the wave-packet is formed from a linear superposition of several plane waves Eq. (1.9) with wavevector k distributed in the range of $2\Delta k$. In a one-dimensional case, the wave-packet can be expressed by

$$\Phi(x,t)=\frac{1}{2\pi}\int_{k_0-\Delta k}^{k_0+\Delta k}\phi(k,t)e^{i(kx-\omega t)}dk\,. \tag{1.20}$$

Since ω relates to k, then we can expand ω at k_0 by

$$\omega=\omega_0+(\frac{d\omega}{dk})_{k_0}\Delta k+\frac{1}{2!}(\frac{d^2\omega}{dk^2})_{k_0}(\Delta k)^2+\cdots \tag{1.21}$$

If we only consider the former two terms in the dispersive relation, *i.e.*

$$\omega=\omega_0+(\frac{d\omega}{dk})_{k_0}\Omega,$$

where $\Omega=\Delta k=k-k_0$. In this case the wave-packet in Eq.(1.20) is now denoted by

$$\Phi(x,t)=\phi(k_0)e^{i(k_0x-\omega_0t)}\int_{-\Delta k}^{\Delta k}d\Omega e^{i\left[x-(d\omega/dk)_{k_0}t\right]\Omega}$$
$$=2\phi(k_0)\frac{\sin\left\{\left[x-(\frac{d\omega}{dk})_{k_0}t\right]\Delta k\right\}}{x-(\frac{d\omega}{dk})_{k_0}t}e^{i(k_0x-\omega_0t)}, \tag{1.22}$$

where the coefficient of $e^{i(k_0x-\omega_0t)}$ is the amplitude of the wave-packet. Its maximum is $2\phi(k_0)\Delta k$, which occurs at x=0, but it is zero at $x=x_n=n\pi/\Delta k\,(n=\pm1,\pm2\cdots)$. Figure1.9 shows that the amplitude of the wave packet decreases with increasing distance of propagation due to the dispersion effect. Hence, the dispersion effect results directly in the damping of the microscopic particle (wave). We can demonstrate that a wave-packet could eventually collapse after some time.This means that the wave-packet is unstable, thus it cannot be used to express the corpuscle feature of microscopic particles in linear quantum mechanics.

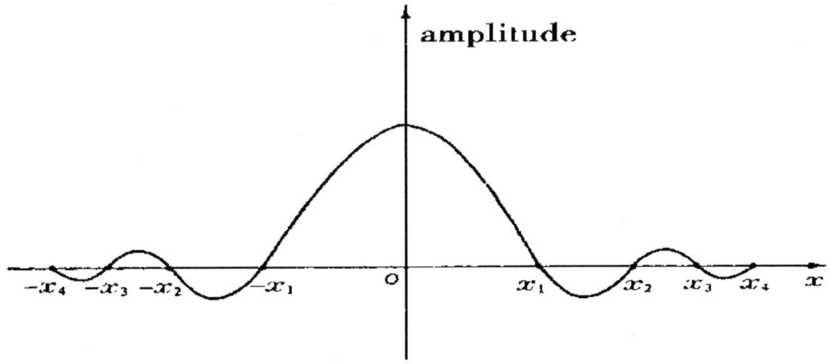

Figure 1.9. The changes of amplitudes of a plane wave propagating along the x-direction.

However, other men again suggest the following Gaussian wave-packet to represent the corpuscle behavior of a microscopic particle

$$\psi(x,t=0)=A_0\exp[-\beta_0^2x^2/2] \tag{1.23}$$

at $t = 0$, where A_0 is a constant.

Although this wave packet is localized at $t = 0$ because $|\psi|^2=|A_0|^2\exp[-\beta_0^2x^2]\to 0$ at $x\to\infty$, the wave packet is also inappropriate to denote the corpuscle feature of the particles because it disperses and always attenuates with time during the course of propagation, that is

$$\psi(x,t)=\frac{1}{\sqrt{2\pi}}\int_{-\infty}^{\infty}\Psi(k)\exp[i(kx-\hbar k^2t/2m]dk$$
$$=\frac{1}{\beta_0\sqrt{\beta_0^{-2}+i\hbar t/m}}\exp[-x^2(\beta_0^{-2}+i\hbar t/m)/2]$$

where $\Psi(k,t=0)=\frac{1}{\sqrt{2\pi}}\int_{-\infty}^{\infty}\psi(x,t=0)e^{ikx}dx$.

In this case,

$$|\psi(x,t)|^2=\frac{1}{\sqrt{1+(\beta_0^2\hbar t/m)^2}}e^{-x^2/\beta_t^2},\ \beta_t=\frac{1}{\beta_0}\sqrt{1+i\beta_0^2\hbar t/m}$$

This result indicates clearly that the Gaussian wave-packet is also unstable and dispersed as time goes by, and its position is also uncertain and varies with the time. Obviously, these results are not the feature of corpuscle. The corresponding uncertainty relation in this case can be obtained:

$$\Delta x \Delta p = \frac{\hbar}{2}\sqrt{1+\beta_0^4\hbar^2 t^2 / m^2} \tag{1.24}$$

where $\Delta x = \frac{1}{2\beta_0}\sqrt{1+\beta_0^4\hbar^2 t^2 / m^2}, \Delta p = \frac{\hbar\beta_0}{\sqrt{2}}$

According to these features, we can affirm that the Gaussian wave-packet still cannot be used to describe and denote the corpuscle property of a microscopic particle. This is another example of intrinsic difficulties of linear quantum mechanics, which implies that solving this problem has always been a challenge in linear quantum mechanics.

On the other hand, the linear Schrödinger equation results in plenty of concepts and relationships,such as the uncertainty relationship between the position and momentum and the mechanical quantities, which are denoted by some average values in an any state, and so on. These results are all contradictory with the wave-corpuscle duality and our traditional understandings.

Therefore, we can conclude that the microscopic particles described by the linear Schrödinger equation or linear quantum mechanics have no wave-corpuscle duality at all. This means that the wave feature of microscopic particles is intrinsic and essential, which cannot be changed in linear quantum mechanics no matter what improvements of descriptions. This is just a defect of quantum mechanics and cannot be complemented and solved in this theory although many improvements were used.We can say that it is only an approximate theory and does not describe correctly and completely the states and properties of the microscopic particles. This compels the theory of linear quantum mechanics to be developed further.

1.4. The Disputations on The Difficulties in Quantum Mechanics

The above difficulties of linear quantum mechanics evoked the continuous contentions and raised doubts further about whether or not the theory is correct among physicists. Actually, if we take a closer look at the history of physics, one would know that not so many fundamental assumptions were required for all physical theories but for linear quantum mechanics, which have eight assumptions.Very clearly, these assumptions of linear quantum mechanics necessarily result in its incompleteness and its limited applicability. The contentions are caused just by these assumptions, or the fundamentals of the linear quantum mechanics, which contain the Heisenberg matrix mechanics, the Schrödinger wave mechanics, Born's statistical interpretation of the wave function, and the Heisenberg uncertainty principle, etc. In other words, the contentions were about how to interpret quantum mechanics, which included the interpretation of the wave-corpuscle duality, probability explanation of the wave function, the difficulty in controlling interaction between measuring instruments and objects being measured, the Heisenberg uncertainty principle, Bohr's complementary (corresponding) principle, single particle versus many particle systems, the problems of microscopic causality and probability, process of measuring quantum states, etc. [59-68].

On the other hand, linear quantum mechanics can describe, in principle, physical systems with many particles, but it is not easy to solve such a system and approximations must be used to obtain some approximate solutions. In doing this, certain features of the system, which could be important, must be neglected. In this case, many enjoyed the successes of linear quantum mechanics, but others questioned whether linear quantum mechanics is the right theory of the real microscopic physical world because of the problems and difficulties it encountered [59-68].

It is known that linear quantum mechanics was built in the 1920s, but the above problems were always topics of heated debates among different views up to now. It was quite exceptional, in the history of physics that so many prominent physicists from different institutions were involved, and the scope of the debate was so wide. The group in Copenhagen School, headed by Bohr, represented the view of the main stream in these discussions [21-22]. As early as the 1920s, there were heated disputes on the statistical explanation and completeness of the wave function arose between Bohr and other physicists, including Einstein, de Broglie, Schrödinger, etc. [69-70].

The issues being debated and problems encountered between them [59-68] can be briefly described and summarized as follows.

(a) The correctness and completeness of linear quantum mechanics were challenged, i.e., is linear quantum mechanics correct? Is it complete and self-consistent? Can the wave-corpuscle duality of microscopic particle systems be completely described by linear quantum mechanics? Do the fundamental hypotheses of linear quantum mechanics contradict each other?

(b) The second problem is the question: is linear quantum mechanicsa dynamic or a statistical theory? Does it describe the motion of a single particle or a system of particles? The dynamic equation seems to be an equation for a single particle, but its mechanical quantities are determined based on the concepts of probability and statistical average. This caused confusion about the nature of the theory itself.

(c) How to describe the wave-corpuscle duality of microscopic particles in linear quantum mechanics? What is the nature of a defined microscopic particle based on the hypotheses of linear quantum mechanics? The wave-corpuscle duality was established by the de Broglie relations [32] and was affirmed by many experiments, as mentioned above. Can the statistical interpretation of wave function correctly describe this property? There are also difficulties in using wave package to represent the particle nature of microscopic particles as elucidated in Section 1.3. Thus, we can confirm that describing the wave-corpuscle duality was a major challenge to linear quantum mechanics.

(d) Was the uncertainty principle due to the intrinsic properties of microscopic particles or a result of the uncontrollable interaction between the measuring instruments and the system being measured?

(e) A particle appears in space in the form of a wave, and it has a certain probability to be at a certain location. However, it is always a whole particle, rather than a fraction of it, being detected in a measurement. How can this be correctly interpreted? Is the explanation of this problem based on wave package contraction in the measurement correct?

Because these issues are important, and they concern the fundamental hypotheses of linear quantum mechanics, many scientists were involved in the debate and contentions. In this discussion, the view of linear quantum mechanics and its interpretation by Bohr and others in the Copenhagen school dominated the debate, but many prominent physicists, including Einstein, respected as the authority, doubted and continuously criticized Bohr's interpretation. This resulted in a life-long dispute between Bohr [21-22] and Einstein, which was unprecedented. Unfortunately, after being debated and disputed for almost a century, there are still no definite answers to most of these questions. Their discussions and disputes may be resolved in three stages. We here introduce and survey some main views of this debate in the following [59-68].

The first stage was during the period from 1924 to1927 when the theory of quantum mechanics had just been established. Einstein proceeded from his own philosophical belief and his scientific goal for an exact description of causality in the physical world, and expressed his extreme unhappiness with the probability interpretation of linear quantum mechanics. In a letter to Born on December 4, 1926, Einstein said, "Quantum mechanics is certainly imposing. But an inner voice tells me that it is not the real thing (der Wahre Jakob). The theory says a lot, but it does not bring us any closer to the secret of the "Old One." I, at any rate, am convinced that He is not playing at dice." Therefore, he intensely doubted and opposed Bohr's interpretations on linear quantum mechanics, thus he debated this problem with Bohr [21-22].

The second stage was from 1927 to 1930. After Bohr had put forward his complementary principle and had done his interpretation as the mainstream interpretation, Einstein was extremely unhappy. His main criticism was directed at the uncertainty relation that Bohr's complementary principle was based upon. At the 5th (1927) and the 6th (1930) International Meetings of Physics at Solway, Einstein proposed two ideal experiments (double slit diffraction and photon box) to prove that the uncertainty relation and formalism of the quantum mechanics contradict each other, and thus to disprove Bohr's complementary principle. But Bohr demolished Einstein's idea each time through resourceful analysis. Since then, Einstein had to accept the logical consistency of quantum mechanics and then turned his criticism to the completeness of the linear quantum mechanics theory.

The third stage was from 1930 until the death of Einstein. The contention during this period is reflected in the debate between Einstein and Bohr over the EPR paradox proposed by Einstein together with Podolsky and Rosen [69]. This paradox concerned the fundamental problem of linear quantum mechanics, i.e., whether it satisfied the deterministic localized theory and the microscopic causality. Since some of the subsequent experiments seem to support the linear quantum mechanics, instead of the Bell inequality, it is necessary to understand the nature of the EPR paradox [69] and the results it brought about. The EPR paradox [69] is briefly described below.

For simplicity but without losing its generality, we assume that the initial relativistic momentum of the pair of particles is $p = 0$ in the system, which is composed of two particles, which move in opposite directions. Then there must be $p_1 = -p_2$ after the two particles interact and depart. However, the magnitude and direction of the momentum of each particle are not known according to physical theory. Assume now that the momentum of particle 1 is measured by a detector and its momentum value $p_1 = +a$ is obtained; the momentum of particle 2 should be determined and it can only be $p_2 = -a$ according to the conservation of momentum in linear quantum mechanics. However, in accordance with the hypothesis of contraction of wave packet in the measuring process in the linear quantum mechanics, the plane wave with momentum $p_1 = a_1$ is "selected" by the detector from wave packet $\phi_1(X_1)$ describing particle 1. In light of the traditional linear quantum mechanics, this process of "spectrum resolution" is due to some kind of "uncontrollable interaction" between the instrument and the wave packet. Under the influence of such an "uncontrollable interaction", the momentum of particle 1 could be $p_1 = a$, or $p_1 = b, \cdots$. However, what is surprising is that there is always $p_2 = -a$ as long as $p_1 = a$ is measured by the detector. This means that this value should be obtained regardless of the measurement on the wave packet $\phi_2(X_2)$ is made or not. In other words, when the wave packet $\phi_1(X_1)$ is measured and contracted, the wave packet $\phi_2(X_2)$ for particle 2 will also be automatically changed and contracted. A series of questions thus arise in this case, such as: what mechanism makes this possible? Does this occur instantaneously, or is it propagating at the speed of light according to the special theory of relativity? How can the wave packet contraction, caused by measurement, automatically guarantee the conservation of momentum? Very evidently, these questions are very difficult to answer [59-68].

Only after careful studies by Einstein and others, were the following conclusions obtained: either the description of linear quantum mechanics was incomplete, or linear quantum mechanics didn't satisfy the criterion of "localization". Einstein always believed that physical phenomena must satisfy the criterion of "localization", i.e. physical quantities cannot propagate with speed greater than the speed of light. Thus, he thought that linear quantum mechanics was an incomplete theory. Due to this remarkable analysis by Einstein, many physicists began to explore the theory of "hidden parameters" of linear quantum mechanics.

The above contentions and queries to linear quantum mechanics by Einstein and others had indeed created quite a stir. Bohr [21-22]had to respond in his own capacity to these queries. In 1935, Bohr published a short essay in Physical Review [21-22], in which he argued that if a system consists of two local particles 1 and 2, and then this system should be described by a wave function $\phi(1,2)$. In this case, the local particles 1 and 2 are no longer mutually independent entities. Even though they are spatially separated at the instant the system is probed, they cannot be considered as independent entities. Thus, there is no basis for statements such as 'the measurement of subsystem 1 could not influence subsystem 2 within the framework of the linear quantum mechanics', thus the idea of Einstein et al., cannot be accepted. Essentially, Bohr was not really against the "paradox" proposed by Einstein and others, but only confirmed that linear quantum mechanics might not satisfy the principle of localization. Bohr further commented that in the final decisive steps of measurement in Einstein's ideal experiment, even though there was no mechanical interference to the system being probed, influence on experimental conditions also existed. Thus, Einstein's arguments could not verify their conclusion that the description of linear quantum mechanics is incomplete [70-72].

Plenty of scientists who closely followed the thought of localization and incompleteness of linear quantum mechanics by Einstein and others thought and believed further that there could exist a hidden variables theory behind linear quantum mechanics, which might be able to explain the probability behavior of microscopic particle. Therefore, the concept of "hidden variables" was proposed soon after linear quantum mechanics was born [73-77]. However, Von Neumann [78] disapproved it, which occurred in his paper in1932. For a long time since then, no one had mentioned this problem. But after the World War II, Einstein repeatedly criticized linear quantum mechanics and suggested again that any actual state should be completely described. Motivated by this thought, Bohm [73-75] put forward the first systematic "hidden variable theory" in 1952. He believed that the statistical characteristics of linear quantum mechanics were due to some "background" fluctuations hidden behind the quantum theory. If the hidden function for a microscopic particle was found then a deterministic description could be established for a single microscopic particle. However, how can the existence of such hidden variables really be proved? In such a case, Bohm [73-75] proposed two experiments to measure the spin correlation of a single proton and the polarization correlation in annihilating radiation of photons, respectively. It was realized later that in Bohm's theory, the single state $\phi(\vec{r},t)$ is essentially a slowly varying state, which describes states of a fluid with random fluctuations. Since the wave function itself cannot have such random fluctuation, a hidden variable could not be introduced. Thus, Bohm's theory, mentioned above, was referred to as a random hidden-variables theory [73-75].

On the other hand, if the motion of particles is considered as a stable Markov process, then a steady state solution of the Schrödinger equation can be obtained from a steady distribution of the Markov chain in this case, and if the Fock-Planck equation was used as the dynamic equation of microscopic particle, a new "hidden variables theories" in linear quantum mechanics can be built. After Bell established the so-called Bell inequality based on Bohm's deterministic "localized variables theory" in 1966, different attempts were made to experimentally verify the correctness of this theory and to further settle the contentions in linear quantum mechanics once and for all. However, the majority of the experiments supported linear quantum mechanics at that time, as mentioned earlier.Meanwhile, it was clear that not all the predictions by linear quantum mechanics could be obtained from the localized hidden variables theory. Thus, the "hidden variable theory" was gradually abandoned.

Therefore, the long contentions between Bohr and Einstein were mainly focused on three issues.

(1)Einstein upheld and adhered to the belief that the microscopic world is no different from the macroscopic world. The particles in the microscopic world are still matter and they really exist regardless of the methods of measurements, there for eany theoretical description of them should also be determinant, instead of probable.

(2) Einstein always thought that thetheory of linear quantum mechanics was not an ultimate and complete theory. He believed that linear quantum mechanics is similar to classical optics, both of them are correct theories based on statistical laws, *i.e.*, when the probability$\left|\phi\left(\vec{r},t\right)\right|^2$ of a microscopic particle at the moment t and position $\vec{r}$ is known, the average value of an observable quantity can be obtained using the statistical method and then compared with experimental results. However, understanding the processes involving a single particle was not satisfactory. Hence, $\phi(\vec{r},t)$ cannot give everything about a microscopic particle system, and the statistical interpretation cannot be ultimate and complete in linear quantum mechanics.

(3) The third issue concerns the physical interpretation of linear quantum mechanics. Einstein was not impressed with it and, made it very clear in a speech at the fifth Selway International Meeting of physics. In an article, "Physics and Reality", published in 1936 in the Journal of the Franklin Institute, Einstein again mentioned that what the wave function,

$\phi(\vec{r},t)$ describes can only be a many-particle system, or an assembly, in terms of the method of statistical mechanics, but under no circumstances couldonly the wave function describe the state of a single particle. Einstein also believed that the uncertainty relation was a result of incompleteness of the description of a microscopic particle by $\phi(\vec{r},t)$ because a complete theory should give precise values for all observable quantities. Einstein also did not accept the statistical interpretation, because he did not believe that electrons possess free will. Thus, Einstein's criticism against linear quantum mechanics was not directed towards the mathematical formalism of linear quantum mechanics, but to its fundamental hypotheses and its physical interpretation. He considered this was due to the incomplete understanding of the microscopic objects. Therefore, Einstein's criticisms were deep, essential, and important, and represent that the difficulties and questions of linear quantum mechanics were quite great and serious.

At the same time, the contradiction between the theory of relativity and the fundamentals of linear quantum mechanics was also a central point of contention. Einstein made the effort to unite the theory of relativity and linear quantum mechanics, and attempted to interpret the atomic structure using field theory. The disagreements on several fundamental issues of linear quantum mechanics by Einstein and Bohr and their followers were deep rooted and worth further study. Although these questions were debated and disputed for more than a century, a consistent conclusion has not been obtained [59-68].

The above descriptions and statements clearly indicate the complexity and seriousness of problems in linear quantum mechanics. This brief review on the disputations between the two great physicists given above should be useful to understand the nature and problems of linear quantum mechanics, and to seek its correct directions and methods of development.

1.5. The Reasons of Difficulty of Linear Quantum Mechanics and Its Direction Eliminating These Difficulties

However, what on earth are the reasons for the difficulties of linear quantum mechanics? This is worth studying deeply and in detail. The reasons we understand are described as follows:

(1).As known, the motions of microscopic particles and their properties are described by the linear Schrödinger equation (1.8) and the Hamiltonian operator of the system, which is represented by

$$\hat{H} = \hat{T} + \hat{V} = \frac{\hbar^2}{2m}\nabla^2 + \hat{V}(\vec{r},t) \tag{1.25}$$

where $\hat{T}$ and $\hat{V}$ are the kinetic and potential operators, respectively, i.e., they consist only of the kinetic and potential operator of particles. This means that the properties of microscopic particles are determined by the kinetic and potential, but the potential can only change the states of microscopic particles, and cannot determine their natures and essences because it is not related to the state or wavefunction of the particle. Therefore, the natures and essences of microscopic particles are only determined by the kinetic term. Thus, there is no force or energy to obstruct and suppress the dispersing effect of kinetic energy in the system, so the microscopic particle disperses and propagates in total space, and cannot be localized at all. This is the main reason why the particles have only a wave feature in linear quantum mechanics.

Evidently, the Hamiltonian in Eq.(1.25) does not represent the practical essences and features of the microscopic particles [57-58,79] because the energy operator of the systems is always associated with the states and properties of the particles in physics, in this case we can affirm that the interaction energies between the particle and background field or among the particles are always related to the wave function of the particles. However, the linear Schrödinger equation(1.8) and the Hamiltonian operator in Eq.(1.25) in linear quantum mechanics have not represented this property.

(2). On the other hand, when the above linear Schrödinger equation is used to solve and explain the quantum mechanical problems of microscopic particles we must incorporate all interactions among the particles or between the particles and the background field, such as the lattices in solids and nuclei in atoms and molecules, including nonlinear and complicated interactions, into the external potential $V(\vec{r},t)$ in Eq.(1.8) using various approximate methods, such as the free electron and average field approximations, Born-Oppenheimer approximation, Hartree-Fock approximation, Thomas-Fermi approximation, and so on [57-58,79]. This implies that we must neglect or freeze or completely blow out the really existent nonlinear and complicated interactions, thus, there are not these interactions, which can balance or suppress the dispersing effect of kinetic energy, in the dynamic equation(1.8) and the Hamiltonian in Eq.(1.25). In this case, it is not strange that the microscopic particles possess only a wave feature, and then the difficulties of the linear quantum mechanics mentioned above occur necessarily [57-58,79].

In general, it is quite difficult to find the real solutions of the linear Schrödinger equation if the above approximate methods are not used, except for the hydrogen atom and molecules and the helium atom as well as others mentioned in Eqs.(1.13)-(1.15).This means that the approximate methods of an average field, etc.,which replaced the real interactions, have been often used to find the quantum mechanical problems of microscopic particles.This sufficiently exposes the approximation and limitations of the linear Schrödinger equation or linear quantum mechanics. Otherwise, it is also very difficult to find out the real solutions of the linear Schrödinger equation in the systems of many-particles and many-bodies except for many approximate methods that were used. Thus, we can judge that linear quantum mechanics is only an approximate theory.

(3). In fact, the linear Schrödinger equation or linear quantum mechanics neglect or erase the real interactions between the microscopic particle and background field or lattices in conceded matters or between the electrons and nucleon in atoms and molecules by freezing the real motions of the background field or lattices and nucleon by means of some approximate interactional potentials and the above approximate methods [57-58,79]. For example, the real interaction between the electrons and lattice is represented by an average periodicity potential in the solid, and the average Coulomb potential $V(\vec{r},t)=-e^2/\vec{r}$ is used to replace the real interaction between the electrons and nucleon in the atoms, in which, thus, the true motions of the lattices and nucleons are completely ignored. In this case, the electron can only disperse and move in waves in the space.This means that linear quantum mechanics have not truly described the motions and properties of microscopic particles in the physical system. This results in that the microscopic particles have only a wave feature and the difficulties of linear quantum mechanics. This is obviously not reasonable and correct. Thus, in the improvement and development of linear quantum mechanics we cannot absolutely fix or neglect the movements of the background field or lattices, and the position of the nucleon in atoms,becauseof real existences of the counteraction effect in these interactions between the particles and the background field or lattices or nucleons. In practice, the motions of the background field or lattices and nucleons are always existent in all systems, which will result directly in the complicated and nonlinear interactions, which can thoroughly change the nature of microscopic particles. Therefore, the true motions of the background field or lattices or nucleons and their interactions with the microscopic particles must be considered in detail, and seriously, in new quantum mechanics. This affirms that the direction of development of quantum mechanics is to establish nonlinear quantum mechanics.

To summarize, the above studied results indicated clearly that there are many considerable difficulties and questions in linear quantum mechanics, which cannot be overcome and eliminated no matter what methods of improvement are used in its own framework. Therefore, these difficulties are intrinsic and essential for quantum mechanics because it is,in essence, a linear, simple, and approximate theory. This means that linear quantum mechanics must be developed forwards. Its direction of development is to establish a nonlinear quantum mechanics, in which two hypotheses of the linear quantum mechanics, i.e., the linearity of this theory and the independence of the Hamiltonian of physical systems on the wave function of the particles must be changed or eliminated, the nonlinear interactions among the microscopic particles or between the particle and the background fields should be added in the Hamiltonian of the systems, and dynamic equations of the particle by considering the movements of all particles and background as well as the interactions between them, in this case the wave-corpuscle duality of microscopic particles,may occur. These problems are described in detail in the following chapters.

REFERENCES

[1] E. Pauling and E.Wilson, *Introduction to quantum mechanics*, 1935.

[2] N. F.Mottand, I.N.Sneddon, *wave mechanics and its applications*, 1948.

[3] D. Bohm, *Quantum theory*, University of London, London, 1954

[4] P. A.M. Dirac, *The principles of quantum mechanics*, Clarendon Press, Oxford, 1947 and 1958.

[5] L. Schiff, *Quantum mechanics*, McGraw-Hill, New York, 1965.

[6] K. Gottfried, *Quantum mechanics*, Vol.I, New York, Benjamin (Fourth printing), 1974.

[7] A. Messianh, *Quantum mechanics*, Vol.II, NorthHolland Publishing Co. Amsterdam, 1970.

[8] E.Merzbachar, *quantum mechanics*, 2 edi. 1970 John Waley &Sons, Inc., 1970.

[9] L. D.Landau, and E.M.Lfshitz, *Quantum mechanics*, Pregamon Press, Oxford, 1958.

[10] D. Bohm. *Quantum theory*, Prentice-Hall, New York,1951.

[11] A. S.Davydov, *Quantum mechanics*, Pergamon Press, Oxford, 1965.

[12] G. L.Trigg, *Quantum mechanics*, D.Van Nostrand Co., Inc., Princeton, 1969.
[13] P. Roman, *Advanced quantum theory*, Addisonn-Wesley, Reading, Massachusetts, 1965.
[14] A. Messiah, *Quantum mechanics*, Interscience, New York, 1961
[15] K.Gottfried, *Quantum mechanics*, W.A.Benjamin, New York, 1966
[16] W.Heisenberg, *The physical principles of the quantum theory*, Chicago University Press, 1930
[17] J. Potter, *quantum mechanics*, North-Holland publishing Co., Amsterdam,1970
[18] S. X. Zhou, *Quantum mechanics* (Chinese), People Education Press, Beijing,1979
[19] T. F.Tuan and E. Gerjuoy, Phys. Rev. 117(1960)756
[20] S. Cheng, C. L. Cocke, V. Frohne, E. Y. Kamber, J. H. McGuire, and Y. Wang, Phys. Rev. A , 47(1993)3923
[21] D. Bohr, *Phys. Rev*. 48 (1935) 696.
[22] D. Bohr and J. Bub, *Phys. Rev*.48(1935) 169.
[23] E. Schrödinger, *Collected paper on wave mechanics*, London: Blackie and Son, 1928.
[24] E. Schrödinger, *Proc. Cambridge Puil. Soc*., 31(1935)555.
[25] W. Heisenberg, *Z. Phys*., 33(1925)879.
[26] W. Heisenberg and H.Euler, *Z. Phys*. 98(1936)714.
[27] M. Born, W. Heisenberg and P. Jorden, *Z. Phys*. 35 (1926) 146.
[28] M. Born and L. Infeld, *Proc. Roy. Soc. A*, 144 (1934) 425.
[29] P. A.Dirac, *Proc. Roy. Soc*. London, A114 (1927) 243.
[30] P.A.Dirac, *The principles of quantum mechanics*, Oxford, 1967.
[31] L. Laudau, *Z. Phys*. 45 (1927) 430.
[32] L.de Broglie, La physique quantique restera-t-elle indeterministe? Gauthier-Villars, Paris,1953.
[33] C.Davisson and L.H.Germer, Diffraction of electron by a crystal Nickel, Phys. Rev. 30(1927) 705.
[34] G.P.Thomson, Experiments on the diffraction of cathode rays, Proceedings of the Royal Society of London. *Series A, Containing Papers of a Mathematical and Physical Character*, 117(778)(1928) 600.
[35] G.P. Thomson, 'On the waves associated with X- rays, and the relation between free electrons and their waves, *Philosophical Magazine*, 7(1929)405.
[36] G.P. Thomson, 'The waves of an electron', *Nature,* 122(1928) 279.
[37] T. F. Tuan and E. Gerjuoy, *Phys. Rev*.117 (1960) 756.
[38] S. Cheng, C. L. Cocke, V. Frohne, E. Y. Kamber, J. H. McGuire, and Y. Wang, *Phys.Rev*. A47 (1993)3923.
[39] I. Reiser, C. L. Cocke, and H. Bräuning, *Phys. Rev*. A 67 (2003) 062718.
[40] D. Misra, H. T. Schmidt, M. Gudmundsson, D. Fischer, N. Haag, H. A. B. Johansson, A. Källberg, B. Najjari, P. Reinhed, R. Schuch, M. Schöffler, A. Simonsson, A. B. Voitkiv, and H. Cederquist, *Phys. Rev. Lett*.102 (2009)153201.
[41] H. T. Schmidt, D. Fischer, Z. Berenyi, C. L. Cocke, M. Gudmundsson, N. Haag, H. A. B. Johansson, A. Källberg, S. B. Levin, P. Reinhed, U. Sassenberg, R. Schuch, A. Simonsson, K. Støchkel, and H. Cederquist, *Phys. Rev. Lett*. 101 (2008) 083201.
[42] L. P. H. Schmidt, S. Schössler, F. Afaneh, M. Schöffler, K. E. Stiebing, H. Schmidt-Böcking, and R. Dörner, *Phys. Rev. Lett*.101 (2008) 173202.

[43] A. L. Landers, E. Wells, T. Osipov, K. D. Carnes, A. S. Alnaser, J. A. Tanis, J. H. McGuire, I. Ben-Itzhak, and C. L. Cocke, *Phys. Rev.* A70 (2004) 042702.

[44] A. Senftleben, T. Pflüger, X. Ren, O. Al-Hagan, B. Najjari,D. Madison, A. Dorn, and J. Ullrich, *J. Phys.* B43 (2010) 081002.

[45] J. S. Alexander, A. C. Laforge, A. Hasan, Z. S. Machavariani,M. F. Ciappina, R. D. Rivarola, D. H. Madison, and M.Schulz, *Phys. Rev.* A78 (2008) 060701.

[46] K. N. Egodapitiya, S. Sharma, A. Hasan, A. C. Laforge,D. H. Madison, R. Moshammer, and M. Schulz, *Phys. Rev.Lett.*106 (2011)153202.

[47] N. Stolterfoht, B. Sulik, V. Hoffmann, B. Skogvall, J. Y.Chesnel, J. Rangama, F. Frémont, D. Hennecart, A. Cassimi,X. Husson, A. L. Landers, J. A. Tanis, M. E. Galassi, and R. D. Rivarola, *Phys. Rev. Lett.* 87 (2001)023201.

[48] S. Chatterjee, D. Misra, A. H. Kelkar, L. C. Tribedi, C. R.Stia, O. A. Fojón, and R. D. Rivarola, *Phys. Rev.* A78 (2008)052701.

[49] N. Stolterfoht, B. Sulik, L. Gulyás, B. Skogvall, J. Y. Chesnel, F. Frémont, D. Hennecart, A. Cassimi, L. Adoui,S. Hossain, and J. A. Tanis, *Phys. Rev.* A67 (2003) 030702.

[50] D. Misra, A. Kelkar, U. Kadhane, A. Kumar, L. C. Tribedi, and P. D. Fainstein, *Phys.Rev.* A74 (2006) 060701.

[51] H. D. Cohen and U. Fano, *Phys. Rev.*150 (1966)30.

[52] M. E. Galassi, R. D. Rivarola, P. D. Fainstein, and N.Stolterfoht, *Phys. Rev.* A66 (2002) 052705

[53] M. E. Galassi, R. D. Rivarola, and P. D. Fainstein, *Phys.Rev.* A70 (2004)032721.

[54] S. F. Zhang, D. Fischer, M. Schulz, A. B. Voitkiv, A. Senftleben, A. Dorn, J. Ullrich, X. Ma, and R. Moshammer. *Phys. Rev. Lett.* 112 (2014) 023201.

[55] J. Ullrich, R. Moshammer, A. Dorn, R. Dörner, L. P. H.Schmidt, and H. Schmidt-Böcking, Rep. *Prog. Phys.* 66 (2003)1463.

[56] Pang Xiao-feng, *Theory of nonlinear quantum mechanics* (Chinese), Chongqing, Chinese Chongqing Press, 1994.

[57] Pang Xiao-feng and Feng Yuan ping, Quantum mechanics in nonlinear systems, New Jersey, World Scientific. Publishing Co., 2005.

[58] M. Ferrero, and A. Van der Merwe, *Fundamental problems in quantum physics*, Kluwer, Academic, Dordrecht, 1993.

[59] M. Ferrero, and A. Van der Merwe, Fundamental problems in quantum physics, Kluwer, Dordrecht, 1995.

[60] M. Ferrero, and A. Van der Merwe, New developments on fundamental problems in quantum physics, Kluwer, Dordrecht, 1997.

[61] R. P. Feynman, *The character of physical law*, MIT Press, Cambridge, Mass, 1998

[62] A. P. French, Einstein, *A Gentenary Volume*, Harvard University Press, Cambridge, Mass, 1979.

[63] C. Garola and A. Ross, *The foundations of quantum mechanics-historical analysis andopen questions*, Kluwer, Dordrecht, 1995.

[64] S. Diner, G. Fargue, Lochak and F. Selleri, *The wave-particle dualism*, D. Riedel, Dordrecht, 1984.

[65] M. Jammer, *The conceptual development of quantum mechanics*, Oxford, 4th edition, 1867.

[66] S. Kobayashi, H. E. Zawa, Y. Muvayama and S. Nomura, eds., *Foundations of quantum mechanics in the light of new technology*, Komiyama, Tokyo, 1990.
[67] T. D. Black, M. M. Nieto, H. S. Pilloff, M. O. Scully and R. M. Sinclair, *Foundations of quantum mechanics*, World Scientific, Singapore, 1992.
[68] A. Einstein, B. Podolsky and N. Rosen, *Phys. Rev.* 47 (1935) 777.
[69] J. S. Bell, *Physics,* 1, (1965) 195.
[70] J. S.Bell, *Rev. Mod. Phys.* 38 (1996) 447.
[71] J. S. Bell, *Speakable and unspeakable in quantum mechanics*, Cambridge University Press, Cambridge, 1987.
[72] D. Bohm, *Quantum theory*, Prentice-Hall Englewood Cliffs, New Jersey, 1951.
[73] D. Bohm, *Phys. Rev.* 85 (1952) 166 and 180.
[74] D. Bohm, and J. Bub, *Rev. Mod. Phys.* 38 (1966) 453.
[75] A. M. Gleason, *J.Math. Mech.* 6 (1957) 885.
[76] A. Daneri, A. Loinger, G. M. Prosperi, *Nucl. Phys.* 33(1962) 297.
[77] J. Von Neumann, *Mathematical foundation of quantum mechanics*, Princeton, N.J, 1955, Chap. IV, VI.
[78] Pang Xiao-feng, *Nonlinear quantum mechanics*(Chinese), Beijing, Chinese Electronic Industry Press, 2009.

Chapter2

PROPOSAL AND BUILD OF DIFFERENT NONLINEAR QUANTUM MECHANICAL MODELS AND THEORIES

About 30 different nonlinear quantum mechanical ideas, models, and theories, including Weinberg's model, and their properties, are briefly stated and reviewed. This also exhibits the necessity to build new nonlinear quantum mechanics because there are many questions in them. In this case, the new nonlinear quantum mechanics proposed by Pang, including the theoretical, mathematical, and experimental foundations, and its characteristics,are depicted in Section 2.6.

2.1. DE BROGLIE'S IDEA FOR THE NONLINEAR QUANTUM MECHANICAL THEORY

In order to overcome the difficulties of linear quantum mechanics, and to further develop the quantum mechanical theory, we must abandon its two hypotheses of the linearity and the independence of Hamiltonian on the wave function of the particles, and establish the theory of nonlinear quantum mechanics by adding the nonlinear interactions into its Hamiltonian and dynamic equations. The first attempt of establishing a nonlinear quantum theory was made by de Broglie, who provided the idea of nonlinear wave theory of microscopic particles, which was described in his book: "the nonlinear wave theory".

In practice, de Broglie [1-3] concluded, through a long period of research, that the present theory of wave motion cannot interpret the relation between particle and wave because this theory was limited to a linear framework from the start. In 1926, he further emphasized that if wave function, $\phi(\vec{r},t)$, is a real field in the physical space, and then the particle should always have a definite momentum and position. Thus de Broglie assumed that $\phi(\vec{r},t)$ describes an essential coupling between the particle and the field, and then we can use this concept to explain the phenomena of interference and diffraction of microscopic particles. In 1927, de Broglie put forward a "dual solution theory" in a paper published in J. de Physique, in which de Broglie proposed that there are two types of solutions in the dynamic equation in quantum mechanics. One is a continuous solution, and it can be represented by $\varphi(\vec{r},t) = \mathrm{Re}^{i\theta(\vec{r},t)}$, which is the Schrödinger wave, resembling that in the linear quantum mechanics, mentioned above, and has only statistical meaning and interpretation and can be normalized. Thus, it does not represent any physical wave. The other type, which is also referred to as a *u* wave and possesses a singularity, is associated with spatial localization of the particle. The corpuscle feature of a microscopic particle is only described by the *u* wave and the position of a particle is determined by a singularity of the *u* wave. After this, de Broglie generalized further the formula of the monochromatic plane wave and stipulated further a rule of associating the particle with the propagation of the wave. Then, the microscopic particle would move inside its wave, according to de Broglie's dual solution theory [1-3]. This suggests that the motion of the microscopic particle, inside its wave, is influenced by a force, which can be derived from a "quantum potential". This quantum potential is proportional to the square of the Planck constant and is dependent on the second derivative of the amplitude of the wave, thus it is a nonlinear force. This force can also be given in terms of the change in the rest mass of the particle. In the case of a monochromatic plane wave, the quantum potential is zero. In the 1950s, de Broglie further improved his "dual solution theory", in which he proposed that the *u* wave satisfies an undetermined nonlinear dynamic equation. These hypotheses led to his own "nonlinear wave theory" [1`-3].

However, de Broglie did not give the exact and concrete nonlinear dynamic equation that the *u* wave should satisfy. Obviously, this model has serious difficulties in describing multi-particle systems and the *s* state of a single-particle. At the same time, the theory also lacked experimental evidence. Thus, even though Einstein supported it, the theory was not taken seriously by the majority and was gradually forgotten.

Although de Broglie's nonlinear wave theory was incorrect, some of his ideas, such as the quantum potential, the *u* wave of nonlinear equation, which is capable of describing a physical particle, provided an inspiration for further development of quantum mechanics.

On the other hand, de Broglie stated and provided that the quantum potential is related to the second derivative of the amplitude of the wave function $\varphi(\vec{r},t) = \mathrm{R}e^{i\theta(\vec{r},t)}$ .This is a challenging idea. Thus Bohm [4-6], who proposed the theory of localized hidden variables in 1952, derived this quantum potential. His results show that the quantum potential is independent of the phase, $\theta(\vec{r},t)$, of the wave function, and is represented in the form of $V = h^2\nabla^2 R / 2mR$, where *R* is the amplitude of the wave, *m* is the mass of the particle, and *h* the Planck constant. With such a quantum potential, Bohm believed [4-6]that the motions of microscopic particles should follow Newton's equation to move. Meanwhile, he thought that a measurement process of the properties of the microscopic particles is always disturbed due to the "instantaneous action" of this quantum potential. The former, however, was less convincing.

However, the quantum potential and nonlinear equations were again introduced in the Bohm-Bub theory proposed in 1966 [6]. They assumed that, in the dual Hilbert space, there exists a dual vector,, $|\varphi_1\rangle$ and $|\delta\rangle$, which satisfies $|\varphi_1\rangle = \sum_n b_n |B_n\rangle$ and $|\delta\rangle = \sum_k \delta_k |B_n\rangle$ respectively, where δ is a hidden variable and satisfies the Gaussian distribution in an equilibrium state. They introduced a nonlinear term in the Schrödinger equation to represent the effect arising from the quantum measurement, and determined the equation containing the nonlinear term based on the relation among the particles, environment, and hidden variables.

Attempts were made to solve the problem concerning the influence of the measuring instruments on the properties of the particles being probed.

de Broglie pointed out [2-3] that the quantum potential can be expressed in terms of the change in the rest mass of the particle and tried to interpret Bohm's quantum potential based on the counteraction of the u wave and domain of singularity. Thus, the quantum potential arises from the interaction between the microscopic particles. It is associated with nonlinear interaction and is able to change the properties of particles. These were encouraging. It seemed promising to make microscopic particles measurable and deterministic by adding a quantum potential with nonlinear effect to the Schrödinger equation, and eventually to have a deterministic quantum theory. A delighted Dirac commented that the results would ultimately prove that Einstein's deterministic or physical view is correct.

In summary, the work of de Broglie and Bohm gave us some good motivations, but their ideas cannot be borrowed indiscriminately. One must go beyond the framework of linear quantum mechanics and look into the nonlinear scheme. To establish a new and correct theory, one must start from the phenomena and experiments which the linear quantum mechanics failed, or had difficulty, to explain, and use completely new concepts and new approaches to study these unique quantum systems. This is the only way to clearly understand the problems in linear quantum mechanics.

2.2. Weinberg's Model for Nonlinear Quantum Mechanics and Some Comments

2.2.1. Weinberg's Idea and Motivation for the Nonlinear Quantum Mechanics

There are many difficulties and problems in linear quantum mechanics as mentioned in Chapter 1, therefore it is necessary to improve it using some new ideas and methods. In this case, it seems long over due to find precision tests of quantum mechanics itself, which would be more stringent than existing tests of specific quantum-mechanical theories. Considering the importance of quantum mechanics in modern science, it is strange how one rarely hears of efforts to test quantum mechanics experimentally with high precision [9-10]. In practice, over the last decades there are plenty of experimental tests of predictions [11-12] that distinguish quantum mechanics from local hidden variable theories, and these experiments have finally ruled out the hidden variable theories, but they do not provide tests of quantum mechanics, itself, to a precision better than about 1%. Weinberg [9-10] thought that this is, in principle, possible to test any specific quantum mechanical theory like quantum electrodynamics. Thus, Weinberg formulated experiments that would show up departures from quantum mechanics itself at a much more miniscule level than departures from any specific theory. In such a case that it should set the upper bounds on these parameters by doing experiments to look for new effects that could arise in the generalized theory, and in this way we could get a sense of how accurate the theory is that we want to test. The trouble is that it is very difficult to find any logically consistent generalization of quantum mechanics. Then the linearity of quantum mechanics is just one good target for the generalization, but the nonlinear terms cannot be arbitrarily added into the Schrödinger equation because the dynamic equation in linear quantum mechanics must exhibit real motions of the microscopic particles. Thus Weinberg proposed some stringent limits on possible nonlinear corrections to linear quantum mechanics by searching for the detuning of resonant transitions, in which a possible nonlinear generalization of linear quantum mechanics can occur, which can be used as a guide in subjecting quantum mechanics to experimental tests for finding a most general such theory and some generalizations of linear quantum mechanics. This means that in order to test the precision of linear quantum mechanics Weinberg has to introduce nonlinearity in this theory. In this case a series of problems occur, for example, what is the form of the nonlinear interaction? Where are the nonlinear interactions from? How do we test the precision of this theory? These problems must be solved before the new theory is built. Weinberg chose one sensitive way to test the linearity of quantum mechanics, i.e., to look for the detuning phenomenon characteristic of any system of nonlinear oscillators, in which the resonant frequency at which a weak external field can strongly perturb a nonlinear system will in general depend on the amplitudes of the various modes excited. So, no matter how we set this frequency, the resonance will be detuned due to the amplitudes change. This will prevent us from being able to drive the system slowly from one mode to another, say in a time T, unless the change in the resonant frequency during this transition is within a natural width of order I /T. This is a good way to look for departures from the linearity of quantum mechanics. In fact, the experimentalists' have already been able to drive hyperfine transitions in ions such as $^9Be^+$ with T as long as several seconds, and with good prospects of increasing this to several minutes. In such a case, these experiments (or related NMR or ESR experiments) can be used to set very stringent limits on any frequency shifts that might arise from nonlinearities in the dynamic equations of the wave function.

Therefore, Weinberg's motivation is to test precisely the linear quantum mechanics by adding some nonlinear interactions into the original quantum theory, to increase the precision of linear quantum mechanics. In order to seek what sort of nonlinearities might actually be expected to show up in such an experiment, and to make all this quantitative, it is essential to formulate some sort of nonlinear generalization of linear quantum mechanics. In accordance with Bollingerand colleagues' idea [11], which proposed a generalized version of linear quantum mechanics that seems physically satisfactory, it is sufficient to consider a discrete system, like the hyperfine states, for which the wave function may be adopted to depend only on a discrete variable k. The time dependence of the wave function can be assumed to be given by a dynamic equation. Based on this idea Weinberg proposed his model of nonlinear quantum mechanics [9-10], which is described as follows.

2.2.2. The Formalism in Weinberg's Nonlinear Model

Based on the above idea Weinberg proposed the nonlinear model for precision tests of linear quantum mechanics [7-8]. In order to understand Weinberg's nonlinear model we here list its basic formalisms, which are described as follows.

(I) Wave Functions Describing the Microscopic Particles

Weinberg still used a wave function ϕ with complex-values to represent the state of a microscopic particle as in ordinary quantum mechanics. For illustrative purposes he takes ϕ

be a function ϕ_k of a discrete variable k that takes values:1, 2, . ..N, but later he considered the cases where the wave function is a function $\Psi(X)$ of a continuous position variable X. Meanwhile, he assumed that the wave functions ϕ_k and $Z\phi_k$ represent the same physical state, where Z is an arbitrary complex number.

(II) Observables of Physical Quantities in This Model

As in ordinary quantum mechanics, the Hermitian operator or matrices A represent observable quantities. At present, he used still real bilinear functions $\phi_k^* A_{kl}\phi_l$ to describe observables in quantum mechanics. However, he generalized these again representing observables by real non-bilinear functions $a(\phi_k^*, \phi_l)$. However, in order to retain the feature of quantum mechanics that, for an arbitrary non-zero complex number Z, the wave functions ϕ_k and $Z\phi_k$ represent the same physical state he required, so that the functions representing observables are (like $\phi_k^* A_{kl}\phi_l$) homogeneous of degree one in both ϕ_k and $Z\phi_k$, i.e.,

$$\phi_k \frac{\partial a}{\partial \phi_k} = \phi_k^* \frac{\partial a}{\partial \phi_k^*} = a \quad . \tag{2.1}$$

Such functions (without a reality condition) form an algebrasum, defined in the obviousway

$$(a+b)(\phi_k^*, \phi_l) = a(\phi, \phi^*) + b(\phi, \phi^*) \tag{2.2}$$

as is multiplication with a complex scalar ξ, then

$$(\xi a)(\phi, \phi^*) = \xi a(\phi, \phi^*) \tag{2.3}$$

On the other hand, the multiplication is defined by

$$a * b = \frac{\partial a}{\partial \phi_k} \frac{\partial b}{\partial \phi_k^*} \tag{2.4}$$

This is an obvious generalization of the matrix multiplication in linear quantum mechanics, if $a = \phi_k^* A_{kl}\phi_l$ and $b = \phi_k^* B_{kl}\phi_l$, then $a * b = \phi_k^* A_{kl} B_{lm}\phi_m$.
One bilinear function will be important here, as in quantum mechanics; it is the norm

$$n = \phi_k^* \phi_k \tag{2.5}$$

which acts as a unit element for the product (2.4):, then it is obtained as follows

$$n * a = a * n = a \tag{2.6}$$

The product (2.4) is distributive, but neither commutative nor associative. From the failure of associativity will follow most of the differences between this formalism and ordinary quantum mechanics.

(III) Symmetries of the Theory for Class of Functions through the Symmetry Transformations

To generalize the linear transformations $\delta\phi_k = i\varepsilon A_{kl}\phi_l$ of quantum mechanics, Weinberg assumed that the change in the wave function associated with an infinitesimal function $a\varepsilon(\phi_k^*, \phi_l)$ is

$$\varepsilon\delta\phi_k = -i\varepsilon\frac{\partial a}{\partial\phi_k^*}. \tag{2.7}$$

Then the change in another function b, induced by $a\varepsilon$ is

$$\varepsilon\delta_a b = -i\varepsilon[\frac{\partial b}{\partial\phi_k}\frac{\partial a}{\partial\phi_k^*} - \frac{\partial b}{\partial\phi_k^*}\frac{\partial a}{\partial\phi_k}]$$

or

$$\delta_a b = i[a,b] = i[a * b - b * a] \tag{2.8}$$

Then this "commutator" in the Weinberg model is obviously antisymmetric:, i.e.,

$$[a, b] = - [b, a]. \tag{2.9}$$

The commutator in Eq.(2.8) satisfies the Jacobi identity

$$[a,[b,c]] + [b,[c,a]] + [c,[a,b]] = 0 \cdot \tag{2.10}$$

Therefore, the symmetry transformations in the Weinberg model meet the Lie algebras relationship as linear quantum mechanics, which is

$$[\lambda_i, \lambda_j] = iC_{ijh}\lambda_h \tag{2.11}$$

(IV) Time Dependence for the Variation of State

Weinberg defined a real "Hamiltonian" function, which is denoted by $h(\phi_k^*, \phi_l)$, and the time dependence of the wave function so that

$$\phi_k(t+\varepsilon) = \phi_k(t) + \varepsilon\delta_h\phi_k(t)$$

or,

$$\frac{d\phi_k}{dt} = -i\frac{\partial h}{\partial \phi_k^*} \tag{2. 12}$$

Weinberg thought that this is a time-dependent nonlinear Schrödinger equation. Equation (2.12) and its complex conjugate show that it is a classical Hamiltonian system, with real q' s and p' sgiven by

$$\phi_k = (q_k + ip_k)/\sqrt{2} \tag{2.13}$$

where the "commutator" in Eq.(2.8) is only proportional to the classical Poisson bracket, it meets also the Jacobi identity in Eq.(2.10), h is the bilinear $h = \phi_k^* H_{kl}\phi_l$,which is called "Hamiltonian function," to distinguish it from the quantum-mechanical Hamiltonian, which is H_{kl} .Any function a of the wave function $\phi(t)$ (and its complex conjugate), that satisfies (2.12), has the time dependence

$$\frac{da}{dt} = -i[a,h] \tag{2.14}$$

which is consistent with that in linear quantum mechanics. Obviously, there are two conserved functions: h itself and the unit element n, which according to Eq.(2.6) commutes with everything because of the failure of associativity. Meanwhile, a product $a * b$ of conserved functions is not necessarily conserved itself, which is different from linear quantum mechanics. In ordinary N-component quantum mechanics, the time-dependence of the wave function is at worst quasi-periodic; it is a sum of N simple harmonic complex terms, in general with incommensuratefrequencies. Many nonlinear Hamiltonian systems display quasi-periodicityof a more general sort; the 2N quantities $\mathrm{Re}\,\phi_k$ and $\mathrm{Im}\,\phi_k$ move on anN-torus, with

$$\phi_k(t) = \sum_{n_1......n_N} L_k(n_1......n_N)\exp[-i\sum_v n_v\omega_v t], \tag{2.15}$$

the difference being that in the nonlinear case the sum runs over all positive andnegative integers and the frequencies ω_k , as well as the constants C_k depend on theinitial conditions.

(V) Galilean Invarianceand Dynamic Equations of Microscopic Particles

Time-translation invariance can be represented in a larger group of symmetries, that of Galilean invariance. In this case,the functions h, p, j, and K,which respectively generate time translations, space translations, rotations, and boosts, are first defined.These quantities satisfy the following commutation relations:

$$[p, h] = [j, h] = [K_i, K_j] = 0\,,\ [K, h] = ip,\ [K_i, p_j] = i\delta_{ij}M$$
$$[j_i, p_j] = i\varepsilon_{ijk}p_k, [j_i, K_j] = i\varepsilon_{ijk}K_k, [j_i, j_j = i\varepsilon_{ijk}j_k \tag{2.16}$$

where M commutes with everything. (All "commutators" are defined as in Eq. (2.8)). It is not easy (though perhaps not impossible) to find realizations of the sub-algebra spanned by j, K, p,and M. For a single free particle of mass m, described by the wave function $\Psi_k(X)$, these are

$$P = -i\int \Psi_k^*(X)\nabla\Psi_k(X)d^3x, j = \int \Psi_k^*(X)[-i(X\times\nabla)\delta_{kl} + S_{kl}]\Psi_l(X)d^3x,$$
$$K = m\int \Psi_k^*(X)X\Psi_k(X)d^3x, M = m\int \Psi_k^*(X)\Psi_k(X)d^3x = mn \tag{2.17}$$

where S isthe Hermitian spin matrice satisfying the algebra S x S = iS.The conventional bilinear kinetic energyof the system is represented by

$$h = -\frac{1}{2m}\int \Psi_k^*(X)\nabla^2\Psi_k(X)d^3x + h' \tag{2.18}$$

Any h' that commutes with j, p, and K provides a realization of the Galileanalgebra.The internal energy h' is the integral over x of ahomogeneous scalar function of $\Psi_k(X)$ and $\Psi_k^*(X)$, i.e.,

$$h' = -\frac{1}{2m}\int d^3xH'(\Psi_k^*(X), \Psi_k(X)) \tag{2.19}$$

where H' satisfies the rotational invariance and homogeneity conditions

$$0 = \frac{\partial H'}{\partial \Psi_k}S_{kl}\Psi_l - \frac{\partial H'}{\partial \Psi_k^*}S_{kl}^*\Psi_l^* \quad H' = \frac{\partial H'}{\partial \Psi_k}\Psi_k = \frac{\partial H'}{\partial \Psi_k^*}\Psi_k^*, \tag{2.20}$$

with total Hamiltonian function given by (2.18) and (2.19). Thus, the dynamic equation of the free particle is the time-dependent Schrödinger equation, which is the form of

$$i\frac{\partial \Psi_k}{\partial t} = \frac{\partial h}{\partial \Psi_k^*} = -\frac{1}{2m}\nabla^2\Psi_k + \frac{\partial H'(\Psi, \Psi^\#)}{\partial \Psi_k^*} \tag{2.21}$$

where H' could be taken in the form $H' = \sum_s \rho^{-s}H_s'$, where the sum runs over integer s = 0, 1, 2, . . ., $\rho(\Psi_k, \Psi_k^*) = \Psi_k^*\Psi_k$, ρ is the density and H_s', is a sum of terms proportional to 1 + s factors of Ψ_k and Ψ_k^*.

Because of the homogeneity (2.20) of the Hamiltonian density H', Weinberg thought the solutions of the time-dependent Schrödinger equation(2.21) can be denoted by separable solutions

$$\Psi_k(x,t) = \varphi(x,t)\psi_k(t) \tag{2.22}$$

with

$$i\frac{\partial\varphi}{\partial t} = -\frac{1}{2m}\nabla^2\varphi$$

and

$$\frac{\partial\psi_k}{\partial t} = \frac{\partial H'(\psi,\psi^\#)}{\partial\psi_k^*} \tag{2.23}$$

However, Weinberg thought that the dynamic equation of the microscopic particles should not choose the following nonlinear Schrödinger equation

$$i\frac{\partial\psi}{\partial t} = -\frac{1}{2m}\nabla^2\psi + \lambda|\psi|^2\psi \tag{2.24}$$

because it lacks the crucial property of homogeneityaccording to which, if ψ is a solution of this equation, then $Z\psi$ is also its solution, where Z is a constant. But he thought the following logarithmic Schrödinger equation proposed by Bialynicki-Birula and Mycielski [12]

$$i\frac{\partial\psi}{\partial t} = -\frac{1}{2m}\nabla^2\psi - b\psi\ln|\psi|^2 \tag{2.25}$$

or the homogeneous non-linear Schrödinger equationsconsidered by Haag[13] and Kibble [14]

$$i\frac{\partial\psi}{\partial t} = -\frac{1}{2m}\nabla^2\psi + \varepsilon\psi A \bullet \nabla Arg\psi \tag{2.26}$$

or homogeneous non-linear Schrödinger equation proposed by Kibble [14]:

$$i\frac{\partial\psi}{\partial t} = -\frac{1}{2m}\nabla^2\psi + \lambda|\nabla\psi|\psi/|\psi|^2 \tag{2.27}$$

to be appropriate to describe the states of free-one-particle because they met all the feature of homogeneity. Shimony [15] suggested further that this equation (2.25) could be tested by neutron interferometry, exploiting the feature that a partial absorption of one of the neutron beams would reduce the value of $|\psi|^2$ and hence affect the subsequent change in phase of the neutron wave function. The experiment of Shull et al., [16] subsequently bound 3.4 x 10^{-13} eV on b. This was subsequently improved to 3 x 10^{-15} eV by measurements of neutron Fresnel diffraction [17]. However, this sort of experiment cannot detect the nonlinearities in local homogeneous equations, where the overall scale of the wave function has no effect on the evolution of the phase[11] because it is possible to observe atomic transitions or precession over times much longer than even a very cold neutron.

(VI) The Separated Systems in This Model

The above discussion of the Galilean group is only suited for a single particle. For the realistic physical systems, which are combined by many separate isolated systems, then the above relationships will be changed. The system with finite number of degrees of freedom is composed of two systems I and II with Hamiltonians $h_1(\psi,\psi^*)$ and $h_2(\varphi,\varphi^*)$, then Weinberg wrote the Hamiltonian of the combined system by [9-10]

$$h_{1+2}(\Psi,\Psi^*)=\sum_l h_1(\psi^{(l)},\psi^{(l)}*)+\sum_k h_2(\varphi^{(k)},\varphi^{(k)}*) \qquad (2.28)$$

where $\psi_k^{(l)}=\varphi_l^{(k)}=\Psi_{kl}$, Ψ_{kl} is the wave function of the combined system with $\mathbf{h_1}$ $+\mathbf{h_2}$. Then dynamic equation of Ψ_{kl} is denoted by

$$i\frac{d}{dt}\Psi_{kl}=\frac{\partial h_{1+2}}{\partial \Psi_{kl}^*}=\frac{\partial h_1(\psi^{(l)},\psi^{(l)}*)}{\partial \psi_k^{(l)*}}+\frac{\partial h_2(\varphi^{(k)},\varphi^{(k)}*)}{\partial \varphi_l^{(k)*}} \qquad (2.29)$$

The Schrödingerequation (2.29) has a solution of $\Psi_{kl}=\psi_k\varphi_l$, where ψ_k and φ_l are, respectively, obtained from the following equations:

$$i\frac{d}{dt}\psi_k=\frac{\partial h_1(\psi,\psi^*)}{\partial \psi_k^*} \text{ and } i\frac{d}{dt}\varphi_l=\frac{\partial h_2(\varphi,\varphi^*)}{\partial \varphi_l^*} \qquad (2.30)$$

The homogeneity of the Hamiltonian functions in this case is expressed by

$$\frac{\partial h_1(\psi^{(l)},\psi^{(l)}*)}{\partial \psi_k^{(l)*}}=\varphi_l\frac{\partial h_1(\psi,\psi^*)}{\partial \psi_k^*},\frac{\partial h_2(\varphi^{(k)},\varphi^{(k)}*)}{\partial \varphi_l^{(k)*}}=\psi_l\frac{\partial h_1(\varphi,\varphi^*)}{\partial \varphi_l^*} \qquad (2.31)$$

Thus the Schrödinger equation (2.29) has a solution of the form of $\Psi_{kl} = \psi_k \varphi_l$, provided that the factors ψ_k and φ_l satisfy the separate Schrödinger equations(2.30) for systems 1 and 2, respectively. This result, that the Schrödinger equation has a solution of the separable form, $\Psi_{kl} = \psi_k \varphi_l$, shows that Eq. (2.28) represents the correct way to combine non-interacting systems.

(VII) The Changes of Basis

Weinberg thought that the commutation relations and the time-dependent Schrödinger equation are left unchanged if we subject the wave function to a transformation $\psi_k \to \psi_k'(\psi, \psi^*)$,which is canonical in the sense that $[\psi_k', \psi_l'^*] = \delta_{kl}, [\psi_k', \psi_l'] = [\psi_k'^*, \psi_l'^*] = 0$ with "commutators" defined by Eq.(2.8). Such canonical transformations can be generated by real functions $f(\psi, \psi^*)$. In this case, the transformed wave function is [9-10]

$$\psi_k' = \overline{\psi}_k(1), \tag{2.32}$$

where

$$i\frac{d}{dt}\overline{\psi}_k(s) = \frac{\partial f(\overline{\psi}(s), \overline{\psi}^*(s))}{\partial \overline{\psi}^*(s)}, \overline{\psi}^*(0) = \psi_k \tag{2.33}$$

In linear quantum mechanics these transformationsare linear, in which the ψ_k is transformed by an element of Sp(2N). On the other hand, as in quantummechanics, Weinbergrequired that functions like have the same homogeneity properties when expressed by ψ_k' and $\psi_k'^*$ as when expressed by means of ψ_k and ψ_k^*. Then ψ_k' must be homogeneous of degree (1,0) in ψ_k and ψ_k^*:

$$\frac{\partial \psi_k'}{\partial \psi_l}\psi_l = \psi_k', \frac{\partial \psi_k'}{\partial \psi_l^*}\psi_k^* = 0 \tag{2.34}$$

Obviously, the generating function $f(\psi, \psi^*)$ must be homogeneous of degree one in both ψ and ψ^*. This means that the norm n is still invariant, i.e.,

$$n = \psi_k^* \psi_k = \psi_k'^* \psi_k'. \tag{2.35}$$

In one word, Weinberg presented a general framework for introducing nonlinear correctionsinto linear quantum mechanics, which can serve as a guide to experiments that would be sensitive to such corrections. In this class of generalized theories described here, the equations thatdetermined the time-dependence of the wave function are no longer linear, but are of Hamiltonian type, although its explicit form has not been given. Also, the wave functions of the particles,which differ by a constant factor, represent the same physical state and satisfy the same time-dependence equations. As a result, there is no difficulty in combining separated subsystems.

2.2.3. Some Problems Discussed by Weinberg's Formalism

(I) The Eigenvalue Problem in Weinberg's Model

It is well known that the state represented by a wave function ψ_k has a definite value α for the observable represented by a matrix A_{kl} , if ψ_k is an eigenvector of A_{kl} , with eigenvalue α in linear quantum mechanics. Weinberggeneralized [9-10] this by assuming that a state represented by ψ_k has a definite value α for the observable represented by a non-bilinear function $\alpha(\psi,\psi^*)$, if and only if

$$\frac{\partial a}{\partial \psi_k^*} = \alpha\psi_k \tag{2.36}$$

and

$$\frac{\partial a}{\partial \psi_k} = \alpha\psi_k^* \tag{2.37}$$

(where as a bilinear $\psi_k^* A_{kl}\psi_l$, this is just the usual eigenvalue condition of quantum). This means that if ψ_k satisfies Eq.(2.36) or Eq.(2.37), then $Z\psi_k$ is also so, where Z is an arbitrary complex number. The α is determined by [9-10]

$$\alpha = \frac{a(\psi,\psi^*)}{n(\psi,\psi^*)} \tag{2.38}$$

where ψ is called the eigenvector satisfying Eq.(2.36). This indicates that α is real if **a** is. In this case, Eq. (2.37) just follows from Eq. (2.36) by taking the complex conjugate. However, for general complex functions, Eq. (2.37) must be taken as a separate condition. If**a**is a symmetry generator, then we can obtain from Eq.(2.36) the relationship of $\delta_a\psi_k = -i\alpha\psi_k$ .This means that ψ is invariant up to a phase under the transformation generated by **a.**In particular, a state with a definite energy value E has a time dependence given by Eqs.(2.12) and (2.36) as $\psi \propto \exp[-iEt]$ just as in linear quantum mechanics. In linear quantum mechanics, the eigenvalues of an observable are denoted by a matrix $\psi_k^* A_{kl}\psi_l / \psi_k^*\psi_k$ .The corresponding variational principle holds, i.e., the eigenvectors of an observable represented by a function $a(\psi,\psi^*)$ are the stationary points of the function

$$\bar{a} = a(\psi,\psi^*)/n(\psi,\psi^*) \tag{3.39}$$

andthe eigenvalues are the values of $\bar{a}$ at these stationary points.Therefore, the function in Eq. (2.39) is the expection value of the observable represented by $a(\psi,\psi^*)$ in the state with wave function ψ. The function $\bar{a}$ in Eq.(2.39) is invariant under multiplication of ψ byan arbitrary complex number Z,

$$\bar{a}(Z\psi, Z\psi^*) = \bar{a}(\psi,\psi^*) \tag{2.40}$$

This manifests that for an N-component wave function ψ_k , $\bar{a}(\psi,\psi^*)$, being defined not on the space of N complex variables, but on the projective space CP^{N-1}, a compact space [7-8].

At the same time, Weinberg's prescriptions were given for determining the states in whichobservables have definite values and for calculating the expectation values of observables forgeneral states, but the calculation of probabilities requires detailed analysis of the method ofmeasurement. A study is presented of various experimental possibilities, including the precession of spinning particles in an external field, experiments of Stern-Gerlach type, and the broadening and detuning of absorption lines, which are described simply as follows.

(II)Spinning Particles in an External Field

Weinberg discussedthe properties of spinning particles (with spatial motion ignored) in various sorts ofweak external field including spin 1, 1/2 and 3/2 that can show departures from linear quantum mechanics.Weinberg examined the properties of a particle of spin 1 or 1/2 in a weak uniform magnetic field and found that all of these are integrable systems with general time-dependence described by Eq.(2.15). He calculated not only the energy eigenvalues but also the characteristic precession frequencies appearing in Eq.(2.15). Weinberg'scalculations were originally carried out for purposes of illustration because the characteristic frequencies play a key role in prediction of the system's spectrum for absorption of radiation.On the other hand Fortson and Chupp[10]pointed out that it was possible to monitor the precession of spinning nuclei coherentlyoverrather long times and measure the characteristic frequencies (or frequency differences) with great precision [9-10]. In this study, Weinberg pointed out that small departures from the linearity of quantum mechanics would show up as a weak dependence of the characteristic frequencies on the initial state of the nuclear spin. Both Fortson and Chupp measured these precision frequencies for Hg^{201} and N^{21},respectively. These nuclei both have spin 3/2, and although the time-dependent Schrodinger equation for spin 3/2 is integrable for a free particle or in a weak magnetic field, it is not integrable for the sort of external field that would be presentin such experiments. Weinberg[9-10] calculated further the precession frequencies for this case by using the method of averaging, and found an interesting dependence on the initial wave functionthrough treating the non-bilinear part of Hamiltonianh as a small perturbation.

For example, in the case of spin 1/2 and a homogeneous magnetic field,its most general Hamiltonian function is bilinear, which is [9-10]

$$h = E_0 n + \mu \psi_m^* (\vec{\sigma}) \psi_l \bullet \vec{B} \tag{2.41}$$

where $\vec{\sigma}$ is the usual Pauli matrice. There is no difference here between linear quantum mechanics and Weinberg's generalized theory because there is no nonlinear interaction. However, for a quadrupole electric field, in linear quantum mechanics a spin of1/2 particle cannot have an interaction of the first order with such a field, but now it can, uniquely in the Hamiltonian function

$$h = E_0 n + Q n^{-1} \psi^+ \sigma_j \psi \psi^+ \sigma_j \psi \left(\frac{\partial E_i}{\partial x_j} + \frac{\partial E_j}{\partial x_i} - \frac{2}{3} \delta_{ij} \nabla . \vec{E} \right) \tag{2.42}$$

where Q is a constant. Corresponding energy eigenvalues are $E_0 + Q\lambda$, where λ runs over the three eigenvalues of E_{ij}. Thus, a free spin 1/2 particle in a generic quadrupole electric field would have three distinct energy values, which are split by a very weak magnetic field into six energy values. If the strength of the magnetic field is increased so that themagnetic dipole term in h is much larger than the electric quadrupole, then the six energy eigenstates will merge into the two expected in linear quantum mechanics. In this case, the time-dependent Schrödinger equation is integrable, but its general solution is complicated. But in the case of axial symmetry, two of the eigenvalues of the quadrupole field tensor E_{ij} are equal. Taking the 3-axis in the direction of the eigenvector for the other eigenvalue, Weinberg wrote the Hamiltonian function [9-10], which is

$$h = n\bar{h}(a) , \tag{2.43}$$

where

$$\bar{h}(a) = Q\lambda_{11}(1-2a)^2 + 4Q\lambda_{\perp} a(1-a) , \quad a = |\psi_2|^2 / n, n = |\psi_1|^2 + |\psi_2|^2 . \tag{2.44}$$

The energy Eigen states and Eigen energy E can also be found by requiring that all components oscillate with the same time-dependent factor exp(-iEt). For all forms of the function $\bar{h}(a)$, there are at least two solutions:

$$\textbf{A: } \psi_1 \neq 0 ,\ \psi_2 = 0 ,\ E = E_A = \bar{h}(0) ,\ \textbf{B: } \psi_1 = 0 ,\ \psi_2 \neq 0 ,\ E = E_B = \bar{h}(1) \tag{2.45}$$

In addition, for sufficiently strong nonlinearities the function $\bar{h}(a)$ has a stationary point at some a_0 andin the range $0 < a_0 < 1$, and, in this case, there is a third class of solutions,

$$\text{C: } a = a_0, \bar{h}(a) = \bar{h}(a_0) \tag{2.46}$$

Weinbergalso studied the nonlinear properties of spins 1 and 3/2 particles using his model and the above method [9-10]. These results are different from those in linear quantum mechanics. Thus, we can use them to inspect the precision of linear quantum mechanics.

(III)The Investigations of Probabilities and Expectation Values in Weinberg's Model

In the historical development of linear quantum mechanics, the interpretation of the wave function by the probabilities occurred somewhat later than the interpretation of eigenvalues as the possible values of observables. Weinberg [9-10] also discussed the eigenvalue problem and its physical interpretation in his generalized version of linear quantum mechanics. As in linear quantum mechanics, hegave the probabilistic interpretation of the wave function. As it is known, in this case there are various ways of formulating the interpretative postulates of linear quantum mechanics. In one common version, it is assumed that the measurement of an observable represented by an operator **A** must give a result equal to one of the eigenvalues of **A**, with a probability equal to the absolute value squared of thecomponent of the wave function along the corresponding eigenvector. This assumption is not appropriate toWeinberg's generalized version of linear quantum mechanics; the analysis of the measurement process using the nonlinear modelindicated that the possible values obtained when we measured some observable in a general state depend on the wave function of that state.

As a natural generalization, Weinberg assumed that in a state represented by a wave function ψ_k the measurement of a set of observables A,B,.... are represented by commuting real homogeneous functions $a(\psi,\psi^*)$, $b(\psi,\psi^*)$,......(commuting in thesense that $a * b = b * a$, etc.) will yield as average values, the expectation values are [9-10]

$$\langle a \rangle_\psi = a(\psi,\psi^*) / n(\psi,\psi^*), \langle b \rangle_\psi = b(\psi,\psi^*) / n(\psi,\psi^*), \tag{2.47}$$

This is just a postulate.

(IV) The Properties of Spectral Lines and Photon Emission in Weinberg's Model

It is known that there are various ways of measuring the spectrum of an atomic or nuclearsystem, in which one can observe the resonant absorption or scattering of individual photons by the system, or, at an opposite extreme, one can observe the effects on the system of a macroscopic oscillating electromagnetic field, containing unlimited numbers of photons. Otherwise, the atomic or nuclear system can be at zero temperature, and hence in the lowest energy Eigen state, or at a high temperature, in which the individual atoms or nuclei are almost uniformly distributed among all possible superposition of states, with only a small excess of lower energy states. Of course, the distributioncan also be non-thermal, as in the measurement of emission spectra, and here also individual atoms or nuclei can or cannot be initially in pure energy eigenstates. Forlinear quantum mechanics, none of these experimental details make much difference in the interpretation of observed spectral frequencies:Weinberg [9-10] got absorption or emission lines at frequencies simply equal to differences of energy eigenvalues. Hegavethat the matters are very different in his nonlinear generalization of quantum mechanics. To explore the various possibilities, hestudied the simplest realistic case and assumed that the atomic or nuclear system may be described by a 2-component wave function ψ_k with k =1 or 2. Also, he assumed that only one normal mode of the electromagnetic field is excited, with frequency ω. The wave function of the system of matter plus radiation is then $\Psi_{k,n}$, where k = 1 or 2, and n= 0, 1, 2,...gives the number of photons in the normal mode. Finally, Weinberg obtained the energy difference between two energy levels, which results in the emission of photons, are related to the average frequency difference.

At the same time, he designed one experiments to detect the changes in the characteristic frequency difference due either to the changes in the initial wave function or the changes in the wave function as it evolves.

(V)The Detuning of Resonant Transition in ^{9}Be Nucleus

Weinberg [9] looked further at the features of detuning resonant transition in ^{9}Be nucleus using his model, in which he proposed to set stringent limits on possible nonlinear corrections to linear quantum mechanicsby searching for the detuning of resonant transitions. A suggested nonlinear generalization of quantummechanics is used to show that such detuning would be expected in the radio frequency (rf) transition in $^9Be^+$ ions that isused to set frequency standards. Measurements at the National Bureau of Standards already set limitsof order 10^{-21}on the fraction of the energy of the Be nucleus that could be due to nonlinear correctionsto quantum mechanics, with good prospects of improving this by2-3 orders of magnitude. In this investigation, Weinberg gave the Hamiltonian of$^9Be^+$ ions [9].

In the absence of the time-varying external field, it is denoted by Eqs.(2.43)-(2.46). If $^9Be^+$ ions act by a small time-varying perturbation $\Delta h = g\psi_2^*\psi_1 \exp[-i\Omega t] + c.c.$, then there is a transition between states A and B, where g is a small coupling parameter, and Ω is a frequency, the time dependence of $\psi_k(t)$ is given by using Eq. (2.24), with $h+\Delta h$ in place of h .These two complex equations can conveniently be boileddown to a single real equation for the action $a = |\psi_2|^2 / n$, then the changes of a is denoted by [9]

$$\frac{da}{dt} = \{4|g|^2 a(1-a) - [\Omega a - \bar{h}(a) + \bar{h}(0)]^2\}^{1/2} \tag{2.48}$$

where $C = \bar{h}(0)$ is an integration constant. In order for the square rootin Eq.(2.48) to not have a negative argument at the ending point a =1, it isnecessary also that

$$\Omega = \bar{h}(a) - \bar{h}(0) = E_B - E_A \tag{2.49}$$

However, the $^9Be^+$nucleus has spin 3/2, if it is in a magnetic field due to its valence electron. In this case, of the absence of a magnetic field, the nuclear Hamiltonian is represented by

$$h_0=\frac{\varepsilon}{n}\sum_{\sigma}\{2\left|\sqrt{3}\psi(\frac{3}{2},\sigma)\psi(\frac{-1}{2},\sigma)-\psi(\frac{1}{2},\sigma)\psi(\frac{1}{2},\sigma)\right|^2+3\left|\psi(\frac{3}{2},\sigma)\psi(\frac{-3}{2},\sigma)-\psi(\frac{1}{2},\sigma)\psi(\frac{-1}{2},\sigma)\right|^2$$
$$+2\left|\sqrt{3}\psi(\frac{1}{2},\sigma)\psi(\frac{-3}{2},\sigma)-\psi(\frac{-1}{2},\sigma)\psi(\frac{-1}{2},\sigma)\right|^2\}$$

(2.50)

where ε is a small coefficient with the dimensions of energy, $n=\sum_{m,\sigma}\left|\psi(m,\sigma)\right|^2$ is the norm, $\psi(m,\sigma)$ is the component of the wave function with a nuclear spin z component equal to m and with quantum numbers for everything else in the problem labeled $\sigma=\pm 1/2$ for $^9Be^+$ions.

If the hyperfine interaction of the $^9Be^+$nucleus with the valence electron's spin, and the interaction of both nucleus and electron are in an external magnetic-field **B**, then ignoring possible non-bilinearterms and taking h as just the expectation value of theusual Hamiltonian of quantum mechanics, then the Hamiltonian of the system may be approximately denoted by [9]

$$h_{QM}=\sum_{m\sigma}\sum_{m'\sigma'}\psi^*(m',\sigma')\psi(m,\sigma)[\mu_e(J_e)_{\sigma\sigma'}.B\delta_{m'm}+\mu_N(J_N)_{m;m}B\delta_{\sigma\sigma'}+k(J_e)_{\sigma\sigma'}(J_N)_{m'm}]$$

(2.51)

where k is the strength of the hyperfine interaction. In the Bollinger's measurement [11], one starts with the $^9Be^+$nucleus in an energy eigenstate with m=-3/2, $\sigma=1/2$ with an rf field tuned to this transition. and drives a single-photon transition to m= --1/2, $\sigma=1/2$. Weinberg chose $\psi_1=\psi(-3/2,1/2)$ and $\psi_2=\psi(-1/2,1/2)$ and dropped other components, Eq.(2.50), then he gained a nonlinear term of $\overline{h}(a)=2\varepsilon a^2$. If the transition of $A\rightarrow B$ is observed to the driven in a time T,then he found out that$|\varepsilon|<2|g|=\pi/T$. from Eq.(2.49).The measurementswere finished [11] with T=1s at a bound on $|\varepsilon|$ of order 10^{-15}eV, which is less than the binding energy per nucleon of the$^9Be^+$nucleus by a factor of order 10^{-21}.This is improved by 1 or 2 orders of magnitude by the reduction of the rf power to lengthen the transition time T, orby use of the Ramsey trick, with several free-precessiontimes of several minutes at various stages in the transition.

2.2.4. Some Comments on the Correctness of Weinberg's Nonlinear Model

Gisin [18]indicated with an example that Weinberg's general framework for introducing nonlinear corrections into linear quantum mechanicsallows for arbitrarily fast communications, in which Weinberg showed how to construct within his framework, an arbitrarilyfast telephone line. In linear quantum mechanicsthere is also "something" that goes arbitrarily fast,since quantum mechanics violates the Bell inequality. But this non-locality does not allow any supraluminal communications.He planned an example of arbitrarily fast communications, which includes the following three steps.

First step. In order to construct the telephone line between the space regions A and B install a source somewhere between them. The source should continuously emit pairs of spin-1/2 particles, with the spins in the singlet state and the spatial coordinates in a state such that one particle travels towards region A and the second particle towards B. He assumed again that the particles move in the y-direction with the irrelevant speed. The line is installed once the flow of particles reaches regions A and B.

Second step. In order to send a message from A to B, the PTT officer in region A rotates a Stern—Gerlach apparatus inthe z-direction and the u-direction. For a single-bit message, the two directions are in the x−z-plane orthogonal to theincoming flow of particles, and are 45^0 from eachother. The way the inhomogeneous magnetic fieldacts on the particles is known from this evidence, in which there are twocounters. For each particle, one of the counters willclick. This process was repeated overmany particles. The flow of the second particles now exists in a mixture of spin states one half in theup state, one half in the down state, depending on thechoice of the A-operator. Note that the two mixturescorrespond to the same density matrix, half the identity, which is also the partial trace over the first spinof the singlet state.

Third step. In order to read the message, the operator in B must be able to distinguish between the two mixtures of spin-1/2 states. A linear evolution, as in linear quantum mechanics, is unable to make such distinctions, hence the impossibility of using quantum correlations to signal arbitrarily fast in linear quantum mechanics. But within Weinberg's generalframework, we have the nonlinear evolutions andthe nonlinear Hamiltonian:

$$h(\psi,\psi^*)=\langle\psi|\sigma_z|\psi\rangle/\langle\psi|\psi\rangle \tag{2.52}$$

where σ_z is the usual Pauli matrix. This is a simplecase of Weinberg's proposal for the interaction of aspin-1/2 and an (external) electric quadrupole. Gisin [18]assumed that the external field whose interaction with the spin-1/2 is described by the Hamiltonian (2.52) is confinedto region B. The evolution equation for the spin-1/2 state corresponding to the Hamiltonian (2.52) reads

$$d\psi/dt=-2i\langle\sigma_z\rangle\sigma_z\psi\text{ , where }\langle\sigma_z\rangle=\langle\psi|\sigma_z|\psi\rangle/\langle\psi|\psi\rangle.$$

If the initial state is either up or down in thez-direction, then the state is stationary. Then the mean value of the "longitudinal polarization"$\langle\sigma_y\rangle$ is always zero. If the initial state is either up or down in the u-direction, then the spinrotates around the z-axis with the same frequency, but the sense of rotation is opposite for the up anddown states. After a time of a quarter of the "Larmorperiod", the spin will have the same positive value of $\langle\sigma_y\rangle$. The B-operator measures then, on many particles, the mean value $\langle \sigma_y \rangle$ and can thus "read" the message sent by the A-operator. The most delicate point in the above example is probably the instantaneous preparation at a distance(second step). Gisin [18] thought that there is evidence for it. The experiments testing quantum mechanics against local hidden variables do not onlyviolate the Bell inequality, but they also agree remarkably well with quantum mechanics. This also supportsthe claim that if one spin of a singlet state pair is "found" to be in the up state, then the other spinis in the down state, for the same direction. MaybeWeinberg, or someone else, will give a different description of what happens in step two, butsomething certainly has to be said, or one accepts the usualinterpretation. A possible way out would be the assumption of spontaneous collapse of quantum correlations, as has been explored in refs. [19-24], but thisis outside Weinberg's framework.

On the other hand, Peres [25]demonstrated that nonlinear variants of Schrödinger's equation in the Weinberg model violate the second law of thermodynamics. He thought that the quantum-mechanical superposition principle asserts that the pure states of a physical system can be represented by rays in a complex linear space. This does not means that the evolution of these rays obeys a linearequation. Then we easily invent nonlinear variants ofSchrödinger's equation' such that if $\phi(0)$ evolves into $\phi(t)$ and $\psi(0)$ evolves into $\psi(t)$, the state represented by $\phi(0)+\psi(0)$ does not evolve into $\phi(t)+\psi(t)$, but intosome other pure state. Thus, Peres demonstrated that possible tests for such nonlinearities in the Weinberg model lead to a violationof the second law of thermodynamics. In this demonstration he considered a mixture of quantum systems representedby a density matrix $\rho=\lambda P_\phi+(1-\lambda)P_\psi (0<\lambda<1)$, where P_ϕ and P_ψ are projectionoperators on the pure states ϕ and ψ ,respectively. Thenon-vanishing eigenvalues of ρ are

$$w_j = 1/2 \pm [1/4 - \lambda(1-\lambda)(1-x)]^{1/2}, x = |\langle \phi | \psi \rangle|^2$$

The entropy of this mixture is $S=-k\sum w_j Lnw_j$ and satisfies dS/dx <0 for any λ. Hence, if the pure quantum states evolve as $\phi(0)\rightarrow\phi(t)$ and $\psi(0)\rightarrow\psi(t)$,then the entropy of the mixture ρ shall not decrease (i.e.that mixture shall not becomeless homogeneous) provided that

$$|\langle \phi(0) | \psi(0) \rangle|^2 \geq |\langle \phi(t) | \psi(t) \rangle|^2 .$$

In fact, if $\langle \phi(0) | \psi(0) \rangle = 0$, then we must have $\langle \phi(t) | \psi(t) \rangle = 0$ because an orthogonal states must remain orthogonal. Then, for a complete orthogonal set ϕ_k , for every ψ we have,

$$\sum_k |\langle \phi_k | \psi \rangle|^2 \equiv 1.$$

If there is some m for which, there is

$$\left|\langle\phi_m(0)|\psi(0)\rangle\right|^2 > \left|\langle\phi_m(t)|\psi(t)\rangle\right|^2,$$

then there must also be some n for which, there is

$$\left|\langle\phi_n(0)|\psi(0)\rangle\right|^2 < \left|\langle\phi_n(t)|\psi(t)\rangle\right|^2.$$

Therefore, the entropy of a mixture of ϕ_n and ψ will spontaneously decrease in a closed system, which violated directly thesecond law of thermodynamics.

If the second law must be retained, then we must have

$$\left|\langle\phi(0)|\psi(0)\rangle\right|^2 = \left|\langle\phi(t)|\psi(t)\rangle\right|^2$$

for ϕ and ψ. Then the phases must be chosen in such a way that,for every ψ,the mapping $\psi(0)\rightarrow\psi(t)$ is either unitary or anti-unitary. The second alternative is ruled out by continuity. In this case, the evolution of quantum states is unitary and the Schrödinger 's dynamic equation mustbe linear.

Therefore, we concluded that there are many contradictions and difficulties in Weinberg's model.

2.2.5. Some Applications of Weinberg's Nonlinear Model

Polchinski [26]thought that Weinberg's model [9-10] mightraise the possibility that the fictitious violation oflocality, which occurs in the Einstein-Podolsky-Rosen(EPR) experiment in linear quantum mechanics [27], mightbecome a real violation in the nonlinear theory. That is,the EPR apparatus might be used to send instantaneoussignals (this issue of information in nonlinear quantum mechanicshas been discussed from a different (thermodynamic) point of view). Thus he studied the relationship of Weinberg's nonlinear quantum mechanical modelwith the EPR paradoxand determined the constraints imposed upon observables by the requirement that transmissionnot occur in the EPR experiment. This leads to adifferent treatment of separated systems than that originally proposed by Weinberg [9-10]. He found that forbidding EPR communication in nonlinear theory necessarily leads to another sort of unusual communication, which occurs between different branches of the wave function [28].In this case, Polchinski considered two widely separated systems, I and II. Thewave function is Ψ_{ij}, where the indices i and j refer tothe two systems, running over 1, . . ., M and 1, . . ., N,respectively. At time t=O the wave function has beenprepared, with correlations between the two systems. Some later time, an observable a_{II} will be measured inthe receiving system II. It is sufficient to consider ordinary linear observables in system II. In the formalism ofWeinberg, linear observables are bilinear functions of $\Psi^+$and Ψ^*, i.e.,

$$a_{II} = \Psi^*_{ij} A_{jl} \Psi_{il} \tag{2.53}$$

where A_{jl} is a Hermitian matrix. At the receiving end, I will assumethat the usual idea of measurement applies. In system I,a signal is sent by turning on a field $\varepsilon(t)$, which couplesto a nonlinear observable $a_I = (\Psi_{ij}, \Psi^*_{kl})$. Then the whole Hamiltonian can be denoted by

$$h = \varepsilon(t) a_I (\Psi_{ij}, \Psi^*_{kl}) \tag{2.54}$$

This gives the equation of motion of a_{II}, which is

$$\frac{da_{II}}{dt} = i\{a_{II}, h\} = \frac{\partial a_{II}}{\partial \Psi_{ij}} \frac{\partial h}{\partial \Psi^*_{ij}} - \frac{\partial a_{II}}{\partial \Psi^*_{ij}} \frac{\partial h}{\partial \Psi_{ij}} \tag{2.55}$$

For a weak field, he found the expectation value of $a_{II}$is simplythe value of a_{II}, and so the condition was made that the measurement made in system II not depend on the field applied in system I, i.e., $\{a_{II}, a_I\} = 0$. In this can he obtained the no-signal condition allowing observation of anyfunction a_{II}, in which theobservables include all the usual linear Hermitian observables, the necessary and sufficient condition for anisolated system not to receive information via EPR correlations is that all observables in the system depend onlyon the density matrix for the system:

$$a_I = h_I [\textstyle\sum_m (\Psi_{im} \Psi^*_{km})] \tag{2.56}$$

However, the observables proposed by Weinberg for separatedsystems,

$$a'_I = \textstyle\sum_m h_I (\Psi_{im}, \Psi^*_{km}) \tag{2.57}$$

are not of this form (except when linear) and therefore allow EPR communication.

Polchinski [26]discussed and showed further that nonlinear observables of the form in Eq. (2.56) lead to unusual communication of another sort.The evolution of the wave function will be coupled to all other possible states. Thus, while the analysis does show that branches are coupled to one another, practical communication between branches may be drowned out by thecoupling to all the other branches of the wave function of the Universe. Finally,he obtainedthat Weinberg's nonlinear quantum mechanics leads either to communication via EPR correlations, or to communication between branches of the wave function.

The communication between branches of the wavefunction seems even more bizarre than faster-than-lightcommunication and consequent loss of Lorentz invariance, but it is not clear that it represents an actual inconsistency. This means that reduction of the wave function never occurs, so that the standard Copenhagen interpretation of quantum mechanics no longer applies. The, many-worlds interpretation of quantum mechanics, in that interpretation, a branch of the wave function, is a subspace in which a macroscopic system in a definite macroscopic state [28,29] becomes the natural one, with communication between the worlds now possible.

Gisin [19,30] has argued that nonlinear quantum mechanics, leads to EPR communication, both in Weinberg'stheory and more generally. He assumed reduction ofthe wave packet — that is, the projection postulate. Aswe have seen, without the projection postulate, EPR communication need no longer occur,provided that the observables in Eq.(2.57) for separated systems are replaced withthe form in Eq.(2.56).

Czachor [31]also showed, by explicit calculation using the form in Eq.(2.57)that EPR communicationoccurs in Weinberg's theory.

To what extent do the results apply to more general nonlinear quantum theories? Weinberg's theory is rather general, carrying over two main assumptions from thelinear theory. The first is that the equation of motion can be put in the form Eq.(2.55). Particularly,"observables"were defined as any quantities, which could be added to $h(\Psi,\Psi^*)$ in Eq. (2.54). This assumption, in any case,seems very well motivated, as it can lead to the usual connection between symmetries and conservation laws, andaid in the interpretation in other ways. The second assumptionis that observables are homogeneous of degree (1, 1) in Ψ and Ψ^*.

However, what are the experimental implications ofthese results? Please note that the communication between branches of the wave function is invalidatmost previous attempts to analyze the experimental consequences of nonlinearities, such as those in Refs. [9-10] and [32] because these analyses ignore previous branching of the wave function and treat macroscopic systems as though they begin in definite macroscopic states. A complete analysis requires consideration of theentire wave function of the Universe and is therefore rather complicated. This manifested that nonlinear effects are very important in some physical phenomena although it is small in some cases. If quantum nonlinearities are observed nonetheless,the thought experiments described herein would seem tobe simple enough to carry out in practice, thereby determining which EPR communication and Everett communication is actually realized.

Jordan[33] discussed and tested also the EPR assumptions without inequalities with two photonsor particles with spin1/2. He found that the inconsistency between quantum mechanics and EPR assumptions about locality and reality can be demonstrated inthe cases as in experiments that test Bell inequalities, but without inequalities, where all the possibilities are all appeared. The measurementchoices are different for the two particles, which are described by virtue of the angles for spin measurements. The calculations was finished by Pauli matrices for a simple example.

At the same time, Abrams[34] studied the relation of Weinberg's nonlinear theory with the polynomial-time solution for problems NP-Complete and #P(including oracle problems) problems. He demonstrated that nonlinear time evolution can be exploited to allow a quantum computer to solve NP-complete and #Pproblems inpolynomial time. He showed how to accomplish this exponential speed up using both the Polchinski and Weinberg models of nonlinear quantum mechanics. He believed that quantum mechanicsis in all likelihood exactly linear. Finally he obtained that if quantum states exhibit small nonlinearities during time evolution, then quantum computers can beused to solveNP-complete and #Pproblems in polynomial time. He provided the algorithms that solve NP-complete and #Poracle problems by exploiting nonlinear quantum logic gates. Using the Weinberg's model as a simple example, the explicit construction of these gates is derived from the underlying physics. He also presented nonlinear quantum algorithmsusing Polchinski type nonlinearities, which do not allow for superluminal communication.

On the other hand, Wodkiewiczet al., [35]derived the nonlinear generalization of the optical Bloch equations with radiative damping using Weinberg's nonlinear wave equation.

As it is known, Weinberg gave a general framework for the nonlinear corrections, which is based on thehomogeneity requirement. Weinberg proposed a nonlinear generalization of the Schrödinger equation involving resonant transitions between only twoatomic levels in its simplest version, as mentioned above. This equation was tested experimentally to set limits on possible corrections to linear quantum mechanics. A hyperfineclock transition of $^9Be^+$has been monitored to test the Weinberg equation. The experiment has set an upperlimit for the nonlinear correction characterized by a coupling constant ε . This upper limit is less than 4 parts in $10^{27}$of the binding energy per nucleon for the $^9Be^+$nucleus [32]. Consecutive experiments performed in ^{22}Ne [36] andin a hydrogen-maser transition [37] have also indicated thatthe impact of the nonlinearity on atomic transitions isvery small.Inview of the experimental results, it showed that ε is verysmall for atomic transitions, thusWodkiewiczet al., investigated only thephysical consequences of the Weinberg theory in a regime, in which coherent transients or spontaneous emissioneffects occur, using comparing the Weinberg's theory with the Fermi nonlinear radiation-reaction wave equation [38] and with Jaynes's neoclassical theory [39].From Weinberg's theory,they derived the nonlinear generalization of the optical Bloch equations with radiativedamping. In the framework of these resonance equations, they further studied the implications of the nonlinearity on coherent transients. The nonlinear constant, ε , which is small if compared to the binding energy per nucleon, can be comparable to the radiative lifetime of an excited state of anion stored in an ion trap. In this case,the nonlinear modification of the optical Bloch equationsled to large-scale observable effects in atomic quantum jumps, resonance fluorescence, or emission lineshapes. These effects will bring the actual upper limit for ε down considerably.

From Eq.(2.25) [9-10] and considering optical application as well as including the radiative decay of the excitedstate and the near-resonant interaction of the two-levelatom with coherent source of radiation [40], Wodkiewiczet al., [35] wrote the following Weinberg optical Bloch equations(WOBE) for the nonlinear wave mechanics, which is the nonlinear equation obtained using the three components of the Bloch vector u, v, and ω

$$\dot{u} = \frac{A}{2}u - \Delta_\varepsilon v - 2\varepsilon\omega v, \dot{v} = \frac{A}{2}v - \Delta_\varepsilon u + 2\varepsilon\omega u + \Omega\omega, \omega = A(1+\omega) - \Omega v \tag{2.58}$$

where A is the standard radiative line width, Ω is the coherent Rabi frequency, and the detuning is defined as $\Delta_\varepsilon = \omega_0 - \omega + 2\varepsilon$, ω_0 is the resonance frequency in the absence of nonlinear corrections.

If spontaneous emissionis neglected in the nonlinearWOBE, the resulting equations are identical to the Fermi or neoclassical equation with A =0. In this case, the Weinberg nonlinearity is mathematically equivalentto the neoclassical Lamb shift. Physically, the two theories are completely different. The neoclassical theory has the nonlinearity purelyelectromagnetic in origin witha dynamical formula for ε . The nonlinear wave mechanicshas no formula for ε and the nature of the nonlinearityis not specified.In the limit of no spontaneous emission, the dynamical predictions of neoclassical theory are identical with theWeinberg nonlinear mechanics.

However,if spontaneous emission is included, then the nonlinearWOBE have the form given by Eqs. (2.58). The nonlinearity changes only the time evolution of the atomic dipole moment d(t) These changes directly affect the frequency spectrum of spontaneous emission. But linear quantum mechanics predicts only a Lorentzian lineshape with a width governed by the spontaneous emissiondecay rate. Because of the non-exponential decay of theatomic dipole moment the frequency spectrum is nolonger Lorentzian. In fact, this spectrum can be obtained exactly through a Fourier transform of the atomicdipole moment [38]. Thus, they obtained that the nonlinear wave mechanics predicts achange of the spontaneous emission power spectrum. The spectrum is non-Lorentzian and nonsymmetric. Deviations from the standard Lorentzian shape are governed by the ratio of the nonlinearity to the radiative decay rate ε /A.

If an atom with radiative decay is driven by an external coherent source of light, then the system will emit the resonance fluorescence photons. The power spectrum of this fluorescence in the case of linear wave mechanics is known. If $\Omega > A$ this spectrum exhibits the Mollowtriplet [41]. If nonlinear corrections are included, the nonlinear WOBE for strong field were unable to be solved. The quantity that we could easily calculate was the steady-state total fluorescent intensity of resonancefluorescence. Then this fluorescence isdirectly related to the atomic inversion $\omega(t)$ .This isin sharp contrast to the pure spontaneous emission; the inversion can be modified by the nonlinearity if the system isdriven by Ω . For nonlinear wave mechanics,we have $\varepsilon \neq 0$ and as a result, the steady-state inversion has to satisfy a cubic equation, which can be solved using the Cardan formulae.

The atomic population of the upper level as a function of the atomic detuning has a well-known Lorentzian line shape in standard wave mechanics. For $\varepsilon \neq 0$, the non-linearity changed the line shape.

As it is well known, this fluorescence is directly related to the atomic inversion. Because the nonlinearity, which is proportional to ε , modifies the resonant frequency, then the shift of the resonant frequency occurs according to the formula in $\Delta_\varepsilon = \omega_0 - \omega + 2\varepsilon$. Thus, they finally gained the steady-state atomic population, which serves as a function of the atomic detuning in weak excitation of Ω =0.5 A and three different values of the nonlinearity ε =0. 1A, ε =0. 5A, and ε =A. Through comparison with the standard quantum mechanical result,they observed a clear asymmetry of the absorption profile, which is shifted to the left of the resonance.The peak values occurred in these cases are power broadened and are shifted due to the nonlinearity.

Therefore, they [35] finally concluded that if spontaneous emission is ignored, the dynamical evolution is governed by elliptic functions instead of sinusoidal functions. Deviations from ordinary quantum mechanics are governed in this case by the ratio Ω / ε .The nonlinear corrections also modified the quantum jumps of fluorescence of ions stored in a trap. For pure spontaneous emission, nonlinear corrections do not change the decay of atomic population but change the spontaneous emission line shape. The line shape becomes non-Lorentzian, the deviations from standard quantum mechanics are governed by the ratio ε /A.The atomic population is influenced by the nonlinearity only if the atom is driven by an external source of radiation. In this case the steady-state resonance fluorescence will be modified by the nonlinear interaction andthe deviations from standard quantum mechanics are governed by a combination of the two ratios Ω / ε and A / ε . In this study, they found that nonlinear modifications of the optical Bloch equations might show a chaos state.This is an interesting result.

2.2.6. Discussion of the Essence of Weinberg's Nonlinear Model

Walsworth et al., [42]demonstrated that Weinberg's nonlinear generalization of quantum mechanics for internal (spin) degrees of freedom [9-10]described, in essence,the properties of systems of composite, or multivalued, spin, such as atoms and molecules, which were obtained from the researches of the generalized Hamiltonian function, energy eigenvalues, and state-vector time dependence. Their investigations are described as follows.

As is pointed by Wigner [43]: "Itis perhaps the most fundamental principle in quantum mechanics that the system of states forms a linear manifold" and "the possibility of a future nonlinear character of the quantummechanics must be admitted of course". Weinberg's nonlinear quantum model provided a framework for the developmentand analysis of experimental tests of the linearity of quantum mechanics. In this theory, as in linear quantum mechanics, physical states are represented by rays (statevectors) in a complex vector space, and physical observables by the (real) generating functions for operators acting on these rays. In this model, however, these operatorsare not linear in general;the distributive property of the operators is loosened, the homogeneity conditionand Galilean invariance were retained. The distributive and homogeneity properties constituted thelinear superposition principle, thus the homogeneity was kept in a realistic theory since it rules out any special normalization of physical states and is essential to a proper treatment of non-interacting separated systems [3]. Weinberg found that nonlinear terms are allowed in the Hamiltonian, and thus the equations of motionof the internal(spin) degrees of freedom of a particle of spin ≥ 1 are changed. The magnitudes of these nonlinear terms are set by parameters, assumed small, such that the generalized theory reduces linear quantum mechanics in the limit so that the parameters go to zero. Therefore, many of the physical features and implications resemble the behaviors of a classical anharmonic oscillator, such as the energy eigenvalues can be shifted; the time dependence of state vectors is not in general an exponential function of a single and fixed frequency; and the resonant frequency of transitions between eigenstates depends on the relative populations of the states, which was exploited experimentally by Bollinger et al., [32]to put a low upper limit on the size of a nonlinear contribution to the quantum mechanics of the spin of the ^{9}Be nucleus and different nuclear systems [36].However, Weinberg's formulation does not address the possibility of nonlinear terms in the Hamiltonian of the translational (spatial) degrees of freedom or deal with the interacting multiple-body problem. Nor does the current form of the theory allow for systems of composite (or multivalued) spin, such as atoms or molecules.

Walsworth et al., [42] considered and studied the last of these problems, in which the construction of a nonlinear multivalued spin Hamiltonianwas found to be a straightforward extension of Weinberg's theory for single-valued spin systems. Most physically significant results, such as the energy eigenvalues and time dependence of state vectors, etc., resemble the single-and multi-valued spin cases.

A notable difference between them is that the nonlinear behavior is allowed for the spin-0 and spin-1/2 states of composite spin systems due to the mixture of state probability among spin manifolds. This new result obtained by them has allowed an experimental upper limit to be placed on the magnitude of nonlinear corrections to the quantum mechanics of the spin of the hydrogen atom [37]. The similarities and differences between Weinberg's modelof single- and multi-valued spins are demonstrated in two example systems: structure of the hydrogen atom (H) and the deuterium atom (D).

In one word, Walsworth et al., [42] extended Weinberg's nonlinear model to treat the systems of composite spin. Among the observable consequences of such nonlinearity is the dependence of the resonant frequency of transitions between the eigenstates on the populations of certain of the system's states. The theory puts no constraints on the magnitudeof the nonlinear effects. Weinberg thought that a scaling of limits on the ε parameters relative to the characteristic energy could vary the spin of the particle under consideration [18,27].

This form of comparison may be correct if nonlinearities are inherent in quantum mechanics and independent of the system. Then linear and nonlinear effects would scale similarly with the characteristic energy of the system, and single-valued spin nuclear systems should exhibit nonlinear behavior of a greater absolute magnitude than composite spin atomic systems. However, if the nonlinearity results from new physical mechanisms, such as aninternal coupling of the states, then the size of the effect depends on the nature of this coupling and the system.

In this case, the observable consequences can onlyexist only if the ε nonlinear parameters vary with the effective particlespin and/or with the constituent particle spins in thecomposite spin case. But the mixed ε 'parameters, which are associated with the significant results, go to zero if the underlying ε parameters have no spin dependence, which equal all nonlinear terms inthe generalized Hamiltonian function, being proportional to the norm n. Therefore, Walsworth et al.'s research result [42]tells us that Weinberg's nonlinear model can only be used to study the observable nonlinear behavior in a composite spin system.

At the same time, Peres [27] and Gisin [18] pointed out and questioned the compatibility of Weinberg's nonlinear model with the second law of thermodynamics and "special relativity" as mentioned above. On the other hand,the author also thinks that Weinberg's model is only a nonlinear quantum mechanical model, and cannot be called a nonlinear quantum mechanical theory because it does not give concrete and explicit dynamic equations, which are satisfied by the microscopic particles, although so-called nonlinear Schrödinger equations, (2.12), or (2.21) and (2.24) were exhibited. In this case, this model did not obtainthe basic properties and natures of motion of microscopic particles, especially since the wave-corpuscle duality, which the microscopic particles must possess, have not been obtained from Weinberg's formalism at all.

In view of the above questions, Weinberg's nonlinear model has not been further researched and developed after 2000. Thus, we can say that nonlinear quantum mechanics and its theory have not yet been established.

2.3. Simple Descriptions of Various Different Nonlinear Quantum Mechanical Ideas and Suggestions

2.3.1. The Geometric Model of Nonlinear Quantum Mechanic Suggestion

As it is known, a geometric description of quantum mechanics, independent of "wave functions" and "complex amplitudes",wasfirstproposed by Birkhoff and Neumann [44] and Piron [45]. The description explores a partial order relation in an idealized set of "yes-no measurements" called"quantum logic". The resulting approach,although mathematically profound, is not physically complete. Meanwhile, the algebraic approach reflecting the physics of operations, which can be performed on statistical ensembles, was also introduced into an axiomatic quantum field theory by Haag and Kastler [46]. On the other hand, Segal [47] and Ludwig [48] provided a so-called "convex" to explore the convex structures of quantum mechanics with special attention concentrated on the convex set of all states (pure and mixed) of a quantum system, which is developed further in [49-57]. The synthesis of the convex and the algebraic approaches has been gradually achieved[46,49-54,58-61]. It brought the complete geometrization of quantum mechanics, including the description of the present formalism of Hilbert spaces in terms of physical axioms [48,53,56,58-61], the general classification of the operation [52,59], the definition of filters as endomorphism of a convex cone of beams [49,54] and finally, the construction of the transition probabilities as affine geometric invariants of a convex set [54].Mielnik [62] thought that the convex set theory may play an important role in quantum physics as the Riemannian geometry in general relativity [54] and showed that the"convex scheme" is flexible enough to comprise nonlinear versions of quantum mechanics in which a nonlinear wave equation plays the role of a Schrödinger equation, such as

$$i\hbar\frac{\partial\phi}{\partial t}=-\frac{\hbar^2}{2m}\nabla^2(|\phi|^2\phi).$$

He generalized(nonlinear)quantum mechanics and tried to build a nonlinear quantum theory, in which a convex scheme of quantum theory is outlined where the states are not necessarily the density matrices in a Hilbert space. The physical interpretation of the scheme is given in terms of generalized “impossibility principles”. The geometry of the convex set of all pure and mixed states (called a statistical figure) is conditioned by the dynamics of the system. With this idea and aim, the geometric description of quantum mechanics based on the convex set theory is provided; the geometry of a system is related with the dynamics, which allows the construction of the convex manifolds of quantum states for systems obeying generalized wave mechanics. Thus,he provided a method of constructing the statistical figures for nonlinear variants of quantum mechanics, where the superposition principle is no longer valid. Examples of that construction are given and its possible significance for the interrelation between quantum theory and general relativity is discussed, some applications of the resulting scheme are indicated,and its relation to other physical theories,was discussed by them.

After this Haas et al., [63]commentedon Mielnik’s generalized quantum mechanics, in which he first judged the nonlinear Schrödinger equations.

$$i\hbar\frac{\partial\phi}{\partial t}=\frac{\hbar^2}{2m}\nabla^2(|\phi|^2\phi)+V|\phi|^2\phi \text{ and } i\hbar\frac{\partial\phi}{\partial t}=\frac{\hbar^2}{2m}\nabla^2\phi+V|\phi|^2\phi+U\phi \tag{2.59}$$

are not appropriate to serve as the dynamic equations in the nonlinear quantum theory because their solutions do not meet the homogeneity condition: $\partial(c\phi)/\partial t=c\partial\phi/\partial t$. They thought that theviolation of the homogeneity condition has the consequence that wave functions with the same shape but differing from the norm represent different pure states. The number $\|\phi\|$ is a constant of motion,which although related to the conservation of the number of systems in the ensemble, is not a normalization factor but influences the ratios of counting rates. This is not a decisive objection against such models but at this stage it appears perhaps more natural to consider wave equations which, although nonlinear, satisfy the above homogeneity condition. Thus, they chose the following nonlinear Schrödinger equation:

$$i\frac{\partial\phi}{\partial t}=-\nabla^2\phi+U\phi+\vec{A}\bullet\nabla S_\phi\phi,...(\hbar=1,m=1/2,\nabla S_\phi=\frac{\phi^*\nabla\phi-\phi\nabla\phi^*}{2i|\phi|^2}) \tag{2.60}$$

to serve as the dynamic equation of the particles because it satisfies the homogeneity condition, where ∇S_ϕ is the gradient of the phase of ϕ , the external agents are a scalar potential U(x, t) and vector potential A(x, t).Therefore, as in Schrödinger theory, the pure states are in one-to-one correspondence with rays in L^2. The availability of the "vector potential" A, as a possible external agent, makes a finer distinction of mixed ensembles possible.

Thus they proved that if $\Phi_i(i=1,2,....n)$ and $\Phi'_j(j=1,2,....n')$ are two different sets of pure states then two ensembles,

$$\omega=\sum_i\lambda_i\Phi_i \text{ and } \omega'=\sum_j\lambda'_j\Phi'_j ,$$

are always distinguishable (unequivalent).Thus,they obtained that the motion by the nonlinear A-term in this model makes mixtures of different sets of pure states distinguishable. As far as only single particle systems are considered one might claim that $|\phi(x)|$ and $\nabla S(x)$ describes the "objective state" of an individual particle. This means not that one achieves a classical theory (where $|\phi(x)|$ and $\nabla S(x)$ are hidden variables of the particle) because problems with such an interpretation would immediately arise as soon as one considers 2-particle systems. Therefore, they argued that in this model the set of (mixed) states is a simplex.It is argued that in Mielnik'smodel [62], the set of (mixed) states is a simplex.

2.3.2. A Neoclassical Model of Nonlinear Quantum Mechanics

Jayneset al., [39] advocated atheory closely related to the Fermi idea of radiation reaction in the mid 1960s. Jaynes's theory does not require electromagneticfield quantization, and dynamical effects like spontaneous decay, which are caused by a nonlinear wave equation.

The best illustration of Jaynes's ideas can be acquired in the framework of only two resonant transitions drivenby a coherent near-resonant radiation. Without radiation reaction, the dynamics of such a two-level atom can be described by the optical Bloch equations involving the three components u, v, and w of the Bloch vector [39].The neoclassical theory of Jaynes leads to the following nonlinear modification of the atomic Bloch equations for a two-level atom irradiated by an external coherentsource of light:

$$\dot{u} = -\Delta v + Auw/2 - 2 \in w, \dot{v} = -\Delta u + Auw/2 + 2 \in uw + \Omega w,$$

$$\dot{w} = A(w^2 - 1)/2 - \Omega v$$

where Ω is the constant coherent Rabi frequency of the external light source, $\Delta = w_0 - w$ is the detuning of theatomic frequency w_0 from the frequency w of the driving light, A is identical to Einstein's coefficient of spontaneous emission, or the standard radiative linewidth, and $\in = 2 \wedge w_0^2 d^2 / 2\pi\hbar c^3$ is the neoclassical and divergent Lamb shift, $\wedge$ is the frequency cutoff parameter, and d is the dipole moment. Note that the nonlinearity in the above Bloch equations has a purely electromagnetic origin.The fundamental difficulties of such a nonlinear theoryhave been recognized by Jaynes, himself, and after almost a full decade of theoretical investigations and experiments, the neoclassical theory has been abandoned.

Although the derivation of the nonlinear Bloch equations was given shortly in the examplewith the neoclassical Jaynes theory [39],Weinberg's theory also cannot directly give predictions equivalent to those based on the neoclassical approach. Thus,Wodkiewicz and Scully [35] compared their semi-classical results with the above results and included further the near-resonant interaction of the two-levelatom with a coherent source of radiation. In these couplings the possible non-bilinear terms emerging from higher-order processes were also ignored due to the nonlinear character of wave mechanics. In this case, the nonlinear quantities chosen by Wodkiewicz and Scully correspond in their notation to: $\varepsilon_2 = 0$ and $\varepsilon_1 = \sqrt{2\hbar\varsigma}$ .This choice leads to the "detuning" ,$\Delta' = \Delta - 2\varsigma$ .If no spontaneous emission is phenomenologically introduced into the Bloch equations, then from Weinberg's model they obtain:

$$\dot{u} = -\Delta' v - 2\varsigma\omega v, \dot{v} = -\Delta' u + \Omega\omega + 2\varsigma\omega u, \dot{\omega} = -\Omega v$$

If we now let $\Delta' = 0$, then it becomes

$$\dot{u} = -2\varsigma\omega v, \dot{v} = \Omega\omega + 2\varsigma\omega u, \dot{\omega} = -\Omega v$$

These equations in their nonlinear quantum mechanical suggestion have an additional conserved quantity, $u^2 + v^2 + u(\Omega / \varepsilon)$, which we shall put equal to 0 in this special case. Differentiating again the last Bloch equation the following equation is obtained

$$\ddot{\omega} = -2\varsigma\omega u - \Omega^2\omega, \dot{\omega} = (2\varsigma^2 - \Omega^2)\omega - 2\varsigma^2\omega^3$$

where Ω is the semi-classical Rabi frequency. This is an explicitly nonlinear equation. Since the results are obtained based on the Weinberg's nonlinear quantum mechanical model, we call it a nonlinear quantum mechanics model.

2.3.3.The Operator Model of Nonlinear Quantum Mechanics

A lot ofpersonssuch as Bohm[64]; Mielnik [62]; Biatynicki-Birula and Mycielski [65]; Davies [66]; Gisin [67];Weinberg [9-10], Wodkiewicz and Scully[34],Haas[63], attempt to incorporate nonlinear operators into linear quantum mechanicsto improve the latter, but the nonlinear operators can cause minor changes of the theory in early researches in Weinberg's model; in which the nonlinearities expected only to help in removing some quantal "paradoxes" or, to test the linear quantum mechanics [9-10]. In this case, it should be stressed that linear quantum mechanics, supplemented by nonlinear operators as representing observables, is not merelyan"improved" or "generalized"quantum mechanics, but should be a totally reconstructed new theory.Then Bugajski [68] thought thatthe nonlinear quantum mechanicshas to be a classical (nonquantal) theory of quantal phenomena.Thus, he [68]verified the dramatic collapse of the elaborate structure of quantum mechanics caused by introducing nonlinear observables, in which such a demonstration demands a theoretical framework, which is sufficiently general to cover quantal and classical theories. This may be calledan "operational approach" to statistical theories (Davies [69]; Ludwig [70]). Bugajski's considerations [68] are obtained based on known mathematical results found by Alfsen [71] and Asimov and Ellis [72].His description of quantal mixed states, as probability measures on pure states, was proposed by Misra [73] and Ghirardi et al. [74].The phenomenon of distinguishing statistical mixtures of quantalpure states by nonlinear operators was demonstrated by Haag and Bannier [63], and the transformations of a physical theory caused by a restriction or an extension of the set of admissible measurements were observed by Holevo [75] and Neumann [76]. Their general conclusion, that nonlinear quantum mechanical suggestionis inevitably a classical theory was announced as early by Bugajski [77,78]], however, in his studies the nonlinear modifications of quantum mechanics is new,and several notions were introduced. These problems are described as follows.

Let V be an order-unit Banach space, elements of the order interval $[0, e] \subset V$, where 0 is the origin and e is the order unit of V, are called effects. The set of effects is endowed with the order inherited from V and two partial operations "+" and "-", defined as the restrictions to [o, e] of the corresponding linear operations over V. The set of effects is convex; its set of extreme effects is denoted Ex [o, e]. The extreme elements of [o, e] are called sharp effects, and the other effects are fuzzy. An observable in V is a vector-valued measure A:B(R) $\rightarrow$ [o, e], where B(R) is the $\sigma-$ field of Borel subsets of the real line R, with A(R) = e. If the range of an observable A is contained in Ex [o, e], which is called sharp; otherwise, it is unsharp (fuzzy if the range of Aconsists of fuzzy effects only). Two effects are called compatible if they both belong to the range of an observable; then two observables are compatible if the set join of their ranges consists of pairwise compatible effects only. The Banach dual of V is a base-norm Banach space V*, the base S of V* is referred to as the set of states of V. S is convex and weak-compact, its extreme elements are referred to as pure states. Thus $W \subset V$ separates the elements of $T \subset V$ if for any $\alpha, \beta \in T$, there is $a \in W$ such that $\alpha(a) \neq \beta(a)$. The interval [o, e] always separatesall states of S. The real number $\alpha(a)(\alpha \in S, a \in [0,e])$ is called the probabilityof occurrenceof the effect **a** in the stateof α. A composition of state α and an observable **A** defines a probability measure on R. S is called a simplex (or a Bauer simplex) if for every point α of S there exists a unique probability measure μ supported by $\overline{ExS}$ (theweak-closure) such that

$$\alpha(a) = \int_{\overline{ExS}} \beta(a) d\mu(\beta),$$

for any $a \in [0,e] \subset V$. If S is a simplex, then ExS =' $\overline{ExS}$, thus S can be identified with $M(\overline{ExS})_1^+$, which is the set of probability measures on $\overline{ExS}$, and V can be identified with $C(\overline{ExS})$, which is the order-unit Banach space of all continuous realfunctions on the compact Hausdorff space $\overline{ExS}$. If and only if S is a simplex, then V is a lattice in its natural ordering and [o, e] is a lattice. If S is a simplex, then V is thought asthe classical. Or else, V is quantal.

In accordance with the above assumptions and rules as well as the nonlinear operators, which are only thenonlinear "observables", the latter are defined as non-bilinear functions on $\mathcal{H}$ x$\mathcal{H}$, where $\mathcal{H}$denotes$a$ separable Hilbert space.Bugajski [68]demonstrated and affirmed thatthe quantum mechanics with nonlinear operators is shown to essentially be a classical theory. In the meanwhile, he also effectively describeda general scheme of delinearization of the linear quantum theory.

2.3.4. Nattermann's Idea of Nonlinear Quantum Mechanics

Nattermann [79] proposed a possible framework for nonlinear quantum theories, into which linear quantum mechanics does not fit as well, and discussed further the notion of "equivalence" in this setting. Finally, he drew attention to persisting severe problems of nonlinear quantum theories. After this study, he thought that the nonlinearity can enter quantum mechanics in various ways, so there are a number of associations a physicist can have with the term "nonlinear quantum mechanics". In quantum field theory, the nonlinearity occurs in the equations of interacting field operators. These equations may be quantizations of nonlinear classical field equations [47] or mathematically tractable models as in φ^4-theory. There are the nonlinear terms for a phenomenological and semi-classical description of self-interactions, e.g., of electrons in their own electromagnetic field [38]. For a complex system, the linear multi-particle Schrödinger equation is often replaced by a nonlinear single-particle Schrödinger equation as in the density functional theory of solid-state physics, in which friction effect on a microscopic level attempts to be included into the nonlinear Schrödinger equations. Many of these approaches incorporate stochastic frictional forces in the nonlinear evolution equation for wave functions [80]. However, what are the fundamental roles of nonlinearity in these phenomena? This is worth to deep study. These problems were investigated by Bialinycki-Birula and Mycielski [81], Weinberg [9-10], and Doebner and Goldin [82-83], as mentioned above. But Bialinycki-Birula and Mycielski [81]re-constructed a model of nonlinear wave mechanics based on Schrödinger equation with nonlinearity $-b\psi \ln|\psi|^2$ .This nonlinearity was selected by assuming the factorization of wavefunctions for composed systems. Its most attractive properties are that they have the lower energy bound and validity of Planck's relation $E=\hbar\omega$ in his assumption. In any number of dimensions, the soliton-like solutions of their equation exist and move in slowly varying fields like classical particles. The Born interpretation of the wave function in linear quantum mechanics is consistent with logarithmic nonlinearity and they tentatively estimate the order of magnitude of the universal constant b.

Nattermann [79] adopted Mielnik's approach [62] for framework of nonlinear quantum theories to develop the main ingredients of a quantum theory with nonlinear time evolutions of wave functions, which is based on a fundamental hypothesis on physical experiments: "all measurements can, in principle, be reduced to a change of the dynamics of the system (e.g., by invoking external fields) and positional measurements". In fact, this point of view was confirmed in scattering experiments, where the localization of microscopic particles is detected asymptotically after interaction. Based on this hypothesis, he built the framework for a nonlinear quantum theory [84], in which a topological space T of wave functions is first built. In the system of one-particle, this topological space is a Hilbert space of square integrable functions, or other function spaces [85]. The time evolutions of wave functionsare given by homeomorphisms of T , $\Phi_t^{(Ext)}$: $T \rightarrow T$,which depend on the time interval t and the external conditions (e.g., external fields) $Ext \in C$. This is its second request. Mielnik's motion group M [62] is the smallest semi-group containing all time evolutions $\Phi_t^{(Ext)}$; close in the topology of pointwise convergence. The third request is that the positional observables X are represented by probability measures on the physical space M, which depend on the wave function $\varphi \in T$, i.e., $X=\{xB \mid B \in B(M)\}$, where

$$xB : T \rightarrow [0,1], \sum_{k=1}^{\infty} xB_k = xB, B = \bigcup_{k=1}^{\infty} B_k,$$

for disjoint $B_k \in B(M)$, where the triple (T ,M, P) is referred to as a quantum system. Using these basic ingredients, the effects and states of the quantum system (T,M, P) can be defined as derived concepts. An effect (or a counter) is a combination of evolutions $T \in M$ and positional measurements $x \in X$.

Mielnik listed a number of nonlinear toy models for his framework [62] and considered finite-dimensional nonlinear systems [86]. At the same time, Haag and Bannier also gave an interesting example of a quantum system with linear and nonlinear time evolutions [63]. Nattermann [79] used the above generalized gauge transformations in order to construct nonlinear quantum systems (H, M, X) with L^2-wave functions that are gauge equivalentto linear quantum mechanics. In this case, he assumed that the time evolution of a nonlinear quantum system is still given by a local, quasi-homogeneous nonlinear Schrödinger equation. In this case, it has to consider strictly local, projective generalized gauge transformations [79] $N_{\gamma t}(\psi_t \exp(i\gamma_t \ln|\psi_t|)$, where γ_t is a time-dependent parameter. Using these transformations and through an iterated process of gauge generalization and gauge closure – similar to the minimal coupling scheme of linear quantum mechanics Nattermann [79] obtained an unified family of nonlinear Schrödinger equations [87-88]:

$$i\partial_t\psi_t = i\sum_{j=1}^{2} \nu_j R_j[\psi_t]\psi_t + \mu_0 V + \sum_{k=1}^{5} \mu_k R_k[\psi_t]\psi_t + \alpha_1 Ln|\psi_t|^2 \psi_t \qquad (2.61)$$

Thus, he finally gave the framework for nonlinear quantum theories that generalizes the usual linear one through the following three remarks.

The first is concerned with the definition of effects and positional observables. Since the realvalued measures were used, then the observables do not allow for an idealization of measurements as in the linear theory, where a projection onto certain parts of the spectrum is possible using the projection-valued measure. Combined subsequent measurements (*histories*) must be described by quite complicated time evolutions. Secondly, we are unable to describe a complete and satisfactory nonlinear theory as yet.

One of the obstacles of quantum mechanical evolution equations comes from the difficulty of the global Cauchy problem for partial differential equations. Whereas, there is a solution of the logarithmic nonlinear Schrödingerequation, in this case [89], which shows that there are only local solutions to non-linearizable Doebner–Goldin equations [90].

Another problem of nonlinear Schrödinger equations in quantum mechanics is the locality of the corresponding quantum theory: EPR-like experiments could indeed lead to superluminal communications, the relevant Gisin-effects [91] could occur if changes of the external conditions inspatially separated regions have instantaneous effects. Since the nonlinear equations considered here are separable, this effect can only occur for entangledinitial wave functions. For the Doebner-Goldin equations, for instance, such effects indeed occur (at least) for certain subfamilies,but are not Galilei invariant [91].

2.3.5. Relativistic Model of Nonlinear Quantum Mechanics

At present, many people are interestedin the establishment of nonlinear quantum mechanics. Different ideas and models were proposed and built based on different motivations mentioned above. apparently well motivated, it becameapparent that nonlinear theories suffer from some prima facieserious difficulties.Svelichny [92] thought thatthere are various typesbut the most notable is the conflict with relativity or causality. Gisin[18,30] and Svetlichny [92] pointed out that the nonlinearity interaction allows us to use EPR-type correlations and the instantaneous nature of state-vector collapse to send a signal across aspace-like interval. Analyzing further, they found that this isa contradiction with relativity [93]. For this problem certain progress were made in order to overcome these difficulties. Because the difficulties stem from the instantaneous state-vector collapse in measurement, then a modification of measurement algorithms could allow for nonlinear processes without superluminal signals. Goldin, Doebner and Nattermann [87,94-95](GDN),as well as Svetlichny [93] argued that nonlinearity property does not lead to superluminal signals that all measurements are in the end expressible in terms of measurements of position, certain nonlinear Schrödinger equations are then observationally equivalent, via a nonlinear "gauge transformation", to the free linear equation. Unfortunately, these theories are still far from understanding the true relation of linearity to relativity.

Thus,Svelichny [92] reconsidered the question of quantum theory without measurements, asthe complete absence of the "measurement process" will eliminate any obstruction tononlinearities from the manifest noncovariance of this process. Of the several "measurementless" theories, the one most adaptable to relativistic considerations is the consistent histories approach [93,96]. He argued that the nonlinearity and relativistic covariance can indeed coexist peacefully and presented further a simple model to support this view. Such a model establishes the logical point and suggests where one should look for experimental evidence.

Svelichny [92] also argued the logical point that the relativistic nonlinear histories without superluminal signals are indeed possible. Because of its *ad hoc* nature,the model presented above cannot be considered realistic. In fact, if the space-time regions involved constitute a time-like chain, then the joint probabilities in the nonrelativistic limit would not be of the type governed by a nonlinear Schrödinger equation.Thus, we have not yet shown that the GDN theories can be obtained as nonrelativistic limits.

All the schemes based on nonlinear histories, which differentiate between space-like and time-like joint probabilities, in principle should exhibit physical effects as one crossesthe light cone. Thus, one can delay the lightrayon one side so that at a certain point the detector events become time-like in a typical photon correlation experiment. In crossingthe light cone, an effect could be present that was not foreseen by the linear theory. Thus, there are strong arguments [93,96] that theories do not suffer such discontinuities at the future light-cone, so itis necessarily linear. Therefore, a true verification of nonlinearity should involve light-cone experiments.

Theories of the type considered here are light cone singularand, for such theories,the notion of non-relativistic limit has to be modified. In usual theories the non-relativistic regime is one that all relevant velocities are small compared to the velocity of light. Hence, in light-cone singular theories one must add the requirement that all relevant space-time intervals be time-like. This further requirement removes the paradox that a causal relativistic theory may have a non-relativistic limit that seems manifestly causal by allowing instantaneous signals through long-range correlations, and may thus remove a major objection to the formulation of nonlinear quantum mechanics.

2.3.6. Logarithmic Nonlinear Interaction Model of Nonlinear Quantum Mechanics

It is well known that many people thought that the quantum mechanics based on the linear Schrödinger equation is only an approximation of the nonlinear theory with the nonlinear Schrödinger equation. The motivation for considering the nonlinear questions is to getmore nonstandard solutions in order to obtain better understanding of the synergism of wave and particle. Bialynicki and Mycielski [81] considered the generalized Schrödinger equation with the additional nonlinear term $F(|\phi|^2)\phi$, where F is some arbitrary function, such as the logarithmic nonlinearity $-b(Ln|\phi|^2)$, b >0. The nonlinear term was selected by assuming the factorization of the wave function for the composed system.The logarithmic nonlinearity is the existence of the lower energy bound and the validity of Planck's relation $E = h\nu = \hbar\omega$. However,they thought that the Born interpretation of the wave function could not be changed. In this theory the estimation of bwas given by the relation b <4 × 10^{-10}eV following the agreement between theory and the observed 2S − 2P Lamb shift in the hydrogen atom. This implies an upper bound to the electron soliton spatial width of 10 μm. Shimony [15] also observed a phase shift by an experiment, which is based on the idea that a phase shift occurs when an absorber is moved from one point to another along the path of one of the coherent split beams in a neutron interferometer. Meanwhile, Shull at. al. [16]also searched for a phase shift with a two-crystal interferometer when an attenuator was moved along the neutron propagation directionin one arm of the interferometer, wherea sheet of Cd with 0.086 mm thick was used for the absorber, from which they gained the upper bound on b of 3.4×10^{-13}eV, which but is the smaller of threeorders of magnitude than that estimated by Bialynicky et al., [81].

On the other hand, Gahler et al., [17] obtained an upper limit on bby searching for variations in the free space propagation of neutrons, where 2nm neutrons were diffracted from an abrupt highly absorbing knife edge at the object position. By comparingthe experimental results with the solutionof the ordinary Schrödinger equation they got b <3 × 10^{-15}eV, which corresponds to an electron soliton width of 3 mm. They obtained also similar results from diffraction of a 100 μm boronwire.

Pardy [97]did the new derivation of logarithmic nonlinearity using Dirac heuristical principle [98] and the Dirac method according to which it is useful to postulate some mathematical requirement in order to get the true information about nature. While the mathematical assumption is intuitive, the consequences have thephysical interpretation. The nonlinear Schrödinger equation of logarithmic nonlinearity can be denoted by [97]

$$i\hbar\frac{\partial\varphi}{\partial t} = -\frac{\hbar^2}{2m}\nabla^2\varphi - b(\ln|\varphi|^2)\varphi + V\varphi \tag{2.62}$$

He found also its solution in the one-dimensional case, which is

$$\varphi(x,t) = c\exp(a/B)\exp[[-B(x-vt+d)^2/4]\exp[-ikx-\hbar\omega t]$$

where $B = 4mb/\hbar^2$, c ,d and a are some constants, and v and m are the velocity and mass of the particle, respectively. In this case he also showed that in the mass limit, i.e.,m $\rightarrow \infty$ he got exactly the delta-function behavior of the probability of finding the particle at point x. This means that there exists the classical motionof a microscopic particle with sufficiently large mass. The nonlinearity of the Schrödinger equation alsosolves the collapse of the wave function and the Schrödinger cat paradox.

2.3.7.Lochanand colleagues' Scheme of Nonlinear Quantum Mechanics

Lochan et al., [99] thought that the knowledge and understanding of linear quantum mechanics are incomplete for the following reasons.(A)The principle of linear superposition was not confirmed experimentally for position eigenstates of objects having more than about a thousand atoms, (B)there is no universally agreed upon explanation for the process of quantum measurement and the observed fact that macroscopic objects are not found in superposition of position eigenstates and (C) the concept of time is classical and hence external to quantum mechanics: there should exist an equivalent reformulation of the theory, which does not refer to an external classical time. Thus,Lochan et al., [99] argued that such a reformulation is the limiting case of a nonlinear quantum theory, with the nonlinearity becoming important at the Planck mass scale. The nonlinearity can provide insights into the previously mentioned problems. In this case, they used a motivated model for a nonlinear Schrödinger equation to indicate that nonlinearity can help in understanding quantum measurement. They obtained that while the principle of linear superposition holds to a very high accuracy for an atomic system, the lifetime of a quantum superposition becomes progressively smaller, as one goes from microscopic to macroscopic objects. This may be used to explain the observed absence of position super positions inmacroscopic objects (lifetime is very small). This means that ongoing laboratory experiments may be used to detect the finite superposition lifetime for mesoscopic objects.

Therefore, Lochan et al., [99] presented the nonlinearity in quantum mechanics, which is inevitable in the mesoscopic domain, and that the linear theory is an excellent approximation in themicroscopic limit. They wished to estimate more precisely for the lifetime of a quantum superposition, and to investigate further the practical feasibility ofeliminating decoherence so that the possible presence of a nonlinearity can be verified or ruled out. This also would be interesting to compare these ideas with the phenomenologicalmodel of continuous spontaneous localization [21, 100]

2.3.8. Fermi's Wave Model of Nonlinear Quantum Mechanics

After the provision of after the linear Schrödinger wave equation in linear quantum mechanics, Fermi proposed his first nonlinear generalization based on a basic goal for describing the mechanism of the dynamical property of radiation due to the charge distributionin an atom. In this model, Fermi [38] proposed his first nonlinear Schrödinger equationusing classical electrodynamics and the radiation-reaction principle combined with the Schrödinger interpretation of the wave function, which is denoted by

$$i\hbar\frac{\partial\psi(\vec{r},t)}{\partial t}=-\frac{\hbar^2}{2m}\nabla^2\psi(\vec{r},t)+V(\vec{r},t)\psi(\vec{r},t)+V_{RR}(\psi*\psi)\psi \quad (2.63)$$

where $V_{RR}=-(2e/3c^3)\dddot{d}.\vec{r}$ is a radiation-reaction potential of a dipole generated by the Schrödinger 's charge distribution $\rho(r)=e\psi*(r)\psi(r)$, here

$$d(\psi*\psi)=e\int d^3r\vec{r}\rho(r)/\int\int d^3r\rho(r) \quad (2.64)$$

Very clearly, the nonlinear Schrödinger equation (2.63) proposed by Fermi satisfies the homogeneity condition imposed by Weinberg [9-10] on all possible generalizations of quantum mechanics, where the nonlinearity is a consequence of the Schrödinger interpretation leading to the formula (2.64) and has a purely electromagnetic origin. Therefore, Fermi's model was not looking for a new nonlinear version of quantum mechanics, but attempted only the use of this little-known generalization of the wave equation to describe and explain the property of radiation of an atom.

Meanwhile, the nonlinear Schrödinger equation (2.63)) has a remarkable property of predicting that the rate of decay of an excited state is given precisely by the Einstein A coefficient with a value of A, which agrees with the one obtained by Dirac in his famous paper on the quantum theory of light absorption and emission. Despite this success, the Fermi model leads to some very serious problems, such as an exact Eigen state cannot radiate spontaneously. Only linear combinations of the excited and deexcited states can decay and this decay is also non-exponential in general. Because these fundamental difficulties Fermi finally had to abandon his nonlinear wave model and become a devoted advocate of quantum electrodynamics. Thus, Fermi's nonlinear wave model has also been put to rest.

2.4. Simple Survey of Other Ideas of Nonlinear Quantum Mechanics

In the article of "Nonlinear quantum mechanics" Reinisch [101] assumed the occurrence of an inhomogeneous phase in the real amplitude and phase of the wave function in the formulation of the Schrödinger problem, which results in a nonlinear description of the dynamics of the field amplitude in the stationary case, which leads to the simple semi-classical-like dynamical interpretation of quantum mechanics for all quantum states.

In the paper "Nonlinear quantum mechanics: a conflict with the Ptolomean structure?" Mielnik [102] extended the argument of Gisin (1989) and showed that the superluminal signals in nonlinear quantum mechanics are not a specific difficulty for Weinberg's (1989) model, but a generic phenomenon in all non-dissipative quantum theories to try to enlarge the unitary motion group, while incorporating the traditional structures of states and observables. The implications exceed the problem of causality.

In the article of "nonlinear quantum mechanics: results and open questions" Goldin [103] proposed a special type of nonlinear modification for the usual Schrödinger time-evolution equation in quantum mechanics, which was motivated by certain unitary representations of the group of diffeomorphisms of physical space, in the framework of either nonrelativistic local current algebra or quantum Borel kinematics. Subsequently, they developed this related approach to nonlinearity in quantum mechanics considerably further, to incorporate theories of measurement, groups of nonlinear gauge transformations, symmetry, and invariance properties, unification of a large family of nonlinear perturbations, and possible physical contexts for quantum nonlinearity. Some of their results and highlights of some open questions are summarized.

Nikolic [104] pointed out that all measurable predictions of classical mechanics could be reproduced from a quantum-like interpretation of a nonlinear Schrödinger equation. The key observation leading to classical physics is the fact that a wave function that satisfies a linear equation is real and positive, rather than complex. This has profound implications on the role of the Bohmian classical-like interpretation of linear quantum mechanics, as well as on the possibilities to find a consistent interpretation of arbitrary nonlinear generalizations of quantum mechanics.

In the article "Is quantum theory intrinsically nonlinear?" Schuch [105] pointed out that in contrast with classical physics, complex quantities have a fundamental physical meaning in quantum physics and action, being essentially the quantized entity, and should be given more attention instead of focusing mainly on Hamiltonians or Lagrangians that have the dimension of energy. Phase and amplitude of the complex quantities in (time-dependent and time-independent) quantum mechanics are not independent of each other but coupled via some conservation law. This coupling can be understood if the systems are described in terms of complex nonlinear Riccati equations. These equations not only enable a connection to the Pythagorean triples, probably the oldest and most abstract `quantization' problem, but also lead to dynamical invariants with the dimension of action. Factorization of the corresponding operator provides generalized creation and annihilation operators, which is also possible for dissipative systems where no conventional Hamiltonian formalism exists. He shows formal similarities with other fields, particularly with nonlinear dynamics.

Slater [106-107] in "Entropy in nonlinear quantum mechanics" point out that Peres [25] and Weinberg [9-10] have debated and determined the definition of entropy in nonlinear quantum mechanics. This issue can be resolved if one adopts an entropy measure consistent with the approach of Guiasu (following work of Wiener and Siegel) in which a probability distribution over the space of wavefunctions is estimated. This is tantamount to rejecting the von Neumann measure (S_N), -Tr rho; ln rho; of the entropy of a density matrix (rho;), which is the basis of Peres' argument. As an authority for such a rejection of S_N, a detailed critique of this measure, given by Band and Park in the mid-1970s, can be cited. At the same time, he discussed further with Peres and Weinberg the problem of finding a suitable definition of entropy in nonlinear quantum mechanics, in which two further observations are made, firstly, by regarding the covariance matrix of the probability distribution over the phase space of wavefunctions as the nonlinear counterpart (P_{NL}) of the (linear) density matrix (P_L) and employing -Tr PNL In PNL, one obtains a limiting transition (as nonlinearities vanish), in which this entropy measure converges to the definition in ordinary quantum mechanics, -Tr P_L In P_L. Secondly, it is argued that Peres' contention on the nonlinear variants of the Schrödinger equation violate the second law of thermodynamics, which is flawed in that it relies upon the entropy of mixing of non-orthogonal states, which as Dieks and van Dijk (1988) have indicated an undefined concept. A proper approach to associating a quantum mechanical entropy with a mixture of a particle into two non-orthogonal statesby first estimating a suitable two-particle density matrix (p) and then employing -Tr p In p- is outlined.

Jordan [108] in "Probing Weinberg's nonlinear quantum mechanics" provided a search for nonlinear representations of propositions made by assuming that the multiplication rule Weinberg (1989) uses to express the Lie bracket as a commutator can also be used to calculate functions representing squares of propositions. For systems described by two-component wave functions, the result is negative: the functions representing nontrivial propositions are all the same as in ordinary quantum mechanics.

As it is known, nonlinear extensions of quantum mechanics are difficult. Nonlinear Schrödinger and von Neumann equations can be justified but do not seem to be allowed in the usual interpretation. Of particular interestare difficulties with multi-particle entangled states. In this case Czachorand Doebner[109] studied the correlation experiments in nonlinear quantum mechanics, in which they show how one can compute multiple-time multi-particle correlation functions in nonlinear quantum mechanics in a way which guarantees locality of the formalism.

In Nonlinear quantum mechanics, complex classical mechanics and conservation laws for closed and open systems, Schuch [110] indicated that phase and amplitude of the complex wave function are not independent of each other, but coupled, which becomes obvious looking at Madelung's hydrodynamic formulation of quantum mechanics. In the time-independent case, this leads to a kind of conservation law that allows for the reformulation of the linear Schrödinger equation in terms of a nonlinear Ermakov equation, which is equivalent to a complex Riccati equation where the quadratic term in this equation explains the origin of the phase-amplitude coupling. A similar conservation law and corresponding nonlinear equations can also be found in the time-dependent case. The gain from the nonlinear formulations emerges when open systems with dissipation and irreversibility are considered. Describing this kind of system by an effective nonlinear Schrödinger equation leads to a modification of the above-mentioned equations with new qualitative effects like Hopf bifurcations.

As it is known, in quantizing a system with a nonlinear Lagrangian the ordering of variables which do not commute cannot be ignored. Owing to this the usual procedure of canonical quantization does not always go through in this process. Thus, there arises an important question as to how one should set up a Hamiltonian quantum theory in such a case. Thus Omote and Sato[111] established the quantum mechanics of a nonlinear systems, in which a procedure is given for constructing a Hamiltonian quantum theory for a system with a nonlinear (classical) Lagrangian. The result indicated that the resulting formalism retains all the symmetry properties possessed by the original Lagrangian. groups that correspond to symmetries of the system are conserved.

Falck, et al., [112] in "The QCD Hamiltonian and nonlinear quantum mechanics" studied the canonical quantization of SU(N) gauge theory in linear, non-covariant gauges. The canonical formalism is first discussed in the classical theory, with special attention to the features involving nonlinearity and the gauge degrees of freedom. The transition to the quantum theory is then performed for an arbitrary linear gauge, using the covariant quantization rules of nonlinear quantum mechanics. They gave the quantum Hamiltonian in the Weyl-ordered form appropriate for the application of the usual Dyson-Wick perturbative techniques; in this case, the additional ordering terms appeared with respect to the classical

Hamiltonian. They discuss further the relation of their results to author's previous data and the relevance of the ordering terms in field theory.

In "Atomic physics tests of nonlinear quantum mechanics" Bollinger et al., [113] showed the atomic physics experiments, which test a nonlinear generalization of quantum mechanics formulated by Weinberg [9-10]. The experiments search for a dependence of hyperfine transition frequencies or nuclear spin precession frequencies on the relative populations of the hyperfine or nuclear spin states. The experiments set limits less than 10 μ Hz on the size of the possible nonlinear contributions to these frequencies. In some cases, this can be interpreted as a limit of less than 10^{-26} on the fraction of binding energy per nucleon that could be due to a nonlinear correction to a nuclear Hamiltonian. The possibility that a nonlinear addition to quantum mechanics violates causality was also discussed this paper.

Svetlichny [114], in "Nonlinear quantum mechanics at the "Planck scale", argued that the linearity of quantum mechanics is an emergent feature at the "Planck scale", along with the manifold structure of space-time. In this regime the usual causality violation objections to nonlinearity do not apply, and nonlinear effects can be of comparable magnitude to the linear ones and still be highly suppressed at low energies. This offered alternative approaches to quantum gravity and to the evolution of the early universe.

Meanwhile, in "Non-linear quantum mechanics and high energy cosmic rays", Svetlichny [115] pointed out that the presence of energetic cosmic rays above the GZK cutoff may be explained by fundamental nonlinearities in quantum mechanics at the Planck level.

In "On Nonlinear Quantum Mechanics, Non-commutative Phase Spaces, Fractal-Scale Calculus and Vacuum Energy" Castro [116] thought that a novel generalized nonlinear Schrödinger equation based on the modifications of Nottale-Cresson's fractal-scale calculus and resulting from the non-commutativity of the phase space coordinates is explicitly derived. The modifications to the ground state energy of a harmonic oscillator yield the observed value of the vacuum energy density. In the concluding remarks,he discussed how nonlinear and nonlocal quantum mechanics wave equations arise naturally from this fractal-scale calculus formalism which may have a key role in the final formulation of quantum gravity.

Brody, et al., [117] in "Stochastic reduction in nonlinear quantum mechanics" pointed out that stochastic extensions of the Schrödinger equation have attracted attention as plausible models for state reduction in quantum mechanics. They formulated a general approach to stochastic Schrödinger dynamics in the case of a nonlinear state space of the type proposed by Kibble (1979). They derived a number of new identities for observables in the nonlinear theory, and they also established general criteria on the curvature of the state space sufficient to ensure collapse of the wave function.

In "On the application of a Gauss transformation in nonlinear quantum mechanics" Ulmer, et al., [118] gave a representation of the delta;-distribution by Gauss functions, which is used to obtain integral transformations with kernel functions of Gaussian type. It is shown that this type of distribution leads to a new type of nonlinear integro-differential equation with Gaussian kernel. The nonlinear term of the cubic Schrödinger equation can be generated by the limiting case, where the Gauss transformation becomes a delta;-distribution. In the domain of positive curvature of the Gaussian function, an approximation method of the nonlinear equation with the Gauss kernel is obtained.

In "Violation of wave-particle complementarityin nonlinear quantum mechanics" Valentini [119] showed, in the context of a double-slit experiment, that small nonlinearities of the Schrödinger equation, considered by Weinberg [9-10], will in general allow a simultaneous measurement of particle paths and interference, unless the nonlinearities are of a special form such that nearby initial states remain so at all later times.

In "Self-tunneling oscillations in nonlinear quantum mechanics and the electron-transfer problem" Etchegoin [120] pointed out that a small quantum system with a nonlinear self-consistent potential might display a spontaneous symmetry breaking of the ground state. This simple result, which can also be interpreted as a spontaneous self-localization of the wave function, leads to the presence of several equivalent ground states, which do not exist simultaneously. There is a virtual degeneracy in the system. The addition of a stochastic perturbation (noise) to such a system provides the possibility of tunneling among the different equivalent ground states. The tunneling process can lock itself into a characteristic frequency related to the dynamics of the system and not contained in the input noise. A self-pulsating quantum system is then generated based on the intrinsic nonlinearity of the dynamics and the presence of noise, which is used as a seed. They inquired into the relevance of these phenomena for electron transfer processes in large biomolecules. By combining nonlinearity and noise, a controlled transfer rate can be achieved in what is effectively a quantum-switch regulated by external perturbations; a mechanism that could be at work in many electron-transfer processes in proteins.

In "A Hilbert space realization of nonlinear quantum mechanics as a classical extension of its linear counterpart" Del Seta, et al., [121]studied the state-effect-probability structure associated with the quantum mechanics of nonlinear (homogeneous, in general non-additive) operators on a Hilbert space for providing a concrete representation of the features of nonlinear quantum mechanics on a Hilbert space, and showing that the properties of the nonlinear version of quantum mechanics have the structure of a classical logic.

Hubsch [122], in "quantum mechanics is either nonlinear or non-introspective", pointed out that the measurement conundrum seems to have always plagued quantum mechanics for so long that impressions of an inconsistency amongst its axioms have spawned. A demonstration that such purported inconsistency is fictitious may then be in order and is presented here. An exclusion principle of sorts emerges, stating that quantum mechanics cannot be simultaneously linear and introspective (self-observing). The nonlinearity of this latter approach allows quantum mechanics to describe the entire measuring process, and to be applied to the entire universe, for which there is no external observer.

In the papers of"Nonlocal-looking equations can make nonlinear quantum dynamics local" Czachor[123] showed a general method for extending a nondissipative nonlinear Schrödinger and Liovill-von Neumann one-particle dynamics to an arbitrary number of particles. He obtained that the dynamics so gained is completely separable, which is the strongest condition one can impose on dynamics of composite systems, which but demand that for all initial states not only can a subsystem not be affected by any action undertaken by an observer in a separated system , but additionally the self-consistency condition should be met. Meanwhile, he obtained that a correct extension to N-particles includes integro-differential equations, which make the theory fully local. Therefore, a much larger class of nonlinearities satisfying the complete separability condition is allowed. This manifested that the locality condition does not single out logarithmic or one-homogeneous nonlinearities. At the same time,in "Nonlinear Schrödinger equation and two-level atoms"He[124]used the same form of atomic nonlinearity by Weinberg [9-10] and Wockiewicz and Scully [35] to obtain some contradictory results concerning an evolution of the atomic inversion w in a two-level atom in Weinberg's nonlinear quantum mechanics. If the atom is initially in a ground state then either the evolution of w (1) can be linear, if one uses a nonlinear generalization of the Jaynes-Cummings Hamiltonian, or (2) is always nonlinear if one uses the nonlinear Bloch equations derived from the nonlinear atomic Hamiltonian function. It is shown that the difference is rooted in inequivalent descriptions of the composite "atom-plus-field" system. The linear evolution of w results from a "faster-than-light communication" between the atom and the field. If one applies a description without the faster-than-light telegraph; then the calculations, based on a suitably modified Jaynes-Cummings Hamiltonian, lead to the same dynamics of w as is found in semi-classical calculations based on Bloch equations. It is also shown that a nonlinear quantum mechanics, based on a nonlinear Schrödinger equation, does not possess a natural probability interpretation.

In "Lie-Nambu and Lie-Poisson structures in linear and nonlinear quantum mechanics" Czachor [125] gained that the space of density matrices in quantum mechanics can be regarded as a Poisson manifold with the dynamics given by certain Lie-Poisson bracket corresponding to an infinite dimensional Lie algebra. The metric structure associated with this Lie algebra is given by a metric tensor, which is not equivalent to the Cartan-Killing metric. The Lie-Poisson bracket can be written in a form involving a generalized (Lie)-Nambu bracket. This bracket can be used to generate a generalized, nonlinear, and completely integrable dynamics of density matrices.

In "Measurement description by means of a nonlinear Schrödinger equation" Grigorenko [126] discussed a new general formulation for nonlinear quantum mechanics, which showed that a nonlinear deterministic evolution with random variables could model a measurement as a dynamical process. Phases of state vectors are treated as random variables of the theory.

In "Nonlinear quantum spin dynamics: symmetries without conservation laws" Jordan [127] presented that spin examples demonstrate the existence of a nonlinear quantum mechanics. The Hamiltonian operator depends on the state. It contains mean values together with operators. Symmetries depend on these mean values. The requirement for a symmetry operator is not simply that it commutes with the Hamiltonian operator. Rotation invariance does not imply that for the angular-momentum operator. The Hamiltonian operator can change as the state changes. There may be no conserved energy. In examples for spin 1/2; and spin 1, the angle between precessing spin and magnetic field is changed. This suggests the existence of the possibilities for new precision tests of linearity.

In "Nonlinear quantum mechanics with non-classical gravitational self-interaction" Popova [128] pointed out that the original approach for the self-consistent inclusion for gravity should be introduced into quantum mechanics of a particle. (There are no connections with second quantization).

The nonstandard action principle is constructed for the stationary situation, i.e., the quantum particle in a stationary state created some non-classical stationary gravitational field and again interacted with it. In this case, the accompanying problem of covariantization of quantum operators occurred is considered in his investigation. The general theory is illustrated by the Newtonian-Schrödinger idea in quasi-classical limiting cases. The levels of applicability of ordinary quantum mechanics and the problems of measurements and interpretation of non-classical gravity were also discussed in his studies. The `uncertainty relations' connecting uncertainties of some `local' parts of curvature and those of the particle's position and momentum were derived. The superposition principle is generalized basedon some approximate actions.

In "nonlinear quantum mechanics with non-classical gravitational self-interaction, nonstationary situation" Popova, et al., [129-130] pointed out that quantum mechanics (first quantization) with self-consistent gravitational interaction, which was previously constructed for the stationary case, is now extended to the general case. The two requirements for such a theory are realized to obtain the theory maximally resembling a classical field theory and to achieve the invariance of the theory under the rescaling transformations of a wave function. The construction is not trivial, because it rejects the variational principles of extremality of any action and involves some principles of smoothed extremality, which gave some relevant equations. In the meanwhile, they condensed further some related problems in the framework of general quantum mechanics with gravitational self-interaction constructed earlier.

They analyzed some the stationary situations. Here, the problem of gauge invariance generated by translations, which do not violate the 3+1 splitting,was also studied. The notions of position and momentum operators were extended to the general case. The uncertainty relations, including the uncertainty of the Ricci tensor and uncertainties of the position and momentum of a particle were obtained. The correspondence between the stationary and non-stationary cases is examined using the variational principles. At the same time, the one-particle and two-particle problems in the Newtonian-Schrödinger idea limit are considered; the latter problem is compared with the standard two-particle quantum problem to demonstrate the advantage of the authors' approach.

In "Nonlinear quantum mechanics as Weyl geometry of a classical statistical ensemble", Castro [131] derived nonlinear relativistic and nonrelativistic wave equations for spin-0 and 1/2 particles. For a suitable choice of coupling constants, the equations become linear, and Weyl gauge invariant in the spin-0 case. The Dirac particle is much more subtle. When a suitable gauge is chosen and, when the Compton wavelength of the particle is much larger than Planck's length, he recovered the standard Dirac equation. Nonlinear corrections to the Schrödinger equation are obtained and these appear as the first-order relativistic corrections to the nonrelativistic Hamilton-Jacobi equation. Consequently, he constructs non-bilinear homogeneous realizations of an approximate Galilean symmetry. He puts also forth the idea that not only a modification of quantum mechanics might be necessary in order to accommodate gravity, but also quantum mechanics itself might have a geometrical origin with Planck's constant as the coupling between matter and curvature.

In "Nonlinear quantum mechanics: two possibilities", Roberts [132] gave a comparative discussion of two possible nonlinear extensions of quantum mechanics. It is argued that the two extensions are mutually consistent but applicable to completely different phenomena and that, consequently, quantum mechanics will be demonstrated to be truly linear only if both extensions are precluded by experiment.

In "Soliton dynamics and nonlinear quantum mechanics", Minelli, et al. [133], studied the soliton solutions of the logarithmic Schrödinger equation in one dimensions with simple potentials and considered tunneling effects.

In "on the application of a Gauss transformation in nonlinear quantum mechanics. II. Problems with spin" Ulmer [134] pointed out that the nonlinear and nonlocal Schrödinger equation with Gaussian kernel as well as the cubic Schrödinger equation generating in the local-limit case can only be applied to spin-0 fields. But this consideration can be also extended to spin-1/2 fields.

Taking account of the non-locality, the nonlinearity of such a field internal spin-orbit coupling must be considered in order to describe transitions between different multiplicities using some special ways. The Pauli equation did not turn out to be a sufficient starting point, as the spin-orbit coupling is incorrectly obtained in a nonrelativistic approach. However, the assumption that the nonlinear term acts as internal potential is sufficient to derive the correct spin-orbit term in the nonlinear and nonlocal case.

It is also possible to obtain an approximation method in the nonlocal domain, based on a self-interacting 3-dimensional harmonic oscillator with spin. In the local limit, where the Gaussian distribution becomes a delta-distribution, the existence of soliton solutions, and the relationship to the superconductivity, were also discussed by them.

2.5. Comment on the Correctness of the Various Nonlinear Quantum Mechanical Ideas and Models

In sections 2.1-2.4 we stated, described, and elucidated the ideas, points of view, suggestions and models of nonlinear quantum mechanics, which can be used to improve and remold the linear quantum mechanics. The results mentioned above showed clearly that these schemes, plans and projections are all different. I estimated that about 30 different ideas, projects, and models of nonlinear quantum mechanics have been proposed. This indicated straightforwardly the severity of difficulties of linear quantum mechanics on the one hand, and its necessity for improvement, on the other hand, because its improvement has no unity scheme up to now. This means that it is highly necessary to establish a correct and unity nonlinear quantum mechanical theory.

However, what is a correct and available nonlinear quantum mechanical theory? To aim at linear quantum mechanics and its difficulties, we believe that a correct nonlinear quantum mechanics should be build on the basis of reliable experiments and theories, or in other words, that nonlinear quantum mechanics must have solid experimental and theoretical fundamentals, in which the dynamic equations of microscopic particles must be determinant and real, which can represent and reflect truly and completely the properties of motion and interactions of microscopic particles. From the dynamic equations, we can give and exhibit the wave-corpuscle duality of these microscopic particles. Or else, they are not correct. These conditions are just the standards of a correct nonlinear quantum mechanical theory, in which there are not second standard as we thought. Thus, we can judge and inspect the validities of the ideas, points of view, suggestions and models of nonlinear quantum mechanics mentioned above using these standards [135-143].

Simple comments and estimations tell us that the above ideas, points of view, suggestions, and models of nonlinear quantum mechanics could not be correct because they did not satisfythe above standards. Just so, these view points, suggestions and models of nonlinear quantum mechanics have only a short lifetime after they are proposed and established and cannot be used to describe correctly the natures and properties of microscopic particles in various different physical systems. The detailed reasons are described as follows.

For example, there is not a determinate and explicit dynamic equation in the Weinberg model, which also did not give the wave-corpuscle duality of microscopic particles. For most of the other nonlinear quantum ideas, review points, and suggestions mentioned above, their experimental and theoretical fundaments, the dynamic equations of microscopic particles and their wave-corpuscle duality were not all given, established, and elucidated further, respectively. Therefore, these nonlinear quantum mechanic ideas, viewpoints, suggestions, and models cannot be referred to as nonlinear quantum mechanics. Thus,it is quite necessary and urgent to establish and develop a correct nonlinear quantum mechanic theory, which can be used to represent and reflect real and complete motion properties of microscopic particle.. However, in the establishment of nonlinear quantum mechanics we cannot absolutely deny completely the correctness of linear quantum mechanics or speaking, cannot fully abandon the latter. This suggests that nonlinear quantum mechanics must be built based on linear quantum mechanics because they should be self-consistent each other, i.e., the nonlinear quantum mechanics will become the linear quantum mechanics when the nonlinear interactions among the microscopic particles in the systems are not considered, and vice versa. This is very important because the linear quantum mechanics was extensively used and obtained a lot of valuable results, which are declared and well known. We can say only that the linear quantum mechanics is an approximate theory, but we cannot absolutely say that the theory is wrong. Therefore, some correct hypotheses and concepts or formulae in linear quantum mechanics must be retained in nonlinear quantum mechanics [135-142].

On the other hand, we should point out that the homogeneity feature of wave function of the microscopic particles, i.e., $d(c\varphi)/dt = cd\varphi/dt$, which is used in the Weinberg model and other ideas and suggestions mentioned above, cannot be excessively emphasized because it is not both a rule the microscopic particle must obey and a representation of the basic features of the particles. The homogeneity is, in essence, requested and obeyed in linear quantum mechanics, namely, it is a necessary result of the linear superposition theorem of wave functions of microscopic particles. One says that it associates with the normalization condition of wave function of microscopic particles, so it holds only in linear quantum mechanics. However, these situations cannot occur in nonlinear quantum mechanics, in which the physical significance of wave function of microscopic particles is different from that in linear quantum mechanics, although they also represent the states of microscopic particles. In practice, the squared of absolute value of the wave function represents the mass density of the microscopic particle at that point in nonlinear quantum mechanics due to the existence of nonlinear interactions, instead of its probability appearing in linear quantum mechanics [135-139]. Therefore, it is not necessary to obey the homogeneity for the wave function and to normalize them further, in other words, the calculation of normalization of wave functions of the particles has not been specially emphasized and required in nonlinear quantum mechanics because the nonlinear interactions accepted by different particles are different in the nonlinear systems, which are different from those in linear quantum mechanics.

2.6. Pang's Nonlinear Quantum Mechanic Theory

2.6.1. Fundamental Principles of Pang's Nonlinear Quantum Mechanics

To aim at the difficulties of linear quantum mechanics and based on the experimental data and theoretical results of macroscopic quantum effects, they can also be described and elucidated by a macroscopic wave function or order parameters and nonlinear dynamic or like-nonlinear Schrödinger equations satisfied by the wavefunction [135-141]. Pang proposed and established the fundamental principlesof nonlinear quantum mechanics [135-149], which are described as follows.

(1) Microscopic particles in a nonlinear quantum system are described by thefollowing wave function,

$$\phi(\vec{r},t)=\varphi(\vec{r},t)e^{i\theta(\vec{r},t)} \tag{2.65}$$

where both the amplitude $\varphi(\vec{r,t})$ and phase $\theta(\vec{r,t})$ of the wave function are all functions of space and time.

(2) In the nonrelativistic case, the wave function $\phi(\vec{r,t})$ satisfies the generalizednonlinear Schrödinger equation (NLSE), i.e.

$$i\hbar\frac{\partial\phi}{\partial t}=-\frac{\hbar^2}{2m}\nabla^2\phi\mp b|\phi|^2\phi+V(\vec{r},t)\phi+A(\phi) \tag{2.66}$$

or

$$\mu\frac{\partial\phi}{\partial t}=-\frac{\hbar^2}{2m}\nabla^2\phi\mp b|\phi|^2\phi+V(\vec{r},t)\phi+A(\phi) \tag{2.67}$$

where μ is a complex number, V is an external potential field, A is a function of $\phi(\vec{r,t})$, and b is a coefficient indicating the strength of the nonlinear interaction.

In the relativistic case, the wave function $\phi(\vec{r,t})$ satisfies the nonlinear Klein-Gordon equation (NLKGE), including the generalized Sine-Gordon equation (SGE) and the ϕ^4 -field equation, i.e.

$$\frac{\partial^2\phi}{\partial t^2}-\frac{\partial^2\phi}{\partial x_i^2}=\beta\sin\phi+\Gamma\frac{\partial\phi}{\partial t}+A(\phi) \tag{2.68}$$

and

$$\frac{\partial^2\phi}{\partial t^2}-\frac{\partial^2\phi}{\partial x_i^2}\mp\alpha\phi\pm\beta|\phi|^2\phi=A(\phi)\quad(i=1,2,3) \tag{2.69}$$

where Γ represents dissipative or damping coefficient of the system, β is a coefficient indicating the strength of nonlinear interaction, α is a constant related to the feature of the system and $A(\phi)$ is a function of $\phi(\vec{r,t})$.

2.6.2. The Basic Properties of Pang's Nonlinear Quantum Mechanical Theory

Quite clearly, Pang's nonlinear quantum mechanics has the following properties.

(1).It is a fully new theory, which is different from not only linear quantum mechanics but also all nonlinear quantum mechanical ideas, suggestions and models including the Weinberg model because itsdynamic equation(2.65) is only the nonlinear Schrödinger equation with cube nonlinear interactions, instead of logarithmic or other nonlinear Schrödinger equations with their dynamic equations in Eqs.(2.26)-(2.29), (2.33), (2.55),(2.59)- (2.60), and (2.62)-(2.63) in the latter.

As it is known, there are the only two fundamental hypotheses of nonlinear quantum mechanics in Eqs.(2.65)-(2.69), which are quite different from linear quantum mechanics in Chapter 1. These dynamic equations, in Eqs.(2.65)-(2.69), included the nonlinear interactions of $b|\phi|^2\phi$, $\beta\sin\phi$ and $\beta|\phi|^2\phi$ .They are just generalizations of the linear Schrödinger and linear Klein-Gordon equations in linear quantum mechanics. Therefore, it is a natural development of linear quantum mechanics, thus the new theory is self-consistent with linear quantum mechanics because it becomes the linear Schrödinger equation and Sine-Gordon equation in linear quantum mechanics only if these nonlinear interactions equal zero. This clearly manifested that the new theory is developed based on linear quantum mechanics. Therefore, it is a new theory.

(2). This new theory has solid theoretical and experimental foundations, its theoretical foundations are the superconductive and superfluid theories because the equation(2.66) resembles the Ginzburg-Landau(GL) equation, which is used to study the motion of electrons in the superconductors, and the Gross Pitaerskii equation, which was used to describe the properties of helium atoms in the superfluid state. Its experimental foundations are just the macroscopic quantum effects, including the superconductive and superfluid phenomena, which exhibit some quantum effect in macroscopic scales [135-143]. Because these dynamic equations all have a soliton solution, thus the mathematical fundament of nonlinear quantum mechanics is the nonlinear partial-differential equation and solitontheory.Then we can affirm that its wave functions and mechanical quantities are no longer the vectors in Hilbert space, but the vectors in Hermitian-symmetry space. Hence, Pang's nonlinear quantum mechanics aretheresults or integrationof superconductivity, superfluidity, and modern soliton theories.It is completely different from linear quantum mechanics described in Chapter 1.

(3) Equation (2.65) indicates the basic form of wave function of states of the microscopic particles, where $\varphi(\vec{r},t)$ is its amplitude and $\theta(\vec{r},t)$ is its phase. Obviously, they all have a determinant physical significance and satisfy different dynamic equations, which can be acquired by inserting Eq. (2.65) into Eq. (2.66). We ascertained in a one-dimensional case that they meet the following two equations:

$$\frac{\partial^2\varphi}{\partial x'^2}-\frac{\partial\theta}{\partial t'}\varphi-(\frac{\partial\theta}{\partial x'})^2\varphi+b\varphi^3=V(x',t)\varphi \qquad (2.70)$$

and

$$\varphi\frac{\partial^2\theta}{\partial x'^2}+2\frac{\partial\varphi}{\partial x'}\frac{\partial\theta}{\partial x'}+\frac{\partial\varphi}{\partial t'}=0 \tag{2.71}$$

respectively, where $x'=x\sqrt{2m}/\hbar$, $t'=t/\hbar$. From Eq. (2.71) we can find the relationof $\theta(x',t')$ with $\varphi(x',t')$, then the solution, $\varphi(x',t')$, of Eq. (2.70) can also be found and, the b and V(x',t') known. If the system is a bulk and uniform material, then $\theta(x',t')$ is a constant, which is independent with x' and t', thus Eq. (2.70) can be represented by

$$\frac{\partial^2\varphi}{\partial x'^2}+b\varphi^3-V(t')\varphi=0 \tag{2.72}$$

This is a time-independent nonlinear Schrödinger equation, which is the same as the time-independent G-L equation in superconductors [150-152]. This demonstrated again that the theory and experiments of the superconductor is only the experimental and theoretical foundations of Pang's nonlinear quantum mechanics. This feature of wave function of the microscopic particles in Pang's nonlinear quantum mechanics does not occur in linear quantum mechanics and other nonlinear quantum mechanic schemes, ideas, and models mentioned above.

(4) The new theory breaks through the independence of Hamiltonian of the systems on the wave functions of microscopic particles and the homogeneity of dynamic equations of microscopic particles as well as the theoretical linearity in linear quantum mechanics. As it is known, the Lagrange density function, $\mathcal{L}$, and Hamiltonian density, $\mathcal{H}$, of the systems corresponding to Eq. (2.66) at $A(\phi)=0$ in Pang's theory can be represented, respectively, by

$$L'=\frac{i\hbar}{2}\left(\phi^*\partial_t\phi-\phi\partial_t\phi^*\right)-\frac{\hbar^2}{2m}\left(\nabla\phi\cdot\nabla\phi^*\right)-V(x)\phi\phi^*+(b/2)\left(\phi\phi^*\right)^2 \tag{2.73}$$

and

$$\mathcal{H}=\frac{i\hbar}{2}\left(\phi^*\partial_t\phi-\phi\partial_t\phi^*\right)-\mathcal{L}=\frac{\hbar^2}{2m}\left(\nabla\phi\cdot\nabla\phi^*\right)+V(x)\phi\phi^*-(b/2)\left(\phi\phi^*\right)^2 \tag{2.74}$$

where L'= $\mathcal{L}$, its momentum density of the particle system is defined as P= $\partial\mathcal{L}/\partial\phi$. Equations (2.73)-(2.74) clearly show that the Lagrange density function and the Hamiltonian density of the systems are all related to the wave function of the state of the particles and involve a nonlinear interaction, (b/2) $(\phi\phi^*)^2$. The above, when compared to the linear quantum theory, thus shows that two major breakthroughs were made in Pang's nonlinear quantum mechanics. One is the linearity of the dynamic equation, and the other is the fundamental hypothesis of independence of the Hamiltonian operator on the wave function of the microscopic particles in linear quantum mechanics. In nonlinear quantum mechanics, the dynamic equations are also a nonlinear function of the wave function, $\phi(\vec{r},t)$, i.e., they are nonlinear partial differential equations. The Hamiltonian operators also depend on the wave function $\phi(\vec{r},t)$. In this respect, Pang's nonlinear quantum mechanics is truly a breakthrough relative to linear quantum mechanics, i.e. it has broken the independence of Hamiltonian of the systems on the wave functions of microscopic particles and the particle's homogeneity of dynamic equations as well as the theoretical linearity in linear quantum mechanics. Therefore, Pang's nonlinear quantum mechanics are completely different from linear quantum mechanics, and the above ideas, suggestions, and models of nonlinear quantum mechanics. Thus, it is a novel theory.

In this case, the natures of the microscopic particles are simultaneously determined by the kinetic and nonlinear interaction terms in dynamic equations of nonlinear quantum mechanics in Pang's theory. The balance between them results necessarily in the form of soliton states of microscopic particles and their motions. This means that the microscopic particles in Pang's nonlinear quantum mechanics have different properties from those in linear quantum mechanics, which will be changed thoroughly, and no longer disperse and may be localized. These are described in detail in Chapter 3.

(5) The physical meanings of the wave function of microscopic particles describe considerable changes. We know from the soliton theory that the dynamic equations in Eqs.(2.66)–(2.69) have all a soliton solution due to the existences of the nonlinear interactions, which can suppress and balance the dispersed effect of kinetic energy because they are related to the wave function of the particles, thus the natures of the microscopic particles represented by the wave function $\phi(\vec{r},t)$ vary to become a soliton or a solitary wave. This means that the wave function $\phi(\vec{r},t)$ no longer represents a linear or dispersive wave occurring in linear quantum mechanics. In this case, the square of the absolute value of the wave function $\phi(\vec{r},t)$ in Eq.(2.65), $|\phi(\vec{r},t)|^2 = |\varphi(\vec{r},t)|^2 = \rho(\vec{r},t)$ is no longer the probability of finding the microscopic particle at a given point in space-time, but rather gives us the mass density of the microscopic particles at that point because $\int|\phi|^2 d\tau = m$, which will be conformed in the following studies, where m is the effective mass of the microscopic particle. Thus, the concept of probability or the statistical interpretation of wave function in linear quantum mechanics is no longer relevant in Pang's nonlinear quantum mechanics. Therefore, the nonlinear interactions and nonlinear quantum mechanics in Pang's theory vary through the natures and rules of motion of microscopic particles. These problems will be described in Chapters 3-4.

At the same time, we should point out that the nonlinear interactions occurring in Eqs.(2.66) -(2.69), delineate and reflect the real motions and interactions of microscopic particles in physical systems.Instead of artificial results, they haveall solid physical mechanisms, which will be described in Chapter 4. This showed clearly that Pang's nonlinear quantum mechanical theory represented and exhibited the true properties of physical systems.

(6) Obviously, the Lagrange density functions in Eq. (2.73) and the Hamiltonian density in Eq. (2.74) have space-time symmetry. That is, L and H remain unchanged under the transformations of $t \to -t$ and $\vec{r} \to -\vec{r}$ or $x \to -x$, $y \to -y$ and $z \to -z$.

Meanwhile, they also possess the $U(1)$ symmetry, i.e., they are unchanged while undergoing the transformations of

$$\phi(\vec{r},t) \rightarrow \phi'(\vec{r},t) = e^{-iQ_j\theta}\phi(\vec{r},t) \tag{2.75}$$

due to the fact that the Hamiltonian density in Eq. (2.74) contains the product of the $\phi_j(\vec{r,t})s$ and the invariance of charge of the system, i.e., $(Q_1 + Q_2 + \cdot\cdot\cdot + Q_n) = 0$, and the facts of $\nabla\phi_j \rightarrow e^{-i\theta Q_j}\nabla\phi_j$ in this case, where Q_j is the charge of the particle, θ is a phase independent of x.

(6) The concept of operator in linear quantum mechanics can still be used in Pang's nonlinear quantum mechanics. However, the majority of them arenot fully linear operators, and thus certain properties of linear operators, such as the conjugate Hermitian of the momentum and the coordinate operators, are no longer required, and $i\hbar\partial/\partial t$ no longer correspond to the energy operator. Instead, nonlinear operators are constructed and used in Pang's nonlinear quantum mechanics. For example, Equation (2.66) may be written as

$$i\hbar\frac{\partial\phi}{\partial t} = \hat{H}(\phi)\phi \tag{2.76}$$

The Hamiltonian operator $\hat{H}$ has a nonlinear dependence on $\phi(\vec{r,t})$ and is given at $A(\phi) = 0$ by

$$\hat{H}(\phi) = -\frac{\hbar^2}{2m}\nabla^2 - b|\phi|^2 + V(\vec{r},t) \tag{2.77}$$

In general, equations (2.66) – (2.69) in the one-dimensional case can be expressed as

$$\phi_{t'} = {d\phi}/{dt'} = K(\phi) \ , \ \phi = \phi(x',t') \tag{2.78}$$

according to the Lax method [153], where $K(\phi)$ is a nonlinear operator or hereditary operator. In an one-dimensional case, a new operator $Q(\phi)$, which is obtained from the generator of the translation group, can be used to produce the vectorial field $K(\phi)$, i.e.

$$K(\phi) = Q(\phi)\phi_{x'} \tag{2.79}$$

and the operator, $Q(\phi)$, is called a nonlinear recursion operator. The nonlinear Schrödinger equation (2.66) at $A(\phi) = 0$ can be generalized as

$$Q(\phi) = -iD + 4i\phi D^{-1}\,\mathrm{Re}(\bar{\phi}) \tag{2.80}$$

where the operator D denotes the derivative with respect to x, and

$$D^{-1}f(x')=\int_{-\infty}^{x'}f(\xi)d\xi .$$

From the hereditary property of $K(\phi)$, we can obtain the following vector field

$$K_n(\phi)=Q(\phi)^n\phi_{x'} \quad \text{n=0,1} \tag{2.81}$$

The equation of motion of the eigenvalue λ' of the recursion operator $Q(\phi)$ maybe expressed as

$$\lambda_{t'}^{'}=\partial\lambda'/\partial t'=K'(\phi)[\lambda'] \tag{2.82}$$

where $K'(\phi)$ is the variational derivative with respect to ϕ and is given by

$$K'(\phi)[\lambda']=\frac{\partial}{\partial\varepsilon}K(\phi+\varepsilon\lambda')\big|_{\varepsilon=0} \tag{2.83}$$

The equation describing the time variation of the recursion operator can be obtained via;

$$\frac{\partial}{\partial t'}Q(\phi)=K'(\phi)Q(\phi)-Q(\phi)K'(\phi)=[K'(\phi),Q(\phi)]. \tag{2.84}$$

This equation is very similar to the Heisenberg matrix equation in linear quantum mechanics. However, both $K'(\phi)$ and $Q(\phi)$ here are nonlinear operators. Therefore, we may refer to Eq. (2.84) as the Heisenberg equation of nonlinear operator in Pang's nonlinear quantum mechanics.

(8) Pang's nonlinear quantum mechanic theory describes an integral system. For ageneral nonlinear equation given in (2.78), we know that $K(\phi)$ is a nonlinear operator of ϕ. According to the Lax method [153], if two linear operators L and $\hat{B}$ corresponding to Eq. (2.78), which depends on ϕ, satisfy the following Lax operator equation

$$iL_{t'}=\hat{B}\hat{L}-\hat{L}\hat{B}=\lfloor\hat{B},\hat{L}\rfloor \tag{2.85}$$

where $t'=t/\hbar$ and $\hat{B}$ is a self-adjoint operator, then the eigenvalue k and Eigen function Ψ of the operator L satisfy the equation:

$$\hat{L}\Psi=k\Psi \tag{2.86}$$

with

$$\Psi(x',t') = \begin{pmatrix} \Psi_1 \\ \Psi_2 \end{pmatrix},$$

but $\hat{B}$ satisfies the equation:

$$\hat{B}\Psi = i\Psi_{t'} \tag{2.87}$$

For the following nonlinear Schrödinger equation in Eq. (2.66) at $V(\vec{r},t) = A(\phi) = 0$

$$i\phi_{t'} + \phi_{x'x'} + b|\phi|^2\phi = 0 \tag{2.88}$$

where $x' = x\sqrt{2m}/\hbar$, using Zakharov and Shabat's method [154-155] we see that the concrete representations of L and $\hat{B}$ for Eq. (2.88) are as follows

$$\hat{L} = i\begin{pmatrix} 1+s & 0 \\ 0 & 1-s \end{pmatrix}\frac{\partial}{\partial x'} + \begin{pmatrix} 0 & \phi^* \\ \phi & 0 \end{pmatrix}, \quad \hat{B} = -s\begin{pmatrix} 1 & 0 \\ 0 & 1 \end{pmatrix}\frac{\partial^2}{\partial x'^2} + \begin{pmatrix} \dfrac{|\phi|^2}{1+s} & i\phi^*_{x'} \\ -i\phi_{x'} & -\dfrac{|\phi|^2}{1-s} \end{pmatrix} \tag{2.89}$$

where $s^2 = (1 - 2/b)$, $\phi(\vec{r},t)$ and satisfies Eq. (2.88). This is just the Lax form for the nonlinear Schrödinger equation (2.88) or (2.66).

In general, many nonlinear dynamic equations can be written either in the Lax form, or in the Hamiltonian form of a compatibility condition for over-determined linear spectral problems. In the former, the nonlinear dynamic equations with soliton solutions are often expressed as (2.78), corresponding the eigenvectors and eigenvalues of the linear operator L and its time evolution are determined by Eqs. (2.86) and (2.87), respectively.

We now assume that the Eigen function of the operator L satisfies the equation:

$$i\Psi_{t'} = \hat{B}\Psi + f(L)\Psi \tag{2.90}$$

where the function $f(L)$ may be chosen according to convenience. In this case, we can write the over-determined system of linear matrix equations as

$$\Psi_{x'} = U(x',t',\alpha)\Psi, \Psi_{t'} = G(x',t',\alpha)\Psi \tag{2.91}$$

where U and G are 2×2 matrices. The compatibility condition of this system is obtained by differentiating the first equation of Eq.(2.91) with respect to t' and the second one with respect to x' and then subtracting one from the other, thus we obtained

$$U_{t'} - G_{x'} - [U, G] = 0,$$

or

$$U_{t'} - G_{x'} = [U, G] \tag{2.92}$$

We here emphasize that the operators U and G depend not only on t' and x' but also on some parameters, α , which are called spectral parameters. Then the condition (2.92) must be satisfied by α . In this case Makhankov et al., [156-157] gave the representations of the operators U and G, which are denoted by

$$U = -i\alpha\begin{pmatrix} (1+s)^{-1} & 0 \\ 0 & (1-s)^{-1} \end{pmatrix} + \begin{pmatrix} 0 & i(1+s)^{-1}\phi^* \\ i(1+s)^{-1}\phi & 0 \end{pmatrix}$$

and

$$G = -i\alpha^2\begin{pmatrix} s(1+s)^{-2} & 0 \\ 0 & s(1-s)^{-2} \end{pmatrix} + i2\alpha\begin{pmatrix} 0 & \dfrac{s\phi^*}{(1+s)(1-s^2)} \\ \dfrac{s\phi}{(1-s)(1-s^2)} & 0 \end{pmatrix} +$$

$$\begin{pmatrix} \dfrac{i(\phi\phi^*)}{1-s^2} & \dfrac{\phi_{x'}}{1+s} \\ \dfrac{\phi_{x'}}{1-s} & -\dfrac{i|\phi|^2}{1-s^2} \end{pmatrix}$$

The presence of a continuous time-independent parameter is a reflection of the fact that the nonlinear Schrödinger equation (2.88) describes a Hamiltonian system with a set of an infinite number of conservation laws, which will be discussed in Chapter 3. In the case of a system with a finite number (N') of degrees of freedom, one can sometimes succeed in finding $2N'$ first integrals of motion between which the Poisson brackets are zero (in this case they are said to be in involution). Such a system is called completely integral. In such a case, equation (2.91) with Eq.(2.92) are referred to as the compatibility condition, and their consequence is the appearance of the nonlinear Schrödinger equation (2.88). This means that, for each solution, $\phi(x', t')$ of Eq.(2.88), there is always a set of basis function Ψ , parameterized by α , which can be obtained through solving the set of linear equations (2.90)– (2.92). Thus, we can conclude that the nonlinear quantum systems described by

Pang's nonlinear quantum mechanics is completely integrable. Zakharov and Faddeev [158] further generalized this concept into the quantum field theory.

(9) The representation of vector nonlinear Schrödinger equationsin Pang's theory. When a system is composed of n same particles in same condition, among which there are nonlinear interactions, the system of microscopic particles can be described by a vectorial nonlinear Schrödinger equation [156-157] according to Eq.(2.66), which can often be written as

$$i\hbar\frac{\partial\phi}{\partial t}=-\frac{\hbar^2}{2m}\nabla^2\phi-b(\bar{\phi}\phi)\phi+V(\vec{r},t)\phi \tag{2.93}$$

where

$$\phi=\begin{pmatrix}\phi^{(1)}\\ \cdot\\ \cdot\\ \phi^{(n)}\end{pmatrix}, \tag{2.94}$$

ϕ is a column vector of n-components, describing the "isospace" state, $\bar{\phi}$ is its complex conjugate vector. The corresponding Lagrangian and Hamiltonian densities of the system are given, respectively, by

$$L'=\frac{i\hbar}{2}\left(\bar{\phi}\phi_t-\phi\bar{\phi}_t\right)-\frac{\hbar^2}{2m}\left(\phi_x\cdot\bar{\phi}_x\right)-V(x)\phi\bar{\phi}+(b/2)\left(\phi\bar{\phi}\right)^2 \tag{2.95}$$

and

$$H'=-\frac{\hbar^2}{2m}\left(\phi_x\cdot\bar{\phi}_x\right)+V(x)\phi\bar{\phi}-(b/2)\left(\phi\bar{\phi}\right)^2 \tag{2.96}$$

where L'= $\mathcal{L}$ and H'= $\mathcal{H}$, $\bar{\phi}=\phi^*\gamma_0$ and γ_0 is a diagonal matrix, and the internal product $(\bar{\phi}\phi)=\phi^\dagger\gamma_0\phi$ is conserved. Such transformations also conserve $\mathcal{L}$ and $\mathcal{H}$. That is, they are symmetrical transformations of the systems, which do not change the vector nonlinear Schrödinger equation (2.93). When γ_0 is also an n-component unit matrix, the transformations form a compact group $U'(n)$.On the other hand, if

$$\gamma_0=\text{diag.}(1,1\cdots 1,-1,-1\ldots.-1), \tag{2.97}$$

which plays the role of isotropic space metric and the dagger sign (†) denotes a Hermitian conjugate, the so-transformation belongs to the non-compact group $U(i, j)$, here "i" is the number of "1", j is the number of "-1"in γ_0. Zakharov and Shabat [154-155] showed that Eq.

(2.93) is completely integrable in the case of the$U(1)$ group for both positive ($b>0$) (the U (1, 0) model) and negative ($b<0$) (the U (0, 1) model) coupling constants. Furthermore, Manakov [159] integrated the vector equation (2.93) for the $U(2)$ group (the $U(2, 0)$ model, $b>0$), and the integrability for the case of the$U(1, 1)$ group was shown by Makhankov [156]. These can be considered as special cases of the general system.

The results mentioned above indicate clearly that Pang's nonlinear quantum mechanics has a completely theoretical formulation and formalism, which are fully different from the ideas, points of view, schemes, suggestions, and models of nonlinear quantum mechanics mentioned above. Therefore, it is new and available. In the following chapters, we mainly studied the properties and rules of Pang's nonlinear quantum mechanics.

REFERENCES

[1] L. de Broglie, "*Non-Linear Wave Mechanics-A Causal Interpretation*," Elsevier, Amsterdam. 1950.

[2] L. de Broglie, *La physique quantique restera-t-elle indeterministe?* Gauthier-Villars, Paris,1953.

[3] L. de Broglie, Interpretation of quantum mechanics by the double solution theory, Annales de la Fondation Louis, 12(4) (1987) 1.

[4] D. Bohm, *Quantum theory*, Prentice-Hall Englewood Cliffs, New Jersey, 1951.

[5] D. Bohm, *Phys. Rev.* 85 (1952) 166 and 180.

[6] D. Bohm, and J. Bub, *Rev. Mod. Phys.*38 (1966) 453.

[7] Wu Ta-You, *Quantum Mechanics*, World Scientific, Singapore, 1986.

[8] A. Aspect, P.Grangier and G.Roger, *Phvs. Rev. Lett.* 49 (1982) 91.

[9] S. Weinberg, *Phys.Rev. Lett.* 62 (1989)485.

[10] S. Weinberg, *Ann. Phys.* (N. Y.) 194 (1989) 336.

[11] J. J. Bollinger, D. Prestage, S. M. Itano, and D. Wineland, *Phys. Rev. Lett.*54 (1985) 1000.

[12] I. Blalynicki-Birula, and J. Mycieski, *Ann. Phys.* (N.Y.) 100 (1976) 62.

[13] R. Haag, "Local Quantum Physics", Springer Verlag, Berlin,1992.

[14] T. Kibble, *Commun. Math. Phys.* 64 (1978) 73.

[15] A. Shimony, *Phys. Rev.* A 20 (1979)394.

[16] C. G. Shull, D. K. Atwood, J. Arthur and M. A. Horne, *Phys. Rev. Lett.* 44 (1980) 765.

[17] R. Gähler, A. G. Klein, and A. Zeilinger, *Phys. Rev.* A23 (1981) 1611.

[18] N. Gisin, *Phys. Lett. A, 143(1990) 1.*

[19] N.Gisin, Phys. Rev. Lett. 53 (1984)1776

[20] Ph. Pearle, *Phys. Rev.* D 13 (1976) 857.

[21] G. C. Ghirardi, A. Rimini and T. Weber, *Phys. Rev.* D 34(1986) 470.

[22] J.S. Bell, in: Schrödinger centenary ofa polymath,Cambridge Univ. Press, Cambridge, 1987 23.L. Diosi, *Phys. Rev.* A 40 (1989) 1165.

[23] T. F. Jordan, Ann. Phys. 225 (1993) 83.

[24] K. E. Eriksson, *Phys. Scr.* 36 (1987) 870.

[25] A. Peres, *Phys.Rev. Lett.* 63 (1989) 1114.

[26] J. Polchinski, *Phys. Rev. Lett.* 66(1991) 397–400.

[27] A. Einstein, B. Podolsky, and N. Rosen, *Phys. Rev.* 47 (1935) 777.
[28] S.Weinberg, *Phys. Rev. Lett.* 63 (1989) 1115.
[29] B. S. DeWitt and N. Graham, *The many worlds Interpretation of quantum mechanics*, Princeton Univ. Press, Princeton, 1973.
[30] N. Gisin, Helv. *Phys. Acta*, 62 (1989) 363.
[31] M. Czachor, "Mobility and nonseparability", Found. *Phys. Lett.*, 4 (1991) 351
[32] J. J. Bollinger, D. J. Heinzen, W. M. Itano, S. L. Gilbert, and D. J. Wineland, *Phys.Rev. Lett.* 63 (1989) 1031.
[33] T. F.Jordan, Phys. Rev. A , 50(1994)62.
[34] D. S. Abrams; *Phys.Rev. Lett.* 81(1998)3992.
[35] K. Wodkiewicz and M. O. Scully, *Phys. Rev.* A,42(1990)5111.
[36] T. E. Chupp and R. J. Hoare, *Phys. Rev. Lett.* 64 (1990) 2261.
[37] R. L. Walsworth, J. F. Silvera, E. M. Mattison, and R. F. C. Vessot, *Phys.Rev.Lett.* 64(1990) 2599.
[38] E. Fermi, Atti della Realeaccademia nazionale dei Lincei. Rendiconti, 5 (1927) 795.
[39] C. R. Stroud, Jr. and E. T. Jaynes, *Phys. Rev.* A 1 (1970) 106.
[40] L. Allen and J. H. Eberly, *Optica1 Resonance and Tao-level Atoms*, Wiley, New York, 1975, Chap. 2.
[41] P. Meystre, and M. Sargent III, *Elements of quantum optics*, Springer, New York, 1989, Chap. 15.
[42] R. L. Walsworth and I. F. Silvera, *Phys. Rev.* A 42(1990)63.
[43] 'E. Wigner, *Ann. Math.* 40 (1939) 149.
[44] G. Birkhoff, von Neumann, J.: *Ann. Math.* 37(1936) 823.
[45] C. Piron, Helv. *Phys. Acta* 37, 439 (1964).
[46] R. Haag, D. Kastler, J. *Math. Phys.* 5(1964) 256.
[47] I.E. Segal, *Ann. Math.* 48 (1947) 930.
[48] G. Ludwig, Z. Natnrforsch. 22a (1967) 1303 and 22a, (1967) 1324.
[49] G: Dähn, Commun. *Math. Phys.*, 9 (1968) 192; 28 (1972)109 and 28(1972) 123.
[50] E.B. Davies, J.T: Lewis, Commun. *Math. Phys.* 17 (1970) 239.
[51] E.B. Davies, Commun. *Math. Phys.*,15 (1969)277; 19 (1970) 83; 22(1971)51.
[52] C.M. Edwards, Commun. *Math. Phys.* 16 (1970) 207; 20(1971)26 ; 24 (1972) 260.
[53] G. Ludwig, Commun. *Math. Phys.*, 13 (1967) 331; 9 (1968)1; 26 (1972) 78.
[54] B. Mielnik, Commun. *Math. Phys.* 15 (1969) 1.
[55] H: Neumann, Commun. *Math. Phys.* 23 (1971) 100.
[56] P. Stolz, Commun.*'Math. Phys.* 11(1969)303 and 23 (1971) 117.
[57] S. Gudder, Commun. *Math. Phys.*, 29 (1973)249.
[58] J. Gunson, Commun. *Math. Phys.* 6(1967) 262.
[59] K. E. Hellwig, K: Kraus, Commun. Math. Phys. 11(1969) 214 and 16 (1970) 142.
[60] J. C. T: Pool, Commun, *Math. Phys.*, 9 (1968) 118.
[61] J. C. T. Pool,: Commun, *Math. Phys.*, 9 (1968) 212.
[62] B. Mielnik, Commun. *Math. Phys.*,37(1974)221.
[63] R. Haas and U. Bannier, Commun. *Math. Phys.* 60 (1978)1.
[64] D.Bohm, An ontological foundation for the quantum theory, in symposium on the foundations of modern physics (Joensuu, 1987), P. Lahti and P. Mittelstaedt, eds., World Scientific, Singapore,1987.
[65] I. Biatynicki-Birnla, and J. Mycir, *Annals of Physics*, 100 (1976) 62.

[66] E. B. Davies, Commun. in *Math. Phys.*, 64 (1979) 191.
[67] N. Gisin, Helv. *Phys. Acta*, 54 (1981) 457.
[68] S. Bugajski, *Inter. J. Theor. Phys.,* 30 (1991) 961.
[69] E. B. Davies, Commun. in *Math. Phys.,* 64 (1979)191-210.
[70] G. Ludwig, Foundations of quantum mechanics I, Springer-Verlag, New York,1983.
[71] E. M. Alfsen, Compact Convex Sets and Boundary Integrals, Springer-Verlag, Berlin, 1971.
[72] L. Asimov and A. J. Ellis, *Convexity theory and Its applications in functional analysis*, Academic Press, London, 1980.
[73] B. Misra, On a new definition of quantal states, in Physical Reality and Mathematical Description, Enz C. P. and Mehra J., eds., D. Reidel, Dordrecht, Holland, 1974.
[74] G. C. Ghirardi, A. Rimini, and T.Weber, Nuovo Cimento B 36 (1976) 97.
[75] A. S. Holevo, *Probabilistic and statistical aspects of quantum theory*, North-Holland, Amsterdam, 1982.
[76] H. Neumann, The size of sets of physically possible states and effects, in Recent Developments in Quantum Logic, Mittelstaedt P. and Stachow E.W., eds., Bibliographisches Institut, Ziirich,1985.
[77] S. Bugajski, Nonlinear extension of "quantum logic", Report to the short conference on Universal Algebra, Bern, Switzerland, 1981.
[78] S.Bugajski, and P. Lahti, *International Journal of Theoretical Physics*, 19 (1980) 499.
[79] P. Nattermann, *Symmetry in Nonlinear Mathematical Physics* 2(1997) 270.
[80] J. Messer Lett. *Math. Phys.*, 2(1978)281.
[81] I. Bialynicki-Birula and J. Mycielski, *Ann. Phys.* (NY), 100(1976) 62.
[82] H. D. Doebner. and G. A. Goldin, *Phys. Lett.* A, 162(1992)397.
[83] H. D.Doebner and G. A.Goldin, *J. Phys. A: Math. Gen.*, 27(1994) 1771.
[84] P. Nattermann, Generalized quantum mechanics and nonlinear gauge transformations, in: Symmetry in Science IX, editors B. Gruber and M. Ramek, Plenum Press, New York, 1997, pp269–280, ASITPA/4/97, quant-ph/9703017.
[85] P. Nattermann, Nonlinear Schrödinger equations via gauge generalization in group21 – physical applications and mathematical aspects of geometry, groups and algebras, V.1., Doebner H.D., Nattermann P. and Scherer W., editors, World Scientific, Singapore, 1997: pp428–432.
[86] B. Mielnik, J. *Math. Phys., 21*(1980) 44.
[87] H. D.Doebner and G.A.Goldin, *Phys. Rev.* A, 54(1996)3764.
[88] H. D. Doebner, G.A.Goldin, and P. Nattermann, *J. Math.Phys.* 40 (1999) 49.
[89] T. Cazenave and A. Haraux, *Ann. Fac. Sc. Toul.*, 11(1980) 21.
[90] H.Teismann, The Cauchy problem for the Doebner-Goldin equation, in Group21 – Physical Applications and Mathematical Aspects of Geometry, Groups and Algebras, V.1., Doebner H.-D., Nattermann P. and Scherer W., editors, World Scientific, Singapore, 1997:,PP 433–438.
[91] W. Lüucke and P. Nattermann, Nonlinear quantum mechanics and locality, in: Symmetry in Science X, editors B. Gruber and M. Ramek, Plenum Press, New York, 1998, ASI-TPA/12/97, quantph/9707055.
[92] G. Svelichny, *Symmetry in nonlinear mathematical physics*, 2(1997)262.
[93] G. Svetlichny, Quantum evolution and space-time structure, in: nonlinear, Deformed and irreversible quantum systems, Proceedings of the international symposium on

mathematical physics, Arnold Sommerfeld Institute, 15-19 August 1994, Clausthal, Germany, editors Doebner H.-D., Dobrev V.K., and Nattermann P.,World Scientific Publ. 1995, 246–251, quant-ph/9512004.
[94] G. A. Goldin, in: Nonlinear, Deformed and Irreversible Quantum Systems, Proceedings of the International Symposium on Mathematical Physics, Arnold Sommerfeld Institute, 15-19 August 1994, Clausthal, Germany, Editors Doebner H.D., Dobrev V.K., and Nattermann P., World Scientific Publ. 1995, 125.
[95] H. D. Doebner, G. A. Goldin and P. Nattermann, in: Quantization, Coherent States, and Complex Structures, Editors Antoine J.-P., Ali S.T., Lisiecki W., Mladenov I.M. and Odzijewicz A., Plenum Press 1995, pp27.
[96] G. Svetlichny, Lorentz covariance and the covering Law, Pontificia Universidade Catolica of Rio de Janeiropre-print MAT.15/95.
[97] M. Pardy, Possible Tests of nonlinear quantum mechanics in: waves and particles in light and matter, ed. by Alwyn van der Merwe and Augusto Garuccio, Plenum Press, New York, 1994.
[98] A. Pais, Playing with equation, the Dirac way preprint RU/86/150 Rockefeller University New York, 1986.
[99] K. Lochan and T. P. Singh, *PRAMANA journal of January Indian Academy of Sciences,* 76 (2011)67-91.
[100] S L Adler and A Bassi,Science 325 (2009) 275.
[101] G. Reinisch,*Physica A* 206 (1994) 229.
[102] B.Mielnik, Phys.Lett. A, 289(2001)1.
[103] G.A.Goldin , Physics of Atomic Nuclei, 71(5(2008) 884.
[104] H. Nikolic, Classical mechanics as nonlinear quantum mechanics, AIP Conference Proc., 962 (2007) 162-7.
[105] D. Schuch, *Physica Scripta*, 87 (2013) 038117.
[106] P. B.Slater, *J.Physics A* 25 (1992) L359.
[107] P. B. Slater, *J. Physics A* 25 (1992)L935.
[108] T. F.Jordan, *Phys. Lett. A, 151*(1990)215.
[109] M. Czachorand H.D. Doebner, *Phys. Lett.* A, 301 (2002) 139.
[110] D. Schuch, *J. Physics: Conference Series*, 361(2012) 012020.
[111] M. Omote and H. Sato, Prog.Theor.l Phys. 47(1972) 1367.
[112] N. K. Falck, A.C. Hirshfeld, Annals Physics, 144 (1982) 34.
[113] J. J. Bollinger, D.J. Heinzen, W.M. Itano, S.L. Gilbert, D.J. Wineland, Santa Fe Workshop of Foundations of Quantum Mechanics, World Scientific publishing, Singapore, 40-6, 1992.
[114] G.Svetlichny, *Inter. J. Theor. Phys.*, 44 (2005) 2051.
[115] G. Svetlichny, *Foundations Phys. Lett.*, 17 (2004)197.
[116] C. Castro, *Foundations Phys.*, 40 (2010) 70.
[117] D.C. Brody, L.P. Hughston, Proceedings of the Royal Society of London, Series A (Mathematical, Physical and Engineering Sciences), 458(2001) 1117-27, 8 May 2002, R. Soc.publishing, London, UK.
[118] W. Ulmer, H. Hartmann, *Nuovo Cimento A*, 47 (1978) 359.
[119] A. Valentini, *Phys. Rev.* A, 42 (1990) 639.
[120] P. Etchegoin, *Chem. Phys.* 297 (2004) 245.
[121] M. Del Seta, G.Cattaneo, *Inter. J. Theor. Phys,,* 39 (2000) 621.

[122] T. Hubsch, *Modern Phys. Lett.* A, 13 (1998)2503.
[123] M. Czachor, *Phys. Rev. A*, 57 (1998)4122.
[124] M. Czachor, Phys. Lett. A 225 (1997) 1.
[125] M. Czachor, *Acta Physica Polonica B*, 27 (1996) 2319.
[126] A.N. Grigorenko, *J. Phys. A*, 28(1995)1459.
[127] T. F. Jordan, *Phys. Rev. A,* 49 (1994)5086.
[128] A. D. Popova, *Inter. J. Modern Phys. A*, 4 (1989)3229.
[129] A. D. Popova, and A.N. Petrov, *Inter. J.Modern Phys. A*, 8 (1993) 2683.
[130] A. D. Popova and A.N. Petrov, *Inter. J. Modern Phys. A*, 8 (1993) 2709.
[131] C. Castro, *Foundations Phys. Lett.* 4 (1991)81.
[132] C. D. Roberts, *Modern Phys. Lett.* A, 5 (1990) 91.
[133] T. A. Minelli, and A. Pascolini, *Nuovo Cimento*, 27 (1980) 413.
[134] W. Ulmer, *Nuovo CimentoA,* 51 (1979)309.
[135] Pang Xiao-feng, Nonlinear quantum mechanics (Chinese), Beijing, Chinese Electronic Industry Press, 2009, P1—285.
[136] Pang, Xiao-feng, The theory of nonlinear quantum mechanics(Chinese), Chinese Chongqing Press,Chongqing, 1994, P23-69.
[137] Pang Xiao-feng, *Frontiers Physics in China*, 3 (2008)413.
[138] Pang, Xiao-feng, Principle and theory of nonlinear quantum mechanics, *Proc. 3thNational Conf. of quantum mechanics*, 1984, p. 27.
[139] Pang, Xiao-feng, *Problems of nonlinear quantum theory*, Chengdu, Sichuan Normal Univ. Press, 1985. p.23-64.
[140] Pang, Xiao-feng, *Proc. IWPM-6* (Shenyang) 1992, p. 83 and Proc. ICPM (Wuhan) 1991,p. 197.
[141] Pang, Xiao-feng, The elementary principle and theory for nonlinear quantum mechanics, Proc. ICNP. Shanghai, 1989, p. 181 and Proc. 4th APPC (Soul Korea) 1990, p. 213.
[142] Pang, Xiao-feng, The theory of nonlinear quantum mechanics, in research of new sciences, eds.Lui Hong, Chinese Science and Tech. Press, Beijing, 1993, p. 16.
[143] Pang, Xiao-feng, The theory of nonlinear quantum mechanics, *Proc. ICIPES*, Beijing 1991,p.123.
[144] Pang, Xiao-feng, *J Potential Science Sin.* 5 (1985) 16.
[145] Pang, Xiao-feng, *Physica Bulletin Sin.* 5 (1983) 6.
[146] Pang Xiao-feng, *Research and Development of World Science and Technology*, 28 (2006) 11.
[147] Pang Xiao-feng, *Research and Development of World Science and Technology*, 24 (2003) 54.
[148] Pang, Xiao-feng, *J. Southwest Inst. Nationalities, Sin.* 23 (1997) 418.
[149] Pang, Xiao-feng, *J. Southwest Inst. Nationalities, Sin.* 24 (1998) 310.
[150] V. I. Ginzberg, and L. D. Landau, *Zh. Eksp. Theor. Fiz.* 20 (1956) 1064.
[151] V. L. Ginzburg, and L. P. Pitayevsky, *Sov. Phys. JETP* 7 (1968) 858.
[152] V. L. Ginzberg, and D. A. Kirahnits, *Problems in high-temperature superconductivity*, Nauka, Moscow, 1977.
[153] P. L. Lax Commun. *Pure Appl. Math.* 21 (1968) 467.
[154] B. E. Zakharov, and A. B. Shabat, *Sov. Phys. JETP* 34 (1972) 62.

[155] B. E. Zakharov, and A. B. Shabat, *Sov. Phys. JETP* 37 (1973) 823.
[156] V. G. Makhankov, and V. K. Fedyanin, *Phys. Rep.* 104 (1984) 1.
[157] V.G. Makhankov, and N. Makhaidiani and O. Pashaev, *Phys. Lett.* A 81 (1981) 161.
[158] V. Zakharov, and I. Fadeleev, *Funct. Analiz* 5 (1971) 18.
[159] S. Manakov, *Sov. Phys. JETP* 38 (1974) 248.

Chapter 3

The Wave-Corpuscle Duality of Microscopic Particles in Nonlinear Quantum Mechanics and Its Experimental Evidences

In order to judge the correctness of Pang's nonlinear quantum mechanical theory we here examine completely and in detail, the wave-corpuscle duality of microscopic particles given by this theory, and verify them further by experiments. In this investigation the wave-corpuscle duality of the solutions of different nonlinear Schrödinger equations, their dynamic features obtained by both nonlinear Schrödinger equations, classical Euler-Lagrange, Hamiltonian equations, and Newton's motion equation.These determinations of the mass, momentum and energy of the particles and corresponding conservation laws of mass, momentum and energy, which display the corpuscle feature, were depicted completely. Except for the corpuscle feature, we elucidated and also checked their wave feature from the solutions of the dynamic equations, which are composed of a bell-type soliton and carrier wave, and their collisions as well as the reflection and transmission on the interface between two mediums, which confirm that their wave feature differs from both the KdV solitary wave and the linear wave. Finally, we used the experimental results of photon soliton in optic fiber and water soliton in water, which are described by the same dynamic equation, as well as the wave-corpuscle duality of microscopic particles depicted in Chapter 1, to affirm the validity of wave-corpuscle duality of microscopic particles depicted by Pang's nonlinear quantum mechanics, which is here called the new nonlinear quantum mechanics. This also confirms the validity of the latter.

3.1. The Wave-Corpuscle Duality of Microscopic Particles Confirmed by the DynamicEquation without Externally Applied Potential

We study now the wave-corpuscle duality of the microscopic particles using the nonlinear Schrödinger equation shown in Eq.(2.66) at $A(\phi)=0$ and V $(x, t) = 0$ in the one-dimensional case, which is denoted [1-13] by

$$i\phi_{t'} + \phi_{x'x'} + b|\phi|^2 \phi = 0, (b > 0) \tag{3.1}$$

We here assume that its solution in Eq.(3.1) is now represented by

$$\phi(x',t') = \varphi(x',t')\exp[i\theta(x',t')] \tag{3.2}$$

Substituting Eq.(3.2) into Eq. (3.1)we get

$$\varphi_{x'x'} - \varphi\theta_{t'} - \varphi\theta_{x'}^2 - b\varphi^2\varphi = 0, (b > 0) \tag{3.3}$$

$$\varphi\theta_{x'x'} + 2\varphi_{x'}\theta_{x'} + \varphi_{t'} = 0 \tag{3.4}$$

where $x' = x/\sqrt{\hbar^2/2m}$, $t' = t/\hbar$. If let $\zeta' = x' - v_e t'$, then equations (3.3)-(3.4) become

$$\varphi_{x'x'} + v_c\varphi\theta_{x'} - \varphi\theta_{x'}^2 - b\varphi^3 = 0 \tag{3.5}$$

$$\varphi\theta_{x'x'} + 2\varphi_{x'}\theta_{x'} - v_e\varphi_{x'} = 0 \tag{3.6}$$

If fixing the time t' and further integrating Eq. (3.6) with respect to x' we can get

$$\varphi^2(2\theta_{x'} - v_e) = A(t')$$

Now let integral constant A(t')=0 in the above equation, then we can get $\theta_{x'} = v_e/2$. Again substituting it into Eq. (3.5), and further integrating this equation we obtain

$$\int_{\varphi_0}^{\varphi} \frac{d\varphi}{\sqrt{Q(\varphi)}} = x' - v_e t' \tag{3.7}$$

where

$$Q(\varphi) = -b\varphi^4/2 + (v_e^2 - 2v_e v_c)\varphi^2 + c'.$$

When c'=0, $v_e^2 - 2v_c v_e > 0$, then $\varphi = \pm\varphi_0$, where $\varphi_0 = [(v_e^2 - 2v_e v_c)/2b]$ is the root of except for $\varphi = 0$. From Eq. (3.6) Pang obtained the solution of Eq. (3.1), which is

$$\varphi(x',t') = \varphi_0 \sec h[\sqrt{\frac{b}{2}}\varphi_0(x' - v_e t')].$$

Pang [1-13] eventually represented the solution of the nonlinear Schrödinger equation (3.1) in the coordinate of (x,t) by:

$$\phi(x,t) = A_0 \sec h\left\{\frac{A_0\sqrt{bm}}{\hbar}\left[(x-x_0)-vt\right]\right\}\times e^{i[mv(x-x_0)-Et]/\hbar} \tag{3.8}$$

where $A_0 = \sqrt{(mv^2/2-E)/2b}$, v is the velocity of motion of the particle, $E = \hbar\omega$.

In practice, the soliton solution of Eq.(3.1) was first obtained by Zakharov and Shabat in the inverse scattering method [14−15]], which was represented as

$$\varphi(x',t') = 2\left(\frac{2}{b}\right)^{1/2}\eta \sec h\left[2\eta(x'-x_0')+8\eta\xi t'\right]\exp\left[-4i\left(\xi^2-\eta^2\right)t'-i2\xi x'-i\theta\right] \tag{3.9}$$

where η is related to the amplitude of a microscopic particle, ξ relates to the velocity of the particle, $\theta = \arg\gamma$, $\lambda = \xi + i\eta$, $x_0' = (2\eta)^{-1}\log(|\gamma|/2\eta)$, γ is a constant.

This solution in both Eq.(3.8) and Eq.(3.9) are all completely different from Eq. (1.9) in linear quantum mechanics, and consists of an envelope andcarrierwaves. Theformer in Eq.(3.8) is

$$\varphi(x,t) = A_0 \sec h\left\{A_0\sqrt{bm}\left[(x-x_0)-vt\right]/\hbar\right\},$$

which is a bell-type, non-topological soliton with an amplitude A_0, but the carrier wave is $\exp\{i[mv(x-x_0)-Et]/\hbar\}$. This solution in Eq.(3.8) is shown in Figure 3.1a. Therefore, the microscopic particles described by nonlinear Schrödinger equation (3.1) are a soliton [16-17]. The envelope φ(x,t) is a slow varying function, which denotes a mass center of the particle; its position is just at x_0, A_0 is its' amplitude, and its width is given by $W' - 2\pi\hbar / A_0\sqrt{2m}$. Thus, the size of the particle is $A_0W' = 2\pi\hbar/\sqrt{2m}$ and is a constant. This shows that the microscopic particle has a determinant size and is localized at x_0. Its form resembles a "wave packet", but its form and size are invariant in propagation process. According to the soliton theory [16-20], the bell-type soliton in Eq. (3.8) can move freely over macroscopic distances in a uniform velocity, *v*, in space-time,retaining its form, energy, and momentum and exhibits a corpuscle property. However, the wave packet in Eq. (1.24) does not have these properties and will be decaying and dispersing with increasing time. Therefore, this soliton solutiondiffers, in essence, from both the plan wave solution in Eq. (1.19) and the wave packet in Eq. (1.24) in linear quantum mechanics.

Just some properties, the vector $\vec{r}$ or x in the representation in Eq.(3.8), have a definite physical significance, and denote the exact positions of the particles at time *t*.Thus, the wave-function $\phi(\vec{r},t)$ or φ(x, t) can exactly represent the states of the particle at the position$x$ at time *t*. These features are consistent with the concept of particles. Thus, the microscopic particles depicted by Eq. (3.1) display an outright corpuscle feature. This result eliminates the contradiction between the first assumption and the results appeared to solve the problems of quantum mechanics in linear quantum mechanics as described in Chapter 1.

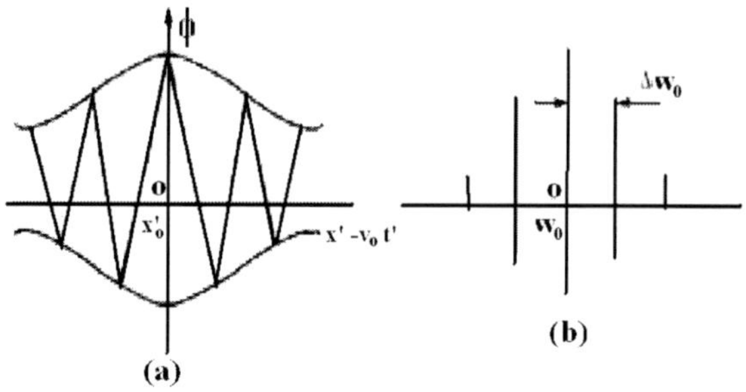

Figure 3.1. The solution of Eq. (3.1) and its features.

Using Eq.(3.9) obtained from the inverse scattering method by Zakharov and Shabat [14-15] or Eq.(3.8) we can also represent the solution of Eq. (3.1) in the coordinate of (x', t') by

$$\phi(x',t') = 2\sqrt{\frac{2}{b}}k\,\mathrm{sec}\,h\left\{2k\left[\left(x'-x_0'\right)-v_e t'\right]\right\}e^{iv_e\left[\left(x'-x_0'\right)-v_c t'\right]/2} \tag{3.10}$$

where v_e is the group velocity of the particle, and v_c is the phase speed of the carrier wave in the coordinate of (x',t'). For a certain system, v_e and v_c are determinants and do not change with time. Meanwhile, we can obtain $2^{3/2}k/b^{1/2} = A_0$, $A_0 = \sqrt{\frac{v_e^2 - 2v_c v_e}{2b}}$.

According to the soliton theory [16-20], the soliton shown in Eq.(3.10) has determinant mass, momentum, and energy, which can be represented by [1-13]

$$N_s = \int_{-\infty}^{\infty} |\phi|^2 dx' = 2\sqrt{2}A_0\ ,\ p = -i\int_{-\infty}^{\infty}\left(\phi^*\phi_{x'} - \phi\phi_{x'}^*\right)dx' = 2\sqrt{2}A_0 v_e = N_s v_e = const\,,$$

$$E = \int_{-\infty}^{\infty}\left[\left|\phi_{x'}\right|^2 - \frac{1}{2}|\phi|^4\right]dx' = E_0 + \frac{1}{2}M_{sol}v_e^2 \tag{3.11}$$

where N_s is just an effective mass of the particles, which is a constant.

Thus, we can confirm that the energy, mass, and momentum of the particle cannot be dispersed in its motion, which concretely embodies the corpuscle features of the microscopic particles.This is completely consistent with the concept of classical particles. This means that the nonlinear interaction, $b|\phi|^2\phi$ in Eq.(2.66), related to the wave function of the particles, balances and really suppresses the dispersion effect of the kinetic term in the dynamic equationmaking the microscopic particles eventually become localized. In this case, the position of the particles, $\vec{r}$ or x, has a determinately physical significance.

However, the envelope of the solution in Eq. (3.8) or (3.10) is a solitary wave. It has a certain wavevector and frequency, as shown in Figure 1(b), and can propagate in space-time, which is accompanied with the carrier wave. Its feature of propagation depends on the concrete nature of the particles. Figure 1(b) shows the width of the frequency spectrum of the envelope $\varphi(x, t)$, which has a localized distribution around the carrier frequency ω_0. This shows that the microscopic particle also has a wave feature.

Thus, we believe that the microscopic particles described by nonlinear quantum mechanics have a simultaneous wave-corpuscle duality. Equation (3.8) or (3.10) and Figure 1a are seen as the most beautiful and perfect representation of the wave-corpuscleduality. This also consists of the de Broglie relation, $E = h\upsilon = \hbar\omega$ and $\vec{p} = \hbar\vec{k}$ ofwave-corpuscle duality and reflect Davisson and Germer's 1927 experimental results of electron diffraction on double seams as well as the traditional concept of particles in physics as shown in Chapter 1. Thus, we have reason to believe the correctness of the nonlinear Schrödinger equation,or nonlinear quantum mechanics, which is used to describe the states and properties of microscopic particles by Pang [1-13].

3.2. The Wave-Corpuscle Duality of the Solutions of the Dynamic Equation with Different Potentials

We can verify that the nature of wave-corpuscle duality of microscopic particles is not changed when varying the externally applied potentials.

As a matter of fact, if $V(x') = \alpha x'$ +c in Eq. (2.66) at $A(\phi) = 0$, where α and c are some constants, in this case Pang [1-13] replaced Eq. (2.70) using,

$$\varphi_{x'x'} - \varphi\theta_{t'} - \varphi\theta_{x'}^2 - b\varphi^2\varphi = \alpha x' + c. \tag{3.12}$$

Now, let

$$\varphi(x',t') = \varphi(\xi), \xi = x' - u(t'), u(t') = -\alpha(t')^2 + vt' + d \tag{3.13}$$

and describes the accelerated motion of $\varphi(x'.t')$ .The boundary condition requires $\varphi(\xi)$ toapproach zero rapidly, then the equation (2.71) can be written as:

$$-\dot{u}\frac{\partial\varphi}{\partial\xi} + 2\frac{\partial\varphi}{\partial\xi}\frac{\partial\theta}{\partial\xi} + \varphi\frac{\partial^2\theta}{\partial\xi^2} = 0 \tag{3.14}$$

where $\dot{u} = \frac{du}{dt'}$. If the $2\,\partial\theta/\partial\xi - \dot{u} \neq 0$, equation (3.14) may be denoted as

$$\frac{\partial\theta}{\partial x'} = \frac{g(t')}{\varphi^2} + \frac{\dot{u}}{2} \tag{3.15}$$

The integration of Eq. (3.15) yields

$$\theta(x',t') = g(t')\int_0^{x'} \frac{dx''}{\varphi^2} + \frac{\dot{u}}{2}x' + h(t') \tag{3.16}$$

where $h(t')$ is an undetermined constant of integration. From Eq. (3.14) Pang [1-13] derived,

$$\frac{\partial\theta}{\partial t'} = \dot{g}(t')\int_0^{x'} \frac{dx''}{\varphi^2} - \frac{g\dot{u}}{\varphi^2} + \frac{g\dot{u}}{\varphi^2}\Big|_{x'=0} + \frac{\ddot{u}}{2}x' + \dot{h}(t') \tag{3.17}$$

From Eq.(3.12) we can obtain

$$\frac{\partial^2\overline{\varphi}}{\partial\xi^2} = \beta\overline{\varphi} - b\overline{\varphi}^3 + g_0^2/\overline{\varphi}^3 \tag{3.18}$$

where β is a real parameter and is defined by with

$$(\alpha x'+c) + \frac{\ddot{u}}{2}x' + \dot{h}(t') + \frac{\dot{u}^2}{4} + \frac{g\dot{u}}{\varphi^2}\Big|_{x'=0} = \overline{V}(\xi) \tag{3.19}$$

Clearly, in the case discussed, $V_0(\xi)=0$. Obviously, it is the solution of Eq. (3.18) when β and g are constant. For large $|\xi|$, we may assume that $\overline{\varphi} \le \beta/|\xi|^{1+\Delta}$, when Δ is a small constant. To ensure that $d^2\overline{\varphi}/d\xi^2$ and $\overline{\varphi}$ approach zero when $|\xi| \to \infty$, only the solution corresponding to $g_0=0$ in Eq. (3.18) remains stable. Therefore we choose $g_0=0$ and obtain the following from Eq. (3.15)

$$\frac{\partial\theta}{\partial x'} = \frac{\dot{u}}{2} \tag{3.20}$$

Thus, from Eq. (3.19) we obtain

$$\alpha x'+c = -\frac{\ddot{u}}{2}x' + \beta - \dot{h}(t') - \frac{\dot{u}^2}{4},\ h(t') = (\beta - v^2/4 - c)t' - \alpha^2 t'^3/3 + v\alpha t'^2/2]\}. \tag{3.21}$$

Substituting Eq. (3.21) for Eqs.(3.17) and (3.19), we obtain

$$\theta = (-\alpha t' + v/2)x' + (\beta - v^2/4 - c)t' - \alpha^2 t'^3/3 + v\alpha t'^2/2] \tag{3.22}$$

Finally, substituting the above into Eq. (3.18), we see that:

$$\frac{\partial^2 \bar{\varphi}}{\partial \xi^2} - \beta\bar{\varphi} + b\bar{\varphi}^3 = 0. \qquad (3.23)$$

When $\beta > 0$, Pang gives the solution of Eq. (3.23), which takes the form

$$\bar{\varphi} = \sqrt{2\beta / b}\,\mathrm{sec}\,h(\sqrt{\beta}\xi)$$

Pang [1-13]finally obtained the complete solution in this condition, which is represented as

$$\phi(x',t') = \sqrt{\frac{2\beta}{b}}\,\mathrm{sec}\,h\left\{\sqrt{\beta}\left[\left(x'-x_0'\right)+(\alpha t'^2 - vt' - d)\right]\right\}\times \\ \exp\{i[(-\alpha t' + v/2)\left(x'-x_0'\right)+(\beta - v^2/4 - c)t' - \alpha^2 t'^3/3 + v\alpha t'^2/2]\} \qquad (3.24)$$

This is a soliton solution. Thus, if V(x')=c, the solution can be represented as

$$\phi(x',t') = \sqrt{\frac{2\beta}{b}}k\,\mathrm{sec}\,h\left\{\sqrt{\beta}\left[\left(x'-x_0'\right)-v(t'-t_0')\right]\right\}\times \exp\{i[v\left(x'-x_0'\right)/2-(\beta - v^2/4 - c)t']\} \quad (3.25)$$

At $V(x') = 2\alpha x'$ and $b = 2$, we can also derive a corresponding soliton solution from the above process. However, in this case Chen and Liu [21-22] adopted the following transformation:

$$\phi(x',t') = \phi'(x',\tilde{t}')\exp[-2i\alpha x'\tilde{t}' + 8i\alpha^2\tilde{t}'^3/3], x' = x' - 2\alpha\tilde{t}'^2, t' = \tilde{t}' \qquad (3.26)$$

to make Eq. (2.66) at $A(\phi) = 0$ becomes

$$i\phi'_{\tilde{t}'} + \phi'_{\tilde{x}'\tilde{x}'} + 2|\phi'|^2\phi' = 0. \qquad (3.27)$$

Thus Chen and Liu [21-22] represented the solution of Eq. (2.66) at $A(\phi) = 0$ and$V(x') = ax'$, $b = 2$ by

$$\phi(x',t') = 2\eta\,\mathrm{sec}\,h\left\{2\eta\left[\left(x'-x_0'\right)+(2\alpha t'^2 - 4\xi't')\right]\right\}\times \exp\{-i[2(\xi' - \alpha t')\left(x'-x_0'\right)+ \\ 4(\xi'^2 - \eta^2)t' + 4\alpha^2 t'^3/3 - 4\xi'\alpha t'^2] + \theta_0\} \qquad (3.28)$$

At the same time, utilizing the above method Pang [1-13] also found the soliton solution of Eq. (2.66) at $A(\phi) = 0$ and $V(x) = kx^2 + A(t)x + B(t)$, which could be represented as:

$$\phi = \varphi(x-u(t))e^{i\theta(x,t)} \tag{3.29}$$

where

$$\varphi(x',t') = \sqrt{\frac{2\beta}{b}}\operatorname{sec}h\left\{\sqrt{\beta}\left[\left(x'-x_0'\right)-u(t')\right]\right\}, \tag{3.30}$$

$$\theta(x',t') = [-2\alpha\sin(2\sqrt{k}t'+\gamma)+u_0(t')/2)](x'-x_0')+\int_0^{t'}\{[-\alpha^2(2\cos(2\sqrt{k}t''+\gamma)+u_0(t'')/2)]^2 -$$
$$B(t'')+[-2\alpha\sin(2\sqrt{k}t''+\gamma)+u_0(t'')/2]\}dt''+\overline{L}t'+\theta_0$$

here $\overline{L}$ is a constant related to A(t').

When $A(t)=B(t)=0$, the solution is still Eq. (3.29), but

$$u(t') = 2\cos(2\sqrt{k}t')+u_0(t'),$$
$$\theta(x',t') = [-2\sqrt{k}\sin(2\sqrt{k}t')+u_0(t')/2)](x'-x_0')+\int_0^{t'}\{[-k(2\cos(2\sqrt{k}t'')+u_0(t'')/2)]^2 +$$
$$[-2\sqrt{k}\sin(2\sqrt{k}t'')+u_0(t'')/2]\}dt''+\theta_0$$

For the case of $V_0(x')=\alpha^2x'^2$ and b=2, where α is constant, Chen and Liu [21-22] assume $u(t')=(2\xi/\alpha)\sin(2\alpha t')$, thus they represent the soliton solution in this condition by:

$$\phi(x',t') = 2\eta\operatorname{sec}h\left\{2\eta\left[\left(x'-x_0'\right)-(2\xi'/\alpha)\sin(2\alpha t')\right]\right\}\times\exp\{i[2\xi'\left(x'-x_0'\right)\cos 2\alpha(t'-t_0')- $$
$$+4\eta^2(t'-t_0')-(\xi'^2/\alpha)\sin[4\alpha(t'-t_0')+\theta_0]\} \tag{3.31}$$

where $2\eta=\sqrt{\beta}$ is the amplitude of microscopic particles, $4\xi'$ is related to its group velocity inEqs. (3.28) and (3.31), and ξ' are the same as ξ in Eq. (3.19).

From the above results we see clearly that these solutions of nonlinear Schrödinger equation (2.66) at $A(\phi)=0$ under influence of different potentials,V(x)=c, $V(x')=\alpha x'$, $V(x')=\alpha x'+c$, $V(x)=kx^2+A(t)x+B(t)$ and $V_0(x')=\alpha^2x'^2$, still consist of envelope and carrier waves, which are analogous to Eq. (3.8), too, some bell-type solitons with a certain amplitude A_0, groupvelocity v_e and phase speed v_c, and have a mass center and determinant amplitude, width, size, mass, momentum and energy.

By inserting these solutions,Eqs. (3.24), (3.25), (3.28),(3.30), and (3.31) into Eq. (3.11), we can determine the effective masses, momentums and energies of these microscopic particles, respectively, all of which have determinant values.

Therefore, we can determine that the microscopic particles described by these dynamic equation(2.66) at $A(\phi)=0$ always possess a wave-corpuscle duality, as shown in Figure 3.1, which are completely different from those in Eqs.(1.9)-(1.18), although they are acted on by different external potentials, in whichthese potentials only change the amplitude, size, frequency, phase, group and phase velocities of the particles, where the velocity and frequencies of some particles are related to time and oscillatory. These results indicate that in Eq. (2.66) at $A(\phi)=0$, the kinetic energy term decides the wave feature of the particles, but the nonlinear interaction determines its corpuscle feature, and their combination results in the wave-corpuscle duality of microscopic particle. In this case, the external potentials only influence the wave form, phase, and velocity of particles, but cannot affect the wave-corpuscle duality. These results eliminated again the contradictions between the first assumption and results that occurred in solving the problems of quantum mechanics described in Chapter 1.They clearly and directly verify the correctness for the nonlinear Schrödinger equationand the necessity of nonlinear quantum mechanics, proposed and built by Pang [1-13], which is used to describe the states and properties of microscopic particles.

3.3. THE LOCALIZED PROPERTY OF MICROSCOPIC PARTICLES IN NONLINEAR QUANTUM MECHANICS

How is the microscopic particles described by the nonlinear Schrödinger equation (2.66) at $A(\phi)=0$ be localized and stabilized? In order to answer this question, Pang [23-25] discusses the localized properties of the microscopic particles. For the nonlinear Schrödinger equations (2.66) at $A(\phi)=0$, its time-independent formin a one-dimensional case is expressed by

$$\frac{\hbar^2}{2m}\frac{\partial^2\varphi'}{\partial x^2}=-b|\varphi'|^2\varphi'+(V-E)\varphi' \tag{3.32}$$

which is obtained through inserting

$$\phi(x',t')=\varphi'(x',t')\exp[iEt'/\hbar] \tag{3.33}$$

into Eq.(2.66) at $A(\phi)=0$. In this case Eq.(3.32) can be represented by

$$E\varphi'=(\hat{H}_0-|b||\varphi'|^2)\varphi'.\text{ with } \hat{H}_0=-\frac{\hbar^2}{2m}\nabla^2+V(\vec{r}) \tag{3.34}$$

If $V(x')$and b are independent as related to t', then the equation (3.32) in a one-dimensional case, may be written as:

$$\frac{\hbar^2}{2m}\frac{\partial^2\varphi'}{\partial x^2}=\frac{d}{d\varphi'}V_{eff}(\varphi') \tag{3.35}$$

where $V_{eff}(\varphi')$ is the effective potential of the system, and can represented by [63-65]

$$V_{eff}(\varphi') = \frac{1}{4} b|\varphi'|^4 - \frac{1}{2}(V - E)|\varphi'|^2 . \tag{3.36}$$

when $V > E$and $V < E$, and the relationship between $V_{eff}(\varphi')$ and φ', is shown as in Figure 3. 2. From this figure we see that there are two minimum values in the effectivepotential, which corresponds to the two ground states of the microscopic particle at b>0, *i.e.*

$$\varphi_0' = \pm\sqrt{(V - E)/b} . \tag{3.37}$$

Equation(3.34) shows clearly that the energy of the microscopic particle in nonlinear quantum mechanics is lowered relative to that in linear quantum mechanics due to the nonlinear interaction, which means that the particle is localized and in a stable state.In this case, its localized or binding energy is just determined by the nonlinear interaction energy $|b||\varphi'|^2$. This phenomenon can be verified from Eqs.(3.36)-(3.37) because the effective potential in Eq.(3.36) is a double-well potential, and its energy is $-(V - E)^2/4b \le 0$ at b>0. This shows that the microscopic particle is really bound and localized due to its negative binding energy. This localization is achieved through repeated reflections on the microscopic particles induced by the double-well potential field. The two ground states limit the energy diffusion, thus the energy of the particle is gathered, a soliton is formed, and the particle is eventually localized [23-25]. Obviously, this is due to the nonlinear interaction because the binding energy and two ground states of the particle are closely related to the nonlinear energy parameter b, in this casethe nonlinear interaction: b $(\varphi'\varphi'^*)^2 \ne 0$ or $b \ne 0$. In other words, once the nonlinear interaction is eliminated, i.e., b=0, then the particles will disperse, cannot be localized and have only a common ground state, which is $\varphi' = 0$. This displays that the nonlinear interaction is the basic reason of thelocalization of the particles, in other words, it plays a determinant and key role in the localized process, i.e., there is not localization of microscopic particle without the nonlinear interaction. This demonstrates and exhibits that linear quantum mechanics cannot give the localized or wave-corpuscle feature of microscopic particle no matter how it changes itsframework due to absence of nonlinear interaction [23-25]. This is just a truth and intrinsic, namely, this conclusion cannot be changed.

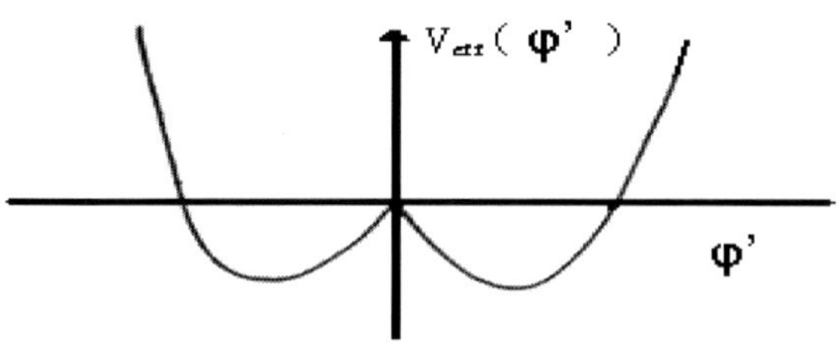

Figure 3.2. The effective potential of nonlinear Schrödinger equation.

However, when $V > 0$, $E > 0$ and $V<E$, or $|V| > E$, $E > 0$ and $V > 0$, for $b > 0$, the microscopic particles are *not* localized by those mechanisms because the systems cannot provide the double-well potential mentioned above in these conditions.

On the other hand, if the nonlinear interaction is repellant (*i.e.*, $b<0$), then equation (2.66) at $A(\phi)=0$ becomes:

$$i\hbar\frac{\partial}{\partial t}\phi+\frac{\hbar^2}{2m}\frac{\partial^2\phi}{\partial x^2}-|b||\phi|^2\phi=V(x,t)\phi. \tag{3.38}$$

Although its effective potential is not a double-well potential, and the bell-type soliton in Eq. (3.8) cannot occur as well, a dark or hole soliton may be formed in this case. Zakhorov and Shabat [14-15] obtained the dark soliton solution of Eq. (3.37) at $V(x) = 0$, which was experimentally observed in optical fiber and discussed in detail in the Bose-Einstein condensation model by Barger et al., [26].

On the other hand, if $V(x, t) = V(x)$ or is a constant, then the equation (3.38) also has a kink soliton solution. Inserting Eq. (3.33) into Eq. (3.38) we can get:

$$\frac{\hbar^2}{2m}\frac{\partial^2\varphi'}{\partial x^2}-|b|\varphi'^3+[E-V(x)]\varphi'=0.$$

If $V$is independent of x and $0<V<E$, the kink solution can be represented by:

$$\varphi'=\frac{\sqrt{2(E-V)}}{|b|}\tanh\left[\sqrt{\frac{2(E-V)}{\hbar^2}}(x-x_0)\right].$$

If, by inserting Eq. (3.33) into Eq. (3.38) we can obtain another quasi-Eigen equation:

$$E\varphi'=(\hat{H}_0+|b||\varphi'|^2)\varphi'.$$

where $\hat{H}_0=-\frac{\hbar^2}{2m}\nabla^2+V(\vec{r})$.

This shows and predicts that the Eigen energy of the particle is larger than that in Eq.(3.34) in this case, i.e., it will lift due to the existence of nonlinear interaction, thus this model describes the nonlinear localization of another kind of particle, such as holes, in nonlinear semiconductors.

3.4. The Stability of Microscopic Particles with the Wave-Corpuscle Duality

Stability is an important property of the particles. The so-called stability of particles happens, in general, when the stability points toward the fact that the particles cannot be dispersed in the process of propagation and interaction. This means that the microscopic particles depicted by linear quantum mechanics are unstable because they can be dispersed easily in motion. However, we ask: "Are the microscopic particles in nonlinear quantum mechanics stable?" And; "How do we verify their stability?"

Usually, three types of stability of microscopic particles are considered in nonlinear quantum mechanics; (1) with respect to perturbation of the initial state, (2) with respect to a perturbation of the dynamic equation governing the system dynamics (structural stability); and (3) with respect to minimal energy state under action of an externally applied field.

In the first case, the problem has been investigated using various approaches (linear and nonlinear). In the linear approximation, the stability problem is usually reduced to an eigenvalue problem of linear equations. In the nonlinear case, it is reduced to the study of Lyapunov inequalities. The study of the structural stability of microscopic particles can be done within the framework of dynamic equations under different types of perturbations. Attempts have been made to construct a general perturbation theory of microscopic particles based on the dynamic equations by using the Green function method and spectral transformation (i.e. a transformation from the configuration space (x, t) to the scattering data space based on the well-known two-time formalism). However, the latter is only suitable for a rather restricted class of perturbation functional. Although some results were obtained in this direction, the studies cannot be considered as complete. In some cases, numerical studies are very effective tools. From the computational point of view, both problems can be investigated within the framework of a unified approach. In the first case, one studies the dynamics, described by an unperturbed dynamic equation, of a perturbed or unperturbed initial state given in the form of a soliton solution, the stability of which is examined. In the second case, the initial state evolution is governed by a perturbed equation. In both cases, the solutions depend on some specific parameters, which are slowlyvaryingfunctions of time.

In the first case, a solution is considered as stable if initial perturbations are not magnified as the initial state evolves with time. In accordance with this definition, weakly radiating soliton-like solutions, these would be solitons, which are not destroyed under initial perturbations, can be considered stable. Obviously, structure-stable solutions are the solutions that conserve their shape for a sufficiently long time. The notion of a "sufficiently long time" is relative to the time scale of physical processes occurring in the systems. These problems are not studied further here. More details on these problems can be found in Makhankov et al.'s publications [27-30].

However, we must ask, "How can the stability of macroscopic particles exposed in an externally applied field be proved?" In this condition, the interactions between the microscopic particles are very complicated, thus it is difficult to define and determine the behavior and stability of each one individually, again using the above strategies of initial and structure stability as well as the collision rule of the particles. Instead, we apply the fundamental work-energy theorem in classical physics: a mechanical system in the state of minimal energy is said to be stable because an external energy must be supplied, in order to change this state. Pang [23-25] demonstrated the stability of the microscopic particles described by the nonlinear Schrödinger equation (2.66) at $A(\phi)=0$, using the minimum energy theorem. In fact, this stability principle is very effective, when microscopic particles are exposed to an externally applied field. This method is outlined in the following by Pang [23-25].

If we let $\phi(x,t)$ represent the field of the particle, and assume that it has derivatives of all orders, and of all integrations, and is convergent and finite, then the Lagrange and Hamiltonian density function, corresponding to the nonlinear Schrödinger equation (2.66), at $A(\phi)=0$ are represented by Eqs.(2.73)-(2.74), respectively, in which the nonlinear interaction, b $(\phi\phi^*)^2$ is involved; the nonlinear interaction can obstruct and suppress the dispersive effect of kinetic energy of microscopic particles in the dynamic equation.

In the general case, the total energy of the system, which can be obtained from Eq.(2.74), is represented by [23-25].

$$E(t') = \int_{-\infty}^{\infty}\left[\left|\frac{\partial\phi}{\partial x'}\right|^2 - \frac{b}{2}|\phi\phi^*|^2 + V(x')|\phi|^2\right]dx' \tag{3.39}$$

However, in this case, b and $V(x,t')$ are not functions of t'.So, the total energy of the systems is a conservative quantity, i.e., $E(t')$=E=const. We can demonstrate that when $x' \to \pm\infty$, the solutions of Eq. (2.66) at $A(\phi)=0$ = 0or $\phi(x,t)$ should tend to zero rapidly, i.e.

$$\lim_{|x'|\to\infty}\phi(x',t') = \lim_{|x|\to\infty}\frac{\partial\phi}{\partial x'} = 0$$

Then $\int_{-\infty}^{\infty}\phi^*\phi dx' = \int_{-\infty}^{\infty}\rho(x')dx' =$ constant, or

$$\frac{\partial}{\partial t'}\int_{-\infty}^{\infty}\phi^*\phi dx' = 0,$$

which denotes the mass conservation of the microscopic particle.

Thus, $\phi^*\phi dx' = \rho(x')dx'$ can be regarded as the mass in the interval of x' to x'+dx'. Thus,the position of mass center of a microscopic particle at $x_0' + v_e t'$ in this case can be represented by

$$\langle x' \rangle = x_g', x_0 = \int_{-\infty}^{\infty}\phi'^* x'\phi dx' / \int_{-\infty}^{\infty}\phi^*\phi dx' \tag{3.40}$$

Since

$$v = v_g = \frac{d}{dt'}\{\int_{-\infty}^{\infty}\phi^* x'\phi dx' / \int_{-\infty}^{\infty}\phi^*\phi dx'\} = \int_{-\infty}^{\infty}(\phi_{t'}^* x'\phi + \phi^* x'\phi_{t'})dx' / \int_{-\infty}^{\infty}\phi^*\phi dx',$$

then, from Eq. (2.66) at $A(\phi)=0$ and its conjugate equation,

$$-i\hbar\frac{\partial\phi^*}{\partial t}=-\frac{\hbar^2}{2m}\nabla^2\phi^*\pm b|\phi|^2\phi^*+V(\vec{r},t)\phi^* \tag{3.41}$$

we can determine that the velocity of mass center of the microscopic particle can be denoted by

$$v_g=d\langle x'\rangle/dt'=-2i\int_{-\infty}^{\infty}\phi^*\phi_{x'}dx'/\int_{-\infty}^{\infty}\phi^*\phi dx' \tag{3.42}$$

However, for different solutions of the same nonlinear Schrödinger equation (3.1), $\int_{-\infty}^{\infty}\phi^*\phi dx'$, $\langle x'\rangle$ and dx'/dt' can have different values. Therefore, it is unreasonable to compare the energy between a definite solution and other solutions. We should compare the energy of one particular solution to that of another, similarsolution. The comparison is only meaningful for many microscopic particle systems that have the same values of

$$\int_{-\infty}^{\infty}\phi^*\phi dx'=K,\ \langle x'\rangle=u \text{ and } d<x'>/dt'=\dot{u}$$

at the same time t_0'. Based on these, we can finally determine the stability of the solutions of Eq. (3.1), for example, Eq. (3.8). Thus, we assume that the different solutions of Eq. (3.1) satisfy the following boundary conditions at a definite time t_0':

$$\int_{-\infty}^{\infty}\phi^*\phi dx'=K,\ \langle x'\rangle\big|_{t'=t_0'}=u(t_0'),\ \frac{d<x'>}{dt'}\bigg|_{t'=t_0'}=\dot{u}(t_0') \tag{3.43}$$

Pang [23-25] assumes the solution of Eq. (3.1) to have the form of Eq. (3.2). Substituting Eq. (3.2) into Eq. (3.39), we obtain the energy formula:

$$E=\int_{-\infty}^{\infty}[(\frac{\partial\varphi}{\partial x'})^2+\varphi^2(\frac{\partial\theta}{\partial x'})^2-\frac{b}{2}\varphi^4+V(x')\varphi^2]dx' \tag{3.44}$$

At the same time, the equation (3.43) becomes

$$\int_{-\infty}^{\infty}\varphi^2dx'=K,\ \frac{\int_{-\infty}^{\infty}x'\varphi^2dx'}{\int_{-\infty}^{\infty}\varphi^2dx'}=u(t_0'),\ \frac{2\int_{-\infty}^{\infty}\varphi^2\frac{\partial\theta}{\partial x'}dx'}{\int_{-\infty}^{\infty}\varphi^2dx'}=\dot{u}(t_0') \tag{3.45}$$

Finding the extreme value of the functional Eq. (3.44) under the boundary conditions of Eq. (3.45) by means of the Lagrange uncertain factor method, we obtain the following Euler equations:

$$\frac{\partial^2\varphi}{\partial x'^2}=\{V(x')+C_1(t_0')C_2(t_0')[x'-u(t_0')]+C_3(t_0')[2\frac{\partial\theta}{\partial t'}-\dot{u}(t_0')]+(\frac{\partial\theta}{\partial x'})^2\}\varphi-b\varphi^3 \quad (3.46)$$

$$\varphi^2\frac{\partial^2\theta}{\partial x'^2}+2\frac{\partial\varphi}{\partial x'}\varphi\frac{\partial\theta}{\partial x'}+2C_3(t_0')\varphi\frac{\partial\varphi}{\partial t'}=0 \quad (3.47)$$

where the Lagrange factors C_1, C_2, and $C_3$are all functions of t'. Nowlet $C_3(t_0')=-\frac{1}{2}\dot{u}(t_0')$.

If $2\frac{\partial\theta}{\partial x'}-\dot{u}(t_0')\neq 0$, from Eq. (3.47) we can get

$$\frac{2}{\varphi}\frac{\partial\varphi}{\partial x'}=\frac{-\partial^2\theta/\partial x'^2}{(\partial\theta/\partial x'-\dot{u}(t_0')/2)}.$$

Integration of the above equation yields

$$\varphi^2=\frac{g(t_0')}{(\partial\theta/\partial x'-\dot{u}(t_0')/2)}$$

or

$$\left.\frac{\partial\theta}{\partial x'}\right|_{t'=t_0'}=\frac{g(t_0')}{\varphi^2}+\frac{\dot{u}(t_0')}{2} \quad (3.48)$$

where g(t$_0$’) is an integral constant. Thus,

$$\theta(x',t')=g(t_0')\int_0^{x'}\frac{dx''}{\varphi^2}+\frac{\dot{u}(t_0')}{2}x'+M(t_0'), \quad (3.49)$$

here, M(t$_0$’) is also an integral constant. Again let

$$C_2(t_0')=\frac{1}{2}\ddot{u}(t_0'), \quad (3.50)$$

and substituting Eqs. (3.47) - (3.50) into Eq. (3.46), we obtain:

$$\frac{\partial^2 \varphi}{\partial (x^{'})^2} = \left\{ V(x^{'}) + \frac{\ddot{u}(t_0)}{2} x^{'} + \left[C_1(t_0) - \frac{\ddot{u}(t_0)}{2} u(t_0) + \frac{\ddot{u}^2(t_0)}{4} \right] \right\} \varphi - b\varphi^3 + \frac{g^2(t_0)}{\varphi^3} \quad (3.51)$$

Letting

$$C_1(t_0') = \frac{u(t_0')\ddot{u}(t_0')}{2} - \frac{\dot{u}^2(t_0')}{2} + M(t_0') + \beta' \quad (3.52)$$

where β' is an undetermined constant, which is a function of t'-independent, and assuming $Z = x' - u(t_0')$, then $\partial^2\varphi / \partial x'^2 = \partial^2 \varphi / \partial Z^2$ is only a function of Z. To make the right-hand side of Eq. (3.51) also be a function of Z, the coefficients of φ , φ^3 and must also be functions of Z, thus, $g(t_0') = g_0 = const$, and

$$V(x') + \frac{\ddot{u}(t_0')}{2} x' + M(t_0') - \frac{u^2(t)}{4} = \widetilde{V}_0(Z)$$

Then, equation (3.51) becomes

$$\frac{\partial^2 \varphi}{\partial (x')^2} = \left\{ \widetilde{V}\left[x' - u(t_0')\right] + \beta' \right\} \varphi - b\varphi^3 + \frac{g^2(t_0')}{\varphi^3} \quad (3.53)$$

Since in the present case, hence equation (3.53) becomes:

$$\frac{\partial^2 \varphi}{\partial (x')^2} = \beta' \varphi - b\varphi^3 + \frac{g^2(t_0')}{\varphi^3} \quad (3.54)$$

Therefore, φ is the solution of Eq. (3.54) for the parameters β'= constant and $g(t_0')$= constant.

For sufficiently large, we may assume that $|\varphi| \le \widetilde{\beta} / |Z|^{1+\Delta}$, where Δ is a small constant. However, in Eq. (3.54) we can only retain the corresponding solution $\varphi(Z)$ to ensure that $Lim_{|\varepsilon| \to \infty} d^2\varphi / dZ^2 = 0$. Thus, equation (2.54) becomes

$$\frac{\partial^2 \varphi}{\partial (x')^2} = \beta' \varphi - b\varphi^3 \quad (3.55)$$

In fact, if $\partial\theta/\partial t' = \dot{u}/2$, then from Eq.(3.53) -(3.54) we can verify that the solution given in Eq. (3.8) satisfies Eq.(3.55). In such a case, it is not difficult to show that the energy corresponding to the solution Eq. (3.8) of Eq. (3.55) has a minimal value under the boundary conditions given in Eq. (3.55).

Thus, we concluded that the solution of the nonlinear Schrödinger equation (3.3),at $A(\phi) = 0$ or microscopic particles with the wave-corpuscle duality in nonlinear quantum system are stable, in such a case, from the minimum energy theorem [23-25].

3.5. The Corpuscle and Classical Characteristics of Motion of Microscopic Particles

3.5.1. The Features of Newton's Motion of Microscopic Particles

Since the microscopic particle described by the nonlinear Schrödinger equation (2.66) at $A(\phi) = 0$ has a corpuscle feature and, as mentioned above, is also quite stable, thus its motion in action of a potential field in space-time should have, in itself, rules of motion. In a number of such cases, Pang [31-32] studied this rule of motion of microscopic particles in depth.

We know that the solution, Eq. (3.8), shown in Figure 1, of the nonlinear Schrödinger equation (3.1), exhibits the behavior of $\partial\phi/\partial x' = 0$ at $x' = x_0'$. Thus we can infer that the position of the soliton is localized at $x' = x_0'$ at t'=0, which is just the position of the mass center of the microscopic particle, and is defined by Eq. (3.40). Its velocity is represented in Eq. (3.41). We now determine the acceleration of the mass center of the microscopic particles and its rules of motion in an externally applied potential.

Now utilizing Eq. (2.66) at $A(\phi)$ = 0and its conjugate equation (3.41) we can obtain:

$$\frac{\partial}{\partial t'}\int_{-\infty}^{\infty}\phi^*\phi_{x'}dx' = \int_{-\infty}^{\infty}\phi_{t'}^*\phi_{x'}dx' + \int_{-\infty}^{\infty}\phi^*(\phi_{t'})\phi_{x'}dx' = i\int_{-\infty}^{\infty}\{\ \phi^*\frac{\partial}{\partial x'}[\phi_{x'x'} + b\phi^*\phi^2$$
$$-V\phi] - [\phi_{x'x'}^* - b\phi(\phi^*)^2 - V\phi^*]\phi_{x'}\}dx' = i\int_{-\infty}^{\infty}\phi^*\frac{\partial V}{\partial x'}\phi dx' \tag{3.56}$$

where $x' = x/\sqrt{\hbar^2/2m}$, $t' = t/\hbar$. We here utilize the following relations and the boundary conditions $\int_{-\infty}^{\infty}(\phi^*\phi_{x'x'x'} - \phi_{x'x'}^*\phi_{x'})dx' = 0, \int_{-\infty}^{\infty}b(\phi^{*2}\phi\phi_{x'} + \phi^*\phi^2\phi_{x'}^*)dx' = 0$ $\int\phi^*\phi dx'$ = constant(or a function of t') and

$$\lim_{|x'|\to\infty}\phi^* x'\frac{\partial\phi}{\partial x'} = \lim_{|x'|\to\infty}\frac{\partial\phi^*}{\partial x'}x'\phi = 0,\ \ \operatorname{Lim}_{|x'|\to\infty}\phi(x',t') = \lim_{|x'|\to\infty}\phi_{x'}(x',t') = 0$$

where

$$\phi_{x'} = \frac{\partial \phi}{\partial x'}, \phi_{x'x'x'} = \frac{\partial^3 \phi}{\partial x'^3}.$$

Thus, we find:

$$\frac{d}{dt'}\int_{-\infty}^{\infty} \phi^* x'\phi dx' = \int_{-\infty}^{\infty} (\frac{\partial \phi^*}{\partial t'} x'\phi + \phi^* x'(\frac{\partial \phi}{\partial t'}) dx' = -2i\int_{-\infty}^{\infty} \phi^* \phi_{x'} dx' \tag{3.57}$$

In the systems, the position of the mass center of the microscopic particle can be represented by Eq. (3.40), thus the velocity of the mass center of the microscopic particle is represented by Eq. (3.41). Then, the acceleration of the mass center of the microscopic particle can also be denoted [31-32] by

$$\frac{d^2}{dt'^2} < x' > = -2i\frac{d}{dt'}\{\frac{\int_{-\infty}^{\infty} \phi^* \phi_{x'} dx'}{\int_{-\infty}^{\infty} \phi^* \phi \, dx'}\} = -2\int_{-\infty}^{\infty} \phi^* V_{x'} \phi^* dx' = -2 < \frac{\partial V}{\partial x'} > \tag{3.58}$$

If ϕ is normalized, i.e., $\int_{-\infty}^{\infty} \phi^* \phi dx' = 1$, then the above conclusions are not changed.

When V=V(x') in Eq. (3.58) is the external potential field experienced by the microscopic particles, then at the mass center $x' = < x' > = x_0'$ we can expand the term :

$$\left\langle \frac{\partial V(x')}{\partial x'} \right\rangle$$

in Eq.(3.58) and take again the expectation value on this term, which can be denoted by

$$\left\langle \frac{\partial V(x')}{\partial x'} \right\rangle = \frac{\partial V(< x' >)}{\partial < x' >} + \frac{1}{2!} < (x' - < x' >)^2 > \frac{\partial^3 V(< x' >)}{\partial < x' >^3}$$

where

$$\Delta_{ij} = \left\langle ((x' - \langle x' \rangle)^2 \right\rangle = \left\langle (x_i' - x_0')(x_j' - x_0') \right\rangle = \left\langle (x_i' - \langle x_i' \rangle)(x_j' - \langle x_j' \rangle) \right\rangle$$

For the microscopic particle described by Eq. (2.66) at $A(\phi) = 0$ or Eq. (3.1), the position of the mass center of the particle is known and determinant, which is just $< x' > = x_0'$ =constant, or 0. Since here we study only the rule of motion of the mass center x_0, which means that the terms contained in $<x'^2>$ are considered and included, then $< (x' - < x' >)^2 >$ =0 can be obtained.

Thus,

$$\left\langle \frac{\partial V(x')}{\partial x'} \right\rangle = \frac{\partial V(<x'>)}{\partial <x'>}$$

Pang [31-32] finally obtained the acceleration of the mass center of the microscopic particle in the nonlinear quantum mechanics, Eq. (3.58), which is denoted as

$$\frac{d^2}{dt'^2} <x'> = -2\frac{\partial V(<x'>)}{\partial <x'>} \tag{3.59}$$

Returning to the original variables, the equation (2.59) becomes

$$m\frac{d^2 x_0}{dt^2} = -\frac{\partial V}{\partial x_0} \tag{3.60}$$

where x_0' =<x'> is the position of the mass center of the microscopic particle.

Equation (3.60) is an example of an equation of motion of mass center of the microscopic particles in nonlinear quantum mechanics. It resembles the Newton-type equation of motion of classical particles, which is a fundamental dynamics equation in classical physics. Thus, it is not difficult to conclude that the microscopic particles, depicted by the nonlinear quantum mechanics,are a property of the classical particle.

The above equation of motion of particles can also derive from Eq. (2.66) at $A(\phi) = 0$ by another method. As it is known, the momentum of the particle depicted by Eq. (2.66) at $A(\phi) = 0$ is obtained from Eq. (3.10) and denoted by

$$P = \frac{\partial L}{\partial \phi} = -i\int_{-\infty}^{\infty} \left(\phi^* \phi_{x'} - \phi_{x'}^* \phi\right) dx'.$$

Utilizing Eq. (2.66) at $A(\phi) = 0$ and Eqs. (3.56)-(3.57) Pang obtained [1-13,31-32]

$$\frac{dP}{dt'} = \int_{-\infty}^{\infty} 2V(x')\frac{\partial}{\partial x'}|\phi|^2 dx' = -2\int_{-\infty}^{\infty} \frac{\partial V}{\partial x'}|\phi|^2 dx' = -2\left\langle \frac{\partial V(x')}{\partial x'} \right\rangle \tag{3.61}$$

where the boundary condition of $\phi(x') \to 0$ as $|x'| \to \infty$ is used. Utilizing again the above result, we can also get that the acceleration of the mass center ofthe particle is the form of

$$\frac{dP}{dt'} = -2\frac{\partial V(x_0')}{\partial x_0'}$$

or

$$m\frac{d^2x_0}{dt^2}=-\frac{\partial V}{\partial x_0} \tag{3.62}$$

where x_0' is the position of the center of the mass of the macroscopic particle. This is the same as Eq. (3.60). Therefore, we can confirm that the microscopic particles in nonlinear quantum mechanics satisfy the Newton-type equation of motion for a classical particle.

3.5.2. Euler-Lagrange and HamiltonianEquation ofMotion of Microscopic Particles

Using the above variables ϕ in Eq. (2.66) at $A(\phi)=0$ and ϕ^* in Eq. (3.41), one can determine the Poisson bracket and further develop the equations describing the motion of microscopic particles in the form of Hamiltonian equations. For Eq. (2.66) at $A(\phi)=0$, the variables ϕ and ϕ^* satisfy the Poisson bracket:

$$\left\{\phi^{(a)}(x),\phi^{(b)}(y)\right\}=i\delta^{ab}\delta(x-y) \tag{3.63}$$

where

$$\left\{A,B\right\}=i\int_{-\infty}^{\infty}\left(\frac{\delta A}{\delta\phi}\frac{\delta B}{\delta\phi^*}-\frac{\delta B}{\delta\phi}\frac{\delta A}{\delta\phi^*}\right)$$

The corresponding Lagrangian density $\mathcal{L}$ in Eq.(2.73) associated with Eq. (2.66) at $A(\phi)$ = 0 can be written in terms of $\phi(x',t')$and its conjugate ϕ^* (x',t') viewed as independent variables.

From Eq. (2.73) Pang [,1-13,31-32] find that:

$$\frac{\partial L'}{\partial\varphi^*}=\frac{i\hbar}{2}\varphi_t-V(x)\varphi+b(\varphi^*\varphi)\varphi$$

and

$$\partial_t(\frac{\partial L'}{\partial\phi_t^*})+\nabla.(\frac{\partial L'}{\partial\nabla\phi^*})=-\frac{i\hbar}{2}\phi_t-\frac{\hbar^2}{2m}\nabla.\nabla\phi$$

whereL' =$\mathcal{L}$, we can thus obtain

$$\frac{\partial L'}{\partial\varphi^*}-\partial_t(\frac{\partial L'}{\partial\varphi_t^*})-\nabla.(\frac{\partial L'}{\partial\nabla\varphi^*})=i\hbar\varphi_t+\frac{\hbar^2}{2m}\nabla^2\varphi-V(x)\varphi+b(\varphi^*\varphi)\varphi \tag{3.64}$$

Through comparison with Eq. (2.66) at $A(\phi) = 0$ Pang arrived at:

$$\frac{\partial L'}{\partial \phi^*} = \partial_t \left(\frac{\partial L'}{\partial \phi_t^*}\right) + \nabla.\left(\frac{\partial L'}{\partial \nabla \phi^*}\right)$$

or

$$\frac{\partial L'}{\partial \phi} = \partial_t \left(\frac{\partial L'}{\partial \phi_t}\right) + \nabla.\left(\frac{\partial L'}{\partial \nabla \phi}\right) \tag{3.65}$$

Equation (3.65) is the well-known Euler-Lagrange equation for this system. This shows that the nonlinear Schrödinger equation amounts to the Euler-Lagrange equation in nonlinear quantum mechanics, in other words, the dynamic equation, or the nonlinear Schrödinger equation, can be obtained from the Euler-Lagrange equation in nonlinear quantum mechanics, ifthe Lagrangian function of the system is known. This is different from linear quantum mechanics, in which the dynamic equation is the linear Schrödinger equation, instead of the Euler-Lagrange equation.

On the other hand, Pang [1-13,31-32] also determined the Hamilton equation of the microscopic particle from the Hamiltonian density of this system in Eq. (2.74). In fact, we can obtain from Eq. (2.74)

$$\frac{\delta H'}{\delta \phi^*} = -\frac{\hbar^2}{2m}\nabla^2\phi + V(x)\phi - b(\phi^*\phi)\phi \tag{3.66}$$

where H' $=\mathcal{H}$. Thus, from Eqs. (2.66) and(2.74) we can give

$$i\hbar\frac{\partial \phi}{\partial t} = \frac{\delta H'}{\delta \phi^*} = -\frac{\hbar^2}{2m}\nabla^2\phi + V(x)\phi - b(\phi^*\phi)\phi$$

Thus,

$$i\hbar\frac{\partial \phi}{\partial t} = \frac{\delta H'}{\delta \phi^*}, \text{ or, } i\hbar\frac{\partial \phi^*}{\partial t} = -\frac{\delta H'}{\delta \phi} \tag{3.67}$$

Equation (3.67) is simply a complex form of the Hamilton equation in nonlinear quantum mechanics. In fact, the Hamilton equation can be represented by the canonical coordinate and momentum of the particle. In this case, the canonical coordinate and momentum are defined by:

$$q_1 = \frac{1}{2}\left(\phi + \phi^*\right), p_1 = \frac{\partial L'}{\partial\left(\partial_t q_1\right)}; \quad q_2 = \frac{1}{2i}\left(\phi - \phi^*\right), \quad p_2 = \frac{\partial L'}{\partial\left(\partial_t q_2\right)},$$

respectively. Thus, the Hamiltonian density of the system in Eq. (2.74) takes the form:

$$\mathcal{H} = \sum_n p_n \partial_{t'} q_n - \mathcal{L}$$

and the corresponding variation of the Lagrangian density, $\mathcal{L}$= L', can be written as:

$$\delta L' = \sum_n \frac{\delta L'}{\delta q_n}\delta q_n + \frac{\delta L'}{\delta(\nabla q_n)}\delta(\nabla q_n) + \frac{\delta L'}{\delta(\partial_{t'} q_n)}\delta(\partial_{t'} q_n) \tag{3.68}$$

From Eq. (3.68), the definition of p_i, and the Euler-Lagrange equation is,

$$\frac{\partial L'}{\partial q_n} = \nabla \cdot \frac{\partial L'}{\partial \nabla q_n} + \frac{\partial p_n}{\partial t}$$

One obtains a variation of the Hamiltonian in the form of

$$\delta\mathcal{H} = \sum_n \int (\partial_{t'} q_n \delta p_n - \partial_{t'} p_n \delta q_n) dx'$$

Thus, one pair of dynamic equations can be obtained and expressed by,

$$\frac{\partial q_n}{\partial t'} = \delta\mathcal{H} / \delta p_n, \quad \frac{\partial p_n}{\partial t'} = -\delta\mathcal{H} / \delta q_n \tag{3.69}$$

This is analogous to the Hamilton equation in classical mechanics and has same physical significance with Eq.(3.67), but the latter is often used in nonlinear quantum mechanics. This result shows that the nonlinear Schrödinger equationfor the dynamics of the microscopic particle can also be obtained from the classical Hamilton equation, if the Hamiltonian of the system is known.

From this study we also seek a new way of finding the equation of motion of the microscopic particles in nonlinear systems, i.e., only if the Lagrangian or Hamiltonian of the system is known, we can obtain the equation of motion of the microscopic particles from the Euler-Lagrange or Hamilton equations. This means that Euler-Lagrange or Hamilton equations and the nonlinear Schrödinger equation are identical for describing the dynamic properties of microscopic particles in nonlinear quantum mechanics. This is a very interesting result. As is known, the Euler-Lagrange equation and the Hamilton equation are fundamental equations in classical theoretical (analytic) mechanics, and were used to describe laws of motion of classical particles. Thus, we can affirm that the microscopic particles evidently possess wave-corpuscle duality in nonlinear quantum mechanics. However, the above results and such methods of finding dynamic equations of microscopic particles are impossible in linear quantum mechanics. Therefore, these results exhibit sufficiently the essential distinction between the nonlinear and linear quantum mechanics.

On the other hand, as can be seen from the de Broglie relation; $E = h\upsilon = \hbar\omega$ and $\vec{p} = \hbar\vec{k}$, representing the wave-corpuscle duality of the microscopic particles in quantum theory, we see that the frequency ω and wave vector $\vec{k}$ can play a role as the Hamiltonian of the system and also as the momentum of the particle, respectively, even in the nonlinear systems and has thus the relation:

$$\frac{d\omega}{dt'} = \left.\frac{\partial\omega}{\partial k}\right|_{x'}\frac{dk}{dt'} + \left.\frac{\partial\omega}{\partial x'}\right|_{k}\frac{\partial x'}{\partial t'} = \mathbf{0}$$

as is the case of the usual stationary media.

From the above result, we also know that the usual Hamilton equation in Eq. (3.67), for the nonlinear systems, remains valid for the microscopic particles. Thus, the Hamilton equation in Eq. (3.67) can be now represented by another form [31-32]

$$\frac{dk}{dt'} = -\frac{\partial\omega}{\partial x'}\Big|_{k} \quad \text{and} \quad \frac{dx'}{dt'} = \frac{\partial\omega}{\partial k}\Big|_{x'} \tag{3.70}$$

And, in the energy picture, where $k = \partial\theta/\partial x'$ is the time-dependent wavenumber of the microscopic particle, its frequency, θ , is the phase of the wave function of the microscopic particles.

3.5.3. Verification of Correctness of Classical Properties of Motion of Microscopic Particles

We now present concrete examples to confirm the correctness of the laws of motion of the microscopic particlesdescribed by the nonlinear Schrödinger equation in nonlinear quantum mechanics mentioned above [1-13]. For example:

(1)For the microscopic particles described by Eq. (2.66) at $A(\phi) = 0$ and V = 0 or constant, of which the solutions are that from Eqs. (3.8) or (3.9),(3.10) and (3.25), respectively, we see that the acceleration of the mass center of microscopic particles is zero based on:

$$m\frac{d^2}{dt^2}\langle x\rangle = -\frac{\partial V(<x>)}{\partial <x>} = 0$$

and, in this case this means that the velocity of the particle is a constant. In fact, if we insert Eq. (3.9) into Eq. (3.42), we can obtain the group velocity of the particle:

$$v_g = d<x'>/dt' = -2i\int_{-\infty}^{\infty}\phi^*\phi_{x'}dx' == \text{constant.}$$

This shows that the microscopic particle moves in a uniform velocity in the space-time relationship, its velocity as the group velocity of the soliton. Thus, the energy and momentum of the microscopic particles are retained in the motion process, with these properties being the same as in classical particles.

On the other hand, if the dynamic equation (3.70) is used from Eq. (3.9), we can see that the acceleration and velocity of the microscopic particles are $\frac{dk}{dt'}=0$ and

$$v_g=\frac{dx'}{dt'}=\frac{\partial\omega}{\partial k}=v_e=-4\xi,$$

respectively,where

$$\omega=-\partial\theta/\partial t'=4(\xi^2-\eta^2)=k^2-4\eta^2,\quad k=\partial\theta/\partial x'=-2\xi,$$
$$\theta=-4\left(\xi^2-\eta^2\right)t'-2\xi x'+\theta_0.$$

For the solution in Eq. (3.25) at V = constant, $\theta=v_e(x'-x'_0)/2-(\beta-v_e^2/4-C)t$, then

$$\omega=(\beta-v_e^2/4-C),\ k=v_e/2,\text{ and thus } v_g=\frac{dx'}{dt'}=\frac{\partial\omega}{\partial k}=-2k=-v_e.$$

These results of the acceleration and velocity of microscopic particles are in agreement with the above data obtained from Eqs. (3.9) and (3.60). This indicates that these moved laws shown in Eqs.(3.59), (3.60), (3.62), (3.67), (3.69), and (3.70) are self-consistent, correct and true in nonlinear quantum systems.

(2)In the case of $V(x')=\alpha x'$, the solution of Eq. (2.66) at $A(\phi)=0$ is Eq. (3.28) by Chen and Liu [21-22], which is also composed of an envelope and carrier wave, the mass center of the particle is at x'_0, which is its localized position. From Eq. (3.60) we can determine that the accelerations of the mass center of the microscopic particle, in this case, is given by

$$\frac{d^2x'_0}{dt'^2}=-2\frac{\partial V\left(\langle x'\rangle\right)}{\partial\langle x'\rangle}=-2\alpha=\text{constant} \tag{3.71}$$

On the other hand, from Eq. (3.28) we know that

$$\theta=2\left(\xi-\alpha t'\right)x'+\frac{4\alpha^2t'^3}{3}-4\alpha\xi t'^2+4\left(\xi^2-\eta^2\right)t'+\theta_0, \qquad (3.72),$$

where ξ is the same as ξ' in Eq. (3.28). Again, utilizing Eq. (3.70) we can find:

$$k = 2(\xi - \alpha t'), \omega = 2\alpha x' - 4(\xi - \alpha t')^2 + (2\eta)^2 = 2\alpha x' - k^2 + (2\eta)^2$$

Thus, the group velocity of the microscopic particle is found via:

$$v_g = \frac{d\overline{x'}}{dt'} = \frac{\partial \omega}{\partial k} = 4(\xi - \alpha t') \tag{3.73}$$

Then its acceleration is given by:

$$\frac{d^2 \tilde{x}'}{dt'^2} = \frac{dk}{dt'} = -2a = \text{cons t an t}, \quad \text{here}(x_0' = \tilde{x}') \tag{3.74}$$

Comparing Eq. (3.71) with Eq. (3.74) we find that they are also same. This indicates that Eqs. (3.60), (3.62), (3.67) (3.69) and (3.70) are correct. In such a case, the microscopic particle moves in a uniform acceleration. This is similar with that of the classical particles in an electric field.

(3) For the case of $V(x') = \alpha^2 x'^2$, which is a harmonic potential, the solution of Eq. (2.66) at $A(\phi)$= 0 in this case is Eq. (3.31) , which was obtained by Chen and Liu [21-22]. This solution also contains an envelope and a carrier wave, and has a mass center. Its position is at x'_0, which is the position of the microscopic particle.

When Eq. (3.60) is used to determine the properties of motion of the particle, in this case, Pang [23-25] found the acceleration of the center of mass of the particle, which is:

$$\frac{d^2 x_0'}{dt'^2} = -4\alpha^2 x_0' \tag{3.75}$$

And, at the same time, from Eq. (3.31) we see that

$$\theta = 2\xi\left(x' - x_0'\right)\cos[2\alpha(t' - t_0')] + 4\eta^2(t' - t_0') - (\xi^2/\alpha)\sin[4\alpha(t' - t_0')] + \theta_0 \tag{3.76}$$

where it is the same as ξ' in Eq. (3.31). And from Eqs. (3.70) and (3.76) we can find that:

$$\begin{aligned} k &= 2\xi\cos 2\alpha\left(t' - t_0'\right), \\ \omega &= 4a\xi x' \sin 2\alpha\left(t' - t_0'\right) - 4\xi^2\cos 4\alpha\left(t' - t_0'\right) - 4\eta^2 \\ &= 2\alpha x'\left(4\xi^2 - k^2\right)^{1/2} - 2k^2 + 4\left(\xi^2 - \eta^2\right), \end{aligned}$$

Thus, the group velocity of the microscopic particle is:

$$v_g = \frac{\alpha x'}{\xi'} \frac{k}{\sqrt{1-k^2/4\xi'^2}} - 2k = 2\alpha(x' - x_0')ctg[2\alpha(t' - t_0')] - 4\xi' \cos(2\alpha(t' - t_0'))$$

While its acceleration is seen as:

$$\frac{dk}{dt'} = \frac{\partial \omega}{\partial x'} |_k = -2\alpha\sqrt{4\xi^2 - k^2} = -4\xi\alpha \sin[2\alpha(t' - t_0')]. \tag{3.77}$$

Since $\frac{d^2\tilde{x}'}{dt'^2} = \frac{dk}{dt'}$, here is $\left(\tilde{x}' = x_0'\right)$, we have

$$\frac{dk}{dt'} = \frac{d^2\overline{x'}}{dt'^2} = -4\xi\alpha \sin[2\alpha(t' - t_0')]$$

and

$$x' = \frac{\xi}{\alpha} \sin[2\alpha(t - t_0')] \tag{3.78}$$

Finally, the acceleration of the microscopic particle is presented as:

$$\frac{d^2\overline{x'}}{dt'^2} = \frac{dk}{dt'} = -4\alpha^2 \overline{x}' \tag{3.79}$$

With equation (3.79) also being the same as Eq. (3.75), we also confirm the validity of Eqs. (3.57), (3.60), (3.62), (3.65), (3.67), and (3.69) - (3.70).

In these cases, the microscopic particle moves in harmonic form and this resembles the result of the motion of classical particles.

From the above studies, we draw the following conclusions [1-13]. (1) The motions of microscopic particles in nonlinear quantum mechanics can be described not only by the nonlinear Schrödinger equation, but also by the Hamiltonian principle, and the Lagrangian and Hamilton equations, its changes of position with changing time satisfying the law of motion of classical particles in both uniform and inhomogeneous states. This not only manifests that the nature of the microscopic particles described by nonlinear quantum mechanics differ completely from those in linear quantum mechanics, *but* sufficiently displays the corpuscle nature of the microscopic particles.

(2) It demonstrates that the external potentials can change the states of motion of the microscopic particles (although it cannot vary its wave-corpuscle duality) and, for example, the particles move with a uniform velocity at V(x')=0 or constant, or in uniform acceleration, which corresponds to the motion of a charged particle in a uniform electric field. But when $V\left(x'\right) = \alpha^2 x'^2$, the macroscopic particles show localized vibrations with a frequency of 2

α and amplitude of ξ/α, thus the corresponding classical vibrational equation is $x' = x_0' \sin\omega t'$, with $\omega = 2a$ and $x_0' = \xi/\alpha$.

The laws of motion of the center of mass of microscopic particles expressed by Eq. (3.60) and Eqs. (3.67) – (3.70), in nonlinear quantum mechanics, are consistent with the equations of motion of macroscopic particles.

The correspondence between a microscopic particle and a macroscopic object shows that the microscopic particles described by nonlinear quantum mechanics have exactly the same moved laws and properties as classical particles. The results not only verify the necessity of development and correctness of nonlinear quantum mechanics, based on the nonlinear Schrödinger equation, also clearly exhibits the limits and approximation of linear quantum mechanics*and* can resolve some of the difficulties of linear quantum mechanics and some of the problems as described in Chapter1. Therefore, the results mentioned above have important significance in physics and nonlinear sciences.

3.6. The Conservation Laws of Motion of Microscopic Particles

As is well known, classical particles satisfy the invariance and conservation laws of mass, energy, momentum and angular momentum, all of which are the elementary and universal laws of matter in nature. Meanwhile, we know from Eq. (3.10) that microscopic particles have a determinant mass, momentum, and energy in nonlinear quantum mechanics. However, whether the mass, momentum and energy of the microscopic particles also have an invariance and conservation in nonlinear quantum mechanics is still in debate. In this section we will study this problem and give further the conservation laws of mass, energy and momentum and angular momentum, among the other features of the microscopic particles depicted by the nonlinear Schrödinger equation (2.66) at $A(\phi) = 0$. Too, we will demonstrate that the microscopic particles also satisfy these conventional conservation laws and have the properties of classical particles.

3.6.1. The Mass, Energy and Momentum of Microscopic Particles and Their Conservations

For the microscopic particles described by the nonlinear Schrödinger equation (2.66) at $A(\phi) = 0$ we can define their number density, number current, densities of momentum and energy as [1-13]

$$\rho = |\phi|^2, p = -i\hbar(\phi^* \phi_x - \phi\phi_x^*), \tag{3.80}$$

where $\phi_x = \frac{\partial}{\partial x}\phi(x,t), \phi_t = \frac{\partial}{\partial t}\phi(x,t)$.

From Eq. (2.66) at $A(\phi) = 0$ and its conjugate equation (3.41) as well as Eqs. (3.8) - (3.9) and (3.80) we obtain [1-13]

$$\frac{\partial p}{\partial t'} = \frac{\partial}{\partial x'}[2(\frac{\partial \phi}{\partial x'})^2 + b|\phi^* \phi|^2 + 2V|\phi|^2 - (\phi^* \frac{\partial^2 \phi}{\partial x'^2} + \phi \frac{\partial^2 \phi^*}{\partial x'^2}) + 2iV(\phi^* \frac{\partial \phi}{\partial x'} - \phi \frac{\partial \phi^*}{\partial x'})]$$

$$\frac{\partial \rho}{\partial t'} = \frac{\partial J}{\partial x'},$$

$$\frac{\partial \in}{\partial x'} = \frac{\partial}{\partial x'}[\rho p + i(\frac{\partial \phi^*}{\partial x'} \frac{\partial^2 \phi}{\partial x'^2} - \frac{\partial \phi}{\partial x'} \frac{\partial^2 \phi^*}{\partial x'^2}) - iV(\phi^* \frac{\partial \phi}{\partial x'} - \phi \frac{\partial \phi^*}{\partial x'})]$$

where $x' = x\sqrt{2m} / \hbar, t' = t / \hbar$ .Thus, we arrive at the following forms for the integral of motion

$$\frac{\partial}{\partial t'} M = \frac{\partial}{\partial t'} \int \rho dx' = 0, \frac{\partial}{\partial t'} P = \frac{\partial}{\partial t'} \int p dx' = 0, \frac{\partial}{\partial t'} E = \frac{\partial}{\partial t'} \int \in dx' = 0, \quad (3.81)$$

These formulae represent only the conservation of mass, momentum, and energy in this case. This shows that the mass, momentum, and energy of the microscopic particles in the nonlinear quantum theory still satisfy the conventional rules for the conservation of matter. We can easily see that the mass, momentum and energy of the particles described by the nonlinear Schrödinger equation(2.66) at $A(\phi) = 0$or (3.1) satisfy the general conservation rules of matter. This means that the microscopic particles described by the nonlinear Schrödinger equation possess a corpuscle feature, thus we have reason to believe that the new theory is correct.

We clearly understand from Eqs.(3.80)-(3.81) the physical significance of the wave function $\phi(x',t')$ in Eq. (2.66) at $A(\phi) = 0$. When $\phi(x',t')$ in truth represents the states and properties of microscopic particles, $|\phi(x',t')|^2$ represents the number, or mass density of particles at certain point in place-time, its integration with respect to x over whole place represents the effective mass of the microscopic particle. Therefore, it no longer represents the probability of the particle occurring at a certain point in linear quantum mechanics. This is only the true essence and physical meaning of the wave function of the particles, $\phi(x',t')$, in Eq. (2.66) at $A(\phi) = 0$ in the nonlinear quantum theory. Therefore, the wave function $\phi(\vec{r},t)$ represents the states of microscopic particles with corpuscle feature, which differs from that in linear quantum mechanics.

3.6.2. General Conservation Features of the Microscopic Particles

In practice, the microscopic particles described by the nonlinear Schrödinger equation (2.66) at $A(\phi)$= 0 include much of the conservation laws, except for the conservation examples shown in Eqs. (3.80) and (3.81). These conservation laws can be found by virtue of the invariance of the action relative to several groups of transformations through the Noether theorem. This problem was investigated by Gelfand and Fomin (1963), and Bulman and Kermel (1989). (See C. Sulem and P. L. Sulem et al.'s book [33]). We first give the Noether theorem, related to nonlinear Schrödinger Equation (2.66) at $A(\phi) = 0$, as shown by their method [33].

To simplify, they [33] introduce the following notations:

$$\bar{\xi}=(t',x')=(\xi_0,\xi_1,\ldots\xi_d),\partial_{t'}=\partial_0,\partial=(\partial_0,\partial_1\ldots\partial_d) \text{ and } \Phi=\left(\Phi_1,\Phi_2\right)=\left(\phi,\phi^*\right).$$

From the Lagrangian in Eq. (2.73) related to the nonlinear Schrödinger equationtheyrepresent the action of the system by

$$S\{\phi\}=\int_{t_0'}^{t_1'}\int L'(\phi,\nabla\phi,\phi^*,\nabla\phi^*)dx'dt' \text{ or}$$

$$S\{\phi\}=\int_D\int_{x'}^{\infty}L'(\Phi,\partial\Phi)d\bar{\xi} \tag{3.82}$$

where L' $=\mathcal{L}$ is the Lagrange density function of the system. Under the action of a transformationT^{ε}, which depends on the parameter ε, they [33] have

$$\xi\rightarrow\xi(\bar{\xi},\Phi,\varepsilon),\Phi\rightarrow\bar{\Phi}(\bar{\xi},\Phi,\varepsilon)$$

where ξ and $\bar{\Phi}$ are assumed to be differentiable with respect to ε. When $\varepsilon=0$, the transformation reduces to the identity. And for the infinitesimally small ε, we have

$$\xi=\bar{\xi}+\delta\bar{\xi},\bar{\Phi}=\Phi+\delta\Phi.$$

At the same time, $\Phi(\bar{\xi})\rightarrow\Phi(\xi)$ by the transformationT^{ε}, and the domain of integration, D,is transformed into $\tilde{D}$, thus:

$$S\{\phi\}\rightarrow S\{\bar{\phi}\}=\int_D\int_{x'}^{\infty}L'(\bar{\Phi},\partial\bar{\Phi})d\xi,$$

where $\tilde{\partial}$ denotes differentiation with respect to ξ . The change in the limit of ε under the above transformation can be expressed as

$$\delta S=\int_D\int_{x'}^{\infty}[L'(\bar{\Phi},\partial\bar{\Phi})-L'(\Phi,\partial\Phi)]d\bar{\xi}+\int_D\int_{x'}^{\infty}[L'(\Phi,\partial\Phi)]\sum_{\nu=0}^{d}\frac{\partial\delta\xi}{\partial\bar{\xi}}d\bar{\xi} \tag{3.83}$$

where we used the Jacobian expansion;

$$\frac{\partial(\xi_0,\cdots\cdot_d)}{\partial(\xi_0,\cdots\xi_d)}=1+\sum_{\nu=0}^{d}\frac{\partial\delta\xi_\nu}{\partial\overline{\xi}_\nu}$$

and $L'(\overline{\Phi},\partial\overline{\Phi})$ in the second term on the right-hand side has been replaced by the leading term $L'(\Phi,\partial\Phi)$ in the expansion. Sulem et al., [33] obtained the following results

$$L'(\tilde{\Phi},\tilde{\partial}\tilde{\Phi})-L'(\Phi,\partial\Phi)=\frac{\partial L'}{\partial\Phi_i}[\tilde{\Phi}_i(\overline{\xi})-\Phi_i(\overline{\xi})]+\frac{\partial L'}{\partial(\partial_\upsilon\Phi_i)}[\tilde{\partial}_\upsilon\tilde{\Phi}_i(\overline{\xi})-\partial_\mu\Phi_i(\overline{\xi})]$$

$$=\frac{\partial L'}{\partial\Phi_i}\partial\Phi_i+\partial_\mu(L'\delta\xi_\upsilon)-L'\frac{\partial\delta\xi_\upsilon}{\partial\xi_\upsilon}+\partial_\upsilon[\frac{\partial L'}{\partial(\partial_\upsilon\Phi_i)}]\delta\Phi_i-\partial_\mu[\frac{\partial L'}{\partial(\partial_\upsilon\Phi_i)}]\delta\Phi_i$$

Equation (3.83) can now be replaced by

$$\delta S=\int_D\int_{X'}^{\infty}\left\{\frac{\partial L'}{\partial\Phi_i}-\frac{\partial}{\partial\xi_\upsilon}[\frac{\partial L'}{\partial(\partial_\upsilon\Phi_i)}]\right\}\delta\Phi_i d\overline{\xi}+\int_D\int_{X'}^{\infty}\frac{\partial}{\partial\xi_\upsilon}[L'\delta\xi_\upsilon+\frac{\partial L'}{\partial(\partial_\upsilon\Phi_i)}\delta\Phi_i]d\overline{\xi} \qquad (3.84)$$

where

$$\delta\overline{\Phi}_i=\overline{\Phi}_i(\xi)-\Phi_i(\overline{\xi})=\partial_\nu\Phi_i\delta\xi_i+\delta\Phi_i(\overline{\xi}),\overline{\partial}_\nu\overline{\Phi}_i(\xi)-\partial_\nu\Phi_i(\overline{\xi})=(\overline{\partial}_\nu-\partial_\nu)\Phi_i(\overline{\xi})+\partial_\nu[\overline{\Phi}_i(\xi)-\Phi_i(\overline{\xi})]$$

$$\partial_\upsilon=\frac{\partial\overline{\xi}_\mu}{\partial\xi_\upsilon}\tilde{\partial}_\mu=\left(\delta_{\upsilon\mu}+\frac{\partial\delta\xi_\mu}{\partial\xi_\upsilon}\right)\tilde{\partial}_\mu=\tilde{\partial}_\upsilon+\frac{\partial\delta\xi_\mu}{\partial\xi_\upsilon}\tilde{\partial}_\mu,$$

$$\frac{\partial}{\partial\overline{\xi}_\nu}(L'\delta\overline{\xi}_\nu)=L'\frac{\partial\delta\overline{\xi}_\nu}{\partial\overline{\xi}_\nu}+\frac{\partial L'}{\partial\Phi_i}\partial_\nu\Phi_i\delta\overline{\xi}_\nu+\frac{\partial^2L'}{\partial(\partial_\mu\Phi_i)}\partial^2_{\mu\nu}\Phi_i\delta\overline{\xi}_\nu,$$

$$\frac{\partial L'}{\partial(\partial_\nu\Phi_i)}\partial_\nu\int_{x'}^{\infty}\delta\Phi_i=\frac{\partial}{\partial\overline{\xi}_\nu}[\frac{\partial L'}{\partial(\partial_\nu\Phi_i)}\delta\Phi_i]-\frac{\partial}{\partial\overline{\xi}_\nu}[\frac{\partial L'}{\partial(\partial_\nu\Phi_i)}\delta\Phi_i]\delta\Phi_i$$

where they [33]use the Euler-Lagrange equation, then the first term on the right-hand side in the equation of δS vanishes. From Eqs.(3.82)-(3.84), Sulem et al. [33], got the following Noether theorem:

(A) If the action, Eq.(3.82), is invariant under the infinitesimal transformation of $\phi\to\phi+\delta\phi$ and $\overline{\xi}\to\overline{\xi}+\delta\overline{\xi}$,where $\overline{\xi}=(t',x_1',....x_d')$,the following conservation law holds

$$\frac{\partial}{\partial\xi_\upsilon}[L'\delta\xi_\upsilon+\frac{\partial L'}{\partial(\partial_\upsilon\Phi_i)}\delta\Phi_i]=0,$$

$$\text{or } \frac{\partial}{\partial \xi_\upsilon}[L^{'}\delta\xi_\upsilon + \frac{\partial L^{'}}{\partial(\partial_\upsilon \Phi_i)}\left(\delta\Phi_i - \frac{\partial \Phi_i}{\partial \zeta_\mu}\delta\xi_\upsilon\right)] = 0 \tag{3.85}$$

in terms of $\delta\hat{\Phi}_i$ definition above.

(B) If the action, Eq.(3.82), is invariant under the following infinitesimal transformation:

$$t' \to \bar{t}' = t' + \delta t'(x', t', \phi), x \to \bar{x}' = x' + \delta x'(x', t', \phi), \phi(x', t') \to \bar{\phi}(\bar{t}', \bar{x}') = \phi(t', x') + \delta\phi(t', x'),$$

then it yields that:

$$\int\left[\frac{\partial L'}{\partial \phi_{t'}}\left(\partial_{t'}\phi\partial_{t'} + \nabla\phi\cdot\delta\bar{x}' - \delta\phi\right) + \frac{\partial L'}{\partial \phi_{t'}^*}\left(\partial_{t'}\phi^*\partial_{t'} + \nabla\phi^*\cdot\delta\bar{x}' - \delta\phi^*\right) - L'\delta t'\right]dx'$$

is a conserved quantity.

For the microscopic particles described by nonlinear Schrödinger equation (2.66) at $A(\phi) = 0$ we can obtain

$$\frac{\partial L'}{\partial \phi_{t'}} = \frac{i}{2}\phi^*, \text{ and } \frac{\partial L'}{\partial \phi_{t'}^*} = -\frac{i}{2}\phi,$$

where L' =$\mathcal{L}$ is given in Eq. (2.73). Thus, the following conservation laws and invariance can be obtained from the above Noether theorem [33].

(a)Invariance under time translation and energy conservation law.

The action demonstrated in Eq. (3.82) is invariant under the infinitesimal time translation

$t' \to t' + \delta t'$ with $\delta x' = \delta\phi = \delta\phi^* = 0$, thus equation (3.85) becomes :

$$\partial_{t'}\left[\nabla\phi\cdot\nabla\phi^* - \frac{b}{2}\left(\phi\phi^*\right)^2 + V(x', t')\phi^*\phi\right] - \nabla\cdot\left(\phi_{t'}\nabla\phi^* + \phi_{t'}^*\nabla\phi\right) = 0$$

This results in the conservation of energy, i.e.,

$$E = \int\left(\nabla\phi\cdot\nabla\phi^* - \frac{b}{2}\left(\phi^*\phi\right)^2 + V(x', t')\phi^*\phi\right)dx' = constant \tag{3.86}$$

(b)Invariance of the phase shift or gauge invariance and mass conservation law is investigated.

It is very clear that the action above, as related to the nonlinear Schrödinger equation, is invariant under the phase shift $\bar{\phi} = e^{i\theta}\phi$, which gives $\delta\phi = i\theta\phi$ with $\delta t' = \delta x' = 0$ for infinitesimal θ. In such a case, equation (3.85) becomes

$$\partial_{t'}|\phi|^2 + \nabla \bullet \left\{ i\left(\phi\nabla\phi^* - \phi^*\nabla\phi\right)\right\} = 0 \qquad (3.87)$$

This means the conservation of mass or number of particles, i.e.

$$N = \int |\phi|^2 dx' = \text{constant}$$

and the continuum equation $\frac{\partial N}{\partial t'} = \nabla \cdot \vec{j}$,where $\vec{j}$ is the mass current density is defined as: $\vec{j} = -i(\phi\nabla\phi^* - \phi^*\nabla\phi)$.

(c) Invariance of space translation and momentum conservation law.

If the above action is invariant under an infinitesimal space translation $x' \to x' + \delta x'$ with $\delta t' = \delta\phi = \delta\phi^* = 0$, then equation (3.85) becomes:

$$\partial_{t'}\left[i\left(\phi\nabla\phi^* - \phi^*\nabla\phi\right) + \nabla\cdot\left\{2\left(\nabla\phi^*\times\nabla\phi + \nabla\phi\times\nabla\phi^* + L'\right)\right\}\right] = 0$$

This results in the conservation of momentum of

$$\vec{P} = i\int(\phi\nabla\phi^* - \phi^*\nabla\phi)dx' = \text{constant} \qquad (3.88)$$

Thus, we then have

$$N\frac{d\langle x'\rangle}{dt} = \int x'\partial_{t'}|\phi|^2 dx' = -\int x'\nabla\left[i\left(\phi\nabla\phi^* - \phi^*\nabla\phi\right)\right]dx' = \int i\left(\phi\nabla\phi^* - \phi^*\nabla\phi\right)dx' = \vec{P} = -\vec{J} = -\int \vec{j}dx'$$

(3.89)

where the center of mass of the microscopic particles, $\langle x' \rangle$,is defined as in Eq.(3.40). Equation (3.89) then becomes the definition of momentum in classical mechanics. It shows clearly that the microscopic particles described by nonlinear quantum mechanic have the feature of classical particles.

(d) Invariance under space rotation and angular momentum conservation laws.

The action shown in Eq. (3.82), is invariant under a rotation of angle $\delta\theta$ around an axis $\vec{I}$ such that $\delta t' = \delta\phi = \delta\phi^* = 0$, and $\delta\vec{x}' = \delta\theta\vec{I}\times\vec{x}'$, thus the following conservation of the angular momentum can be obtained:

$$\vec{M} = i\int \vec{x}'\times\left(\phi^*\nabla\phi - \phi\nabla\phi^*\right)dx'$$

In addition to the above, Sulem et al., [33]also derived another invariance of the nonlinear Schrödinger equation from the Noether theorem for the nonlinear Schrödinger equation (2.66) at $A(\phi) = 0$.

(e).The Galilean Invariance.

If the action shown is an invariant under the Galilean transformation:

$$x' \to x'' = x' - vt', t' \to t'' = t', \phi(x',t') \to \phi''(x'',t'') = \exp[-i(vx''/2 + \vec{v}.\vec{v}t''/2)]\phi(x'',t'') \quad (3.90)$$

then the equation (2.66) at $A(\phi) = 0$ can also retain its invariance. For an infinitesimal velocity, v, there are $\delta\phi = \phi(x',t') - \phi(x',t') = -(i/2)vx'\phi(x',t')$, $\delta x' = vt, \delta t' = 0$ in this case.

After integration over the space variables, equation (3.85) leads to the conservation law shown in Eq. (3.89), which implies that the velocity of the center of mass of the microscopic particles is a constant in this case. This clearly exhibits that the microscopic particles have a particulate nature.

3.7. THEPROPERTIES OF MOTION AND COLLISION OF MICROSCOPIC PARTICLES WITH ATTRACTIVE NONLINEAR INTERACTIONS

3.7.1. The Properties of a Collision of Microscopic Particles

As it is known, the collision among the particles is the most effective method to verify and display whether or not the microscopic particles have the corpuscle feature. Therefore, we give considerably attention to the properties of collision of microscopic particles described by the nonlinear Schrödinger equation.Just so, Pang studied [34-36] the properties of propagation and collision of microscopic particles described by the nonlinear Schrödinger equation in Eq.(2.66) at $A(\phi) = 0$ in the fourth–order Runge-Kutta method [37-38].

For the purpose we divide Eq.(2.66) at $A(\phi) = 0$ at b>0in the one-dimensional case into the following two-equations

$$i\hbar\frac{\partial\phi}{\partial t}+\frac{\hbar^2}{2m}\frac{\partial^2\phi}{\partial x^2}=\chi\phi\frac{\partial u}{\partial x}, \tag{3.91a}$$

$$M(\frac{\partial^2 u}{\partial t^2}-v_0\frac{\partial^2 u}{\partial x^2})=\chi\frac{\partial}{\partial x}\left(|\phi|^2\right). \tag{3.91b}$$

We think that in this case, Eqs. (3.91) may describe the features of motion of the studied particle, and another particle (such as phonon), or background field (such as lattice), with mass M, and velocity v_0, respectively, where u is the characteristic quantity of another particle (as phonon) or of vibration (such as, displacement) of the background field. The coupling effect between the two modes of motion is caused by the interactionof deformation of the background field with the studied particle–background field, such as, dipole-dipole interaction, χ is just the coupling coefficient between them and represents the change of interaction energy between the studied particle and background field due to a unit variation of the latter or field. The relation between the two modes of motion can be obtained from the latter in Eq.(2.91b) and represented by

$$\frac{\partial u}{\partial x}=\frac{\chi}{M(v^2-v_0^2)}|\phi|^2+A$$

If inserting this representation into Eq.(3.91) yields only the nonlinear Schrödinger equation (2.66) at V(x)=constant, where there is a nonlinear coupling coefficient in this model, $V(x)=A\chi$, A is an integral constant. In order to use the fourth–order Runge-Kutta method [37-38] to solve numerical equation (3.91) we must further discretize them, in this case, theequations (3.91) are [34-36] now denoted by

$$i\hbar\dot{\phi}_n(t)=\varepsilon\phi_n(t)-J[\phi_{n+1}(t)+\phi_{n-1}(t)]+(\chi/2r_0)[u_{n+1}(t)-u_{n-1}(t)]\phi_n(t) \tag{3.92a}$$

$$M\ddot{u}_n(t)=W[u_{n+1}(t)-2u_n(t)+u_{n-1}(t)]+(\chi/2r_o)[|\phi_{n+1}|^2-|\phi_{n-1}|^2] \tag{3.92b}$$

where the following transformation relation between continuous and discrete functions are used

$\phi(x,t)\to\phi_n(t)$ and $u(x,t)\to u_n(t)$, $u_{n\pm1}(t)=u_n(t)\pm r_0\frac{\partial u_n(t)}{\partial x}+\frac{1}{2!}r_0^2\frac{\partial^2 u_n(t)}{\partial x^2}\pm....$,

$$\phi_{n\pm1}(t)=\phi_n(t)\pm r_0\frac{\partial\phi_n(t)}{\partial x}+\frac{1}{2!}r_0^2\frac{\partial^2\phi_n(t)}{\partial x^2}\pm....$$

where $\varepsilon=\hbar^2/mr_0+A\chi, J=\hbar^2/2mr_0^2, W=Mv_0^2/r_0^2$, r_0 is the distance between two neighboring lattice points. If using transformation: $\phi_n\to\phi_n\exp(i\varepsilon t/\hbar)$ we [35] can eliminate the term $\varepsilon\phi_n(t)$ in Eq. (3.92a). Again making a transformation: $\phi_n(t)\to a_n(t)=a(t)r_n+ia(t)i_n$, then Eq. (3.92) become

$$\hbar \dot{a}r_n = -J(ai_{n+1} + ai_{n-1}) + (\chi / 2r_0)(u_{n+1} - u_{n-1})ai_n \tag{3.93a}$$

$$-\hbar \ \dot{a}i_n = -J(ar_{n+1} + ar_{n-1}) + (\chi / 2r_0)(u_{n+1} - u_{n-1})ar_n \tag{3.93b}$$

$$\dot{u}_n = y_n / M \tag{3.93c}$$

$$\dot{y}_n = W(u_{n+1} - 2u_n + u_{n-1}) + (\chi / 2r_o)(ar_{n+1}^2 + ai_{n+1}^2 - ar_{n-1}^2 - ai_{n-1}^2) \tag{3.93d}$$

$$|a_n|^2 = |ar_n|^2 + |ai_n|^2 = |\phi_n|^2 \tag{3.94}$$

where ar_n and ai_n are real and imaginary parts of a_n. Equation (3.93) can determine the states and behaviors of the particle. Their solutions can be found from the four equations. There are four equations for one structure unit. Therefore, for the quantum systems constructed by N structure units there are 4N-associated equations. When the fourth-order Runge-Kutta method [37-38] is used to numerically calculate their solutions we must further discretize them, in which n is replaced by j, let the time be denoted by n, and the step length of the space variable is denoted by h in the above equations. An initial excitation is required in this calculation, which is chosen as,

$a_n(o)$=ASech $[(n-n_0)\ (\chi / 2r_0)^2/4JW]$,

where A is the normalization constant) at size n, for the applied lattice, $u_n(0)=y_n(0)=0$. In the numerical simulation, it is required that the total energy and the norm (or particle number) of the system must be conserved. The system of units, ev for energy, $\overset{0}{A}$ for length, and ps for time, are proven to be suitable for the numerical solutions of Eqs.(3.93)-(3.94). The one-dimensional system is composed of N units and fixed, where N is chosen to be N=50, and a time step size of 0.0195 is used in the simulations. Total numerical simulation is performed through data parallel algorithms and MALAB language.

If the values of the parameters, $M, \varepsilon, J, W, \chi$ and r_0 , in Eq.(3.93) are appropriately chosen we can calculate the numerical solution of the associated equation (3.93) by using the fourth-order Runge-Kutta method, thus the changes of $|\phi_n(t)|^2 = |a_n(t)|^2$, which is "probability" or number density of the soliton occurring at the nth structure unit, with increasing time and position in time-place can be obtained. This result is shown in Figure 3. 3, which shows that the amplitude of the solution can retain constancy in the motion process, i.e., the solution of Eq.(3.92) or Eq.(2.66) at $A(\phi) = 0$ and V(x)=constant is very stable while in motion. According to the soliton theory[5,13,16] we can obtain that Eq.(3.93) hasan exact soliton solution, which has a feature of classical particles.

In order to verify the corpuscle feature of soliton solution of the nonlinear Schrödinger equation (3.93) we study further their collision property in accordance with the soliton theory [5,13,16]. Thus we further simulated numerically the collision behaviors of two soliton solutions of Eq.(2.66) atV(x) = $\varepsilon = \hbar^2 / mr_0 + A\chi$ =constant using the fourth-order Runge-Kutta method. This process, resulting from two solitons colliding head-on from opposite directions, are set up from opposite ends of the channel, as shown in Figure 3.4, where the above initial conditions simultaneously motivate from the opposite ends of the channels. From this figure, we see clearly that the initial two solitons, having clock shapes separating 50 unit spacings in the channel, collide with each other at about 8ps and 25 units. After this collision, the two solitons in the channel go through each other without scattering and retain their clock shapes to propagate toward and separately along its own channels. The collision properties of the solitons described by the nonlinear Schrödinger equation (2.66) at $A(\phi) = 0$ are the same as those obtained by Zakharov and Shabat [59-60]. Clearly, the property of collision of the solitons of Eq.(2.66) at $A(\phi) = 0$ is the same as the rules of collision of macroscopic particles. Thus, we can conclude that the microscopic particles described by nonlinear Schrödinger equations in nonlinear quantum mechanics have a corpuscle feature.

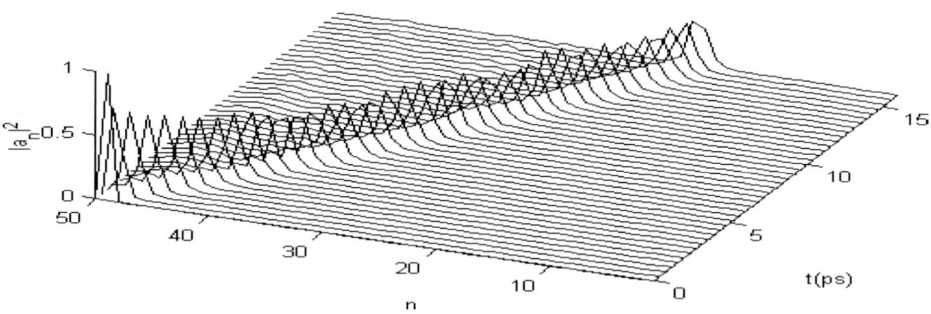

Figure 3.3. The features of propagation of the solutions of Eqs. (3.93)-(3.94).

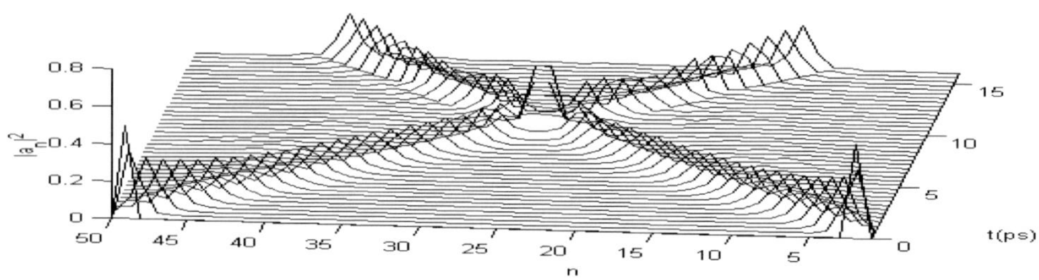

Figure 3.4. The features of a collision of microscopic particles described by Eqs.(3.93)-(3.94).

However, we see clearly that there is a wave peak with large amplitude in the colliding process in Figure 3.4. Obviously, this is a result of complicated superposition of solitary waves of two solitons. Thus, we obtained that this result displays the wave feature of the particles. Therefore, the collision process shown in Figure 3.4obviously represents that the microscopic particles have a both a corpuscle and a wave feature, which is due to the nonlinear interaction $b|\phi|^2 \phi$. Therefore, the collision process of the microscopic particles in nonlinear quantum mechanics shows both an obvious corpuscle feature and a wave feature, i.e., wave-corpuscle duality.

3.7.2. The Change of Features of Microscopic Particles after the Collision

Zakharov and Shabat [14-15] discussed the properties of the collision of two particles as depicted by the most sampled nonlinear Schrödinger equation (3.1) at b=1>0 and b< 0 based on the inverse scattering method and Zakharov and Shabat equation [14-15], which was established based on Lax representations in Eq.(2.85) [39]. They found the change of properties of the soliton solutions after collision. In the studies of b=1>0, they first gave this single soliton solution of this equation, which is presented in Eq.(3.9),where $2\sqrt{2}\eta$ is the amplitude of the soliton, $2\sqrt{2}\xi$ denotes its velocity, is the eigenvalues of the linear operator L in Eq.(3.1).

In such cases, they further found the N-soliton solutionof Eq.(3.1) and thus studied the collision features of two solitons in the system. Here we adopt results of their research to explain the rules and properties of collision between the microscopic particlesdepicted by Eq.(3.1) in nonlinear quantum mechanics. The translations of the mass center and phase of each particle after this collision were obtained and are represented, respectively, by

$$x_{0m}^{'+} - x_{0m}^{'} = \frac{1}{\eta_m} \prod_{p=m+1}^{N} \left| \frac{\lambda_m - \lambda_p}{\lambda_m - \lambda_p^*} \right| < 0, \text{ and } \theta_m^+ - \theta_m = -2 \prod_{p=m+1}^{N} \arg\left(\frac{\lambda_m - \lambda_p}{\lambda_m - \lambda_p^*} \right)$$

and where η_m and λ_m are, respectively, constants related to the amplitude and eigenvalue of mth particles,. The equations show that, after collision of two particles moving with different velocities and amplitudes, the shift of position of the mass center of the particles and their variation of phase are a constant. The collision process of the two particles can be described as in the above equation and is shown as follows. Before the collision and in the case of $t' \to -\infty$, the slowest soliton is in the front while the fastest at the rear, they collide with each other at t'=0, and after the collision and $t' \to \infty$, they are separated and the positions are reversed. Thus, Zakharov and Shabat [14-15]discovered that, as the time t varies from $-\infty$ to ∞, the relative change of mass center of two particles, $\Delta x_{0m}^{'}$, and their relative change of phases, $\Delta\theta_m$, can, respectively, be formularized as:

$$\Delta x_{0m}^{'} = x_{0m}^{'+} - x_{0m}^{'-} = \frac{1}{\eta_m} \left(\sum_{k=m+1}^{N} \ln \left| \frac{\lambda_m - \lambda_p}{\lambda_m - \lambda_p^*} \right| - \sum_{k=1}^{m-1} \ln \left| \frac{\lambda_m - \lambda_p}{\lambda_m - \lambda_p^*} \right| \right) \tag{3.95a}$$

and

$$\Delta\theta_m = \theta_m^+ - \theta_m = 2\prod_{k=1}^{m-1} \arg\left(\frac{\lambda_m - \lambda_p}{\lambda_m - \lambda_p^*} \right) - 2\prod_{k=m+1}^{N} \arg\left(\frac{\lambda_m - \lambda_p}{\lambda_m - \lambda_p^*} \right) \tag{3.95b}$$

Equation (3.95) can be interpreted by assuming that the microscopic particles collide pairwise and every microscopic particle collides with others. In each paired collision, the faster microscopic particle moves forward by a measure of $\eta_m^{-1}\ln\left|\left(\lambda_m-\lambda_k^*\right)/\lambda_m-\lambda_k\right|$, $\lambda_m>\lambda_k$, and the slower one shifts backwards by a measure of $\eta_k^{-1}\ln\left|\left(\lambda_m-\lambda_k^*\right)/\lambda_m-\lambda_k\right|$. The total shift is equal to the algebraic sum of their shifts during the paired collisions. Thus, there is no effect at all because of multi-particle collisions. In other words, in the collision process, each time the faster particle moves forward by a measure of phase shift, the slower one shifts backwards by a measure of phase shift. The total shift of the particles is equal to the algebraic sum of those of the pair during the paired collisions.

The situation is the same with the phases. This rule of collision of the microscopic particles described by the nonlinear Schrödinger equation (3.1) is the same as that of classical particles, or speaking in another way, they also meet the collision laws of macroscopic particles (i.e., during the collision these microscopic particles interact and exchange their positions in the space-time trajectory as if they had passed through each other). After the collision, the two microscopic particles may appear to be instantly translated in space and/or time but otherwise unaffected by their interaction. This translation is called a phase-shift as mentioned above.

In one dimension, this process results from two microscopic particles colliding head-on from opposite directions, or in one direction between two particles with different amplitudes or velocities. This is possible because the velocity of a particle depends on the amplitude. The two microscopic particles surviving a collision, completely unscathed,clearly demonstratethe corpuscle feature of the microscopic particles. This property separates the microscopic particles (solitons) described by nonlinear quantum mechanics from the particles in linear quantum mechanical regime. Thus, this demonstrates again the classical feature of the microscopic particles in nonlinear quantum mechanics.

3.7.3. The Changes of the Features of Microscopic Particles in the Collision Process

In the above study, we take note only of the changes of phase shift of the two particles after the collision by analyzing the changes in the mass center and phase of the solutions of Eq.(3.1) from $t'\to-\infty$ to $t\to\infty$, but not the process. Thus, we cannot clearly understand the true properties of collision of the soliton solutions of Eq.(3.1). Obviously, the collision process is very complicated as shown in Figure 3.4. Desem and Chu [40-42] have paid attention to the details of this problem. Using the above features and following Zakharov and Shabat's approach [14-15], Desem and Chu [40-42] gave a solution corresponding to two discrete eigenvalues $\lambda_{1,2}$ for the two interacting microscopic particles in the process of collision, which is represented by:

$$\phi(x',t')=\frac{|\alpha_1|\cosh(a_1+i\theta_1)e^{i\theta_2'}+|\alpha_2|\cosh(a_2+i\theta_2)e^{i\theta_1'}}{\alpha_3\cosh(a_1)\cosh(a_2)-\alpha_4\left[\cosh(a_1+a_2)-\cos(A')\right]} \tag{3.96}$$

where $\theta'_{1,2} = 2\left[2\left(\eta_{1,2}^2 - \xi_{,2}^2\right)t' - x'\xi_{1,2}\right] + \left(\theta'_0\right)_{1,2}$, $A' = \theta'_2 - \theta'_1 + \left(\theta_2 - \theta_1\right)$,

$\alpha_{1,2} = 2\eta_{1,2}\left(x' + 4t'\xi_{1,2}\right) + \left(x'_0\right)_{1,2}$,

$$\left|\alpha_{1,2}\right|e^{i\theta_{1,2}} = \pm\left\{\left[\frac{1}{2\eta_{1,2}} - \frac{\eta}{\left(\Delta\xi^2 + \eta^2\right)}\right] \pm i\frac{\Delta\xi}{\left(\Delta\xi^2 + \eta^2\right)}\right\}$$

$\alpha_3 = \frac{1}{4\eta_1\eta_2}, \alpha_4 = \frac{1}{2\left(\eta^2 + \Delta\xi^2\right)}$, $\lambda_{1,2} = \xi_{1,2} + i\eta_{1,2}, \Delta\xi = \xi_2 - \xi_1, \eta = \eta_1 + \eta_2$, η and ξ are the same as those in Eq. (3.9), and represent the velocities and amplitudes of the microscopic particles, $\left(x'_0\right)_{1,2}$ the positions, and $\left(\theta'_0\right)_{1,2}$ the phases. They are all determined by the initial conditions. Of particular interest, here is an initial pulse waveform,

$$\phi(0, x') = \sec h\left(x' - \bar{x}'_0\right) + \sec h\left(x' + \bar{x}'_0\right)e^{i\theta} \tag{3.97}$$

which represents the motion of two microscopic particles into the system at t=0. Equation (3.97) will evolve into two particles described by Eq. (3.96). The interaction between the two microscopic particles given in Eq. (3.97) can therefore be analyzed through the two-particle function in Eq. (3.96).

Given the initial separation of x'_0, the phase difference θ between the two microscopic particles, and the eigenvalues $\lambda_{1,2}$,a_0,and θ'_0 can be evaluated by solving the Zakharov and Shabat equation [14-15]. For this solution, Eq. (3.97) is used as the initial condition. Substituting the eigenvalues obtained into Eq. (3.96), we then obtain the description of the interaction between the two microscopic particles.

The two microscopic particles described by Eq. (3.97) interact through a periodic potential in t' and $\cos A'$ term. The period is given by $\pi/\left(\eta_2^2 - \eta_1^2\right)$. The propagation of two microscopic particles with the initial conditions $\theta = 0$, $\xi_1 = \xi_2 = 0$, $\left(\theta'_0\right)_1 = \left(\theta'_0\right)_2 = 0$, obtained by Desem and Chu [40-42] are shown in Figure3.5. The two microscopic particles are initially separated but coalesce into one microscopic particle at $A' = \pi$. Then they separate and revert to the initial state with separation x'_0 at $A' = \pi$, and so on. An approximate expression for the separation between the microscopic particles as a function of the distance along the system can be obtained provided the two microscopic particles are well resolved in such a case (Gordon [43-44], Karpman et al., [45-47]).

Assuming that the separation between the particles is sufficiently large, one can obtain the separationas $\Delta x' = \ln\left[(2/a)\left|\cos\left(at'\right)\right|\right], a = 2e^{-\bar{x}'_0}$ .Thus, the period of oscillation is approximately given by $t'_p = (\pi/2)\exp[\bar{x}'_0]$ (see Blow and Doran 1983; Gordon 1983 [44]).

Meanwhile, Tan et al., [48] also discussed the features of collision of solitons obtained from the nonlinear Schrödingerequation, their collision phenomenon is similar to the above results.

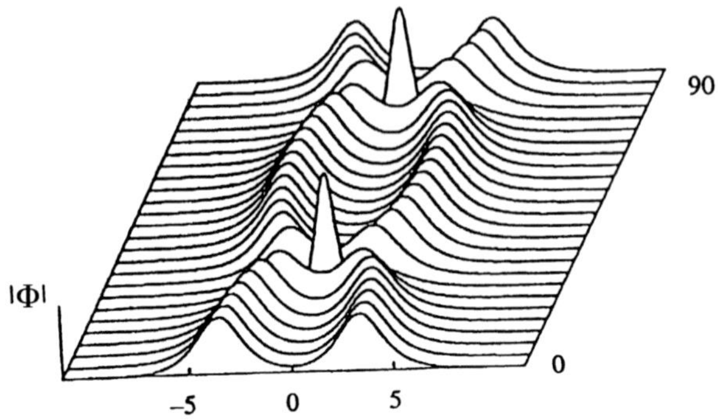

Figure 3.5. Interaction with two equal amplitude microscopic particles. Initial microscopic particles separation=3.5 pulse width (pw) [Re.40-47].

3.8. The Features of Collisions of Microscopic Particles under Action ofRepulsiveNonlinear Interaction

3.8.1. The Properties of Collision of Microscopic ParticleswithRepulsive Nonlinear Interaction

It is shown, in the case of repulsive nonlinear interaction, i.e.,$b< 0$ in the equation (3.1) still has a soliton solution. When $\lim_{x\to\infty}\left|\phi(x',t')\right|^2 \to$ constant, and $\lim_{x\to\infty}\phi_{x'}=0$, the solutions are then dark (hole) solitons, in contrast to the bright soliton,mentioned above,when $b> 0$. The bright soliton was observed experimentally in focusing on fibers with a negative dispersion, but the hole soliton solution was also observed in defocusing fibers with normal dispersion effect by Emplit et al., [49-52]. In practice, this result is an empty state without matter in the microscopic world. Hence $b>0$corresponds to an attraction between Bose particles, and $b<$ 0corresponds to a repulsive interaction between them. Thus, reversing the sign of b not only leads to changes in the physical picture of the phenomenon described by the nonlinear Schrödinger equation(3.1), but also requires considerable restructuring of the mathematical formalism for its solution. Thus, the solution of the nonlinear Schrödinger equation must be again analyzed. Also, the collision rules of the microscopic particles in the case of $b< 0$ must be studied separately, which we obtainedby using the inverse scattering method explored by Zakharov and Shabat [14-15].

We dealt with the direct scattering problem of Eq.(3.1) with b<0, in the Zakharov and Shabat equation [14-15], which works as long as factors ($1-s^2$) and (1-s) are replaced by (s^2-1) and (s-1), respectively, the relation of scattering data of $a(\lambda)$ and $b'(\lambda)$, $|a(\lambda)|^2+|b(\lambda)|^2=1$, is replaced by $|a(\lambda)|^2-|b'(\lambda)|^2=1$. The corresponding scattering data are now:

$$b'(\lambda,t') = b'(\lambda,0)e^{i4\lambda\varsigma_j t'}, .a(\lambda,t') = a(\lambda,0),$$
$$c_j(t') = c_j(0)e^{i4\lambda_j \upsilon_j t'}, \varsigma_j = i\sqrt{1-\lambda_j^2} = i\upsilon_j \quad (3.98)$$

The above equations enable us to calculate the scattering matrix at an arbitrary instant of time from the initial data. This yields $\phi(x',t')$ at an arbitrary instant of time. The integral equations (Marchenko equations) can be reconstructed to determine the "potential" $q(x',t') = \phi(x',t')/(s^2-1)$ from the scattering data. Of particular physical interest is the state $|\phi(x',t')|^2 \to$ constant and $\phi_{x'} \to 0$ as $x' \to \pm\infty$, which corresponds to the propagation of a wave through a condensate of constant density. In such a case, Zakharov and Shabat [14-15] obtained the soliton solution of Eq. (3.1) at b< 0 by the inverse scattering method, which is representedby:

$$\sqrt{\frac{b}{2}}\phi(x',t') = \frac{(\lambda - i\upsilon)^2 + e^{2\nu(x'-x_o')-2\lambda t'}}{1+e^{2\nu(x'-x_o')-2\lambda t'}}, \nu = \sqrt{1-\lambda^2} \quad (3.99)$$

This soliton can move with a constant velocity. Thus

$$\frac{b}{2}|\phi(x',t')|^2 = 1 - \frac{\nu^2}{\cosh^2[\nu(x'-x_0' - 2\lambda t')]} \quad (3.100)$$

The parameter λ characterizes the amplitude and velocity of the microscopic particles, and x_0' is the position of its mass center at t'=0, where $d(\ln\mu)/dt' = 2\lambda\nu$ or $\ln\mu = 2\nu(x_0' + 2\lambda t')$.

We can verify the eigenvaluecorresponding to a soliton moving with velocity 2λ and that the collision between these microscopic particles satisfies the rule for b>0 in Eq. (3.1). Thus, we consider the interaction of two microscopic particles (solitons) with velocities $2\lambda_2$ and $2\lambda_1$. To this end, it is necessary to obtain the corresponding two-soliton solution.

Owing to the complexity of explicit formula for the solution, here we limit our discussion to the asymptotic behavior as $t' \to \pm\infty$. Following the same procedure as above; Zakharov and Shabat[14-15] show that as $t' \to \pm\infty$, the two-soliton solution breaks up into individual solitons:

$$\phi(x',t') \to \phi_0(x'-2\lambda_1 t', \lambda_1, x_1'^+) + \phi_0(x'-2\lambda_2 t', \lambda_2, x_2'^+), t' \to +\infty$$
$$\phi(x',t') \to \phi_0(x'-2\lambda_1 t', \lambda_1, x_1'^-) + \phi_0(x'-2\lambda_2 t', \lambda_2, x_2'^-), t' \to -\infty$$

The scattering of the microscopic particles by each other results in the displacement of their mass centers of $\delta x_1' = x_1'^+ - x_1'^-, \delta x_2' = x_2'^+ - x_2'^-$. To determine these displacements, we note first the discontinuity in the phase of wave function of a microscopic particle (soliton) with velocity $2\lambda_1$, which can be expressed as:

$$\theta_i' = \arg\phi(-\infty) - \arg\phi(+\infty) = 2\tan^{-1}(v_i / \lambda_i)$$

Then the total phase discontinuity for N eigenvalues is $\theta' = \sum_{k=1}^{N} \theta_k'$, which does not depend on the relative positions of the microscopic particles. Zakharov and Shabat [14-15] ultimately gave the displacements of the two microscopic particles after the collision,which were:

$$\delta x_1' = \frac{1}{2v_1}\ln\left(\frac{1}{\left|a_2(\lambda_1, iv_1)\right|^2}\right) = \frac{Y}{2v_1},\ \delta x_2' = \frac{1}{2v_1}\ln\left(\left|a_1(\lambda_2, iv_2)\right|^2\right) = -\frac{Y}{2v_2}, \tag{3.101}$$

where $a_{1,2}(v,\lambda) = \left[\frac{iv + \lambda - iv_{1,2} - \lambda_{1,2}}{iv + \lambda + iv_{1,2} + \lambda_{1,2}}\right]$ and $Y = \ln\left[\frac{(\lambda_1 - \lambda_2)^2 + (v_1 + v_2)^2}{(\lambda_1 - \lambda_2)^2 + (v_1 - v_2)^2}\right]$. (3.102)

Thus, the microscopic particle, which has the greater velocity, acquires a positive shift, and the other a negative shift. Their behavior is one of repelling each other, as in classical particles. From (3.101) they get:

$$v_1\delta x_1' + v_2\delta x_2' = 0.$$

Tsuzuki also obtained this relation directly from Eq.(3.1) for $b< 0$by analyzing the motion of the mass center of a Bose gas. It can be interpreted as the conservation of mass center of the microscopic particles during the collision. This shows sufficiently the classical feature of microscopic particles described by Eq.(3.1) with $b<0$.

The collision of many particles can similarly be studied by the above method. The result obtained shows that in this case, the slowest soliton is in the front while the fastest is at the rear. The faster microscopic particle tracks the slower microscopic particle, they collide with each other at t'=0, and after the collision and $t' \to \infty$, they are reversed. In studying the collision process, Zakharov and Shabat obtained the total displacement of a particle, which is equal to the sum of the displacements in individual collisions.Regardless of the details of the collisions, it is

$$\delta_j = x_j'^{+} - x_j'^{-} = \sum_{i=1} \delta_{ij} \tag{3.103}$$

where $\delta_{ij} = sign(\lambda_i - \lambda_j)\frac{1}{2v_i}\ln\left[\frac{(\lambda_i - \lambda_j)^2 + (v_i + v_j)^2}{(\lambda_i - \lambda_j)^2 + (v_i - v_j)^2}\right]$.

From the above studies we see that collisions of many microscopic particles can be described by the nonlinear Schrödinger equation (3.1), with both $b > 0$or $b < 0$, the results showed that they satisfy the rules of classical physics. This sufficiently demonstrates further the corpuscle feature of microscopic particles in the nonlinear quantum mechanics.

3.8.2. The Detailed Mechanism and Rules of Collision of Microscopic ParticleswithRepulsive Nonlinear Interaction

In the following, we describe a series of laboratory and numerical experiments dedicated to the research of the detailed structure, mechanism and rules of the collision of the microscopic particles as described by the nonlinear Schrödinger equation (3.1) with b<0. The properties and rules of such a collision of particles (solitons) depicted by Eq.(3.1) at $b < 0$ have first been studied by Aossey et al., [53].

Asossey et al., [53] represented the hole-particle or dark spatial soliton of Eq.(3.1) at $b < 0$ as:

$$\phi(x',t') = \phi_0\sqrt{1 - B^2 \operatorname{sech}^2(\xi')}e^{\pm i\Theta(\xi')} \tag{3.104}$$

where

$$\Theta(\xi') = \sin^{-1}\left[\frac{B\tanh(\xi')}{\sqrt{1 - B^2 \operatorname{sech}^2(\xi')}}\right], \xi' = \mu(x' - \upsilon_t t')$$

Here, $B$is a measure of the amplitude ("blackness") of the solitary wave (hole or dark soliton) and may have a value between –1 and 1, and where $\upsilon_t$is the dimensionless transverse velocity of the particle center, and $\overline{\mu}$ is the shape factor of the particle. The intensity (I_d) of the solitary wave (or the depth of the irradiance minimum of the dark soliton) is given by $B^2\phi_0^2$. Aossey et al.showed that the shape factor $\overline{\mu}$ and the transverse velocity υ_t are related to the amplitude of the particles, which can be obtained from the nonlinear *Schrödinger* equation via fiber optics, which is:

$$\overline{\mu}^2 = n_0|n_2|\overline{\mu}_0^2 B^2\phi_0^2, \text{ and } \upsilon_t \approx \pm\sqrt{(1 - B^2)\frac{|n_2|\phi_0^2}{n_0}}$$

where $n_0$and $n_2$are the linear and nonlinear indices of refraction for the optical fiber material, where $|n_2|\phi_0^2 << n_0$ is assumed. Asossey etal. [93]represented the individual phase shifts of two particles in the collision process by:

$$\delta x_j' = \sqrt{\frac{n_0}{|n_2|\phi_0^2}}\frac{1}{2\mu_0 n_0 B_j}\ln\left[\frac{\left(\sqrt{1-B_1^2}+\sqrt{1-B_2^2}\right)^2+(B_1+B_2)^2}{\left(\sqrt{1-B_1^2}+\sqrt{1-B_2^2}\right)^2+(B_1-B_2)^2}\right] \tag{3.105}$$

The interaction of the microscopic particles can be investigated numerically by using a split-step propagation algorithm, which was derived by Skinner et al. [54-56], and Thurston et al.[57], in order toclosely predict experimental results. The results of a simulated collision between two equi-amplitude microscopic particles are shown in Figure 3.6, which are similar to that of general microscopic particles (bright solitons), shown in Figure 3.1. We note that the two particles interpenetrate each other, while retaining their shape, energy, and momentum, but experience a phase shift at the point of collision. In addition, there is also a well-defined interaction length in z,along the axis of time t,which depends on the relative amplitude of two collidingmicroscopic, particles. This case also occurs in the collision of two solitons of KdVequation[16-20,58],which is denoted by

$$\phi_t + \phi\phi_x = \mu\phi_{xxx} \tag{3.106}$$

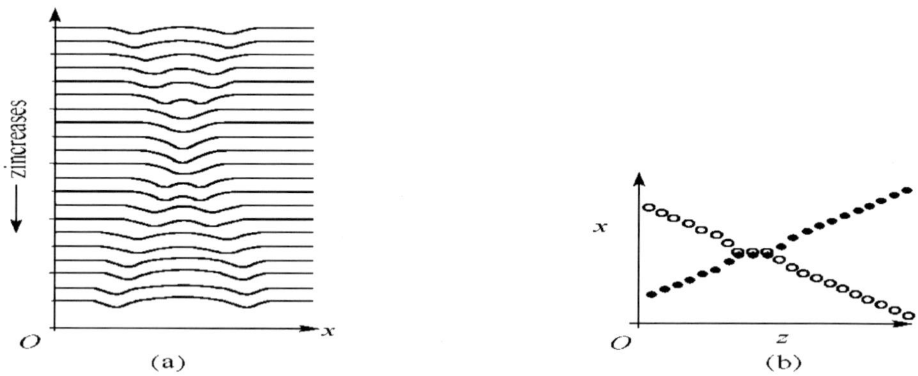

Figure 3.6. Numerical simulation of an overtaking collision of equi-amplitude dark particles. (a) Sequence of the waves at equal intervals in the longitudinal position. (b) Time-of-flight diagrams of the signal [Res.54-57].

Its soliton solution takes the form:

$$\phi(x,t) = A_j\, sech^2[(A_j/12\mu)^{1/2}(x-vt)] \tag{3.107}$$

Cooney et al., [59-62] studied the overtaking collision to interpret the experimental results on the microscopic particles described by the nonlinear Schrödinger equation (3.1) at b<0 and KdV solitons.

The model is based on the fundamental property of solitons and says that two microscopic particles can interact and collide, but survive the collision and remain unchanged. Rather than using the exact functional form of sechξ for microscopic particles described by Eq.(3.1), the microscopic particles are represented by rectangular pulses with an amplitude$A_j$and a width $W_j$where the subscript $j$denotes the jth microscopic particles. An evolution of the collision of two microscopic particles is shown in Fig.3.7 (a).

In this case, Aossey et al., [53]considered two microscopic particles with different amplitudes. The details of what occurs during the collision need not concern us here other than to note that the microscopic particles with the larger-amplitude has completely passed through the one with the smaller amplitude. In regions that can be considered external to the collision, the microscopic particles do not overlap, as there is no longer aninteraction between them. The particles are separated by a distance, $D = D_1 + D_2$, after the interaction. This manifests itself in a phase shift in the trajectories depicted in Figure 3.7(b). This was noted in the experimental and numerical results. The minimum distance is given by the half-widths of the two as microscopic particles, $D \geq W_1/2 + W_2/2$. Therefore,

$$D_1 \geq \frac{W_1}{2} \text{ and } D_2 \geq \frac{W_2}{2} \tag{3.108}$$

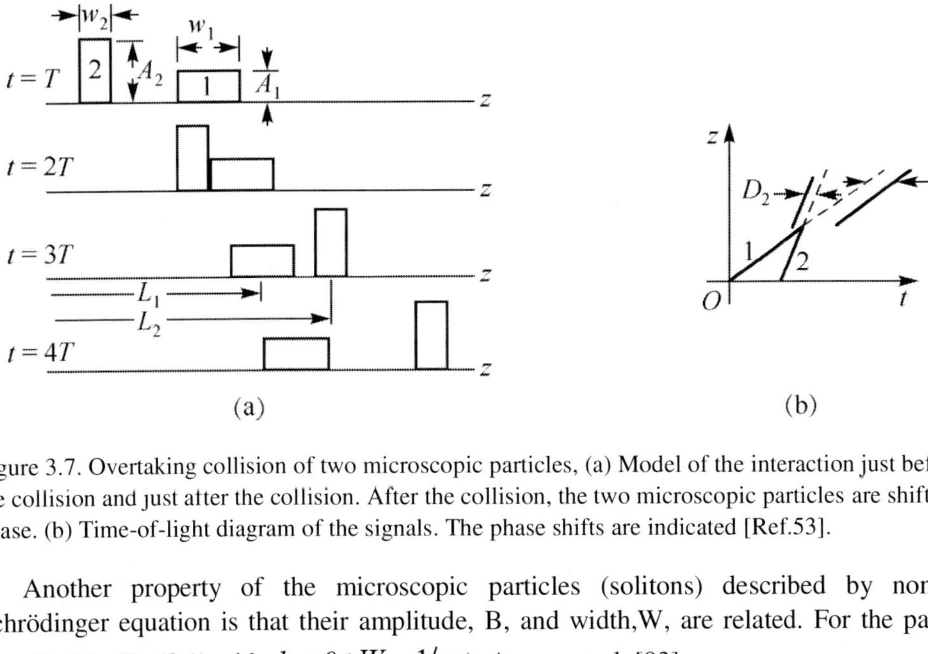

Figure 3.7. Overtaking collision of two microscopic particles, (a) Model of the interaction just before the collision and just after the collision. After the collision, the two microscopic particles are shifted in phase. (b) Time-of-light diagram of the signals. The phase shifts are indicated [Ref.53].

Another property of the microscopic particles (solitons) described by nonlinear Schrödinger equation is that their amplitude, B, and width,W, are related. For the particles described by Eq.(3.1) with $b < 0$ ($W \approx 1/\mu$), Aossey et al.,[93]gave

$$B_j W_j = \text{constant} = K_1 \tag{3.109}$$

Using the minimum values in Eq.(3.108), we find that the ratio of the repulsive shifts for the particles described by Eq. (3.1) with $b< 0$ is given by:

$$\frac{D_1}{D_2} = \frac{B_2}{B_1} \tag{3.110}$$

Results obtained from simulation of this kind of particle are presented in Figure 3.8(a), wherein the solid line in the figure corresponds to Eq.(3.110).

In addition to predicting the phase shift that results from the collision of two particles, the model also allows us to estimate the size of the collision region and/or the duration of the collision. Each microscopic particle depicted in Figure 3.7 travels with its own amplitude-dependent velocity v_j . For the two microscopic particles to interchange their positions during a time ΔT , they must travel a distance L_1 and L_2,

$$L_1 = \upsilon_1 \Delta T \text{ and } L_2 = \upsilon_2 \Delta T \tag{3.111}$$

The interaction length must then satisfy the relation

$$L = L_2 - L_1 = (\upsilon_2 - \upsilon_1)\Delta T \ge W_1 + W_2 \tag{3.112}$$

Equation (3.111) or (3.112) can be written in terms of the amplitudes of the two particles. For the particles depicted by Eq.(3.1) at b<0, combining Eqs.(3.108) and (3.112), Aossay et al., [53]obtained:

$$L \ge K_1 \left[\frac{1}{B_1} + \frac{1}{B_2} \right] \tag{3.113}$$

In Figure 3.9(a), the results for the particles described by Eq.(3.1) at b<0 are presented. The dashed line corresponds to Eq. (3.113) with $B_2 = 1$ and $K_1 = 6$. The interaction time (solid line) is the sum of the widths of the two particles (soliton), minus their repulsive phase shifts, and multiplied by the transverse velocity of particles 1. Since the longitudinal velocity is a constant, this scales as the interaction length. From the figure, we see that the theoretical result obtained using the above simple collision model is in ample agreement with that of the numerical simulation.

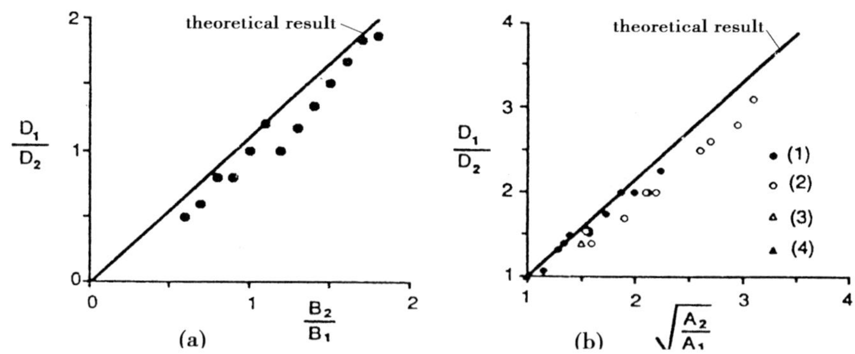

Figure 3.8. Summary of the ratio of the measured phase shifts as a function of the ratio of amplitudes. (a) The particle in Eq.(3.1) at b<0, the solid line corresponds to Eq.(3.105). (b) KdV solitons, the data are from (1) this experiment, (2) Zabusky et al., [64], (3) Lamb's [65],and (4) Ikeziet al.'s [63] results. The solid line corresponds to Eq.(3.1) [Re.43].

The discussion presented above and the corresponding formulae reveal the mechanism and rule for the collision between microscopic particles depicted by nonlinear Schrödinger equation (3.1) at b<0.

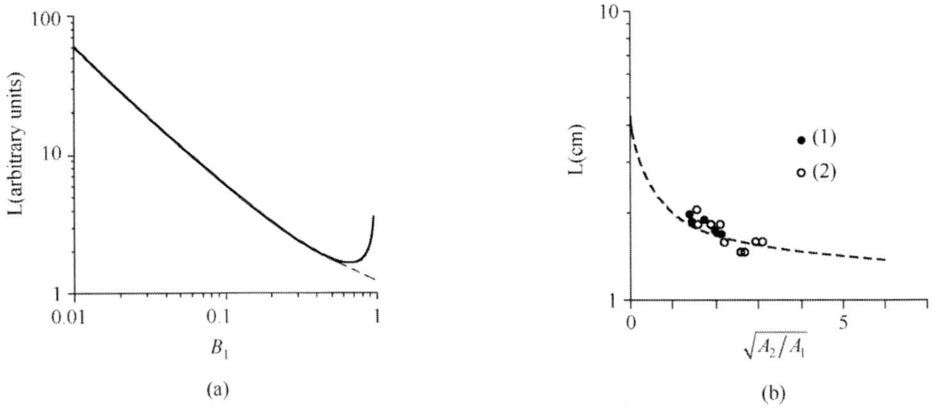

Figure 3.9. Summary of the measured interaction length as a function of the amplitudes. (a) The particles described by Eq.(3.1) at b<0, the dashed line corresponds to Eq. (3.113) with $B_2 = 1$ and $K_1 = 6$. (b) KdV solitons, the symbols represent (1) experiment results of Ikezi et al. [63] and (2) numerical results of Zabusky et al. [64]. The dashed line corresponds to Eq. (3.115) with $K_2 = A_1 = 1$ [Re.53].

In order to verify the validity of this simple collision model, Aossey et al., [53]studied the collision of the solitons using the exact form of $sec\,h^2\xi$ for the KdV equation (3.107)[16-20,58] and the collision model shown in Figure 3.7. For the KdV soliton, they found from Eqs. (3.107) that:

$$A_j\left(W_j\right)^2 = \text{cons}\tan t = K_2 \text{ and } \frac{D_1}{D_2} = \frac{W_1/2}{W_2/2} = \sqrt{\frac{A_2}{A_1}} \tag{3.114}$$

where A_j and W_j are the amplitude and width of the *j*th KdV soliton, respectively.

Corresponding to the above, Aossey et al., obtained

$$L \geq K_2\left(\frac{1}{\sqrt{A_1}} + \frac{1}{\sqrt{A_2}}\right) = \frac{K_2}{\sqrt{A_1}}\left(1 + \sqrt{\frac{A_1}{A_2}}\right) \tag{3.115}$$

for the interaction length.

Aossey et al., [53]compared their results for the ratio of the phase shifts as a function of the ratio of the amplitudes for the KdV solitons, with those obtained in the experiments of Ikezi [63], Taylor and Baker [66], and those obtained from numerical work of Zabusky and Kruskal [64] and Lamb [65], as shown in Figure 3.8(b). The solid line in Figure 3.8(b) corresponds to Eq. (3.1). Results obtained by Aossey et al., [53] for the interaction lengths are shown in Figure 3.9(b) as a function of amplitudes of the colliding KdV solitons. Numerical results (which were scaled) from Zabusky and Kruskal are also shown for comparison. The dashed line in Figure 3.9(b) corresponds to Eq. (3.115), with $A_1 = l$ and $K_2 = l$.

Since the theoretical results obtained by the collision model based on macroscopic bodies in Figure 3.7 are consistent with experimental data for the KdV soliton, shown in Figures 3.8(b) and 3.9(b), it is reasonable to believe the validity of the theoretical model of collision of microscopic particles (soliton) presented above, and results shown in Figures 3.8(a) and 3.9(a) for theparticles described in the nonlinear Schrödinger equation (3.1), which are obtained using the same model as that shown in Figure 3.7. Thus, the above colliding mechanism for the microscopic particles clearly shows the corpuscle feature of the microscopic particles described by the nonlinearequationin the nonlinear quantum systems.

3.9. The Properties of Propagation of Microscopic Particles at Interfaces of Media

3.9.1. The Wave Natures of Microscopic Particles Described By Nonlinear Quantum Mechanics

As mentioned above, microscopic particles in nonlinear quantum mechanics also have the wave property in addition to the corpuscle property. This wave feature can be conjectured based on the following reasoning:

(1) Equations(2.66) at or Eq.(3.1) are wave equations and their solutions, Eqs.(3.8) - (3.10),(3.24)- (3.25), are solitary waves having the features of traveling waves. A solitary wave consists of a carrier wave and an envelope wave, has a certain amplitude, width, velocity, frequency, wave vector, and so on, although the latter are different from the classical waves or the de Broglie waves in quantum mechanics.

(2) The solitary waves have reflection, transmission, scattering, diffraction, and tunneling effects, just as those of classical waves or the de Broglie waves in linear quantum mechanics. The diffraction and tunneling effects of microscopic particlescould be studied. In this section, we only consider the reflection and transmission of the microscopic particles at an interface.

The propagation of microscopic particles (solitons) in a nonlinear non-uniform media are different from those in uniform media. The non-uniformity can be due to a physical confining structure or two nonlinear materials being juxtaposed. One could expect that a portion of microscopic particles that was incident upon such an interface from one side would be reflected and a portion would be transmitted to the other side due to its wave feature.

Lonngrenet al., [67-70]observed the reflection and transmission of solitons in plasma consisting of a positive ion and a negative ion interface, and numerically simulated the phenomena at the interface of two nonlinear materials. To illustrate the rules of reflection and transmission of microscopic particles, here we discuss the work of Lonngren, et al.

3.9.2.Reflection and Transmission Effects of Microscopic Particles atInterface of the Media

Lonngren et al. [67-70], numerically simulated the behaviors of solution of the nonlinear Schrödinger equation (3.1). They found that the signal had the property of a soliton. These results are in agreement with numerical investigations of similar problems by Aceves et al., [71-72].

A sequence of pictures, obtained by Lonngren et al. [67-70],at uniform temporal increments of the spatial evolution of the signal, are shown in Figure 3.10. From this figure, we note that the incident wave propagating toward the interface between the two nonlinear media splits into a reflected and transmitted soliton at the interface. Thus, wethink that the microscopic particles described by the nonlinear Schrödinger equation (3.1) also possess this feature. Thus, we can use the following results of reflection and transmission of the soliton in Eq.(3.1) at interface to exhibit those microscopic particles depicted by same equation of dynamics.

From the numerical values used in producing the figure, the relative amplitudes of the incident, the reflected and the transmitted solitons can be deduced. In this calculation, Lonngren et al.[69-70],assumed that the energy that is carried by the incident particle is all transferred to either the transmitted or reflected microscopic particle and none is lost through radiation. Thus,

$$E_{\text{inc}} = E_{\text{ref}} + E_{\text{trans}} \tag{3.116}$$

Lonngrenet al., [69-70] introduced here a characteristic impedance $Z_c$to represent the feature of propagation of the particles and further, gave approximately the energy of the soliton by

$$E_j = \frac{A_j^2}{Z_c} W_j , \tag{3.117}$$

where the subscript $j$refers to the incident, reflected or transmitted microscopic particles. The amplitude of the microscopic particle is A_j and its width is W_j. Hence, equation (3.116) can be written as

$$\frac{A_{\text{inc}}^2}{Z_{\text{cI}}} W_{\text{inc}} = \frac{A_{\text{ref}}^2}{Z_{\text{cI}}} W_{\text{ref}} + \frac{A_{\text{trans}}^2}{Z_{\text{cII}}} W_{\text{trans}} \tag{3.118}$$

For the microscopic particle described in the nonlinearSchrödinger equation (3.1) we can find thefollowing relation from Eqs. (3.8) -(3.10)

$$A_j W_j = \text{constant} \tag{3.119}$$

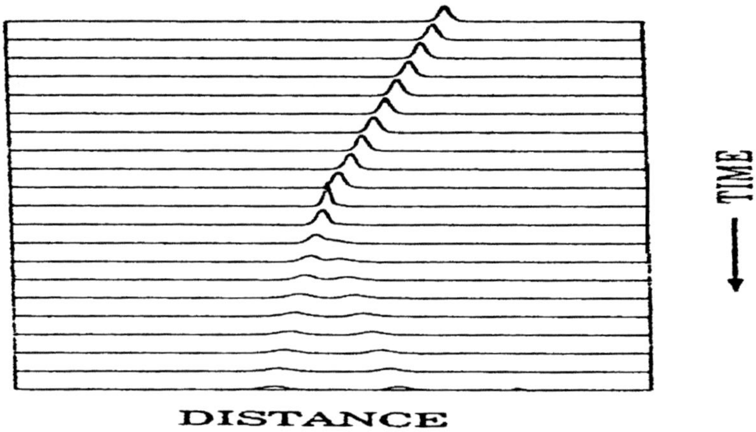

Figure 3.10. Simulation results showing the collision and scattering of an incident soliton in Eq. (3.1) (top above) onto an interface by Lonngren et al.The peak nonlinear refractive index change is 0.67%ofthelinearrefractive index for the incident solitons (particles), and the linear offset between the two regions isalso 0.67% [67-70].

Thus they obtained the following relation between the reflection coefficient $R=A_{ref}/A_{inc}$ and the transmission coefficient $T = A_{trans} / A_{inc}$

$$1 = R + \frac{Z_{cI}}{Z_{cII}} T \tag{3.120}$$

for the microscopic particle described by Eq.(3.1) in nonlinear quantum mechanics. This is consistent with the results obtained by the numerical simulation for the solitons described by the nonlinear Schrödinger equation (3.1).

3.9.3. The Wave Features of KDV Soliton and the Behavior of Linear Wave

To clearly understand the above features of the reflection and transmission of the microscopic particles at the interface, Lonngren et al. [67-70],periments of the KdV soliton. Inthe experiment, they detected the signal, which had the characteristics of a KdV soliton. They show a sequence of pictures taken from a small probe at equal spatial increments, starting initially in a homogeneous plasma region and then propagating into an inhomogeneous plasma sheath adjacent to a perturbing biased object and found that the probe first detects the incident soliton and then, sometime later, the reflected soliton. The signals are observed, as expected, to coalesce together as the probe passed through the point where the soliton was actually reflected. Beyond this point where the density started to decrease in the steady-state sheath, a transmitted soliton was observed. From this experiment, the relative amplitudes of incident, the reflected and the transmitted solitons can be deduced. For the KdV solitons shown Eq. (3.107), they found that the relation of its amplitude (A_j)and width ($W_j= (A_j / 12\mu)^{1/2}$) is

$$A_j W_j^2 = \text{constant} \tag{3.121}$$

Thus Lonngren et al., got:

$$1 = R^{3/2} + \frac{Z_{eI}}{Z_{eII}} T^{3/2} \tag{3.122}$$

This result is consistent with the experimental results of KdV solitons by Nishida[73-74] and computed data obtained by Andersen et al. [75-77]. Meanwhile, we see from Eqs.(3.120) and (3.122) that the relations between the reflection and the transmission coefficients are different for the microscopic particle in nonlinear quantum mechanics and the KdV soliton. Obviously, this is due to differences of properties of propagation between the microscopic particle described by the nonlinear Schrödinger equation and KdV soliton.

The above rule of propagation of the microscopic particles, described by nonlinear quantum mechanics, is also different from that of linear waves in classical physics. Lonngren et al. [67-70],found that a linear wave obeyed the following relation:

$$1 = R^2 + \frac{Z_{eI}}{Z_{eII}} T^2 \tag{3.123}$$

This can be also derived from Eq. (3.118) by assuming the linear waves. The width of the incident, reflected and transmitted pulses, W_j, will be the same. For the linear waves we can obtain

$$R = \frac{Z_{cII} - Z_{cI}}{Z_{cII} + Z_{cI}} \text{ ,and } T = \frac{2Z_{cII}}{Z_{cII} + Z_{cI}}, \tag{3.124}$$

which also satisfy Eq. (3.123).

Obviously, equation (3.123) is different from Eq. (3.120). These results in Eqs.(3.120), (3.122)-(3.123) indicate clearly that the microscopic particles described by the nonlinear Schrödinger equation have a wave feature, but its property is different from that of both KdV soliton and linear classical waves or the de Broglie waves in linear quantum mechanics.

3.10. ThePropertieS of QuantumFluctuation in Nonlinear Quantum Systems

3.10.1. TheQuantization of Nonlinear Schrödinger Equation

Similar to linear quantum mechanics, wave function in nonlinear quantum theory can also be quantized. The commonly used methods are the direct quantized method, the canonical quantization method, and the path integration method. However, because of the nonlinear effect, the actual procedure of quantization becomes much more complicated. We here quantize the nonlinear Schrödinger equation using Lai et al.'s method [78,79]. The main steps for the direct quantization are outlined below.

The nonlinear Schrödinger equation (3.1) in which b is replaced by 2b, is now given by

$$i\frac{\partial}{\partial t'}\phi(x',t') = -\frac{\partial^2}{\partial x'^2}\phi(x',t') + 2b\phi^*(x',t')\phi(x',t')\phi(x',t') \tag{3.125}$$

in a one-dimensional case, where $t' = t/\hbar$, $x' = x\sqrt{2m}/\hbar$. We can quantize Eq. (3.125) inthe Fourier transform space by defining the Fourier transform of the above field quantities, which can be approximately represented by:

$$b(q,t') = \frac{1}{\sqrt{2\pi}}\int_{-\infty}^{+\infty}\phi(x',t')e^{iqx'}dx' \ , \tag{3.126}$$

thus we can obtain the equation of motion for the amplitude in the Fourier transformation of space from Eq. (3.125), which is:

$$i\frac{\partial}{\partial t'}b(q,t') = -q^2 b(q,t') + 2b\int dq_1 dq_2 b^*(q_1,t')b(q_2,t')b(q+q_1-q_2,t') \tag{3.127}$$

The field envelope can be normalized so that it represents the microscopic particle at time, *t'*. In the light of quantum mechanics [78,79], *b*(q,*t'*) can be identified with the microscopic particle annihilation operator at time *t*, $\hat{b}(q,t')$,and $b^*(q,t')$ can be identified with the creation operator $\hat{b}^+(q,t')$. Note that the second term on the right-hand side of Eq.(3.127) denotes the third-order nonlinearity, which is the form of a convolution. This is because, for a broadband field, one has to integrate over the Fourier-transform space. Obviously, the creation and annihilation operators satisfy the following commutation relations

$$[\hat{b}(q',t'),\hat{b}^+(q,t')] = \delta(q'-q) \text{,} \quad [\hat{b}(q',t'),\hat{b}(q,t')] = [\hat{b}^+(q',t'),\hat{b}^+(q,t')] = 0 \tag{3.128}$$

Thus the quantized equation for Eq. (3.127) can be obtained and represented by

$$i\frac{\partial}{\partial t'}\hat{b}(q,t') = -q^2\hat{b}(q,t') + 2b\int dq_1 dq_2 \hat{b}^+(q_1,t')\hat{b}(q_2,t')\hat{b}(q+q_1-q_2,t') \tag{3.129}$$

Evidently, equation (3.129) can be obtained by the following equation:

$$i\frac{\partial}{\partial t'}\hat{b}(q,t') = [\hat{b}(q,t'),\hat{H}] \tag{3.130}$$

with

$$\hat{H}=[\int q^2 \hat{b}^{+}(q,t')\hat{b}(q,t')dq+b\int \hat{b}^{+}(q,t')\hat{b}^{+}(q_1,t')\hat{b}(q_2,t')\hat{b}(q+q_1-q_2,t')dqdq_1dq_2] \quad (3.131)$$

By virtue of defining new field operators as the inverse Fourier transforms of the annihilation and creation operators and the application of the inverse Fourier transformation (3.129), and via Lai et al. [78,79], we obtain the quantum nonlinear Schrödinger equation

$$i\frac{\partial}{\partial t'}\hat{\phi}(x',t') = -\frac{\partial^2}{\partial x'^2}\hat{\phi}(x',t') + 2b\hat{\phi}^{+}(x',t')\hat{\phi}(x',t')\hat{\phi}(x',t') \quad (3.132)$$

The operators $\hat{\phi}(x',t')$ and $\hat{\phi}^{+}(x',t')$ are simply the annihilation and creation operators of microscopic particles at a point x' and time t', they satisfy the commutation relations:

$$[\hat{\phi}(x'',t'),\hat{\phi}^{+}(x',t')]=\delta(x'-x''),[\hat{\phi}(x'',t'),\hat{\phi}(x',t')]=[\hat{\phi}^{+}(x'',t'),\hat{\phi}^{+}(x',t')]=0 \quad (3.133)$$

By means of Eq. (2.133), Lai et al., wrote the equation (3.132) as:

$$i\frac{\partial}{\partial t'}\hat{\phi}(x',t')=[\hat{\phi}(x',t'),\hat{H}] \quad (3.134)$$

with

$$\hat{H} =[\int \hat{\phi}^{+}_{x'}(x',t')\hat{\phi}_{x'}(x',t')dx'+ b\int \hat{\phi}^{+}(x',t')\hat{\phi}^{+}(x',t')\hat{\phi}(x',t')\hat{\phi}(x',t')dx'] \quad (3.135)$$

The quantum nonlinear Schrödinger equation (3.132) is the operator evolution equation of a nonlinear quantum system with the quantum Hamiltonian in Eq. (3.135). Since the quantum nonlinear Schrödinger equation can be derived from the quantum Hamiltonian, it is hence a well-defined operator equation.

Equations (3.132)–(3.135) form the representation of dynamic equations of nonlinear quantum mechanics in the Heisenberg representation and correspond to the non-relativistic case. In the Schrödinger representation, the above question is denoted in terms of the time evolution of a state $|\Phi\rangle$ of the system

$$i\frac{d}{dt'}|\Phi\rangle = \hat{H}_s|\Phi\rangle \quad (3.136)$$

with

$$\widehat{H}_s = [\int \hat{\phi}^{+}_{x'}(x')\hat{\phi}_{x'}(x')dx' + b\int \hat{\phi}^{+}(x')\hat{\phi}^{+}(x')\hat{\phi}(x')\hat{\phi}(x')dx'] \tag{3.137}$$

where $\hat{\phi}(x')$ and $\hat{\phi}^{+}(x')$ are the field operators in the Schrödinger representation and satisfy the following commutation relation

$$[\hat{\phi}(x''), \hat{\phi}^{+}(x')] = \delta(x' - x'') \; [\hat{\phi}(x''), \hat{\phi}(x')] = [\hat{\phi}^{+}(x''), \hat{\phi}^{+}(x')] = 0 \tag{3.138}$$

In the Heisenberg representation,the quantum nonlinear Schrödinger equation (3.132)can be solved by the quantum inverse-scattering method. In the Schrödinger representation, equation (3.136) could be solved via Bether's ansatz. Lai et al. [78,79], expanded the quantum state in the Fock space and substituted it into Eq.(3.136). The result is a wave-function equation that has many degrees of freedom (like equations in many-particle systems). However, for the quantum nonlinear Schrödinger equation (3.132), this wave-function equation (3.136) is in a simple form and can be solved analytically.

Lai et al. [78,79],expanded the quantum state of the system in Fock space by

$$|\Phi\rangle = \sum_n c_n \int \frac{1}{\sqrt{n!}} g_n(x'_1 ..., x'_n, t')\phi^{+}(x'_1)...\phi^{+}(x'_n)dx'_1 ...dx'_n |0\rangle \tag{3.139}$$

This shows that the state $|\Phi\rangle$ is a superposition of states produced from the vacuum state by creating particles at points x'_1, x'_2, · · ·, $x'_n$with the weighting functions g_n. However, the microscopic particles are Bosons, thus $g_n$should be a symmetric function of x'_j, in this case $c_n$and g_n should satisfy the following normalization conditions

$$\sum_n |c_n|^2 = 1 \tag{3.140}$$

$$\int |g_n(x'_1, ...x'_n, t')|^2 dx'_1 ...dx'_n = 1 \tag{3.141}$$

Inserting Eqs. (3.139) and (3.137) into Eq. (3.136) and using Eq. (3.138), Lai et al., obtained the following equation for $g_n(x'_1, \cdots, x'_n, t')$

$$i\frac{d}{dt'} g_n(x'_1, ...x'_n) = [-\sum_{j=1}^{n} \frac{\partial^2}{\partial x'^2_j} + 2b \sum_{1 \le i \le j \le n} \delta(x'_j - x'_i)] g_n(x'_1, ...x'_n, t') \tag{3.142}$$

Clearly, this is the Schrödinger equation for a one-dimensional Boson system with a function interaction. The t-dependent relation in Eq.(3.142) can be factored out by assuming a solution of the form

$$g_n(x'_1, \cdots x'_n, t') = g_n(x'_1, \cdots x'_n) e^{iE_n t'} \tag{3.143}$$

where the equation satisfied by $g_n(x'_1, \cdots, x'_n)$ can be represented as follows:

$$[-\sum_{j=1}^{n}\frac{\partial^2}{\partial x'^2_j}+2b\sum_{1\le i\le j\le n}\delta(x'_j-x'_i)]g_n(x'_1,\cdots x'_n)=E_n g_n(x'_1,\cdots x'_n) \quad (3.144)$$

This shows that the quantum nonlinear Schrödinger equation is equivalent to the evolution equation of a one-dimensional Boson system with a $\delta-$function interaction. It is surprising that the quantum nonlinear Schrödinger equation can be solved exactly. It was first solved by Bether's ansatz and then by the quantum inverse-scattering method.

3.10.2. The Effects of Quantum Fluctuation of Microscopic Particles in the Nonlinear Quantum Systems

As it is known in this quantum system, there are both conservation quantities of the particle numbers

$$\hat{N}=\int\phi^{+}(x')\phi(x')dx' \quad (3.145)$$

and the total momentum

$$\hat{P}=i\int\left[\frac{\partial}{\partial x'}\phi^{+}(x')\phi(x')-\phi^{+}(x')\frac{\partial}{\partial x'}\phi(x')\right]dx' \quad (3.146)$$

Kartner and Boiven [80]proved the results of these calculations so that the boson number and the momentum operator commute, then we see that the common eigenstates of $\hat{H}$, $\hat{P}$ and $\hat{N}$ exist in such a case.

In the case of a negative of the nonlinear interaction constant$b$,in the quantized nonlinear Schrödinger equation, the interaction between the bosons is attractive and the Hamiltonian has bound states. A subset of these bound states is characterized solely by the eigenvalues of $\hat{N}$ and $\hat{P}$. Kartner and Boiven [80] denoted the wave functions of these states by:

$$g_{n,p'}=N_{n,}\exp\left(ip'\sum_{j=1}^{\infty}x'_j+\frac{b}{2}\sum_{1\le i,j<n}^{\infty}\left|x'_i-x'_j\right|\right), \quad (3.147)$$

where $N_n=\sqrt{\frac{(n-1)!|b|^{n-1}}{2\pi}}$. Thus

$$g_n(x_1,\cdots,x_n,t)=\int dp'g_n(p')g_{np}(x_1,\cdots,x_n,t)e^{-iE(n,p')t}, \quad (3.148)$$

where $f_n = \sqrt{f(p')}e^{-inp'x_0}$, and $f(p') = \frac{c\phi\{-(p'-p_0')^2/|2(\Delta p')^2|\}}{\sqrt{2\pi(\Delta p')^2}}$

Using $g_{n,p}$ given in Eq. (3.147),they found that $|n,P\rangle$ decays exponentially with separation between an arbitrary pair of bosons. In essence, it describes an n-quantum soliton moving withmomentum $P = \hbar np'$ and energy

$$E(n,p') = np'^2 - |b|^2(n^2-1)n/12.$$

By construction, the quantum number p' in this wave function is related to the momentum ofthe center of mass of the $n$interacting bosons. The corresponding position operator is now definedas

$$\hat{X} = \lim_{\varepsilon\to 0}\int x'\hat{\phi}^+(x')\hat{\phi}(x')dx'\left(\varepsilon+\hat{N}\right)^{-1}$$

The limit of $\varepsilon \to 0$ is introduced to regularize the position operator for the vacuum state.

We are interested in the quantum fluctuations of Eqs.(3.145)-(3.146) for a state in Eq.(3.139), with a large average Boson number and a well-defined mean field. Kartnerand Boiven [80]decomposed these operators in its mean values and the remainder, which is responsible for the quantum fluctuations.

$$\hat{\phi}(x) = \left\langle\Phi'(0)\left|\hat{\phi}^+(x')\right|\Phi(0)\right\rangle + \hat{\phi}_1(x'), \left[\hat{\phi}_1(x'),\hat{\phi}_1^+(x'')\right] = \delta(x'-x''), \left[\hat{\phi}_1(x'),\hat{\phi}_1^+(x'')\right] = 0 \qquad (3.149)$$

Since the field operator $\hat{\phi}$ is is time-independent in the Schrödinger representation, Kartner and Boiven [80] chose $t = 0$ for definiteness. Inserting Eq. (3.149) into Eqs. (3.145)-(3.146) and (3.148) and neglecting terms of a second and higher order in the noise operator, Kartner et al. [80], obtained that:

$$\hat{N} = n_0 + \Delta\hat{n}, n_0 = \int dx'\left(\left\langle\hat{\phi}^+(x')\right\rangle\left\langle\hat{\phi}(x')\right\rangle\right), \Delta\hat{n} = \int dx'\left(\left\langle\hat{\phi}^+(x')\right\rangle\hat{\phi}_1(x')\right) + c.c.,$$

$$\hat{P} = \hbar n_0 p_0 + \hbar n_0 \Delta\hat{p}', p_0 = \frac{i}{n_0}\int dx'\left\langle\hat{\phi}_{x'}^+(x')\right\rangle\left\langle\hat{\phi}(x')\right\rangle, \Delta\hat{p}' = \frac{i}{n_0}\int dx'\left\langle\hat{\phi}_{x'}^+(x')\right\rangle\left\langle\hat{\phi}_1(x')\right\rangle + c.c.,$$

$$\hat{X} = x_0'(1-\frac{\Delta\hat{n}}{n_0}) + \Delta\hat{x}', x_0' = \frac{1}{n_0}\int dx'x'\left\langle\hat{\phi}^+(x')\right\rangle\left\langle\hat{\phi}(x')\right\rangle, \Delta\hat{x}' = \frac{1}{n_0}\int dx'x'\left\langle\hat{\phi}_{x'}^+(x')\right\rangle\left\langle\hat{\phi}_1(x')\right\rangle + c.c.,$$

where $\Delta\hat{x}'$ is the deviation from the mean value of the position operator, and $\Delta\hat{n}, \Delta\hat{p}'$, and $\Delta\hat{x}'$ are linear in the noise operator. Because the third- and fourth-order correlators of $\hat{\phi}_1$ and $\hat{\phi}_1^+$ are very small, they can be neglected in the limit of a large n_0 .Note that $\Delta\hat{p}'$ and $\Delta\hat{x}'$ are conjugate operators, thus, in order to complete this set they introduced a variable conjugate to $\Delta\hat{n}$,

$$\Delta\hat{\theta} = \frac{1}{n_0}\int dx'\left\{i\left[\hat{\phi}^+(x') + x'\left\langle\hat{\phi}^+_{x'}(x')\right\rangle\right] - p_0 x'\left\langle\hat{\phi}^+_{x'}(x')\right\rangle\right\} \times \hat{\phi}_1(x') + cc$$

As is known, if the propagation distance is not too large, the mean value of the particle is given to the first order by the classical soliton solution

$$\left\langle\hat{\phi}(x')\right\rangle = \phi_{0,n_0}(x',t')\left[1 + O\left(\frac{1}{n_0}\right)\right]$$

with

$$\phi_{0,n_0}(x',t') = \frac{n_0\sqrt{|b|}}{2}\exp\left[i\Omega_{nl} - ip_0'^2 t' + ip_0'(x' - x_0') + i\theta_0\right] \times \mathrm{sech}\left[\frac{n_0|b|}{2}(x' - x_0' - 2p_0' t')\right],$$

and where the nonlinear phase shift is $\Omega_{nl} = n_0^2|b|^2 t'/4$.

If $p_0' = x_0' = \theta_0 = 0$, we obtain the following for the fluctuation operators in the Heisenberg representation,

$$\Delta\hat{n}(t') = \int dx'\left[g_{-n}(x')^* F'_{nl} + c.c\right], \Delta\hat{\theta}(t') = \int dx'\left[g_{-\theta}(x')^* F'_{nl} + c.c\right],$$

$$\Delta\hat{p}'(t') = \int dx'\left[g_{-p'}(x')^* F'_{nl} + c.c\right], \Delta\hat{x}'(t') = \int dx'\left[g_{-x'}(x')^* F'_{nl} + c.c\right],$$

with $F'_{nl} = e^{i\Omega_{nl}}\hat{\phi}_1(x',t')$,and the set of adjoint functions:

$$g_{-\theta}(x') = \frac{i\sqrt{|b|}}{2}\left[\sec\mathrm{h}(x'_{n0}) + x'_{n0}\frac{d}{dx'_{n0}}\sec\mathrm{h}(x'_{n0})\right], g_{-n}(x') = \frac{n_0\sqrt{|b|}}{2}\sec h(x'_{n0}),$$

$$g_{-p'}(x') = -\frac{in_0\sqrt{|b|^3}}{4}\frac{d}{dx'_{no}}\sec\mathrm{h}(x'_{n0}), g_{-x'}(x') = \frac{1}{n_0\sqrt{|b|}}x'_{n0}\sec\mathrm{h}(x'_{n0}),$$

and where

$$x'_{no} = \frac{1}{2} n_0 |b| x'$$

Thus, Kartner et al. [80] obtained that

$$\langle \Delta\hat{n}_0^2 \rangle = n_0, \; \langle \Delta\hat{\theta}_0^2 \rangle = \frac{0.6075}{n_0}, \; \langle \Delta\hat{p}'^2_0 \rangle = \frac{1}{3n_0\tau_0^2}, \; \langle \Delta\hat{x}'^2_0 \rangle = \frac{1.645\tau_0^2}{2n_0}, \tag{3.150}$$

where $\tau_0^2 = 2/n_0|b|$ is the width of the microscopic particle. The uncertainty products of the Boson number and phase, momentum, and position are, respectively,

$$\langle \Delta\hat{n}_0^2 \rangle \langle \Delta\hat{\theta}_0^2 \rangle = 0.6075 \geq 0.25, n_0^2 \langle \Delta\hat{p}'^2_0 \rangle \langle \Delta\hat{x}'^2_0 \rangle = 0.27 \geq 0.25, \tag{3.151}$$

However, the quantum fluctuation of the particle cannot be white because interaction between the particles results in correlations between them in this case. Thus, Kätner and Boivin [80] assumed a fundamental soliton state with a Poissonian distribution for the boson number $p' = \mathrm{n}_0^{\mathrm{n}} e^{-n_0} / \mathrm{n}!$ and a Gaussian distribution for the momentum Eq. (3.148) with a width $\langle \Delta p'^2_0 \rangle = n_0 |b|^2 / 4\mu$, where μ is a parameter of the order of unity compared to n_0. They finally obtained the minimum uncertainty values of:

$$\langle \Delta\hat{\theta}_0^2 \rangle = \frac{0.25}{\langle \Delta\hat{n}^2 \rangle} = \frac{0.25}{n_0} \left[1 + O\left(\frac{1}{n_0} \right) \right], \tag{3.152}$$

and

$$\langle \Delta\hat{x}'^2_0 \rangle = \frac{0.25}{\langle n_0^2 \rangle} = \frac{0.25\mu\tau_0^2}{n_0} \left[1 + O\left(\frac{1}{n_0} \right) \right] \tag{3.153}$$

up to an order of $1/n_0$, for the corresponding initial fluctuations in soliton phase and timing. Thus, at t'=0 the fundamental soliton with the given Boson number and momentum distributions is a minimum uncertainty state in the four collective variables, the Boson number, phase, momentum, and position, up to the terms of $O(1/n_0)$, which are of the form:

$$\langle \Delta\hat{n}_0{}^2 \rangle \langle \Delta\hat{\theta}_0^2 \rangle = 0.25 \left[1 + O\left(\frac{1}{n_0} \right) \right] \tag{3.154}$$

and

$$n_0^2\left\langle \Delta p'^2_0\right\rangle\left\langle \Delta \hat{x}'^2_0\right\rangle = 0.25\left[1+O\left(\frac{1}{n_0}\right)\right] \tag{3.155}$$

These are the uncertainty relations arising from the quantum fluctuations in the nonlinear quantum theory of microscopic particles, described by the nonlinear Schrödinger equation. They are minimum uncertainty relations. Therefore, we conclude that that the uncertainty relationship of the particles takes the minimum values in the nonlinear quantum theory regardless of whether the state is coherent or squeezed, or a system is classical or quantum. This minimum uncertainty relationship between the mechanical quantities clearly displays the wave-corpuscle duality of microscopic particles in the nonlinear quantum theory.

3.11. The Experimental Confirmation of Wave-Corpuscle Duality and Soliton Motion of Microscopic Particles in Nonlinear Quantum Mechanics

3.11.1. Methods That Experimentally Verify the Wave and Corpuscle Natures of Particles

The theoretical predictions related to the wave-corpuscle duality, localization,and soliton motionof microscopic particles,described by the nonlinear quantum mechanics, mentioned above,should be subject to experimental verification. Only if these questions are verified and solved can we affirm the validity of the nonlinear quantum mechanics, or else, it is not a correct theory. Therefore, the experimental evidences play an important role in its validity.In practice, we provided many experimental evidences of wave-corpuscle duality of microscopic particles including the electron, neutron, and He atom in Chapter 1, which are enough to support and verify the validity of the results of the wave-corpuscle duality of microscopic particles obtained from Pang's nonlinear quantum mechanics, mentioned above. We here provided again some experimental evidences, which were obtained from macroscopic experiments, for the correctness of the wave-corpuscle features of microscopic particles.It is fortunate that the dynamic equation in nonlinear quantum mechanics, and the cubically nonlinear Schrödinger equation, can be used to describe the movements of macroscopic objects or fluids.

In order to verify this point, we now write the wave function of the microscopic particles in Eq. (2.66) at $A(\phi)=0$ as $\phi=\sqrt{\rho}e^{i\theta}$ in terms of the phase and amplitude, where $\rho=|\phi|^2=|\varphi|^2$. This transformation is usually called Madelung'stransformation. After substitution in the nonlinear Schrödinger equation (3.1) and the separation of the real and imaginary parts of the equation, one obtains (after rescaling the time by a factor of 2):

$$\rho_{t'}+\nabla.(\rho\nabla\theta)=0, or, \rho_{t'}+\nabla.\vec{j}=0 \text{ with } \vec{j}=\rho\nabla\theta \tag{3.156}$$

and

$$\theta_{t'} + \frac{1}{2}|\nabla\theta|^2 - \frac{b}{2}\rho = \frac{1}{2\sqrt{\rho}}\Delta\sqrt{\rho} \qquad (3.157)$$

Taking as a density of macrofluid and θ as a hydrodynamic potential, from this interpretation of $\rho = |\phi|^2$, we know clearly that the meaning of $|\phi|^2$ is, in truth, far from the concept of probability in linear quantum mechanics. It essentially represents the mass density of the microscopic particles at a certain point in the place-time in the nonlinear quantum theory.

If Eq. (3.156) is integrated over x' one finds the integral of motion N = with $\vec{j} \to 0$, when $x' \to \pm\infty$, N is then the mass of the microscopic particle, and Eq. (3.156) is nothing but the mass conservation for the microscopic particle. It is, in essence, a continuum equation of fluid hydrodynamics that occurs in macrofluid hydrodynamics, equations (3.156) –(3.157) with b <0, and can be identified with the equations for an irrotational barotropic gas. Thus, the equation (3.156) is also called the fluid-dynamical form of the nonlinear Schrödinger equation. This clearly shows the classical features of the nonlinear Schrödinger equation, i.e.,the motions of microscopic particles in nonlinear quantum mechanics resemble the movements of classical particles in macrofluid hydrodynamics. Therefore, these discussions not only enhance our understanding of the meaning of the wave function of the microscopic particle, $\phi(x',t')$, in nonlinear quantum mechanics, but also points out the direction and method for inspecting and verifying the above nature and properties of microscopic particles, described by nonlinear quantum mechanics.

On the other hand, we already know that the light soliton occurred in the optical fiber, which is formed through the nonlinear self-focus mechanism of light, and can be described by the nonlinear Schrödinger equation.

In view of the two phenomena we here use the properties of the water-soliton and light-soliton (or photon-soliton),obtained from the experiments,to verify the correctness of nonlinear behaviors or localizationfeatures of microscopic particles described by the nonlinear Schrödinger equation.

3.11.2. TheDynamic Equation of Non-propagating Solitary Wave in Water and Its Properties

Wu, et al., [81-82], in 1984, first found non-propagating solitary waves in water.The form and outline of the solitary wave is shown in Figure 3.11. Subsequently, Cui and Pang et al. [83-87],experimentally conformed this phenomenon and further studied the properties of the solitary wave.

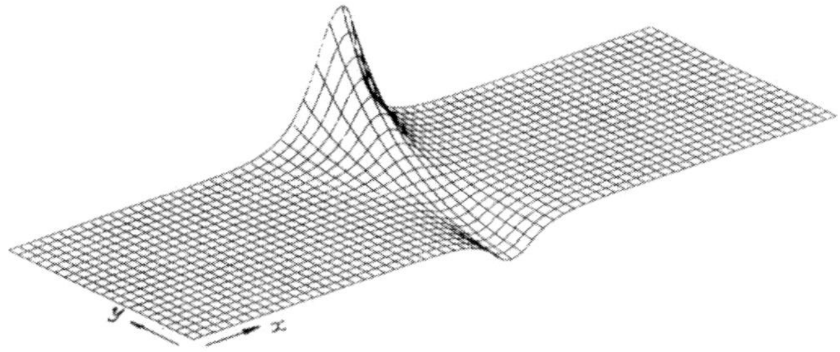

Figure 3.11. The non-propagating solitary wave generated in water trough obtained by Wu et al., [81].

Miles et el [88-92], Larraza et al., [93-94] and Pang et al., [95-96]theoretically demonstrated that this phenomenon can be very well described by a nonlinear Schrödinger equation. Larrazaet al., [93-94]derived the concept that the surface wave, in water troughs, satisfies the following nonlinear Schrödinger equation,

$$C^2\frac{d^2\phi}{dx^2}+(\omega_1^2-\omega^2)\phi+A|\phi|^2\phi=0, \tag{3.158}$$

and, based on the velocity potential of the fluid, ϕ, satisfies the Laplace equation and the corresponding boundary conditions.

For and $k\phi>1.022$, the soliton solutionof Eq.(3.158) is in the form:

$$\phi=\sqrt{\frac{2(\omega_1^2-\omega^2)}{A}}\sec h\left[\frac{(\omega_1^2-\omega^2)x}{C^2}\right] \tag{3.159}$$

where

$$C^2=\frac{g}{2k}\left[\overline{T}+kd(1-\overline{T}^2)\right], A=\frac{1}{8}k^4(6\overline{T}^4-5\overline{T}^2+16-9\overline{T}^{-2}),$$

and d is the depth of water, where $k=\pi/\tilde{b}$, and $\tilde{b}$ and is the width of the water troughs, $\overline{T}=\tanh(kd)$, $\omega_1^2=gk\overline{T}$, and where g is the acceleration due to gravity, and finally, where ω is the frequency of the externally applied field.

A slightly different form of the nonlinear Schrödinger equation was obtained by Miles et al., [88-92] in a similar phenomenon, and is given by:

$$B\phi_{xx}+(\beta+A|\phi|^2)\phi+v\phi^*=0 \tag{3.160}$$

and it has the following soliton solution:

$$\phi = e^{i\theta} \sec h(\sqrt{\frac{A'}{2B}} X), \tag{3.161}$$

at $v = 0$, where,

$$B = \bar{T} + kd(1 - \bar{T}^{2}), A' = \frac{1}{8}(6\bar{T}^{4} - 5\bar{T}^{2} + 16 - 9\bar{T}^{-2}), \beta = \frac{(\omega^2 - \omega_1^2)}{2\varepsilon\omega_1^2}, r = \frac{\omega^2 a_0}{\varepsilon g}, \omega_1 = \sqrt{gk \tanh(kd)}, X = 2\sqrt{\varepsilon \bar{T}} kx,$$

a_0 and θ are constants, and ε is a small and positive scaling parameter.

Further, Pang et al., [95-96] obtained the nonlinear Schrödinger equation of motion of the water soliton,obtained from its mechanism of generation and properties of this phenomenon, which is represented by

$$\frac{\partial^2 \phi_n(x)}{\partial x^2} - K^2 \phi_n(x) + r' \left|\phi_n(x)\right|^2 \phi_n(x) = 0 \tag{3.162}$$

in x direction. This equation has a solitonsolution.Thus, the velocity potential of the water in this case can be denoted by

$$\phi = B_n \phi_n(x) \cos(k_n y) \frac{\cos\left[k_n'(z+d)\right]}{\cos(k_n^1 d)}, \tag{3.163}$$

where the movement of water wave in x direction is denoted by

$$\phi_n(x) = B_n \sqrt{\frac{2K^2}{r'}} \sec h[\sqrt{K^2}(x - x_0)] \tag{3.164}$$

with andwhere $\omega_0^2 = \beta'/\rho$, and β' represent the surface tension constant. Then ρ is the density of the water, α is the nonlinear coefficient of the surface water, and B_n is a constant.

From Eqs.(3.158)-(3.164), we see that the water soliton satisfies the time-independent nonlinear Schrödinger equation, and has a bell-shaped form, which is similar with Figure 3.1 and Eqs.(3.8)-(3.9), i.e., they are all bell-type solitons. Therefore, their dynamic equations and outlines are the same as those of microscopic particles in nonlinear quantum mechanics.

Thus, we can use the properties of water soliton obtained from the experiments to verify the correctness of the wave-corpuscle duality and soliton motion of microscopic particles in nonlinear quantum mechanics, mentioned above. These experimental results of the water soliton are described as follows.

3.11.3. Experimental Results of Nonlinear Features of Non-Propagating Water Solitons

The experimental apparatus used by Cui et al., [83-87] is quite simple. It included an organic glass trough of 38 cm long and 2.0-5.0 cm wide, which was filled with various liquids to a depth of d=2.0-5.0 cm and was put on a vibrational plane-station. A cone loudspeaker was driven by a low frequency (7-15Hz) signal in the vertical direction, and another signal (12-25Hz) in the horizontal direction, with a power amplifier and appropriate measuring instruments.

The water trough was placed on the vibrational plane station, which was driven by the power amplifier. During the initial periods, when the frequency of the driving signal was below a certain threshold, only a few small ripples were observed and but no significant wave generation took place on the surface of the water trough. In this case, the density of the water molecules was uniform.However, above the threshold of the driving frequency, if a nonlinear initial disturbance was again applied to the surface of water trough, then a parametrically excited wave occurred at half the driven frequency. As large numbers of water molecules assembled, they got together to form a bell-type, non-propagating solitary wave on the surface of the water in the trough, due to the nonlinear interaction among the water molecules arising from the surface tension (as shown in Figure 3.12). Thus, the soliton of the same type as the soliton of the nonlinear Schrödinger equation in nonlinear quantum mechanics appears on the surface of the water in this case.

An obvious peculiarity of this phenomenon, is that the density of the water molecules where the the soliton occurs are much larger than that in other parts of the water trough (as shown in Figure 3.13), therefore, we infer that the water solitonis formed through gathering of the water molecules, then its mechanism of form can be called*self-localization*of water molecules due to the nonlinear interaction, which arises from the changes of the surface tension,in this case.

With different liquids, and different initial disturbances, and proper vibrations, one or two, or even more solitary waves can be observed on the water surface. However, the shapes of these solitary waves are different for different liquids. If the driving amplitude is further increased, several solitary waves can also occur.

From these experiments, we can obtain the following conditions for forming the solitary waves:

(1) The liquid must have a certain depth.
(2) The coefficient of surface tension of the liquid must be smaller than a criticalvalue; its upper boundary depends on the characteristics of the liquid.
(3) The vertical driving frequency is about twicethe intrinsic frequency, thehorizontalfrequency is about the same as the inherent frequency, and the driving amplitude must be within a certain range in this experiment.
(4) There must be a nonlinear initial excitation.
(5) The channel should not be too wide.

Another interesting distinguishing behavior is that there is a fixed position in the x direction at which the peak of the wave occurs, and the solitary wave formed does not propagate but oscillates in the y direction. Also, the frequency of oscillation is about the same as the intrinsic frequency. In the x direction, the profile of the soliton is a bell-shaped curve, which is similar to Figure 3.1. In glycerin-water, the soliton form fits the expression $\phi = 1.7 \sec h(x/1.25)$ cm.

The measured profiles of the soliton for the other four liquids are shown in Figure 3.14, and they are expressed in the order of increasing surface tension. Curve I is that of an ideal fluid whose surface tension is zero. Curve V is the wave of glycerin-water. In these experiments, it is clear that the amplitude of the soliton increases and its width decreases with increasing the surface tension. If the coefficient of the surface tension is greater than its upper bound, then no soliton can be formed.

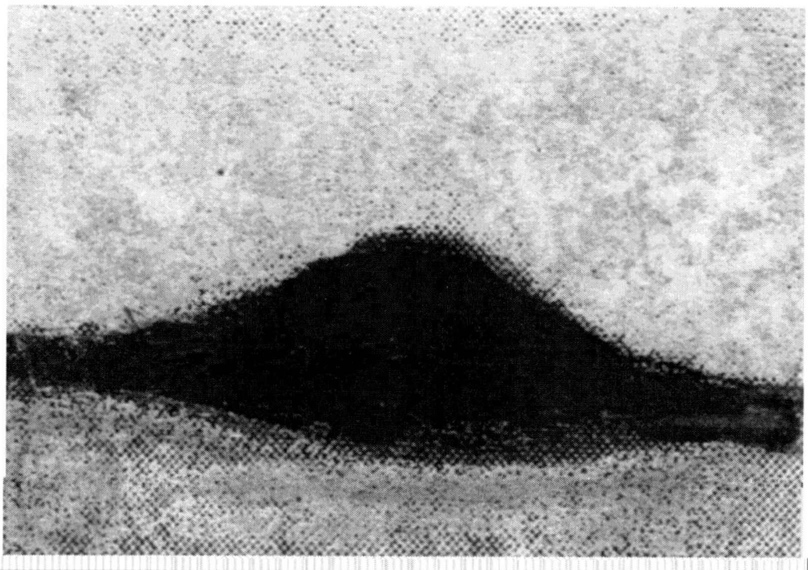

Figure 3.12. The bell-type non-propagating surface water soliton occurred in the water trough.

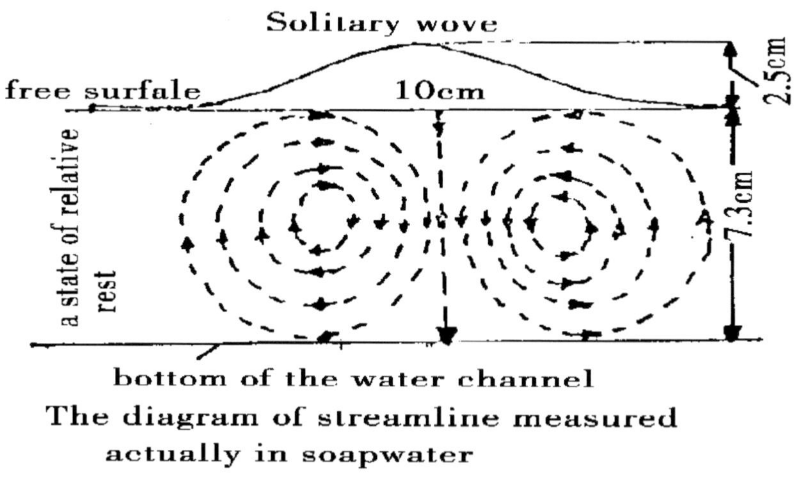

Figure 3.13. The change of density of water and the illustration of the direction of the motion of water molecules.

The surface water soliton can be moved under the influence of external forces. For example, when they are pushed with a little oar, or when they are blown by the wind, or when saltwater is added into the channel to increase the surface tension, then the soliton moves toward greater surface tension. If the channel is titled to a slope of 0.05, the soliton moves toward the shallow side, and the height of the solitary wave is reduced.As soon as the soliton reaches the point where the depth of water is the minimum, then the motion of the soliton stops. Meanwhile, the motion of the soliton can often be prevented by another soliton, which is likely out of phase and has small amplitude. These dynamic properties of water soliton are seen to be the same as those of classical particles and the microscopic particles in nonlinear quantum mechanics. Thus, these experimental resultscan be regarded as an experimental verification of the dynamic features of the microscopic particle in the nonlinear Schrödinger equation,mentioned above.

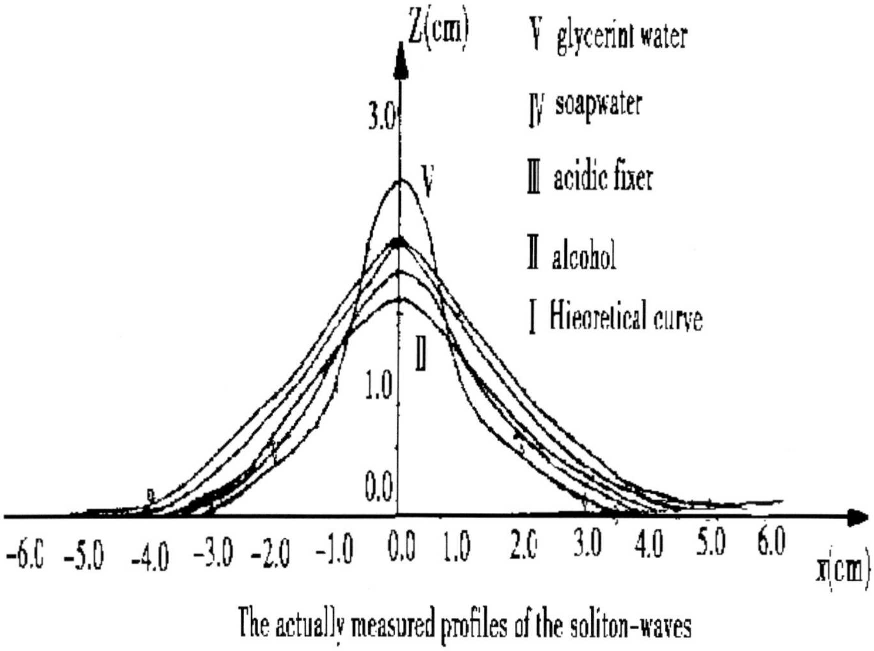

Figure 3.14.Measured profiles of water solitary waves.

Figure 3.15. Collision between two solitons that are out-of-phase.

The following two types of interactions, or collisions, between two solitons may be distinguished. (1) Out-of-phase solitons that repel each other and do not merge as shown in Figure 3.15, (2) Interactions between two solitons, which are in-phase, which show a cyclic process of attraction→merging→separation→and re-attraction as the driving amplitude is increased, as shown in Figure 3.16. These properties of collision between the water solitons are the same as those obtained by nonlinear quantum mechanics for microscopic particles. Thus, the collision properties of microscopic particles in nonlinear quantum mechanics, in which the forms of the microscopic particles can be retained after a collision, are experimentally confirmed.

In the case of a horizontal driving force, the driving frequency required for form of the soliton increases when the depth of the liquid increases. If the channel is rotated horizontally, by up to 50 degrees while the driving signal remains the same, then the watersoliton remains at its original place. Similar phenomena can be observed in nonrectangular shaped channels, such as, in round, ring, or trapezoid channels, and V- or X-shaped channels.

At the same time, the solitons can be easily generated and are very stable if the liquid is magnetized,their collision process between two solitary waves can also be easily observed in the magnetized liquid.

These properties of water solitary waves and their rules of motion, shown in Figures 3.12-3.16, are similar with Eqs.(3.60) and(3.62) and Figures3-3-3.4 and 3.6-3.7 in nonlinear quantum mechanics, respectively. This means that the results of nonlinear quantum mechanics, the wave-corpuscle duality and soliton motions of microscopic particles, are again confirmed by some experiments.

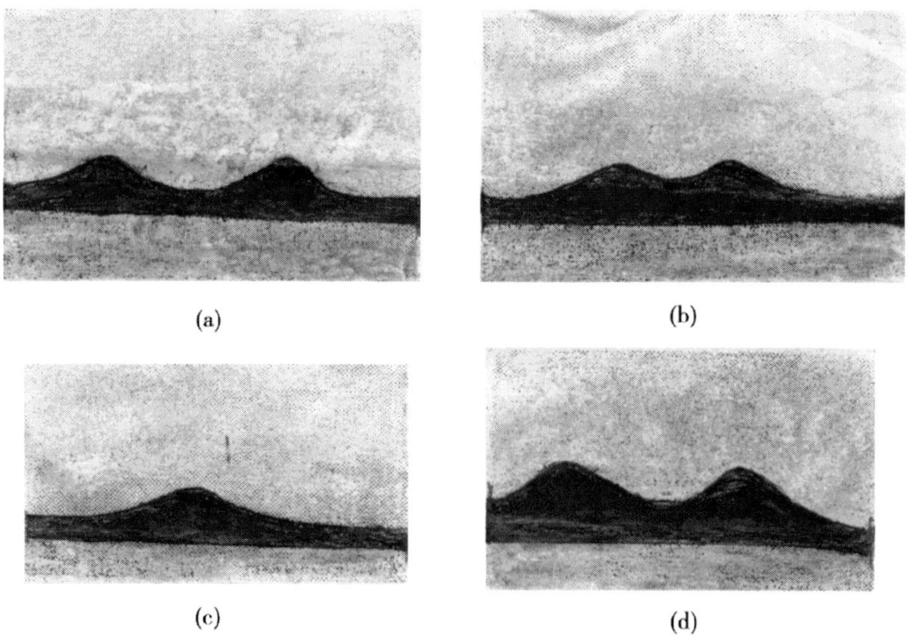

Figure 3.16.The collisionproperties between two water solitons, which are in-phase.

3.11.4. Experimental Evidence Provided by theOptical Solitons in Fiber

(1). Some Experimental Results

On the other hand, Hasegawa et al.'s experiment results [97-100]show that the light transmission in optical fibers satisfies the nonlinear Schrödinger equation, and that the light can be self-focused to form a light soliton that was obtained by nonlinear interactionor the Kerr effect in optical fibers [101].Its form is also similar with Figure 3.1. These phenomena can be clearly observed experimentally, the properties of the light soliton are measured experimentally. Thus, we can also use the experimental results of the light soliton to verify the validity of wave-corpuscle duality of microscopic particles described in nonlinear quantum mechanics. In the following, we summarize the features of the light soliton.

In order to experimentally verify the propagation of a soliton in an optical fiber, it is necessary to generate a short optical pulse with sufficiently high power and to use a fiber, which has a loss rate less than 1 dB/km. Also, it is required that the spectral width of the laser, which is narrower than the inverse of the pulse width in time, be applied. In other words, it requires the generation of a pulse with a narrow spectrum, similar to that which, in 1980, Mollenauer et al., [102-106], at Bell Laboratories, succeeded in using experimentally, for the first time, to verify the soliton transmission in an optical fiber.

This was achieved by using an F^{2+}color center laser, which is pumped by an Nd:TAG laser. Utilizing a fiber with a relatively large cross-section (10^{-6} cm^2) and a length of 700 meters, they transmitted an optical pulse with a 7 ps width and measured the shape of the output pulse by means of autocorrelation. Their experimental results are shown in Figure 3.17. The pulse shape is measured by the autocorrelation at the output side of the fiber for different input power levels. It can be seen clearly from Figure 3.17 that, while the output pulse width increases for a power below the threshold of 1.2W, it continuously decreases for an input power above 1.2W. The appearance of two peaks in the case of an input power of 11.4W is a result of a phase interference of three solitons, which were produced simultaneously in this case.

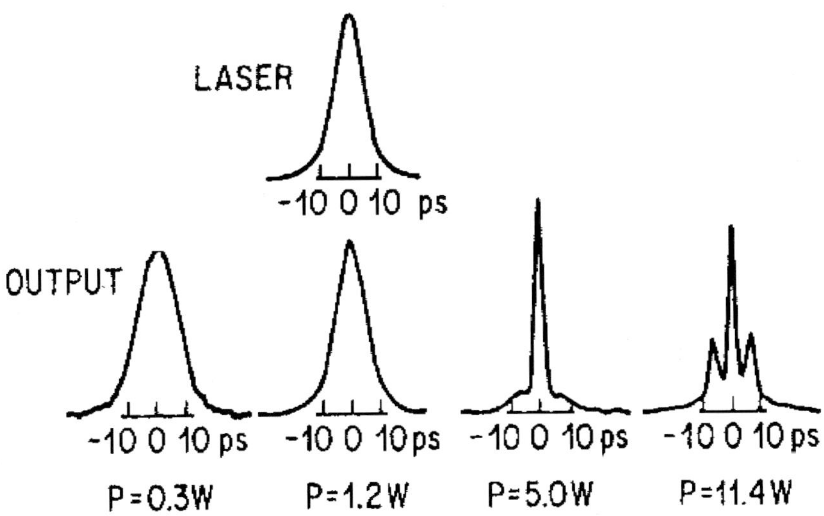

Figure 3.17.Experimental observation of light [97].

This is consistent with results of numerical calculations obtained by Hasegawa et al., [97-100]. The periodic behavior of the higher order solitons was confirmed by Stolen et. al. [107],in a later experiment. Most experimental observations of optical solitons were achieved by using the color center laser because of the need to produce the Fourier-transform limited pulse. However, optical solitons have also been successfully observed using other types of lasers, such as the dye laser or laser diodes. But, in these experiments, it has been necessary to control frequency chirping and to have a narrow spectrum for a well-defined soliton.

In 1987, Mitschke and Mollenauer [108-110] experimentally observed the interaction between optical solitons. Both repulsion and attraction regimes were observed and, depending on the soliton phase differences in the experiments, solitons were generated using a soliton laser. The pulses first passed through a Michelson interferometer, giving a pair of pulses. The length of one arm of the interferometer permitted a change in the distance between the pulses from zero up to several picoseconds. The soliton phase difference was also measured. In this experiment they used a polarization preserving 340 m low-loss (>0.3 dB/km) fiber with D=14.5 ps/(nm·km) at an operating wavelength of 1.52μm. The pulse duration was ≈1ps, therefore the fiber length was 10 soliton periods. Such a long fiber enables the researchers to clearly observe interactions between two well-separated soliton pulses. The results of the measurements are shown in Figure 3.18 for both repulsive and attraction regimes [111-112]. As seen below, the result is different from that predicted by the theory of optical solitons in the case of attractive interaction, and in the region where there is a small initial distance between the pulses.

Theoretical calculations [97-100] indicated that the two interacting optical solitons pass through each other. This is seen in Figure 3.18, illustrated by the oscillating structure of the theoretical curve. The experimental data deviate somewhat from such a feature. In the unstable region, the attractive force becomes repulsive, because of the influence of the Raman self-frequency shift.

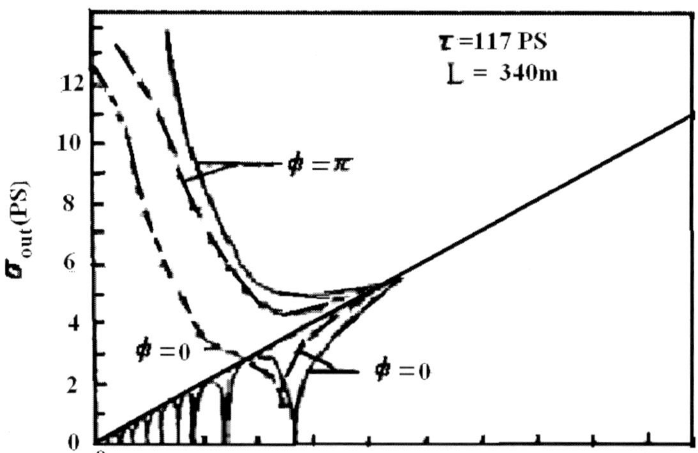

Figure 3.18. Shown is the pulse separation at an output of (σ_{out}) vs. the pulse separation at input (σ_{in})in a 340m fiber.The solid curve: a theoretical result, dashed line: experimentaldata.Without interaction, all points would fall on the 45^0 line [Re.108-110].

Mollenauer and Smith [102-106] discovered that between optical solitons, separated at long distance (more than 1000 km), there exists a long-range phase-independent interaction. Dianov etal. [112-113] showed that the electrostrictional mechanism may be responsible for the observed anomalous interaction. Reynaud and Barthelemy [114] observed interaction of spatial solitons between two soliton beams. These results demonstrated the existence of a mutual interaction between two close propagating optical soliton beams, which depends on the relative phase, and the distance between the beams [111-112].

In addition, Vysloukh and Cherednik [115-116] developed an effective method based on a combination of the inverse scattering theory and numerical methods to describe soliton interaction, and to follow the evolution of an arbitrary pulse. The method is useful for studying dynamics of multi-soliton states, and for studying the interaction of optical solitons in nearly integrable systems. These experimental results of soliton collisions are consistent with that of the microscopic particles described by the nonlinear Schrödinger equation,mentioned above.

(2) Theoretical Investigation

From Figure 3.17, we see that a bell-type soliton, which is the same as that in Figure 3.1, occurred in the optical fiber. This exhibits clearly that the light wave with dispersive feature was self-localized or self-focused as a stable soliton described by the nonlinear Schrödinger equation due to the nonlinear interaction, which is caused by the Kerr effect in the light fiber. This makes us believe that microscopic particles in nonlinear quantum mechanics can be self-localized or self-focused as a soliton with wave-corpuscle duality under action of nonlinear interactions. In this case, it is very necessary to theoretically deduce the dynamic equation of light soliton, in light fiber, using the nonlinear quantum mechanical theory.

Pang [5-6,117-118]performed the deductionof thenonlinear dynamic equation of the motion of the photons or propagation of light wavesin the fiber, which is described as follows:

As is known, the refractive index of the medium, n, is defined by the ratio of the speed oflight, c, in the vacuum to the phase velocity of the wave in the medium, i.e., $n=c/v_c=ck/\omega$. The refractive index of a dielectric medium is related to the frequency of the incident wave in the dispersive mediums. This phenomenon is called the dispersive effect, which arises from molecular structure of the medium. In the dispersive medium, the quantities, such as $\partial n/\partial\omega$, $\partial^2 n/\partial\omega^2$ etc., always have finite values. Thus, Pang[5-6,117-118]expands the wave frequency $\omega(=ck/n)$ around the wave vector k_0 for the carrier wave shown in Figure 3.1 by:

$$\omega-\omega_0=\frac{\partial\omega}{\partial k}\Big|_{k_0}(k-k_0)+\frac{1}{2!}\frac{\partial^2\omega}{\partial k^2}\Big|_{k_0}(k-k_0)^2+\frac{1}{3!}\frac{\partial^3\omega}{\partial k^3}\Big|_{k_0}(k-k_0)^3+\cdots \qquad (3.165)$$

This process shows the change of the frequency of the modulated wave because its wave vector, k, deviates from the carrier wave vector k_0.

Due to the fact that lightwave envelope function, $\vec{E}(x',t')$, whichis the electric field of the light wave in fiber, where $x'= x/\sqrt{\hbar^2/2m}$, $t'=t/\hbar$, in Figure 3.1, is a slowly varying function in both *x'* and *t'*, we can perform a Fourier transform on this function using the Fourier space variables $\Delta k(=k-k_0)$,because of a small wave vector shift from the carrier wave vector, and $\Delta\omega(=\omega-\omega_0)$, which denotes the small frequency shift, in the following form:

$$\overline{E}(\Delta k,\Delta\omega)=\frac{1}{2\pi}\int_{-\infty}^{+\infty}\int_{-\infty}^{+\infty}E(x',t')e^{i(\Delta kx'-\Delta\omega t')}dt'dx' \tag{3.166}$$

The corresponding inverse transform is:

$$E(x',t')=\frac{1}{2\pi}\int_{-\infty}^{+\infty}\int_{-\infty}^{+\infty}\overline{E}(\Delta k,\Delta\omega)e^{-i(\Delta kx'-\Delta\omega t')}d(\Delta k)d(\Delta\omega)$$

From Eqs. (3.165) and (3.166), we see that that $\partial E/\partial t'$ and $\partial E/\partial x'$ can be identified as the Fourier transform of $i\Delta\omega\overline{E}$ and $-i\Delta k\overline{E}$, respectively. Making use of these, we can rewrite $\Delta\omega$ and Δk as $-i\partial/\partial t'$ and $i\partial/\partial x'$, respectively. In terms of these operators, equation (3.165) can then be written as

$$-i\frac{\partial}{\partial t'}=i\omega'\frac{\partial}{\partial x'}-\frac{\omega''}{2!}\frac{\partial^2}{\partial x'^2}-\frac{\omega'''}{3!}\frac{\partial^3}{\partial x'^3}+\cdots \tag{3.167}$$

Operating Eq. (3.167) on the envelope function $\vec{E}(x',t')$ and keeping terms up to the order of the second derivative with respect to the carrier frequency, Pang got

$$i(\frac{\partial}{\partial t'}+\omega'\frac{\partial}{\partial x'})E-\frac{\omega''}{2!}\frac{\partial^2 E}{\partial x'^2}=0 \tag{3.168}$$

where $\omega'=(\partial\omega/\partial t')\big|_{k_0}$, $\omega''=(\partial^2\omega/\partial k^2)\big|_{k_0}$. The group velocity of the envelope-modulated wave is defined by: $v_g=\partial\omega/\partial k=\omega'$, $\omega''=\partial v_g/\partial k$. This shows that ω'' is given by the wavevector dependence of the group velocity of the wave. Therefore, ω'' expresses the dispersion feature of the group velocity of the wave. If $\omega''=0$, then the solution of Eq. (3.168) can be represented by an arbitrary function of $x-v_gt$, *i.e.*, $\varphi(x-v_gt)$. This shows that the wave envelop of the microscopic particle propagates at the group speed. Furthermore, we may introduce a new coordinate system which moves with the group velocity, $\tau = \varepsilon^2 t', \xi = \varepsilon(x' - \omega' t')$, where $\varepsilon = \Delta k_0 / k_0$ is a small quantity designating the width of the spectrum. Then, equation (3.168) becomes

$$i\frac{\partial E}{\partial \tau} - \frac{\omega''}{2}\frac{\partial^2 E}{\partial \xi^2} = 0 \tag{3.169}$$

Equation (3.169) is a dispersive equation. It shows that the wave envelope of the photon deforms the propagation due to the group velocity dispersion. If a short optical pulse is transmitted through a medium, then $\partial^2 \vec{E}/\partial x'^2$ may become large, and inversely proportional to the square of the photon width. Thus, the form of the pulse distorts during the transmission. In the fiber with a certain cross-sectional size, which is approximately equal to the wavelength of the photon, the group velocity dispersion ω'' is decided by the property of the medium itself. For a shorter wavelength, ω'' is the position, *i.e.*, the group velocity then increases with increasing the wave vector. But at a longer wavelength, ω'' is negative and the group velocity is a decreasing function of the wave vector. The domain of the wavelength corresponding to positive ω'' is called a normal dispersion region, while that corresponding to the negative ω'' is called an anomalous dispersion region. If one chooses the wavelength that corresponds to the minimum loss of the fiber, then the group dispersion ω'' becomes negative and the photon shape distorts considerably due to the change of the group velocity dispersion [97-100]. However, for optical fiber the nonlinear interaction can change the property of propagation of the phonon. Kerr discovered that the refractive index of a fiber increases in proportion to the square of the electric field $\vec{E}(x',t')$, which is called the Kerr effect and it is caused by a deformation of the electron orbits in the fiber due to the applied electric field. It can be expressed by $n = n_0 + n_2|\vec{E}|^2$ [97 100]), where n_2 is called the Kerr coefficient in the case of a fiber. Obviously, this is a nonlinear effect because n is a nonlinear function of the electric field $\vec{E}(x',t')$) [5-6,117]. (For a grass fiber, n_2 is about $1.2\times10^{-22}\ m^2/V^2$. For a typical optical fiber with a $60\mu m^2$ effective cross-sectional area and $10^6\ V/m$ electric field, produced by importing an optical pulse with 100 mW optical power, the refractive index increases by a factor of about 10^{-10}). Because of this effect, the frequency of the light wave in the medium occurs and causes the change by a factor of $-n_2|E|^2(\omega/c) = -n_2|E|^2(2\pi/\lambda n_0)$. When the photon propagates in the medium, it experiences this effect. Thus, it is reasonable to add the nonlinear interaction into the above equation of motion in Eq. (3.169) [5-6,117]. With this, the equation (3.169) becomes

$$i\frac{\partial E}{\partial t'} - \frac{\omega''}{2!}\frac{\partial^2 E}{\partial x'^2} + b|E|^2 E = 0 \tag{3.170}$$

where $b = 2\pi n_2 \rho' / n_0 \bar{\lambda}$, where $\bar{\lambda}$ is the wavelength of light and ρ' denotes the reduction factor due to the fact that the intensity of the light wave varies in the cross-section of the material and takes a value approximately constant (or 1/2 for the fiber) in most cases.

Evidently, equation (3.170) is a nonlinear Schrödinger equation, which is similar with Eq.(3.1). Its solutions are some bell-type solitons as shown in Figure 3.17, which are also similar to Figure 3.1. In this model, the nonlinear interaction results from the interaction of the light wave, or photonswith a Kerr effect, which makes the photons self-focus a soliton with wave-corpuscle duality. This is the self-focus mechanism of generation of nonlinear interaction. Just so, the aboveresults can serve as the experimental evidences of the wave-corpuscle duality of microscopic particles described the nonlinear quantum mechanics, or nonlinear Schrödingerequation.

In fact, the dynamic equation of the light soliton in fiber was first obtained by Hasegawa [97-100], in which the dispersive effect of the medium and the Kerr effect ofthe optical fiber were simultaneously considered. Hasegawa obtained a nonlinear equation for the propagation of the electric field $\vec{E}(x,t)$ of the light due to the nonlinearity of thefiber, which is represented by:

$$i\frac{\partial \vec{E}}{\partial \xi} - \frac{k''}{2}\frac{\partial^2 \vec{E}}{\partial \tau^2} + g\frac{\left|\vec{E}\right|^2 \vec{E}}{\varepsilon^2} = 0 \tag{3.171}$$

This is also a nonlinear Schrödinger equation, which is similar to Eq.(3.170) and Eq.(3.1), except for some parameters, where $\vec{E} = \mathrm{Re}\{\vec{E}(x,t)\exp[i(k_0 x - \omega_0 t)]\}$ is the electric field of light wave, the envelope function $\vec{E}(x,t)$ is a slowly varying function of time and space, which indicates that the frequency spectrum of electric field $\vec{E}(x,t)$ has a localized structure around the carrier frequency ω_0.

Briefly, the above experimental resultsof water-soliton and light-soliton are enough and sufficient to verify and check the validity of the wave-corpuscle duality of microscopic particles described in the nonlinear Schrödinger equation, thus we conclude that the new nonlinear quantum theory established, based on the nonlinear Schrödingerequation, is correct and real, and is confirmed by the experiments.

REFERENCES

[1] Pang, Xiao-feng, *Physica B*, 403 (2008)4292.

[2] Pang, Xiao-feng, The theory of nonlinear quantum mechanics: in Research of New Sciences, eds. Lui Hong, Beijing, *Chin. Science and Tech.* Press, 1991, pp16-20.

[3] Pang, Xiao-feng,*Frontiers of physics in China,* 3, (2008) 413.

[4] Pang, Xiao-feng and Feng Yuan-ping, Quantum Mechanics in Nonlinear Systems, Singapore: World Scientific Publishing Co. 2005.

[5] Pang, Xiao-feng. *Nonlinear quantum mechanics*, Beijing, Chinese Electronic Industry Press 2009.

[6] Pang, Xiao-feng. *The Theory of nonlinear quantum mechanics*, Chongqing, Chinese Chongqing Press, 1994.

[7] Pang, Xiao-feng. Establishment of nonlinear quantum mechanics, *Research andDevelopment of World Science and Technology*, 28(2006) 11.

[8] Pang, Xiao-feng, *Research and Development of World Science and Technology*, 24(2003) 54.

[9] Pang, Xiao-feng. Features of motion of microscopic particles in nonlinear systems and nonlinear quantum mechanics in Sciencetific Proceding-Physics and Others, Beijing: Atomic Energy Press, 2006.

[10] Pang, Xiao-feng. *Physica B*, 405(2010) 2317.

[11] Pang, Xiao-feng. *Far EastJournal of Mechanical Engineering and Physics*, 2 (2010)1.

[12] Pang, Xiao-feng, *Modern Physics*, 1 (2011)1-16.

[13] Pang, Xiao-feng. *Soliton physics*, Sichuan Sci. and Tech. Press Sin., Chengdu,2003.

[14] V. E.Zakharov, & A. B. Shabat, *Sov. Phys.-JETP*, 34(1972) 62.

[15] V. E..Zakharov, & A. B. Shabat, *Sov. Phys.-JETP*, 37(1973)823.

[16] Guo bai-lin. &Pang Xiao-feng. *Solitons,* Chinese Science Press, Beijing, 1987.

[17] A. S.Davydov, *Solitons in molecular systems*, D. Reidel Publishing, Dordrecht,1985.

[18] P. G.Drazin, & R. S. Johnson, Solitons, an introduction, Cambridge Univ. Press, Cambridge, 1989.

[19] P.E.Zhidkov, Korteweg-De, *Vries and nonlinear Schrödinger equation: qualitative theory,* Springer, Berlin, 2001.

[20] M. J.Ablowitz, & H.Segur, Solitons and the inverse scattering transform, *SIAM,*1981.

[21] H. H.Chen, & C. S.Liu, solitons in nonuniform media, Phys. Rev. Lett., 37 (1976)693.

[22] H. H.Chen, *Phys. Fluids*, 21(1978)377.

[23] Pang, Xiao-feng, *Mod. Phys. Lett*., B23(2009) 939.

[24] Pang, Xiao-feng. *Chin. Phys. Lett*., 10(7) (1993)437.

[25] Pang, Xiao-feng, *Chin. Phys. Lett*., 10 (1993)570 .

[26] S. Bargur, K.Bongs, S. Dettmer, W. Ertmer, & K. Sengstock, *Phys.Rev. Lett*., 83(1999) 5198.

[27] V.Makhankov, & V. Fedyanin, *Phys. Scr*., 20(1979)552.

[28] V.Makhankov, *Comp. Phys. Com*., 21(1980) 1.

[29] V. G.Makhankov, & V. K. Fedyanin, *Phys. Rep*., 104(1984) 1.

[30] V. G.Makhankov, *Phys. Rep*., 35(1978) 1.

[31] Pang, Xiao-feng, *Physica B,* 404(2009) 3125.

[32] Pang Xiao-feng. *Mod. Phys. Lett*. B, 23 (2009) 939.

[33] C.Sulem, & P. L. Sulem, *The nonlinear Schrödinger equation: self-focusing and wavecollapse,* Springer, New York, 1999.

[34] Pang, Xiao-feng,*Nature Sciences*, 2(2007) 42.

[35] Pang, Xiao-feng, *Int. J. Nonlinear science and numerical Simulation*, 11(2010) 1069.

[36] Pang, Xiao-feng, *Phys. Rev*. E, 62(2000) 6989.

[37] J. Stiefel, *Einfuhrung in die Numerische Mathematik*, Teubner Verlag, Stuttgart, 36(1965)123.

[38] K. E.Atkinson, *An Introduction to Numerical Analysis*, Wiley, New York Inc. 1987.
[39] P. L. Lax, *Comm.Pure and Appl.Math.*, 21 (1968)467.
[40] C.Desem, &, P. L. *Chu, Opt. Lett.* 12(1987)349.
[41] C.Desem, & P. L.Chu, *IEE.Proc.-J. Optoelectron*, 134(1987)145.
[42] C.Desem, & P. L.Chu, *Soliton-soliton interactions in optical solitons*, ed. J. R. Taylor, Cambridge University Press, Cambridge, 107(1992)351.
[43] J. P.Gordon, Theory of the soliton self-frequency shift, *Opt. Lett.*,11(1986)662.
[44] J. P.Gordon,*Opt. Lett.,* 8(1983)596.
[45] V.Karpman, *Phys. Lett.* A, 71 (1979)163.
[46] V.Karpman, & E. Maslov, *Sov. Phys.-JETP* 46(1977) 281
[47] V. I.Karpman, & V. V. Solovev, *Physica D* 3 (1981)487.
[48] B. Tan, & J. P.Bord, *Phys. Lett.* A, 240 (1998)282.
[49] P. Emplit, M. Haelterman, & J. P. Hamaide, *Opt. Lett.,* 18(1993) 1047.
[50] P. Emplit, M.Haelterman, R.Kashyap, & M. D.Lathouwer, *IEEE Photon.* Technol. Lett., 9 (1997) 1122.
[51] P.Emplit, J. P.Hamaide, F. Reynaud, C.Froehly, & A.Barthelemy, *Opt. Commun.*, 62(1987) 374.
[52] V.Karpman, & E. Maslov, *Sov. Phys.-JETP* 48 (1978) 252.
[53] D. W.Aossey, S. R.Skinner, J. T. Cooney, J. E.Williams, M. T.Gavin, D. R. Amdersen, & K. E.Lonngren, *Phys. Rev.*, A45 (1992) 2606.
[54] S. R.Skinner, G. R. Allan, D. R.Andersen, & A. L.Smirl, *IEEE Quantum Electron.* QE-27(1991)2211.
[55] G. R.Allan, S. R.Skinner, D. R.Anderson, & A. L. Smirl, *Optics Letters*, 16 (1991) 156.
[56] S. R.Skinner, G. R. Allan, D. R. Andersen, & A. L. Smirl, Baltimore (Opt. Soc. Am.),1991.
[57] R. N. Thurston, & A. M.Weiner, *J. Opt. Soc. Am.,*B8(1991)471.
[58] V. E.Zakharov, S. V. Manako, S. P. Noviko, & I. P.Pitayersky, *Theory of solitons*, Plenum, New York. 1984.
[59] J. L. Cooney, D. W. Aossey, J. E. Williams, M. T. Gavin, S. K Hyun, Y. C.Hsu, A. Scheller, & K. E.Lonngren,*Plasma Sources Sci. Technol.*, 2(1993)73.
[60] J. L. Cooney, M. T.Gavin, & K. E.Lonngren, *Phys. Fluids*, B3(1991)2758.
[61] J. L. Cooney, M. T.Gavin, J. E.Williams, D. W. Aossey, & K. E. Lonngren, *Phys.Fluids,* B3 (1991) 3277.
[62] J. L.Cooney, D. W.Aossey, J. E.Williams, & K. E.Lonngren, *Phys. Rev. E*, 47(1993)564.
[63] H. Ikezi , R. J.Taylor, & D. R. Baker, *Phys. Rev. Lett.*, 25 (1970)11.
[64] N. J. Zabusky, & M. D.Kruskal, *Phys. Rev. Lett.*, 15(1965)240.
[65] G. L. Lamb, *Elements of soliton theory*, Wiley Interscience, New York, 118(1980)125.
[66] J.P.Taylor, Optical solitons: Theory and experiment, Cambridge Univ. Press, Cambridge. 1992.
[67] K. E.Lonngren, & A. C.Scott, *Soliton in action*, Academic, New York, 1978,pp153.
[68] K. E.Lonngren, D. R. Andersen, & J. L.Cooney, *Phys. Lett.*, A156(1991)441.
[69] K. E.Lonngren,*Plasma Phys.,* 25(1983) 943.
[70] K. E. Lonngren, M .Khaze, E. F.Gabl, & J. M.Bulson, *Plasma Phys.*, 24 (1982)1483.
[71] A. B.Aceves, J. V. Moloney, & A. C. Newell, *Phys. Rev.*, A39(1989)1809.
[72] A. B.Aceves, J. V.Moloney, & A. C.Newell, *Phys. Rev.,* A39(1989)1839.

[73] Y.Nishida, *Phys. Fluids*, 27(1984)2176.
[74] Y. Nishida, K.Yoshida, & T.Nagasawa, *Phys. Lett.*, A131(1988)437.
[75] D. R.Andersen, & J. J. Regan, *J. Opt. Soc. Am.*, B6(1989) 1484.
[76] D. Anderson,*Phys. Scr.*, 20(1979) 345.
[77] G. A., Jr.Swartzlander, D. R. Andersen, J. J.Regan, H. Yin, & A. E.Kaplan, *Phys. Rev.Lett.*, 66(1991) 1583.
[78] Y. Lai, & H. Haus, *Phys. Rev.*, A 40(1989)844.
[79] Y. Lai, & H.Haus, *Phys. Rev.*A,40(1989) 854.
[80] F. X Kartner,& L.Boivin, *Phys. Rev.,* A53(1996).454.
[81] R. J.Wu, R . Keolian, and I.Rudnik, *Phys. Rev. Lett.* 52 (1984) 1421.
[82] R. J.Wu and I. Rudnick, *Phys. Rev.Lett.* 55 (1985)204.
[83] Cui Hong-nong and Yan zhi-qun, J.Xiangtan Univ. *Sinica,* ni:4 (1986) 27.
[84] Cui Hong-nong, *Advances in hydrodynamic* 3 (1988) 46.
[85] Cui Hong-nong, Pang Xiao-feng, *Advances in Hydrodynamic* 6 (1991) 18.
[86] Cui Hong-nong and Yan zhi-qun, *J. Hydrodynamics*, 3, (1988) 43.
[87] Cui Hong-nong, Yang, Xue-qun, Pang, Xiao-feng & Xiang, Long-wan, *J. Hydrodynamics,* 6(1991)18.
[88] J. W. Miles, *J.Fluid Mech.* 148(1984) 451.
[89] J. W. Miles, *J.Fluid Mech.*, 186(1988) 119.
[90] J. W. Miles, *J.Fluid Mech.*, 189(1988) 287.
[91] J. W. Miles, and Henderson, *Ann. Rev. Fluid Mech.* 22 (1990) 143.
[92] Henderson and J. W.Miles, *J.Fluid Mech.* 22(1991) 449.
[93] A. Larraza and S. Putterman, *J.Fluid Mech.* 148(1984) 445.
[94] A. Larraza and S. Putterman, *Phys. Lett. A* 103 (1984) 15.
[95] Pang Xiao-feng and Cui Hong-nong, A theory for non-propagating solitary wave, *Proc. of 4th APPC*, Korea 1990, pp154.
[96] Pang Xiao-feng, Feng Yuan-ping, *Quantum mechanics in nonlinear systems*, World Scientific Publishing Co., New Jersey, 2005.
[97] A. Hasegawa, *Optical Solitons in Fiber*, Berlin, Springer, 1989.
[98] A. Hasegawa, *Appl. Opt.,* 23(1984) 3302.
[99] A. Hasegawa, *Opt. Lett.*, 8 (1983) 342.
[100] A. Hasegawa, *Opt. Lett.,* 9 (1984) 288.
[101] R. N.Thurston, & A. M Weiner, *J. Opt. Soc. Am.*, B8 (1991)471
[102] L. F. Mollenauer, & K. Smith, *Opt. Lett.*, 13(1988) 675
[103] L. F.Mollenauer, J. P.Gordon, & M. N.Islam, *IEEE. J. Quantum Electrons.*, 22(1986) 157.
[104] L. F.Mollenauer, R. H. Stolen, & M. N. Islam, *Opt. Lett.,* 10(1986) 229.
[105] L. F.Mollenuaer and D.M. Bloom,*Opt. Lett.*, 4 (1979) 247.
[106] L. F.Mollenuaer, R. H. Stolen, & J. P. Gordon, *Phys. Rev. Lett.*, 45 (1980) 1095.
[107] R. H.Stolen & C. Lin, *Phys. Rev.* A17 (1978)1448.
[108] F. M. Mitschke, & L. F.Mollauer, *Opt. Lett.,* 11(1986).659.
[109] F.Lin Mollenauer,P James Gordon, solitons inoptical fibers, Elsevier Academic Press, 2006.
[110] F. M. Mitschke, & L. F. Mollenauer, *Opt. Lett.,* 12 (1987) 355.
[111] F.Abdullaev, Darmanyan, S. & P. Khabibullaev, *Optical soliton*, Springer, Berlin, 1994.

[112] E. M.Dianov, P. V.Mamyshev, A. M.Prokhorov, & S. V.Hernikov, *Opt.Lett.*,14(1990) 18.

[113] E. M.Dianov, A. V. Luchnikov, A. N. Pilipetski, & A. N. Arodumov, *Opt. Lett.*, 15 (1990)314.

[114] F. Reynaud, & A. Barthelemy, *Europhysics Lett.,* 12(1990) 401.

[115] V. A. Vysloukh, A. V. Ivanov, & I. V. Cherednik, & Izv.Vysch, *Uchebn. Zaved.* Radiofiz, 30 (1987) 980.

[116] V. A. Vysloukh, & I. V. Cherednik, Dokl. Akad. Nauk SSSR, 289 (1986) 336.

[117] Pang xiao feng, the behavior and properties of microscopic particles in nonlinear systems in *Handbook of solitons: Research, Technology and application*, Nove Science Publishers, Inc., New York, 2009, pp185-254.

[118] Pang Xiao feng, Establishment of nonlinear quantum mechanics and changes of property of microscopic particles as well as their experimental evidences in Contemporary Research in quantum systems, ed. Zoheir Wzziane, Nove Publishers, New York, 2014, pp207-338.

Chapter 4

THE DIFFERENCES OF NONLINEAR QUANTUM MECHANICS FROM LINEAR QUANTUM MECHANICS AS WELL AS THE EFFECTS AND MECHANISM OF GENERATION OF NONLINEAR INTERACTION

In this chapter, we stated in detail the distinctions between the new nonlinear quantum mechanics and linear quantum mechanics, which involve the significances and properties of wave function and mechanical quantities, the solutions of dynamic equations, uncertainty relation and methods solving dynamic equations, and so on, due to the nonlinear interactions in the dynamic equations. Our research points out that these nonlinear interactions are produced from the mechanisms of self-interaction, self-condensation, self-focus, self–trapping and self-localization in different systems caused by the interactions among the moved particles or between moved particles and moved background fields, such as lattice in solids and nucleons in atoms, or the special features of structures of the systems, such as anisotropy, double-well or many-well potentials and the nonlinear feature of structure. These results sufficiently display the validity,practicability, and availability of the new theory, and approximation of linear quantum mechanics (LQM).

4.1. THE VARIATIONS OF PHYSICAL SIGNIFICANCES AND NATURES OF THE WAVE FUNCTIONS OF MICROSCOPIC PARTICLE

We know clearly from the nonlinear and linear quantum mechanics that the states and properties of microscopic particles are all represented by a wave function, but their physical significances and natures of the wave functions are fully different. Therefore, we here state and discuss this problem.

4.1.1. PhysicalSignificances and Natures of the Wave Functions inLinear Quantum Mechanics

It is known thatin linear quantum mechanics [1-3], the wave function $\phi(\vec{r},t)$ in Eq.(1.8) represents a wave, its absolute square, $|\phi(\vec{r},t)|^2$, and denotes the probability of this particle, which occurs at the position $\vec{r}$ and time *t*.The wave function of the state of the microscopic particle satisfies a linear superposition principle, i.e., the superposition of two state wave functions of a particle still represents a state of the particle,the superposed state is denoted as:

$$|\phi\rangle = C_1|\phi_1\rangle + C_2|\phi_2\rangle \tag{4.1}$$

where $|\phi_1\rangle$ and $|\phi_2\rangle$ denotes two states of the particle, and C_1 and C_2 are constants relating to its states. This linear superposition principle of states of the microscopic particleis is due to the linear characteristics of the operators. Thus, the Hilbert space in which the quantum mechanics is defined is a linear space. This is why the theory is referred to as linear quantum mechanics. Therefore, such a superposition is different from that of classical waves [1-3]. This superposition principle results because the wave function must satisfy the homogeneity rule, i.e., if $\phi(\vec{r},t)$ is a solution of this equation, then $Z\phi(\vec{r},t)$ is also its solution, where Z is a constant, namely, they are homogeneous, which is a feature of linear quantum mechanics.

Correspondingly, any wave function, $\phi\left(\vec{r},t\right)$, or state vector $|\phi\rangle$, can be also expanded in terms of the eigenvectors ϕ_L or $|\phi_L\rangle$ of a Hermitian operator $\hat{L}$, i.e.,

$$\phi(\vec{r},t) = \sum_L C_L \phi_L(\vec{r},t) \tag{4.2}$$

or

$$\left|\phi\left(\vec{r},t\right)\right\rangle = \sum_L \langle\phi_L|\phi\rangle|\phi_L\rangle \tag{4.3}$$

,

where $C_L = \langle\phi_L|\phi\rangle$ is the wave function in representations of *L*. If the spectrum of *L* is continuous, then the summation of it, in Eqs.(4.2)-(4.3), should be replaced by an integral: $\int dL \cdot\cdot\cdot$. Clearly, equation (4.3) resembles Eq.(4.2), which can then be regarded as a superposition of many eigenstates $|\phi_L\rangle$.

Meanwhile, we can also regard that equation(4.2)shows a projection of the state vector of a system onto those of its subsystems, and thus constitutes the foundation of a transformation of the state vector $|\phi\rangle$ between different representations in this theory. Then the probability of getting the value *L'* in a measurement of *L* in the state $\phi\left(\vec{r},t\right)$ is $|C_{L'}|^2 = |\langle\phi_{L'}|\phi\rangle|^2$ in the case of discrete spectrum, or $\left|\langle\phi_{L'}|\phi\rangle\right|^2 dL$ in the case of continuous spectrum. This is a basic assumption concerning measurements of mechanical quantities in linear quantum mechanics.

At the same time, in linear quantum mechanics, no new state should occur if a pair of identical particles is exchanged in a system, which consists of some identical particles. This means that the wave function $|\phi\rangle$ of the particles in this system meets

$$\hat{P}_{kj}|\phi\rangle = \lambda|\phi\rangle, \tag{4.4}$$

where $\lambda = \pm 1$ and $\hat{P}_{kj}$ is an exchange operator. This is called the identical theorem of particles, in which the wave function of the particle in a system consisting of identical particles is either symmetric, ϕ_s, ($\lambda = +1$), or antisymmetric, ϕ_a, ($\lambda = -1$), and this feature cannot also be changed with increasing time, it is determined only by the nature of the particles, where the symmetric wave function describes a boson, but the antisymmetric wave function denotes a fermion [1-3].

These results are just the properties of wave function of the state of microscopic particles in linear quantum mechanics.

4.1.2. Physical Significances and Natures of the Wave Functions in the New Nonlinear Quantum Theory

On the contrary,in Pang's nonlinear quantummechanics, the states of microscopic particles are, in general, described by Eq.(3.2); it satisfies the nonlinear Schrödinger equation(2.66) at $A(\phi)=0$ or Eq.(3.1), which, in this case, possesses both wave and corpuscle features as mentioned in Eqs.(3.8)-(3.10).Its absolute square, $|\phi(\vec{r},t)|^2 = \rho(\vec{r},t)$, no longer denotes the probability of finding the microscopic particle at a given point in space-time, but the mass density of the microscopic particles at that point.Thus, we canfind out the particle number or the effective mass of the particle from $\int_{-\infty}^{\infty}|\phi|^2 d\tau = N$ as shown in Eq.(3.11). Then, the concept of probability of the particle does not exist; the difficulty of statistical interpretation for the wave function of the particle in linear quantum mechanics is abandoned in the nonlinear quantum theory. In this case,the hypothesis of independence of the Hamiltonianoperatorof the systems on states of microscopic particles, as mentioned in Eq. (2.74), and the linearity hypothesis of quantum mechanics are also broken through.

On the other hand, the superposition principle of wave function in Eq.(4.1) no longer holds in the new nonlinear quantum theory; a new superposition principle of wave function can be established. This means that the nonlinear interaction not only changes the process of superposition, but also gives rise to the diversificationof its forms. In practice, various different superposition principles for the nonlinear Schrödinger equation were built [4, 5]. Here we gave only the superposed rules of wave function for the equation (3.1) built by Konopelchenko [6-10].

The derivation ofthe superposition rules of statesof microscopic particlesin Konopelchenko's investigation is described as follows.

According to the Lax method [11], if two linear operators L and $\hat{B}$, corresponding to Eq. (3.1), which depend on $\phi_n(t)$, satisfy the following Lax operator equation

$$i\hat{L}_{t'} = \hat{B}\hat{L} - \hat{L}\hat{B} = [\hat{L}, \hat{B}] \tag{4.5}$$

which is a self-adjoint operator, then the eigenvalue $k$and eigenfunction Ψ of the operator L satisfy the equations(2.86)-(2.87), and their concrete representations depend on the nonlinear Schrödinger equation. For the nonlinear Schrödinger equation (3.1), the representations of L for Eq. (4.5), can be represented by Eq.(2.89), where

$$s^2 = (1 - 2/b) \tag{4.6}$$

with

$$t' = t/\hbar \text{ and } x' = x\sqrt{2m}/\hbar \tag{4.7}$$

where, $\phi(\vec{r}, t)$ in Eq.(2.89) satisfies Eq. (3.1). This is just the Lax form for the nonlinear Schrödinger equation (3.1).

In accordance with Zakharov and Shabat's method [12-13], from Lax representations in Eqs. (2.86) - (2.87) and the nonlinear Schrödinger equation (3.1), we can represent Ψ in Eq. (2.87) in a one-dimensional case by

$$\Psi = S\exp(-i\bar{k}x')\psi \tag{4.8}$$

where,

$$S = \begin{pmatrix} 0 & (1-s)^{1/2} \\ (1+s)^{1/2} & 0 \end{pmatrix}, \quad \bar{k} = \frac{k}{1-s^2} = bk/2 \tag{4.9}$$

Inserting Eq. (4.9) and in Eq. (2.89) into Eq. (2.86) the Zakharov-Shabat (ZS) equation [12-13]can be obtained as

$$\psi_{1x'} - q\psi_2 = -i\lambda\psi_1 \tag{4.10a}$$

$$\psi_{2x'} + q^*\psi_1 = i\lambda\psi_2 \tag{4.10b}$$

where

$$\lambda = \bar{k}s\,, \quad q = \frac{i\phi}{(1-s^2)^{1/2}} = i(\frac{b}{2})^{1/2}\phi\ , \tag{4.11}$$

Since S in Eq.(4.9) is not a unitary operator, then Eq. (4.10) is also *not* self-conjugate. Obviously, equation (4.10) can be represented by the following matrix form:

$$i\psi_{x'} + \Phi\psi = \lambda\sigma_3\psi \tag{4.12}$$

where

$$\sigma_3 = \begin{pmatrix} 1 & 0 \\ 0 & -1 \end{pmatrix}, \quad \Phi = \frac{1}{\sqrt{1-s^2}}\begin{pmatrix} 0 & \phi \\ \phi^* & 0 \end{pmatrix}, \quad \psi = \begin{pmatrix} \psi_1 \\ \psi_2 \end{pmatrix} \tag{4.13}$$

Φ is not a Hermitian operator. We represent Eq.(4.10) by the following form:

$$\frac{\partial}{\partial x'}\begin{pmatrix} \psi_1 \\ \psi_2 \end{pmatrix} = -i\lambda\begin{pmatrix} 1 & 0 \\ 0 & -1 \end{pmatrix}\begin{pmatrix} \psi_1 \\ \psi_2 \end{pmatrix} + \begin{pmatrix} 0 & q(x',t') \\ h(x',t') & 0 \end{pmatrix}\begin{pmatrix} \psi_1 \\ \psi_2 \end{pmatrix} \tag{4.14}$$

where function h(x',t') = - q*(x',t') in Eq.(4.10b). Konopelchenko [6-10] has re-written Eq.(4.14) as

$$\frac{\partial Q(x',t')}{\partial t'} = -2i\Pi(P^+)RQ \tag{4.15}$$

here

$$Q = \begin{pmatrix} 0 & q \\ h & 0 \end{pmatrix}, \quad R = \begin{pmatrix} 1 & 0 \\ 0 & -1 \end{pmatrix}, \quad P^+ = \frac{1}{2}iR\frac{\partial}{\partial x'} + \frac{1}{2}i[Q(x'), \int_{-\infty}^{x'} dy'[Q(y'), R^*]]$$

and $\Pi(P^+)$ is a mermorphic function, λ is a constant.

In the above,

$$\psi(x',t') = \begin{pmatrix} \psi_1 \\ \psi_2 \end{pmatrix}$$

is a two component Jost function, from Eq.(4.11) we know that q (x', t') and $h(x', t')$ are wave functions denoting the states of microscopic particles, and satisfy Eq. (3.1).

For the sine-Gordon equation, where $\gamma = A(\phi) = 0$, corresponding Lax equations can be also denoted by Eq. (4.14); but the representations of q(x', t') and h(x', t') differ somewhat.

According to a study by Calogero et al., [14-15], the Backlund transformation (BT) corresponding to Eq. (4.15) from the infinite dimensional group Q to Q' for the systems can be denoted by

$$\sum_{i=1,2} B_i(P^+)(N_i'Q'-QN_i') = 0 \tag{4.16}$$

where , $N_1' = \begin{pmatrix} 1 & 0 \\ 0 & 1 \end{pmatrix}$, $N_2' = \begin{pmatrix} 0 & 0 \\ 0 & 1 \end{pmatrix}$

and $B_1(P^+)$ and $B_2(P^+)$ are entire functions. The operator P^+ has the following property:

$$P^+\phi = \frac{i}{2}R\frac{\partial\phi}{\partial x'} + \frac{i}{2}\int_{-\infty}^{x'} dy'[R\phi(y')Q'(y') - Q(y')R\phi(y')]Q'(x') + \frac{1}{2}Q(x')\int_{-\infty}^{x'} dy'[R\phi(y')Q'(y') - Q(y')R\phi(y')]$$

Konopelchenko [6-10] considered a discrete BT in Eq. (4.16), namely,

$$B_1(P^+) = \prod_{i=1}^{n_1}(P^+ - \lambda_i)\ ,\ B_2(P^+) = \prod_{j=1}^{n_2}(P^+ - \mu_j)$$

here λ_i and μ_j are arbitrary constants and n_1 and n_2 are integers. Then, the discrete BT may be expressed as

$$B = \prod_{j=1}^{n_2} B_{\mu j}^{(2)} \prod_{i=1}^{n_1} B_{\lambda i}^{(1)} \tag{4.17}$$

where $B_\lambda^{(1)}$ and $B_\mu^{(2)}$ are elementary Backlund transformations (EBT), is $EBT\ B_\lambda^{(1)}$ is the BT in Eq. (4.16) at $B_1 = P^+ - \lambda$ and $B_2 = 1$ (λ is an constant), EBT $B_\mu^{(2)}$ is the BT in Eq. (4.17) at $B_1 = 1$ and $B_2 = P^+ - \mu$ (μ is a constant). The concrete expressions of the EBTs can be presented as in

$$B_\lambda^{(1)}(Q \to Q') : -i\frac{\partial q'}{\partial x'} - \frac{1}{2}q'^2 h + 2\lambda q' + 2q = 0\ ,\quad -i\frac{\partial h}{\partial x'} + \frac{1}{2}h^2 q' - 2\lambda h - 2h' = 0 \tag{4.18}$$

$$B_\mu^{(2)}(Q \to Q') : -i\frac{\partial q}{\partial x'} - \frac{1}{2}q^2 h + 2\mu q + 2q' = 0\ ,\quad -i\frac{\partial h'}{\partial x'} + \frac{1}{2}h'^2 q - 2\mu h' - 2h = 0 \tag{4.19}$$

the EBT $B_{\lambda}^{(1)}$ and $B_{\mu}^{(2)}$ commute with each other, i.e.,

$$B_{\lambda}^{(1)}\ B_{\mu}^{(2)} = B_{\mu}^{(2)}\ B_{\lambda}^{(1)} \tag{4.20a}$$

and

$$B_{\mu}^{(2)}\ B_{\lambda}^{(1)} = B_{\lambda}^{(1)}\ B_{\mu}^{(2)} = 1 \tag{4.20b}$$

Konopelchenko described the relation of four solutions of Eq. (4.15), (q_0, h_0), (q_1, h_1), (q_2, h_2), and (q_3,h_3), which are related through the commutativity of EBT $B_{\lambda}^{(1)}$ and $B_{\mu}^{(2)}$,respectively, from Eqs. (4.18) - (4.19), which are as follows [6-10]

$$q_3 = q_0 + 2(\lambda-\mu)/(h_2/2+2/q_1) \tag{4.21a}$$

$$h_3 = h_0 + 2(\mu-\lambda)/(q_1/2+2/h_2). \tag{4.21b}$$

This is simply the nonlinear superposition formula for the four wave functions of the state of microscopic particles described by Eq. (3.1). Only if three solutions (q_0, r_0), (q_1, r_1), and (q_2, r_2) of Eq. (4.15) are known, can we then find the fourth solution (q_3, h_3) via Eq. (4.21).

In this way, we can find an infinite number of solutions of Eq. (4.15). Thus, let us start from the trivial solution

$$Q_{(00)} = \begin{pmatrix} 0 & 0 \\ 0 & 0 \end{pmatrix}$$

and apply all possible discrete BTs in Eq. (4.21) on $Q_{(00)}$. In such a case Konopelchenko [6-10]gave us the following solutions for the entire system corresponding to the vertex (n_1 n_2)from $Q_{(00)}$

$$Q_{(n_1,n_2)} = \prod_{j=1}^{n_2} B_{\mu j}^{(2)} \prod_{i=1}^{n_1} B_{\lambda i}^{(1)} Q_{(00)} \tag{4.22}$$

The solutions $Q_{(0n)}$ and $Q_{(n0)}$ can be found from Eqs. (4.18) and (4.19) and are given by

$$q_{(0n)} = 0\,, h_{(0n)} = \sum_{l=1}^{n} \exp[2i\Pi(\mu_l)t'+2i\mu_l(x'-\bar{x}_{0l}')] \tag{4.23}$$

$$q_{(n0)} = \sum_{l=0}^{n} \exp[-2i\Pi(\lambda_l)t'-2i\lambda_l(x'-x_{0l}')]\,,\ h_{(n0)} = 0 \tag{4.24}$$

where x_{01}' and $\bar{x}_{01}$' are arbitrary constants. Equations (4.23)-(4.24) are also solutions of the linear equation

$$\frac{\partial Q(x',t')}{\partial t'} = -2i\Pi(\frac{1}{2}iR\frac{\partial}{\partial x'})RQ \tag{4.25}$$

which is the linear part of Eq. (4.15). Finally Konopelchenko [6-10] gave us the general nonlinear superposition formulae of states for nonlinear Schrödinger equation (3.1) corresponding to Eq. (4.10).They are:

$$h_{(n_1+1,n_2+1)} = h_{(n_1,n_2)} + 2(\mu-\pi)/(q_{(n_1+1,n_2)}/2 + 2/h_{(n_1,n_2+1)}) \tag{4.26}$$

and

$$q_{(n_1+1,n_2+1)} = q_{(n_1,n_2)} + 2(\lambda-\mu)/(h_{(n_1,n_2+1)}/2 + 2/q_{(n+1,n_2)}) \tag{4.27}$$

Using Eqs. (4.26) - (4.27), any arbitrary solutions such as $Q_{(n1,n2)}$ may be found from Eqs. (4.23) - (4.24). For example, if $Q_{(00)}$, $Q_{(10)}$, and $Q_{(01)}$ are known, then $Q_{(11)}$ can be obtained, i.e.,

$$q_{11} = 2(\lambda_1-\mu_1)\left\{\frac{1}{2}\exp[-2i\Pi(\mu_1)t'+2i\mu_1(x'-\bar{x}'_{01})] - 2\exp[2i\Pi(\lambda_1)t'-2i\lambda_1(x'-x'_{01})]\right\}^{-1}$$

$$h_{(11)} = 2(\mu_1-\lambda_1)\left\{\frac{1}{2}\exp[-2i\Pi(\lambda_1)t'-2i\lambda_1(x'-\bar{x}'_{01})] + 2\exp[-2i\Pi(\mu_1)t'+2i\mu_1(x'-\bar{x}'_{01})]\right\}^{-1} \tag{4.28}$$

Similarly, $Q_{(12)}$ may be obtained from $Q_{(01)}$, $Q_{(11)}$, and $Q_{(02)}$ using Eqs. (4.26)-(4.27),

$$q_{(12)} = 2(\lambda_1-\mu_2)/(h_{(02)}/2+2/q_{(11)}), \quad h_{(12)} = h_{(01)} + 2(\mu_1-\lambda_1)/(q_{(11)}/2+2/h_{(02)}) \tag{4.29}$$

where $q_{(11)}$ is given in (4.28). Other solutions, such as, $Q_{(31)}$, $Q_{(32)}$, etc., can also be obtained from Eqs.(4.26) - (4.29). We can see that an arbitrary solution$Q_{(n1n2)}$ is a simple function of $q_{(10)}$, $q_{(20)}$,…, $q_{(n10)}$ and $h_{(10)}$, $h_{(02)}$, …. $h_{(0n2)}$. Thus, the nonlinear superposition principle for Eqs.(4.26)–(4.27) for the nonlinear Schrödinger equation(3.1) gives a simple algebraic structure for the infinite number of solutions of (4.22). Solutions $Q_{(11)}$, $Q_{(22)}$, and $Q_{(nn)}$, are the soliton-type solutions. They are relatedto each other through the Backlund transformation in Eq. (4.23). If one of them is known, others can be obtained using Eq.(4.25).

Therefore, the superposition principle in Eqs.(4.26)-(4.27) of the wave function of microscopic particles, described by the nonlinear Schrödinger equation (3.1), are in essencedifferent from that in Eq.(4.1); it is both quite complicated and difficult to operate.

The above variation sufficiently exhibits that the nonlinear quantum theory is a new theory, which is different from linear quantum mechanics.

4.2. THE CHANGES OF MEANING AND PROPERTY OF MECHANICAL QUANTITIES FOR THE MICROSCOPIC PARTICLES

4.2.1. The Meaning and Property of Mechanical Quantities in Linear Quantum Mechanics

It is well known that the mechanical quantities of microscopic particles are represented by an operator, including the coordinates and momentum in linear quantum mechanics, but its value in anystate $|\phi\rangle$ is given by the average values of the operator[1-3], such as the value of physical quantity A in the state $|\phi\rangle$ is denoted by

$$\langle \hat{A} \rangle = \langle \phi | \hat{A} | \phi \rangle / \langle \phi | \phi \rangle , \tag{4.30}$$

or

$$\langle \hat{A} \rangle = \langle \phi | \hat{A} | \phi \rangle$$

if ϕ *is* normalized. Then the possible values of $\hat{A}$ can be obtained from these average values. This means that there are many possible values for a mechanical quantity in a state. This hypothesis exhibits the statistical feature in the depiction of motion of microscopic particles in linear quantum mechanics. Therefore, in order to obtain a precise valueof the mechanical quantity *A*, we must find a state such that $\overline{\langle \Delta A \rangle^2} = 0$, where

$$\overline{\langle \Delta A \rangle^2} = \langle \hat{A}^2 \rangle - \langle \hat{A} \rangle^2 .$$

This results in the following eigenequation for the operator $\hat{A}$,

$$\hat{A}\phi_L = A\phi_L \text{ or } \hat{A}|\phi_L\rangle = A|\phi_L\rangle \tag{4.31}$$

where ϕ_L or $|\phi_L\rangle$ is its eigenfunction, A is its eigenvalue. Thus, we can determine the eigenvalues of the operator $\hat{A}$ and the corresponding eigenfunctions ϕ_L from Eq.(4.31). This means that only if this state is the eigenstate of the operator,then its eigenvalue can be determinate.

At the same time, an observable physical quantity, for instance the coordinate *X*, the momentum*P,* and the energy *E,*are represented by a Hermitian operator whose eigenvalues are real. Their eigenfunctions and eigenvalues could be also obtained from Eq.(4.31).

Also, these eigenfunctions, corresponding to different eigenvalues, are orthogonal to each other. Thus, all eigenstates of a Hermitian operator constitute an orthogonal and complete set, $\{\phi_L\}$.

If two quantities, *A* and *B*, satisfy the Poisson brackets,

$$\{A,B\}=\sum_n\left(\frac{\partial A}{\partial p_n}\frac{\partial B}{\partial q_n}-\frac{\partial A}{\partial q_n}\frac{\partial B}{\partial p_n}\right)$$

where q_n and p_n are generalized coordinates and momentum in classical mechanics, respectively, then the corresponding operators, $\hat{A}$ and $\hat{B}$, meet the following commutation relationship:

$$\left[\hat{A},\hat{B}\right]=(\hat{A}\hat{B}-\hat{B}\hat{A})=-i\hbar\left\{\hat{A},\hat{B}\right\} \tag{4.32}$$

in quantum mechanics, where $i=\sqrt{-1}$ and $\hbar$ is the Planck's constant. This is just the correspondence principle in linear quantum mechanics. It shows that the values allowed for a mechanical quantity are quantized, and thus the name “quantum mechanics”. Based on this principle, the following Heisenberg uncertainty relation,for a set of conjugate mechanical quantity,can be obtained, which is denoted by [1-3]

$$\overline{(\Delta\hat{A})^2}\,\overline{(\Delta\hat{B})^2}\geq\left|\hat{C}\right|^2/4 \tag{4.33}$$

where $i\hat{C}=[\hat{A},\hat{B}]$ and $\Delta A=(\hat{A}-\langle\hat{A}\rangle)$. For example, the uncertainty relation of position and momentum is $\Delta x\Delta p\geq h/2$. Therefore, we should know clearly that the Heisenberg uncertainty relation is a necessary result of linear quantum mechanism, instead of the result arising from the measures of mechanical quantities.

4.2.2. The Meaning and Property of Mechanical Quantitiesin the Nonlinear Quantum Theory

On the contrary, we see from Eq.(2.74) that the Hamiltonian operator or Lagrangian and Hamiltonian functions in the nonlinear quantum systems are related to the wave function of macroscopic particles, which is thoroughly different from that in Eq.(1.25) in linear quantum mechanics. Thus, they are some nonlinear operators of the wave function.This means that we have thoroughly broken through the hypothesis of the independence of Hamiltonian operators of the systems on states of microscopic particles, forsaking the above hypothesis of linear quantum mechanics due to the nonlinear interaction.

In fact, the concept of the operator in linear quantum mechanics can still be used in nonlinear quantum theory. However, the majority of them are no longer linear operators, and thus certain properties of linear operators, such as the conjugate Hermitian of the momentum and the coordinate operators, are no longer required; $i\hbar\partial/\partial t$ no longer corresponds to the energy operator. Instead, nonlinear operators are constructed and used in the nonlinear quantum theory. For example, Equation (3.1) may be written as

$$i\hbar\frac{\partial\phi}{\partial t}=\hat{H}(\phi)\phi \tag{4.34}$$

where the Hamiltonian operator $\hat{H}$ is just a nonlinear operator, which depends on and is given by

$$\hat{H}(\phi)=-\frac{\hbar^2}{2m}\nabla^2-b|\phi|^2+V(\vec{r},t) \tag{4.35}$$

From Eqs.(4.34)-(4.35) we can deduce the Heisenberg matrix equation in nonlinear quantum mechanics, which is denoted in Eq.(2.84), where$K'(\phi)$ and $Q(\phi)$here are nonlinear operators. Thus, we may regardEq. (2.84) as the Heisenberg equation of the nonlinear operator in nonlinear quantum mechanicaltheory.

Otherwise,

$$\int_{-\infty}^{\infty}\phi^* x\phi d\tau\,,\ \frac{\partial}{\partial t}\int_{-\infty}^{\infty}\phi^* x\phi d\tau\,,\ \text{and}\ \int\phi^*\hat{H}\phi dx\ \text{ or}$$

in the nonlinear quantum theory, are no longer some average values of the physical quantities in linear quantum mechanics, but represent the position, velocity and acceleration of the mass center and energy of the microscopic particles, respectively, and have determinant values as shown above.

Thus,the presentations of mechanical quantities in the nonlinear quantum theory and their physical significances appear considerably in the variations relative to those in linear quantum mechanics. This has solved the difficulty arising from the average values, which are the possible values of mechanical quantities in linear quantum mechanics. However, the microscopic particlesin the nonlinear quantum theory have determinant mass, momentum and energy and obey universal conservation laws of mass, momentum, energy, and angular momentum as mentioned in Chapter 3, which cannot be obtained from linear quantum mechanics.

The variations mentioned above exhibit sufficiently that the nonlinear quantum theory is a new theory, which is completely different from the original quantum mechanics.

4.3. THE CHANGES OF UNCERTAINTY RELATION FOR MICROSCOPIC PARTICLES IN TWO THEORIES

4.3.1. The Correct Form of the Uncertainty Relation in Linear Quantum Mechanics

As is known, the microscopic particle does not have a determinant position, and it always disperses in total space in a wave form in linear quantum mechanics, hence, the position and momentum of the microscopic particles cannot be simultaneously determined. This last feature, being the well-known uncertainty relation, as mentioned above. The uncertainty relation is an important formula, and a serious problem in linear quantum mechanics, and one that has troubled many scientists over time. This may be an intrinsic property of microscopic particles, or an artifact of linear quantum mechanics, or a problem of the measuring instruments, which results in a long-lasting controversy in physics. Obviously, it is closely related to the basic features of microscopic particles, as mentioned above. When the nonlinear quantum theoryhas been established, in which the nature of the microscopic particles have considerable variations relative to that in linear quantum mechanics, then we expect that the uncertainty relation in the nonlinear quantum theory, including its physical significance and representative form, will be changed relative to those in linear quantum mechanics. These problems are here studied.

The uncertainty relation in linear quantum mechanics is shown in Eq.(4.33).However, is it a correct form?In order to determine its correctness we here again derive systematically the correct representation of the uncertainty relation [1-3]. In this case, we introduce the following function

$$I(\varsigma')=\int\left|\left(\varsigma'\Delta\hat{A}+i\Delta\hat{B}\right)\psi(\vec{r},t)\right|^{2}d\vec{r}\geq 0$$

or

$$\bar{\hat{F}}(\varsigma')=\int\psi^{*}(\vec{r},t)\hat{F}\left[\hat{A}(\vec{r},t),\hat{B}(\vec{r},t)\right]\psi(\vec{r},t)d\vec{r} \tag{4.36}$$

In the coordinate representation, $\hat{A}$ and $\hat{B}$ are operators of two physical quantities, for example, position and momentum, or energy and time, and satisfy the commutation relation. $\left[\hat{A},\hat{B}\right]=i\hat{C}$, $\psi\left(\vec{r},t\right)$ and $\psi^{*}\left(\vec{r},t\right)$ are wave functions of the microscopic particle satisfying the Schrödinger equation (1.8), and its conjugate equation, respectively. In Eq.(4.36) we can denote the operator by

$$\hat{F}=\left(\Delta\hat{A}\varsigma'+\Delta\hat{B}\right)^{2},$$

where $\Delta\hat{A}=\hat{A}-\overline{A}$, $\Delta\hat{B}=\hat{B}-\overline{B}$, $\overline{A}$ and $\overline{B}$ are average values of the physical quantities in the state denoted by $\psi(\vec{r},t)$, clearly $\hat{F}$ is an operator of physical quantity related to $\overline{A}$ and $\overline{B}$, ζ' and is a real parameter. After some simplifications, thus from Eq. (4.36)we can derive:

$$I=\overline{F}=\overline{\Delta\hat{A}^2}\zeta'^2+2\overline{\Delta\hat{A}\Delta\hat{B}}\zeta'+\overline{\Delta\hat{B}^2}\geq 0$$

or

$$\overline{\Delta\hat{A}^2}\zeta'^2+\overline{\hat{C}}\zeta'+\overline{\Delta\hat{B}^2}\geq 0 \tag{4.37}$$

Using mathematical identities, this can be written as

$$\overline{\Delta\hat{A}^2}\,\overline{\Delta\hat{B}^2}\geq\frac{\overline{\hat{C}}^2}{4} \tag{4.38}$$

The above representation is the uncertainty relation, which is often used in linear quantum mechanics. From the above derivation we see that the uncertainty relation was obtained based on the fundamental hypotheses of linear quantum mechanics, such as the properties of operators of the mechanical quantities, the state of particles as represented by the wave function, which satisfies the linear equation (1.8), the concept of average values of mechanical quantities and the commutation relations and eigenequation of operators. Therefore, we can conclude that the uncertainty relation in Eq. (4.38) is a result of linear quantum mechanics.

Since linear quantum mechanics only describes the wave nature of microscopic particles, the uncertainty relation is a result of the wave feature of microscopic particles, and it inherits the wave nature of microscopic particles. This is why its coordinate and momentum cannot be determined simultaneously. This is an essential interpretation for the uncertainty relation Eq. (4.38) in linear quantum mechanics. It is not related to measurement, but is closely related to basic assumptions of linear quantum mechanics. In other words, if linear quantum mechanics is used to describe the states of microscopic particles, then it is necessary to obtain the uncertainty relations for the microscopic particles. However, Eq.(4.33) or Eq.(4.38) is not the correct form; its correct form can be obtained, which is described as follows.

In fact, equation (4.37) can be written in the following form [16,17-22]:

$$\hat{F}=\overline{\Delta\hat{A}^2}\left(\zeta'+\frac{\overline{\Delta\hat{A}\Delta\hat{B}}}{\overline{\Delta\hat{A}^2}}\right)^2+\overline{\Delta\hat{B}^2}-\frac{\left(\overline{\Delta\hat{A}\Delta\hat{B}}\right)^2}{\overline{\Delta\hat{A}^2}}\geq 0 \tag{4.39}$$

or

$$\overline{\Delta\hat{A}^2}\left(\zeta'+\frac{\bar{\hat{C}}}{4\overline{\Delta\hat{A}^2}}\right)^2+\overline{\Delta\hat{B}^2}-\frac{\left(\bar{\hat{C}}\right)^2}{4\overline{\Delta\hat{A}^2}}\geq 0 \tag{4.40}$$

This shows that $\overline{\Delta\hat{A}^2}\neq 0$, and shows that if $\left(\overline{\Delta\hat{A}\Delta\hat{B}}\right)^2$ or $\bar{\hat{C}}^2/4$ is not zero, we cannot obtain Eq.(4.38) and $\overline{\Delta\hat{A}^2}\,\overline{\Delta\hat{B}^2}>\overline{\left(\Delta\hat{A}\Delta\hat{B}\right)^2}$ because when $\overline{\Delta\hat{A}^2}=0$, Eq. (4.40) does not hold. Therefore, $\overline{\left(\Delta\hat{A}^2\right)}\neq 0$ is a necessary condition for the uncertainty relation Eq. (4.38), $\overline{\Delta\hat{A}^2}$ can approach zero, but cannot be equal to zero. Therefore, in linear quantum mechanics, the right uncertainty relation should take the form:

$$\overline{\Delta\hat{A}^2\Delta\hat{B}^2}>\left(\bar{\hat{C}}^2\right)^2/4 \tag{4.41}$$

4.3.2. The Uncertainty Relation of Microscopic Particles in Nonlinear Quantum Mechanics

We now return to calculate the uncertainty relation of the microscopic particles described by the nonlinear quantum mechanics. In such a case, the microscopic particles are a soliton and have a determined position and momentum. Thus, we have reason to believe that the uncertainty relation in this theory should havea different form relative to Eq. (4.41) in linear quantum theory. Pang [34-40,131] derived this relation for position and momentum of a microscopic particle depicted by the nonlinear Schrödinger equation (3.1) with a solution, ϕ_s, as given in Eq. (3.8). The function $\phi_s\left(x',t'\right)$ is a square integral function localized at $x_0'\neq 0$ in the coordinate space. The Fourier transform of this function is given by

$$\phi_s\left(p',t'\right)=\frac{1}{\sqrt{2\pi}}\int_{-\infty}^{\infty}\phi_s\left(x',t'\right)e^{-ip'x'}dx' \tag{4.42}$$

Using Eq.(3.8), then the Fourier transform is explicitly represented as:

$$\phi_s\left(p,t'\right)=\sqrt{\frac{\pi}{2}}\operatorname{sech}\left[\frac{\pi}{4\sqrt{2}\eta}\left(p'-2\sqrt{2}\xi\right)\right]\exp\{i4(\eta^2+\xi^2-p'\xi/2\sqrt{2})t'-i\left(p'-2\sqrt{2}\xi\right)x_0'+i\theta\} \tag{4.43}$$

It shows that $\phi_s\left(p',t'\right)$ is also localized at p in momentum space. Equations (3.8) and (4.43) show that if the microscopic particle is localized in the shape of a soliton in coordinate space, then it is also localized in momentum space [17-22].

For convenience, we introduce here the normalization coefficient B_0 in Eqs. (3.8) and (4.43), then obviously $B_o^2 = \eta / 4\sqrt{2}$ .The position of the mass center of the microscopic particle, $\langle x' \rangle$, and its square, are given by:

$$\langle x' \rangle = \int_{-\infty}^{\infty} x' \left|\phi_s(x')\right|^2 dx', \quad \langle x'^2 \rangle = \int_{-\infty}^{\infty} x'^2 \left|\phi_s(x')\right|^2 dx'. \tag{4.44}$$

We thus find that:

$$\langle x' \rangle = 4\sqrt{2}\eta A_0^2 x_0', \quad \langle x'^2 \rangle = \frac{A_0^2 \pi^2}{12\sqrt{2}\eta} + 4\sqrt{2} A_0^2 \eta x_0'^2 \tag{4.45}$$

respectively. Similarly, the momentum of the mass center of the microscopic particle, $\langle p' \rangle$, and its square, $\langle p'^2 \rangle$, are given by:

$$\langle p' \rangle = \int_{-\infty}^{\infty} p' \left|\hat{\phi}_s(p')\right|^2 dp', \quad \langle p'^2 \rangle = \int_{-\infty}^{\infty} p'^2 \left|\hat{\phi}_s(p')\right|^2 dp' \tag{4.46}$$

which yield:

$$\langle p' \rangle = 16 A_0^2 \eta \xi, \quad \langle p'^2 \rangle = \frac{32\sqrt{2}}{3} A_0^2 \eta^3 + 32\sqrt{2} A_0^2 \eta \xi^3 \tag{4.47}$$

The standard deviations of position $\Delta x' = \sqrt{\langle x'^2 \rangle - \langle x' \rangle^2}$ and momentum are given [17-18] by:

$$\begin{aligned} (\Delta x')^2 &= A_0^2 \left[\frac{\pi^2}{12\eta} + 4\eta x_0'^2 \left(1 - 4\sqrt{2}\eta A_0^2\right) \right] = \frac{\pi^2}{96\eta^2}, \\ (\Delta p')^2 &= 32\sqrt{2} A_0^2 \left[\frac{1}{3}\eta^3 + \eta\xi^3 \left(1 - 4\sqrt{2}\eta A_0^2\right) \right] = \frac{8}{3}\eta^2, \end{aligned} \tag{4.48}$$

respectively. Thus, Pang [17-22] obtains the uncertainty relation between position and momentum for the microscopic particle depicted by the nonlinear Schrödinger equation in Eq. (3.1)

$$\Delta x' \Delta p' = \frac{\pi}{6} \tag{4.49}$$

This result is not related to the features of the microscopic particle (soliton), depicted by the nonlinear Schrödinger equation, because Eq. (4.49) has nothing to do with characteristic parameters of the nonlinear Schrödinger equation. πin Eq. (4.49) comes from the integral coefficient. For a quantized microscopic particle, and πin Eq. (4.49) should be replaced by $\pi\hbar$, because Eq. (4.42) is replaced by:

$$\phi_s(p,t) = \frac{1}{\sqrt{2\pi\hbar}} \int_{-\infty}^{\infty} dx \phi_s(x,t) e^{-ipx/\hbar}. \tag{4.50}$$

Thus, the corresponding uncertainty relation of the quantum microscopic particle is given by [17-22]

$$\Delta x \Delta p = \frac{\pi\hbar}{6} = \frac{h}{12} \tag{4.51}$$

This uncertainty principle also suggests that the position and momentum of the microscopic particle can be simultaneously determined to a certain degree. It is possible to roughly estimate the degree of the uncertainty of these physical quantities. If it is required that $\phi_s(x',t')$ in Eq. (3.8) or if $\phi_s(p',t')$ in Eq.(4.43) satisfies the admissibility condition *i.e.*, $\phi_s(0) \approx 0$, we choose Eq.(3.8). (In fact, in such a case we can get $\phi_s(0) \approx 10^{-7}$, thus the admissibility condition can be satisfied). We then get and $\Delta p' \approx 19.893$, according to (4.50) and (4.51). This result shows that the position and momentum of the microscopic particles in nonlinear quantum mechanics could be determined simultaneously within a certain approximation, that one of these cannot approach infinity.

Also, the uncertainty relation in Eq. (4.51) or Eq. (4.49) differs from the in Eq. (4.50) in linear quantum mechanics. However, the minimum value has not been obtained from both the solutions of the linear Schrödinger equation and neither has the experimental measurement been done up to now, except for the coherent and squeezed states of the microscopic particles. Therefore, we can draw the conclusion that the minimum uncertainty relationship is a result arising from the nonlinear effect, instead of a linear effect, which embodies concretely the wave-corpuscle duality of microscopic particle [17-22].

From this result, we see that when the microscopic particles satisfy $\Delta x \Delta p > \hbar/2$, then their motions obey the laws of linear quantum mechanics, and the particles are waves. When the uncertainty relationship of $\Delta x \Delta p = \hbar/12 \ or \ \pi/6$ is satisfied, the microscopic particles should be described by nonlinear quantum mechanics, and has a wave-corpuscle duality. If the position and momentum of the particles meets $\Delta x \Delta p = 0$, the particles have only a corpuscle feature, i.e., they are the classical particles. Therefore, the minimum uncertainty relation in Eqs. (4.49) and (4.51) clearly exhibit the wave-corpuscle duality of microscopic particles described by nonlinear quantum mechanics, a state which also bridges the gap between the classical and linear quantum mechanics. This is a very interesting result in physics.

4.3.3. The Uncertainty Relation in the Coherent States

Infact, we can represent one-quantum coherent state of the harmonic oscillator [20-23] by

$$|\alpha\rangle = \exp\left(\alpha\hat{b}^{+} - \alpha^{*}\hat{b}\right)|0\rangle = e^{-\alpha^2/2}\sum_{n=0}^{\infty}\frac{\alpha^n}{\sqrt{n-1}}\hat{b}^{+n}|0\rangle$$

in the number representation, which is a coherent superposition of a large number of microscopic particles (quanta). Thus:

$$\langle\alpha|\hat{x}|\alpha\rangle = \sqrt{\frac{\hbar}{2\omega m}}\left(\alpha+\alpha^{*}\right),\quad \langle\alpha|\hat{p}|\alpha\rangle = i\sqrt{\hbar m\omega}\left(\alpha-\alpha^{*}\right)$$

and

$$\langle\alpha|\hat{x}^2|\alpha\rangle = \frac{\hbar}{2\omega m}\left(\alpha^{*2}+\alpha^2+2\alpha\alpha^{*}+1\right),\quad \langle\alpha|\hat{p}^2|\alpha\rangle = \frac{\hbar\omega m}{2}\left(\alpha^{*2}+\alpha^2-2\alpha\alpha^{*}-1\right)$$

where

$$\hat{x} = \sqrt{\frac{\hbar}{2\omega m}}\left(\hat{b}+\hat{b}^{+}\right),\quad \hat{p}=i\sqrt{\frac{\hbar\omega m}{2}}\left(\hat{b}^{+}-\hat{b}\right),$$

and $\hat{b}^{+}\left(\hat{b}\right)$ is the creation (annihilation) operator of microscopic particle (quantum), α and $\alpha*$ are examples of unknown functions, ω is the frequency of the particle, and $m$is its mass. . Thus, we can get:

$$\left(\Delta x\right)^2 = \frac{\hbar}{2\omega m},\quad \left(\Delta p\right)^2 = \frac{\hbar\omega m}{2},\quad \langle\Delta x\rangle^2\langle\Delta p\rangle^2 = \frac{\hbar^2}{4} \tag{4.52}$$

and $\frac{\Delta x}{\Delta p} = \frac{1}{\omega m}$, or $\Delta p=(\omega m)\Delta x$

This is a minimum uncertainty relationship for the coherent state.

For the squeezed state of the microscopic particles:

$$|\beta\rangle = \exp\left[\beta\left(b^{+2}-b^2\right)\right]|0\rangle,$$

which is a two quanta coherent state, we can find that:

$$\langle\beta|\Delta x^2|\beta\rangle=\frac{\hbar}{2m\omega}e^{4\beta},\langle\beta|\Delta p^2|\beta\rangle=\frac{\hbar m\omega}{2}e^{-4\beta}$$

using a similar approach as the above. Here β is the squeezed coefficient [20-23] and $|\beta|<1$. Thus,

$$\Delta x\Delta p=\frac{\hbar}{2},\frac{\Delta x}{\Delta p}=\frac{1}{m\omega}e^{8\beta},$$

$$\text{or } \Delta p=\Delta x(\omega m)e^{-8\beta} \tag{4.53}$$

This shows that the squeezed state also meets the minimum uncertaintyrelationships,that the momentum of the microscopic particle (quantum) is squeezed in the two-quanta coherent state when compared to that in the one-quantum coherent state. On the other hand, the minimum uncertainty relationship of coherent state is not changed with the variation of time. In fact, according to the quantum theory, the coherent state of a harmonic oscillator at time $t$can be represented by:

$$|\alpha,t\rangle=e^{-i\hat{H}t}|\alpha\rangle=e^{-i\hbar\omega\left(\hat{b}+\hat{b}/2\right)t}|\alpha\rangle=e^{-i\hbar\omega t/2-|\alpha|^2/2}\sum_{n=0}^{\infty}\frac{\alpha^n e^{-i\hbar n\omega t}}{\sqrt{n!}}|n\rangle$$
$$=e^{-i\hbar\omega t/2}\left|\alpha e^{-i\hbar\omega t}\right\rangle,$$

where $|n\rangle=(b^+)^n|0\rangle$.

This shows that the shape of a coherent state can be retained during its motion, which is the same as that of a microscopic particle (soliton) in nonlinear quantum mechanics. The mean position of the particle in the time-dependent coherent state is:

$$\langle\alpha,t|x|\alpha,t\rangle=\langle\alpha|e^{iHt/h}xe^{-iHt/h}|\alpha\rangle=\left\langle\alpha\left|x-\frac{it}{h}[x,H]+\frac{(-it)^2}{2!\hbar^2}[[x,H],H]+\cdots\right|\alpha\right\rangle= \tag{4.54}$$
$$\left\langle\alpha\left|x+\frac{pt}{m}-\frac{1}{2!}t^2\omega^2x+\cdots\right|\alpha\right\rangle=\left\langle\alpha\left|x\cos\omega t+\frac{p}{m\omega}\sin\omega t\right|\alpha\right\rangle=\sqrt{\frac{2\hbar}{m\omega}}|\alpha|\cos(\omega t+\theta)$$

where

$$\theta=\tan^{-1}\left(\frac{y}{x}\right),\ x+iy=\alpha,\ [x,H]=\frac{i\hbar p}{m},\quad [p,H]=-i\hbar m\omega^2x.$$

Then, comparing Eq. (4.54) with the solution of a classical harmonic oscillator:

$$x=\sqrt{\frac{2E}{m\omega^2}}\cos(\omega t+\theta),\quad E=\frac{p^2}{2m}+\frac{1}{2}m\omega^2x^2$$

we find that they are similar with:

$$E=\hbar\omega\alpha^2=\langle\alpha|H|\alpha\rangle-\langle 0|H|0\rangle,\quad H=\hbar\omega\left(b^+b+\frac{1}{2}\right).$$

Thus, we can say that the mass center of the coherent state-packet indeed obeys the classical laws of motion, which are also the same as the laws of motion of microscopic particles described by the nonlinear Schrödinger equation discussed in Eqs. (3.55) - (3.59).

At the same time, we can also obtain

$$\langle\alpha,t|p|\alpha,t\rangle=-\sqrt{2m\hbar\omega}|\alpha|\sin(\omega t+\theta),\langle\alpha,t|x^2|\alpha,t\rangle=\frac{2\hbar}{\omega m}\left[|\alpha|^2\cos^2(\omega t+\theta)+\frac{1}{4}\right]$$

$$\langle\alpha,t|p^2|\alpha,t\rangle=2m\hbar\omega\left[|\alpha|^2\sin^2(\omega t+\theta)+\frac{1}{4}\right]$$

Thus, a time-dependent uncertainty relationship can be obtained and represented as:

$$[\Delta x(t)]^2=\frac{\hbar}{2\omega m},\quad [\Delta p(t)]^2=\frac{1}{2}m\omega\hbar,\Delta x(t)\Delta p(t)=\frac{h}{2} \tag{4.55}$$

This is the same with Eq. (4.53). It shows that the minimal uncertainty principle for the coherent state is retained at all times, *i.e.*, the uncertainty relation does not change with time, *t*. This result is also the same with that in nonlinear quantum mechanics, but we do not see this rule in linear quantum mechanics.

The above results show that both one-quantum and two-quanta coherent states satisfy the minimal uncertainty principle. This is the same as that of microscopic particles in nonlinear quantum mechanics. This means that coherent and squeezed states are a nonlinear quantum state, and that the coherence and squeezing of quanta are a kind of nonlinear quantum effect. Just so, are the states of microscopic particles described by the coherent states in the nonlinear quantum systems, such as the Davydov's wave functions [24] , both $\mathrm{ID}_1>$ and $\mathrm{ID}_2>$, and Pang's wave function of exciton-solitons [25-28], in protein molecules and acetanilide; the wave function of proton transfer in hydrogen-bonded systems and the BCS's wave function in superconductors [29], etc., which are all the coherent states.

Hence, the coherence of particles does not belong to the systems described by linear quantum mechanics, because the coherent state cannot be obtained by superposition of linear waves, such as plane waves, a de Broglie wave, or a Bloch wave. Then the minimal uncertainty relation of Eq. (4.51), as well as that of Eqs.(4.53) and (4.55) are only applicable to microscopic particles described by nonlinear quantum mechanics. Thus, it reflects the wave-corpuscle duality of the microscopic particles.

Also, the above results indicate not only the essences of nonlinear quantum effects of the coherent state or squeezing state, but also that the minimal uncertainty relationship is an intrinsic feature of nonlinear quantum mechanics systems, including both the coherent and squeezing states.

Pang et al., [25-28]also calculated the uncertainty relationship and quantum fluctuations and studied their properties in nonlinear electron-phonon systems based on the Holstein model by a new Ansatz including the correlations among the one-phonon coherent and two-phonon squeezing states and the polaron state proposed by Pang. Many interesting results were obtained, such as the minimum uncertainty relationship, which is correct and related to the properties of the microscopic particles, which will be studied in detail in Chapter 7. The results enhanced the understanding of the significance and essences of the minimum uncertainty relationship.

First of all, we give the quantum fluctuation effects of the coherent state. The mean number of quanta in a coherent state is attained by:

$$\bar{n}=\langle\alpha|\hat{N}|\alpha\rangle=\langle\alpha|\hat{b}^{+}\hat{b}|\alpha\rangle=\alpha^{2},\quad \langle\alpha|\hat{N}^{2}|\alpha\rangle=|\alpha|^{4}+|\alpha|^{2}$$

Therefore, the fluctuation of the quantum in the coherent state is:

$$\Delta n=\sqrt{\langle\alpha|\hat{N}^{2}|\alpha\rangle-\left(\langle\alpha|\hat{N}|\alpha\rangle\right)^{2}}=|\alpha|.$$

which leads to $\frac{\Delta n}{\bar{n}}=\frac{1}{|\alpha|}<<1$.

It is thus obvious that the fluctuation of the quantum in the coherent state is very small. This means that the coherent state is quite close to the feature of solitons and solitary waves, which are stable in the nonlinear quantum mechanics. Therefore, the above variation of the uncertainty relationship exhibitsclearly that the nonlinear quantum theory is a new theory, which is completely different from the linear quantum mechanics.

4.4. The Variations of Natures of Microscopic Particles Arising from the Nonlinear Interaction

It is known from Eqs.(1.9)-(1.18), the microscopic particlesonly have a wave nature, not a corpuscle nature, which is an intrinsic nature of microscopic particles described by the linear Schrödinger equation and cannot be changed in linear quantum mechanics.

However, the natures of microscopic particles described by the nonlinear Schrödinger equation were changed. They have both a wave nature and a corpuscle nature, i.e., wave-corpuscle duality. This means that the natures of microscopic particles have essential variations in the nonlinear quantum theory relative to those in linear quantum mechanics. These problems were discussed deeply and in detail, as mentioned above inchapter 3. In this investigation the wave-corpuscle duality of microscopic particles are clearly exhibited, and their properties are summarized as follows.

(1).The solutions of the nonlinear Schrödinger equation with different potentials of V(x)=c, $V(x')=\alpha x'$, $V(x')=\alpha x'+c$,and in Eq.(2.66) at $A(\phi)=0$ all possess a wave-corpuscle duality as described in Chapter 3 [18-22], which is completely different from those in Eqs.(1.13)-(1.15) in linear quantum mechanics. We see from these studies that the solutions of the nonlinear Schrödinger equation in these cases can be always represented by the form of $\phi(x,t)=\varphi(x,t)\exp(\theta(x,t))$, i.e., it consists of an envelope and carrier waves, where $\varphi(x,t)$ is a bell-type soliton, which has a determinant amplitude, width, size, group velocity v_e, mass, momentum, and energy, thus it represents the mass center of the microscopic particle, which always exists, although the externally applied potentials were frequently changed, which can be demonstrated from the results inEqs. (3.8), (3.24), (3.25), (3.28),(3.30), and (3.31). These results clearlyexhibit the corpuscle nature of microscopic particles. But $\exp(\theta(x,t))$ is a carrier wave with a phase speed v_c, thus it displays the wave nature of the particle. Their organic combination and synthesis as shown in Figure 3.1 make the microscopic particles localize and have a wave–corpuscle duality, which the microscopic particles in the linear quantum mechanics do not have. Obviously, this nature is due to the nonlinear interaction because the dispersion effect of kinetic energy terms in Eq. (2.66) is now suppressed by the nonlinear interaction, thus the wave-corpuscle duality of the microscopic particle appears. In this case, the external potentials only influence the wave form, phase, and velocity of the particles, but cannot vary the wave-corpuscle duality. Thus, we can confirm that the nonlinear Schrödinger equation (2.66) at $A(\phi)=0$ really shows the wave-corpuscle duality of microscopic particles, i.e., it is a correct equation, which can be used to describe correctly the states of microscopic particles in the nonlinear quantum systems. These results indicated clearly the necessity and correctness of building and developing the nonlinear quantum theory based on the nonlinear Schrödinger equation; in other words, it manifested that Pang's nonlinear quantum mechanics is correct [18-22].

(2).The localization and stability of microscopic particles described by the nonlinear Schrödinger equationwere justified, which are proved by the extreme theory of effective potentialand the minimum theory of energy of the system, respectively. The former indicated that the microscopic particle is in the binding state with a negative energy in this case, the latter exhibits that the particle described by the nonlinear Schrödinger equation has minimal energy. Thus, the microscopic particles described by the nonlinear Schrödinger equation are localized and stable. Just so, the microscopic particles possess a corpuscle nature as described in Chapter 3 because macroscopic particles in classical physicsalso have these properties.

(3).The corpuscle feature of microscopic particles described by the nonlinear Schrödinger equation, such as the determinant mass, momentum, and energy to satisfy the conservative laws of mass, momentum, and energy, were demonstrated using the equation's solutions, formulae of the mass, momentum, and energy and the Noether theorem, respectively. This means that the nonlinear Schrödinger equation is a correct dynamic equation of microscopic particles in nonlinear quantum mechanics.

(4).From the nonlinear Schrödinger equation, we derived the Newton's law of motion, and the Hamilton and Euler-Lagrange equations of microscopic particles. This means that the nonlinear Schrödinger equation, Newton's equation of motion, and the Hamilton and Euler-Lagrange equations can be used simultaneously to describe the dynamic properties of macroscopic particle. In other words, Newton's equation of motion, the Hamiltonequation,the

Euler-Lagrange equations,andthe nonlinear Schrödinger equation are identical in their description of motion of microscopic particles in the nonlinear quantum theory. This is a singular and unique peculiarity of nonlinear quantum theory, which does not appear in both linear quantum mechanics and classical mechanics. It is well known that Newton's equation of motion, and the Hamilton and Euler-Lagrange equations were frequently used to describe the laws of motion of macroscopic particles in classical mechanics. This implies that the nonlinear Schrödinger equation and corresponding microscopic particles possess a classical feature.

(5).We argued the properties and rules of collision of microscopic particles in the nonlinear theory, using the results of shifts of mass center and phase obtained from the inverse scattering method of the fourth–order Runge-Kutta number simulation method and an analytic way for the nonlinear Schrödinger equation, with positive and negative nonlinear interaction coefficients. The results obtained show that the collision of microscopic particles satisfies the collision rule of classical particles, but there is a high peak in the collision process, which is a result of superposition of solitary waves of two microscopic particles. This implies that collision of the microscopic particles exhibited that they have both corpuscle and wave features.

(6).We found out the minimal uncertainty relation of position and momentum of microscopic particles, described by the nonlinear quantum mechanics, by calculating the uncertainty of position and momentum by the traditional method. We studied further the features of quantum fluctuation of microscopic particles described by the quantum nonlinear Schrödinger equation and got that the uncertainties of quantum fluctuations of the position and momentum of quantum microscopic particles meet the minimal uncertainty relation. These results denoted that nonlinear quantum mechanics is not both linear quantum and classical mechanics; it bridges the gap between the classical and linear quantum mechanics. This is a very interesting result in physics.

(7).We also examined the wave nature of microscopic particles described by the nonlinear Schrödinger equation, in which we found that the solutions of the nonlinear Schrödinger equation, included the wavevector, frequency, velocity and amplitudein its envelope and carrier, and observed the reflection, transmission and diffractionphenomena of microscopic particles at interfaces, but its rules of change are different from those of both classical wave, KdV solitary wave, and linear wave. Thus, we concluded that microscopic particles described by nonlinear quantum mechanics are not both a classical wave, KdV solitary wave and a linear wave.

The above studied results showed clearly that the rules of motion and natures of microscopic particles described by the nonlinear Schrödinger equation are, in essence, different from those in linear quantum mechanics and classical mechanics. It is truly a new theory and bridges the gap between classical mechanics and linear quantum mechanics. Therefore, Pang's nonlinear quantum mechanics has considerable physical significance in science.

4.5. THE ALTERNATIONS OF EIGENVALUE PROBLEMS IN TWO THEORIES

4.5.1. In Linear Quantum Mechanics

The eigenequation for operator $\hat{A}$ is denoted by Eq.(4.31),where ϕ_L or $|\phi_L\rangle$ is its eigenfunction, and A is its eigenvalue. Thus, we can determine the eigenvalues of the operator $\hat{A}$ and the corresponding eigenfunctions from Eq.(4.31).

Owing to the fact that the observable physical quantities are represented by the Hermitian operator [1-3], then they have real eigenvalues. Their eigenfunctions are also orthogonal to each other,i.e., $\int \phi_1^* \phi_2 d\tau = 0$, where ϕ_1 and ϕ_2 are the eigenfunctions corresponding to the eigenvalues A_1 and A_2, respectively.

These eigenfunctions constituted an orthogonal and complete set, $\{\phi_n(x)\}$ or $\{\phi_\lambda\}$. Then an arbitrary wave function $\phi(x)$ can be denoted by the superposition of the complete set, $\{\phi_n(x)\}$ or $\{\phi_\lambda\}$, i.e., $\phi(x) = \sum_n c_n \phi_n(x)$ for discrete eigenvalues, or $\phi(x) = \int c_\lambda \phi_\lambda(x) d\lambda$ for continuous eigenvalues, where $\phi_\lambda(x)$ is the eigenfunction of Hermitian operator $\hat{A}$ with continuous eigenvalue λ.

Meanwhile, the eigenfunctions of the Hermitian operator also have a closure feature, i.e., $\sum_n \phi_n^*(x')\phi_n(x) = \delta(x'-x)$ for discrete eigenvalues and $\int \phi_\lambda^*(x')\phi_\lambda(x) d\lambda = \delta(x'-x)$ for continuous eigenvalues. These results are just the features of the eigenvalue problem of operators of mechanical quantities in linear quantum mechanics.

4.5.2. In the Nonlinear Quantum Theory

The above eigenvalue theory holds no longer in the nonlinear quantum theory due to the existence of nonlinear interaction, in which there is still an eigenvalue problem for the operators, but their properties are changed. We here give mainly the method solving the eigenvalues and eigenfunction of the Hamiltonian operator, which is very useful and significant in the nonlinear quantum theory. For example, the nonlinear Schrödinger equation (2.66) can be written as [18-22]:

$$i\hbar \frac{\partial \phi}{\partial t} = \hat{H}(\phi)\phi$$

where $\hat{H}(\phi) = -\frac{\hbar^2}{2m}\nabla^2 - b|\phi|^2 + V(\vec{r},t)$

is just the Hamiltonian operator of the system. Its eigenequation can be denoted as

$$\hat{H}(\varphi')\varphi' = E\varphi' \tag{4.56a}$$

or

$$E\varphi'(\vec{r}) = -\frac{\hbar^2}{2m}\nabla^2\varphi'(\vec{r}) + V(\vec{r})\varphi'(\vec{r}) - b\left|\varphi'(\vec{r})\right|^2 \varphi'(\vec{r}), \tag{4.56b}$$

where

$$\varphi'(\vec{r}) = \phi(\vec{r},t)e^{iEt/\hbar} \text{ (or } \phi(\vec{r},t) = \varphi'(\vec{r})e^{-iEt/\hbar}\text{)} \tag{4.57}$$

its eigenfunction with eigenenergy E [18-22]. Obviously, the Hamiltonian operator can be expressed as

$$\hat{H} = -\frac{\hbar^2}{2m}\nabla^2 + V(\vec{r}) - b\left|\varphi'(\vec{r})\right|^2 = \hat{H}_0 - b\rho(\vec{r}) \tag{4.58}$$

where $\hat{H}_0 = -\frac{\hbar^2}{2m}\nabla^2 + V(\vec{r})$, and $\rho(\vec{r}) = \left|\varphi'(\vec{r})\right|^2 = \left|\phi(\vec{r},t)\right|^2$

Compared with that in Eq.(1.10), in linear quantum mechanics, we find that the eigenenergy of the Hamiltonian operator, or the system, which can be represented as $E = E_0 - b\rho(\vec{r})$ where E_0 is the eigenenergy of $\hat{H}_0$ in the the linear case of the system in linear quantum mechanics, is reduced about $b\rho(\vec{r})$ relative to a linear case due to the nonlinear interaction in the nonlinear quantum theory. The reduced energy, $b\rho(\vec{r})$, is just the formed energy or binding energy of the soliton, which has come from the microscopic particle under action of nonlinear interaction $b\left|\phi(\vec{r},t)\right|^2 \phi(\vec{r},t)$. This result shows clearly that the microscopic particles are the solitons and have a wave–corpuscle duality, only if the nonlinear interaction is notequal to zero, or else, they have only a wave feature, i.e., there is no localization of microscopic particles without nonlinear interaction, no matter how the potentials in Eqs.(1.8)-(1.10) change. This is a very important conclusion in physics and nonlinear science; it embodies and exhibits clearly the importance of nonlinear interaction in description of the nature of microscopic particles. This is the linearSchrödinger equation and linear quantum mechanics must be replaced by nonlinear Schrödinger equation and nonlinear quantum theory for the wave-corpuscle duality of microscopic particles. This also indicatesclearly the reasons why linear quantum mechanics cannot give the wave-corpuscle duality of the microscopic particle.

On the other hand, the eigenenergy and eigenfunction of the Hamiltonian operator can be obtained from associatively solving the time-independent nonlinearSchrödinger equation (4.56) with $E = E_0 - b\rho(\vec{r})$ and

$$E=\frac{1}{r_0}\int[-\frac{\hbar^2}{2m}\varphi'^*\nabla^2\varphi'(x)+V(\vec{r})\varphi'^*\varphi'^2-b/2|\varphi'|^2\varphi'^*\varphi']dx$$
$$=\frac{1}{r_0}\int[\frac{\hbar^2}{2m}|\nabla\varphi'|^2+V(\vec{r})|\varphi'|^2-b/2|\varphi'|^2|\varphi'|^2]dx \tag{4.59}$$

which is gained by left-multiplying both sides of Eq. (4.56) by φ'^* and integrating it with respect to x. Therefore, the method solving the eigenenergy and eigenfunction of the Hamiltonian operator are different from those in linear quantum mechanics, mentioned above, due to the nonlinear interaction. This is the new way to find the eigenfunctions and eigenvalues of the Hamiltonian operator of a single particle in the nonlinear quantum theory.

4.5.3. Eigenenergies of HamiltonianOperator in the Systems with Many Models of Motion

It is, in general, quite difficult to calculate the eigenfunctions and eigenvalues of the Hamiltonian operator of the systems with many particles or many models of motion using the above method in nonlinear quantum mechanics.

An alternative method is to quantize the Hamiltonian of the systems, which is expressed in the particle number or second quantization picture through the quantization of the wave functions by the creation and annihilation operators [30-36].

In this method, the first step is to split the dynamic equation in order to express the motions of many modes or many particles in the systems. Pang [30-36] replaced the nonlinear Schrödinger equation (2.66) at $A(\phi)=0$ of a single particle by a discrete form:

$$i\hbar\frac{\partial\phi_j}{\partial t}=-\frac{\hbar^2}{2mr_0^2}(\phi_{j+1}-2\phi_j+\phi_{j-1})-b|\phi_j|^2\phi_j+V(j,t)\phi_j,(j=1,2,3...J) \tag{4.60}$$

This method displays how the continuous field is replaced by a lattice field, where r_0 is a space between two neighboring lattice points, j labels the position of lattice points, and J is the total number of lattice points in the lattice field. Thus a discretely vector nonlinear Schrödinger equation for the many particles or many modes of motion can be represented as:

$$(i\hbar\frac{\partial}{\partial t}-\frac{\hbar^2}{mr_0^2}-V(j,t))\bar{\phi}=\varepsilon M\bar{\phi}-b\,diag.(|\phi_1|^2,|\phi_2|^2....|\phi_\alpha|^2)\bar{\phi} \tag{4.61}$$

where $\bar{\phi}(j,t)$ is the column vector, i.e., $\bar{\phi}(j,t)=Col.(\phi_1,\phi_2...\phi_\alpha)$. Equation (4.61) is a vector nonlinear Schrödinger equation with α modes. In Eq. (4.61), b is a nonlinear parameter, M= [M_{nl}]is a $\alpha\times\alpha$ real symmetric dispersion matrix, $\varepsilon=\hbar^2/2mr_0^2$. Here, n and l are integers denoting the number of modes of motion. The Hamiltonian and the particle number corresponding to Eq. (4.61) are represented, respectively, by

$$H=\sum_{n=1}^{\alpha}\left(\hbar\omega_0|\phi_n|^2-\frac{1}{2}b|\phi_n|^4\right)-\varepsilon\sum_{n\neq l}^{\alpha}M_{nl}\phi_n\phi_l\,, \tag{4.62a}$$

and

$$N=\sum_{N=L}^{\alpha}|\phi_n|^2 \tag{4.62b}$$

where $\hbar\omega_0=\hbar^2/2mr_0^2+V(j,t)$.

We have assumed that $V(j,t)$ are independent of j and t . In the canonical second quantization theory, the complex amplitudes (ϕ_n^* and ϕ_n) can be replaced by boson creation and annihilation operators ($\hat{B}_n^+$ and $\hat{B}_n$) in the number representation according to the quantized method [37-40]. If $|m_n>$ is an eigenfunction of a particular mode, then:

$$\hat{B}_n^+|m_n\rangle=\sqrt{m_n+1}\,|m_n+1>,\hat{B}_n\,|m_n>=\sqrt{m_n}\,|m_n-1>\text{ and }\hat{B}_n\,|0>=0.$$

Since no particular ordering is specified in Eq. (4.62), thus we can denote them using the averages:

$$|\phi_n|^2\rightarrow\frac{1}{2}(\hat{B}_n^+\hat{B}_n+\hat{B}_n\hat{B}_n^+)$$

and

$$|\phi_n|^4\rightarrow\frac{1}{6}(\hat{B}_n^+\hat{B}_n^+\hat{B}_n\hat{B}_n+\hat{B}_n^+\hat{B}_n\hat{B}_n^+\hat{B}_n+\hat{B}_n^+\hat{B}_n\hat{B}_n\hat{B}_n^+ +\hat{B}_n\hat{B}_n^+\hat{B}_n\hat{B}_n^+ +\hat{B}_n\hat{B}_n\hat{B}_n^+\hat{B}_n^+ +\hat{B}_n\hat{B}_n^+\hat{B}_n^+\hat{B}_n)$$

with the Boson commutation rule:

$$\hat{B}_n\hat{B}_n^+-\hat{B}_n^+\hat{B}_n=1\,,$$

Equation (4. 62) then becomes:

$$H=\sum_{n=1}^{\alpha}[\left(\hbar\omega_0-\frac{1}{2}b\right)(\hat{B}_n^+\hat{B}_n+1/2)-\frac{1}{2}b\hat{B}_n^+\hat{B}_n\hat{B}_n^+\hat{B}_n]-\varepsilon\sum_{n\neq l}^{\alpha}(M_{nl}\hat{B}_n^+\hat{B}_l) \tag{4.63}$$

$$\hat{N}=\sum_{n=1}^{\alpha}(\hat{B}_n^+\hat{B}_n+\frac{1}{2}) \tag{4.64}$$

From now on, we will use the notation [m_1 , m_2 ,... m_α] to denote the products of number states $|m_1>|m_2>...|m_\alpha>$. Thus, stationary states of the vector nonlinear Schrödinger equation (4.60) must be eigenfunctions of both $\hat{N}$ and $\hat{H}$. Consider any m-quantum state (i.e., the mth excited level, m= $m_1+m_2+...m_j$), with m < α . An eigenfunction can be established by

$$|\phi_m\rangle = C_1\,[\mathrm{m},0,0,\ldots,0]+\ldots+C_2\,[0,\mathrm{m},0,0,\ldots,0]+\ldots+$$
$$C_i\,[0,0,0,\ldots,\mathrm{m}\,,]+\ldots+C_{i+1}\,[\mathrm{m}-1,\mathrm{m},0,0,\ldots,0]+\ldots+C_p\,[0,0,0,\ldots,0,1,1\ldots,1]. \quad (4.65)$$

The number of terms in Eq. (4.65) is equal to the number of ways that m quanta can be placed on α sites, and is given by P= $\frac{(m+\alpha-1)!}{m!(\alpha-1)!}$. The wave function $|\phi_m>$ in Eq. (4.65) is an eigenfunction of $\hat{N}$ for any values of C_α'. Thus, we are free to choose these coefficients so that:

$$\hat{H}\,|\phi_m>=\mathrm{E}|\phi_m>. \quad (4.66)$$

To solve equation (4.66) requires that the column vector C=Col. ($C_1, C_2,\ldots C_p$) satisfies the matrix equation:

$$|\,\mathrm{H}-\mathrm{IE}\,|\mathrm{C}=0 \quad (4.67)$$

where H is a pxp symmetric matrix with real elements. I is a pxp identity matrix, E is just the eigenenergy we need. Then equation (4.66) is the eigenvalue equation of the quantum Hamiltonian operator (4.63) of the systems. We can find the eigenenergy spectra E_m of the systems from Eq. (4.67), if the values of the parameters, ε and ω_0, are given. Scott et al., [37-40] and Pang et al., [30-36] and Chen et al., [41-49] used this method to calculate the energy-spectra of nonlinear excitations (quanta) for many nonlinear systems, for example, small molecules or organic molecular crystals and biomolecules. The results obtained were consistent with experimental data.

For the system with two degrees of freedom, we now discuss the eigenenergy spectra and eigenfunctions of the system with two degrees of freedom (two modes of motion), or two of the same particles by the above method, which were studied by Scott et al., [37-40] and Pang [30-36]. In this case, Eq. (4.61) takes the form

$$(i\hbar\frac{d}{dt}-\hbar\omega_0)\begin{pmatrix}\phi_1\\ \phi_2\end{pmatrix}+\begin{pmatrix}b|\phi_1|^2 & \varepsilon\\ \varepsilon & b|\phi_2|^2\end{pmatrix}\begin{pmatrix}\phi_1\\ \phi_2\end{pmatrix}=0 \quad (4.68)$$

The corresponding $\hat{N}$ and $\hat{H}$ in such a case are given, respectively, by

$$\hat{H} = \left(\hbar\omega_0 - \frac{1}{2}b\right)\hat{N} - \frac{1}{2}b(\hat{B}_1^+\hat{B}_1\hat{B}_1^+\hat{B}_1 + \hat{B}_2^+\hat{B}_2\hat{B}_2^+\hat{B}_2) - \varepsilon(\hat{B}_1^+\hat{B}_2 + \hat{B}_1\hat{B}_2^+) \tag{4.69}$$

$$\hat{N} = \hat{B}_1^+\hat{B}_1^+ + \hat{B}_2^+\hat{B}_2^+ + 1 \tag{4.70}$$

We seek the eigenfunction for both the operators. For the ground state, $|\Phi_0\rangle = |0\rangle|0\rangle$, which is the product of the ground state wavefunctions of the two degrees of freedom, we have:

$$\hat{N}|\Phi_0\rangle = |\Phi_0\rangle,\ \hat{H}|\Phi_0\rangle = E_o|\Phi_0\rangle, E_0 = \hbar\omega_0 - b/2$$

For the first excited state, we have

$$|\Phi_1\rangle = C_1|1\rangle|0\rangle + C_2|0\rangle|1\rangle$$

With $|C_1|^2+|C_2|^2 = 1$. Thus, $\hat{N}|\Phi_1\rangle = 2|\Phi_1\rangle$. From there we can get:

$$\begin{pmatrix} (2\hbar\omega_0 - E_1) - b/2 & \varepsilon \\ \varepsilon & (2\hbar\omega_0 - E_1) - b/2 \end{pmatrix}\begin{pmatrix} c_1 \\ c_2 \end{pmatrix} = 0$$

The first excited state splits into a symmetric and an antisymmetric state, which have corresponding eigenwave functions and Eigenenergies:

$$|\Phi_{1s}\rangle = \frac{1}{\sqrt{2}}(|1\rangle|0\rangle + |0\rangle|1\rangle), E_{1s} = E_0 + \hbar\omega_0 - b - \varepsilon, |\Phi_{1a}\rangle = \frac{1}{\sqrt{2}}(|1\rangle|0\rangle - |0\rangle|1\rangle), E_{1a} = E_0 + \hbar\omega_0 - b + \varepsilon$$

respectively. For the second and third or n^{th} excited states, corresponding eigenwave functions and eigenenergies can be also found in accordance with the above methods.

4.6. The Eigenvalue Problems ofNonlinear SchrödingerEquation

4.6.1. Representations of Eigenequations

As it is known, the nonlinear dynamic equations can, in general, be written either in the Lax form or in the Hamilton form of a compatibility condition [11]. From the Lax equation (4.5) we know that two linear operators $\hat{L}$ and $\hat{B}$, corresponding to Eq.(3.1), satisfy Eq.(4.6) and Eq.(4.7),respectively, i.e., Eq.(4.6) is the eigenequation of the linear operator L ,in which k is its eigenvalues, the evolution of time of the eigenfunction is determined by $\hat{B}$ through Eq. (4.7). From Eq.(4.8) we know that the eigenvalues of L depend on the solution ϕ of nonlinear Schrödinger equation (3.1). This means that to find the eigenequation of $\hat{L}$ is related to Eq.(3.1).

Thus, we may say that, finding the eigenequation of $\hat{L}$ may be called the eigenvalue problem of the nonlinear Schrödinger equation(3.1), which has the feature of the Galilei invariance in the nonlinear quantum theory. Obviously, solving the eigenvalue problem of the nonlinear Schrödinger equation is finally decided by Eq.(4.6) or the Zakharov-Shabat (ZS) equation(4.12) [11-13]. Satsuma and Yajima [50]studied, in detail, the properties of Eq.(4.12) or (4.6). We now discuss the properties of the eigenvalue problem of the nonlinear Schrödinger equation (3.1), which were obtained by Satsuma and Yajima [50].

In fact, if we differentiate Eq. (4.6) and multiply the resulting equation by *i*, we can arrive at:

$$i\left(\Psi\frac{dk}{dt'}+k\frac{d\Psi}{dt'}\right)=i\left(\hat{L}\Psi_{t'}+\frac{\partial\hat{L}}{\partial t'}\Psi\right)=iL\Psi_{t'}+\left[\hat{B}\hat{L}-\hat{L}\hat{B}\right]\Psi=\hat{L}\left(i\Psi_{t}-\hat{B}\Psi\right)+k\hat{B}\Psi$$

where Ψ corresponds to ψ in Eq.(4.12). From the above, we get $i\Psi\dfrac{dk}{dt'}=0$ due to $\Psi\neq 0$. Thus, k or λ is a time-independent eigenvalue, and Eq. (4.6) is simply the eigenequation corresponding to the nonlinear Schrödinger equation (3.1).

In such a case, the eigenequation corresponding to the nonlinear Schrödinger equation (3.1) is the Zakharov-Shabat equation (4.11) or Eq. (4.12), its eigenvalue is k orλ, through Eq. (4.12), the eigenvalue k orλ can also be determined, and itseigenfunction satisfies Eqs. (4.6) - (4.7) or Eqs. (4.12) - (4.13). Thus, the soliton solution of the nonlinear Schrödinger equation (3.1), corresponding to these eigenvalues and eigenfunctions, can be found using the inverse scattering method and is denoted by Eq.(3.8). We now study the eigenfunction of the nonlinear Schrödinger equation (3.1) and its properties using Satsuma and Yajima's method [50].

4.6.2. Features of Eigenfunctions

(1).Orthonormal features: Clearly, if the eigenfunction satisfied the boundary condition, $\psi = 0$ at $|x| \to \infty$, its eigenvalues are discrete and denoted as $\lambda_1, \lambda_2, ..., \lambda_N$, with the corresponding eigenfunctions denoted by $\psi_1, \psi_2, ..., \psi_N$. For a given eigenfunction, $\psi_n(x')$, the eigenequation (4.12) reads

$$i\frac{d\psi_n(x')}{dx'} + \Phi(x')\psi_n(x') = \lambda_n\sigma_3\psi_n(x'), (n = 1, 2, ..., N) \quad (4.71)$$

where Satsuma and Yajima [50] expressed $\Phi(x')$ by Pauli's spin matrices σ_1 and σ_2, it is

$$\Phi(x') = \Re\,[\phi(x')]\sigma_1 - \Im\,[\phi(x')]\sigma_2 \quad (4.72)$$

Multiplying Eq. (4.71) by σ_2, from the left, and transposing the resulting equation, Satsuma and Yajima [159] resulted in

$$-i\frac{d\psi_m^T}{dx'}\sigma_2 - \psi_m^T\Phi^*\sigma_2 = i\lambda_m\psi_m^T\sigma_1 \quad (4.73)$$

where the superscript T denotes a transpose.
Multiplying the above equation by ψ_n from right and Eq. (4.71) by $\psi_m^T\ \sigma_2$ from the left and subtracting one from the other, Satsuma and Yajima obtained the following equation

$$(\lambda_n - \lambda_m)\int_{-\infty}^{\infty}\psi_m^T\sigma_1\psi_n dx' = 0$$

The boundary condition, $\psi_n, \psi_m \to 0$ as $|x'| \to \infty$, is used in this calculation. Thus, the following orthonormal condition can be obtained:

$$\int_{-\infty}^{\infty}\psi_m^T\sigma_1\psi_n dx' = \delta_{nm} \quad (4.74)$$

(2)Symmetry features. Satsuma and Yajima [50] further demonstrated that Eq. (4.71) has the following symmetry properties.

If satisfies $\phi(-x') = \phi^*(x')$, then replacing x' by - x' in Eq. (4.71) and multiplying it by σ_2 from the left, we get:

$$i\frac{d}{dx'}[\sigma_2\psi_n(-x')]+\Phi(x')[\sigma_2\psi_n(-x')]=\lambda_n\sigma_3[\sigma_2\psi_n(-x')]$$

Since $\sigma_2\psi_n(-x')$ is also an eigenfunction associated with λ_n, its behavior resembles that of $\psi_n(x')$ in the asymptotic region, i.e.,

$$\sigma_2\psi_n(-x')\to 0 \text{ as } |x'|\to\infty,$$

thus ψ_n has the following symmetry $\sigma_2\psi_n(-x')=\delta\ \psi_n(x')$, or

$$\psi_n(-x')=\delta\ \sigma_2\psi_n(-x'),(\ \delta=\pm 1) \tag{4.75}$$

Therefore, if $\phi(-x')=-\phi^*(x')$, then $\psi(x')$ satisfies the symmetry property $\psi_n(-x')=\delta\sigma_1\psi(-x')$ with $\delta=\pm 1$.

If $\phi(x')$ is a symmetric (or antisymmetric) function of x', i.e., $\phi(-x')=\pm\phi(x')$, then $\psi_n^{(s)}(x')=\sigma_1\psi*(-x')$ is the eigenfunction belonging to the eigenvalue $-\lambda_n^*$, and $\psi'^{(a)}_n(x')=\sigma_2\psi*(-x')$ is the eigenfunction belonging to the eigenvalue λ_n^*.

The suffix s (or, a) to the eigenfunction $\psi_n^{;}$ indicates that ϕ is symmetric (or antisymmetric). Since $\phi(-x')=\phi\ (x')$, replacing x' with - x' in Eq. (4.71) and taking complex conjugate, they get:

$$i\frac{d}{dx'}[\sigma_1\psi*(-x')]+\Phi(x')[\sigma_1\psi*(-x')]=-\lambda_n^*\sigma_3[\sigma_1\psi^*{}_n(-x')] \tag{4.76}$$

Compared with Eq. (4.71), the above equation implies that $-\lambda_n^*$ is also an eigenvalue and the associated eigenfunction $\psi_n^{s}(x')$ is simply replaced by $\sigma_1\psi^*{}_n(-x')$, with an arbitrary constant. For $\phi(-x')=-\phi\ (x')$, the same conclusion is also obtained by replacing σ_1 with σ_2 in the above derivations. If $\phi(x')$ is real, then the above symmetry property yields:

$$\psi'^{(s)}_n(-x')=\sigma_1[-\delta\sigma_2\psi_n^*(-x')]=\delta\sigma_2\psi'^{(s)}_n(x')$$
$$\psi'^{(a)}_n(-x')=\sigma_2[-\delta\sigma_1\psi_n^*(-x')]=-\delta\sigma_1\psi'^{(a)}_n(x')$$

i.e., $\psi'^{(s)}_n(x')$ having the same parity as $\psi_n(x')$, while $\psi'^{(a)}_n(x')$ has the opposite.

When $\phi(-x')=-\phi\ (x')$ and λ_n are purely imaginary ($\lambda_n=-\lambda_n^*$), the eigenvalues corresponding to the positive and negative parity eigenfunctions degenerate.

If ϕ is a real non-antisymmetric function of x', it can be shown that:

$$\psi_n^*(x') = i\delta\sigma_3\psi_n(x') \tag{4.77}$$

where $\delta = \pm 1$. Because $\Re(\lambda_n) = 0$, from the complex conjugate of Eq. (4.71), one can arrive at: $\psi_n^*(x') \propto \sigma_3\psi_n(x')$. Substituting Eq. (4.77) into the normalization condition in Eq. (4.73), one then has $\delta = \pm 1$. If the eigenvalue of Eq. (4.12) is real, i.e., $\lambda = \xi$ is real, then,

$$i\frac{d\psi}{dx'} + \Phi\psi = \xi\sigma_3\psi \tag{4.78}$$

and the adjoint function of $\psi, \overline{\psi} = i\sigma_2\psi^*$, is also a solution of Eq.(4.78), i.e.,

$$i\frac{d\overline{\psi}}{dx'} + \Phi\overline{\psi} = \xi\sigma_3\overline{\psi}$$

From this and Eq. (4.78), Satsuma and Yajima obtained the following:

$$\frac{d}{dx'}(\psi^+\psi) = \frac{d}{dx'}(\overline{\psi}^+\psi) = \frac{d}{dx'}(\psi^+\overline{\psi}) = \frac{d}{dx'}(\overline{\psi}^+\overline{\psi}) = 0 \tag{4.79}$$

Using the above boundary conditions, they found that the solutions of Eq.(4.12) , $\psi_1(x',\xi)$, and $\overline{\psi}_2(x',\xi)$ satisfy the following relations:

$$\psi_1^+\psi_1 = \psi_2^+\psi_2 = \overline{\psi}_2^+\overline{\psi}_2 = 1, \overline{\psi}_2^+\psi_2 = \psi_2^+\overline{\psi}_2 = 0$$

From $\psi_1 = a(\xi)\overline{\psi}_2 + b(\xi)\psi_2$, they got a= $\overline{\psi}_2^+\psi_1$ and b= $\overline{\psi}_2^+\psi_1$, where $\psi_1(x',\xi) = \begin{pmatrix}1\\0\end{pmatrix}e^{-i\xi x'}$, as $x' = -\infty$ and $\psi_2(x',\xi) = \begin{pmatrix}0\\1\end{pmatrix}e^{+i\xi x'}$, $\overline{\psi}_2(x',\xi) = \begin{pmatrix}1\\0\end{pmatrix}e^{-i\xi x'}$ as $x' = \infty$.

However, if a real (not antisymmetric) initial value is considered, the microscopic particle does not decay into moving solitons, but can form bound states of solitons pulsating with the proper frequency.

4.6.3. Properties of the Eigenvalues

From the eigenequation (4.12), we know that it is invariant under the Galilei transformation. As a matter of fact, if we substitute the following Galilei transformation:

$$\phi'(\tilde{x},\tilde{t}) = e^{ivx' - iv^2/2}\phi(x',t'), \tilde{x} = x' - vt', \tilde{t} = t' \tag{4.80}$$

into Eq. (4.12), then Φ is transformed into

$$\Phi'(\tilde{x}) = \begin{pmatrix} e^{i\theta/2} & 0 \\ 0 & e^{-i\theta/2} \end{pmatrix} \Phi(x') \begin{pmatrix} e^{-i\theta/2} & 0 \\ 0 & e^{i\theta/2} \end{pmatrix} \tag{4.81}$$

where $\theta = vx' - \frac{1}{2}v^2 t' + \theta_0$, the θ_0 is the initial phase, v is the group velocity of microscopic particles. If the eigenfunction $\psi(x')$ is transformed as:

$$\psi'(\tilde{x}) = \begin{pmatrix} e^{i\theta/2} & 0 \\ 0 & e^{-i\theta/2} \end{pmatrix} \psi(x') \tag{4.82}$$

then Eq. (4.12) becomes

$$i\psi'_{x'} + \Phi'\psi' = (\lambda - \frac{v}{2})\sigma_3\psi' \tag{4.83}$$

This result, by Satsuma and Yajima [50], shows that in a reference frame that is moving with velocity v, the eigenvalue of the particles is reduced to v/2 compared with that in the resting frame. It is known that the velocity of the microscopic particle described by Eq. (3.1) is given by $2\Re(\lambda_k)$. However, when it is constant, i.e., $\theta = \theta_0$, the eigenvalue is unchanged because v=0. This means that the nonlinear Schrödinger equation (3.1) is invariant under the gauge transformation of $\phi' = e^{i\theta_0}\phi(x')$.

If the initial value takes the form of $\phi = e^{ivx'}R(x')$, where $R(x')$ is a real, but not antisymmetric function of x', then all the eigenvalues have the common real part, -v/2.

This can be easily shown by the Galilei transformation. In fact, when Satsuma and Yajima [50] arrived at the following solution, $\phi(x', t' = 0) = e^{ivx'}R(x')$, it could be seen that it does not decay to the series of solitons moving with the different velocities, but forms a bound state. In this case, the real parts are common to all the eigenvalues, i.e., the relative velocities of the solitons vanish.

If the wave function ϕ in Eq. (4.12) undergoes a small change, i.e., $\phi \to \phi' = \phi + \Delta\phi$ the corresponding change in Φ is given by

$$\Delta\Phi = \begin{pmatrix} 0 & \Delta\phi \\ \Delta\phi^* & 0 \end{pmatrix}.$$

Then, λ_n and ψ_n changes as $\lambda_n + \Delta\lambda_n$ and $\psi_n + \Delta\psi_n$, respectively. To the first order in the variation, Eq. (4.71) becomes:

$$[i\frac{d}{dx'}+(\Phi-\lambda_n\sigma_3)]\Delta\psi_n+(\Delta\Phi-\Delta\lambda_n\sigma_3)\psi_n=0$$

Multiplying the above equation by $\psi_n^T\sigma_2$ from the left and integrating it with respect to x' over ($-\infty,\infty$), Satsuma and Yajima [50] got:

$$\begin{aligned}\Delta\lambda_n &= -i\int_{-\infty}^{\infty}\psi_n^T\sigma_2\Delta\Phi\,\psi_n dx' \\ &= -\int_{-\infty}^{\infty}\psi_n^T\Re(\Delta\phi)\sigma_3\psi_n dx' + i\int_{-\infty}^{\infty}\psi_n^T\Im(\Delta\phi)\psi_n dx'\end{aligned}$$

If ϕ is a real and non-antisymmetric function of x', Eq. (4.78) holds and

$$\Delta\lambda_n = \delta <n|\Im(\Delta\phi)\sigma_3|n> + i\delta <n|\Re(\Delta\phi)|n> \tag{4.84}$$

Equation (4.84) indicates that if $<n|\Im(\Delta\phi)\sigma_3|n>\neq 0$, the perturbation makes the real part of the finite eigenvalue appear. That is, for the initial value, $\phi(x')+\Delta\phi(x')$, the solution of the nonlinear Schrödinger equation (3.1) breaks up into moving solitons with velocity $2\Re(\Delta\lambda_n)$. If ϕ is real and is either a symmetric or an antisymmetric function of x', the above symmetry properties of eigenvalues of the nonlinear Schrödinger equation (3.1) lead to

$$<n|\Im(\Delta\phi(x'))\sigma_3|n> = -<n|\Im(\Delta\phi(-x))\sigma_3|n> \tag{4.85}$$

Therefore, if $\Im(\Delta\phi)$ is a symmetric function, then $<n|\Im(\Delta\phi)\sigma_3|n>$ vanishes, i.e., $\Re(\Delta\lambda_n)=0$, and the soliton bound state does not resolve into moving solitons even in the presence of the perturbation $\Delta\phi$.

Satsuma and Yajima [50] also obtained the shifts of the eigenvalues of Eq. (4.82) under the double-humped initial values,

$$\phi(x',t'=0)=\phi_0(x'-x_0')+e^{i\theta_0}\phi_0(x'+x_0'),$$

where ϕ_0 is a real and symmetric function of x', x_0' ,where ϕ_0 is real.

From the above investigations, we understand that the eigenvalues and eigenequations of nonlinear Schrödinger equation are very complicated and display different properties.

4.7. THE VARIATIONS OF METHOD SOLVING THEDYNAMIC EQUATIONS IN TWO THEORIES

4.7.1. In the Linear Quantum Mechanics

As it is known,when the linear Schrödinger equation is used to solve and explain the quantum mechanical problems we must incorporate all interactions among particles or between particles and the background field, such as the lattices in solids and nuclei in atoms and molecules, including nonlinear and complicated interactions, into the external potential $V(\vec{r},t)$ in Eq. (1.8).

In this case, we must use various approximate methods to solve the linear Schrödinger equation, such as, the free electron and average field approximations, Born-Oppenheimer approximation, Hartree-Fock approximation, Thomas-Fermi approximation, and thevariational method, and so on [1-3], in which the perturbed method and the Wentael, Kramers and Brillouin (W.K.B.)method were often used in practical calculations. We here simply debated these methods [1-3].

In the perturbed method of independent time, we assume that the perturbed Hamiltonian operator is denoted by $\hat{H}'$, and then thecorrespondent eigenequation in this case should be denoted by

$$\hat{H}\phi = (\hat{H}_0 + \beta\hat{H}')\phi = E'\phi, \tag{4.86}$$

where $\hat{H}_0\phi_0 = E_0\phi_0$, β is a small perturbed parameter. In this case, we can represent the eigenfunction and eigenvalue, respectively, by

$$\phi = \phi_0 + \beta\phi_1 + \beta^2\phi_2 ...., E' = E_0 + \beta E_1 + \beta^2 E_2 ...., \tag{4.87}$$

Substituting Eq.(4.87) into Eq.(4.86) we can obtain many equations, ϕ_1 and E_1, ϕ_2 and E_2 …, respectively. Thus we can gain the representations of ϕ_1 and E_1, ϕ_2 and E_2 … from these equations, respectively. Inserting these representations again, into Eq.(4.87) we finally give the perturbation solutions of Eq.(4.86), which can be expressed by [1-3]

$$\phi = \phi_0 + \beta\phi_1 + \beta^2\phi_2 .... = \phi_0 + \sum_n \frac{H'_{on}\phi_n}{E_0 - E_n} + \sum_m [\sum_n \frac{H'_{mn}H'_{0n}}{(E_0 - E_n)(E_0 - E_m)} - \frac{H'_{mn}H'_{0m}}{(E_0 - E_n)^2}]\phi_m - \frac{1}{2}\sum \frac{|H'_{0n}|^2}{(E_0 - E_n)^2}]\phi_0 + ..., \tag{4.88}$$

and $E' = E_0 + \beta E_1 + \beta^2 E_2 .... = E_0 + H'_{on} + \sum_n \frac{|H'_{0n}|^2}{E_0 - E_n} + ...,$ (4.89)

where $H'_{li} = \int \phi_l^* \hat{H'} \phi_i d\tau$ is a matrixelement, here $l, i = 1, 2, 3, ...m, n$.

If E_0 is degenerated with k degeneracy, i.e., it has k eigenstates of $\phi_{01}, \phi_{02}, \phi_{03}, \cdots \phi_{0k}$, then we should use the degenerated perturbation theory to find the perturbed solution of Eq.(4.86). In this case, the zero order approximate wave-function, and the first order approximate eigenvalues, can be obtained from the following matrix equation:

$$\sum_{i=1}^{k} (H'_{li} - E_1 \delta_{li}) c_i^{(0)} = 0, l = 1, 2, 3, ...k, \phi_0 = \sum_{i=1}^{k} c_i^{(0)} \varphi_{0i} \quad (4.90)$$

In the time-dependent perturbation theory, there is $\hat{H} = (\hat{H}_0 + \hat{H'}(t))$ in this case, where $\hat{H}_0 \phi_n = E_n \phi_n$. In the linear Schrödinger equation,

$$i\hbar \frac{\partial \Psi}{\partial t} = \hat{H}(t)\Psi, \quad (4.91)$$

the wave function Ψ can be denoted by

$$\Psi = \sum_n a_n(t) \Phi_n, \quad (4.92)$$

where $\Phi_n(x,t) = \varphi_n(x) \exp[-iE_n t / \hbar]$. Inserting Eq.(4.92), into Eq.(4.91), we get [1-3]

$$i\hbar \frac{\partial a_m(t)}{\partial t} = \sum_n a_n(t) H'_{mn} \exp[i\omega_{mn} t], \omega_{mn} = (E_m - E_n) / \hbar \quad (4.93)$$

Let now $a_n(t) = a_n^{(0)} + \beta a_n^{(1)} + a_n^{(2)} + ...,$ (4.94)

and inserting Eq.(4.94) into Eq.(4.93) we can obtain that zero order approximates coefficient $a_n^{(0)}$, which is time-independent. First order approximate coefficient $a_n^{(1)}$ can be obtained through solving the equation:

$$i\hbar \frac{\partial a_m^{(1)}(t)}{\partial t} = \sum_n H'_{mn} \delta_{nk} \exp[i\omega_{mn} t] = H'_{mk} \delta_{nk} \exp[i\omega_{mk} t]$$

Then the first order coefficient of transition from nth state to mth state is a form of

$$a_m^{(1)}(t) = \frac{1}{i\hbar}\int H'_{mk}\exp[i\omega_{mk}t]dt\,.$$

In the W.K.B. approximation method, the time-independent linear Schrödinger equation of Eq.(1.8) can be re-written as

$$\frac{\partial^2\phi(x)}{\partial x^2} + \frac{2m}{\hbar^2}[E - V(x)]\phi(x) = 0, \tag{4.95}$$

When $\kappa^2(x) = \frac{2m}{\hbar^2}[E - V(x)] > 0$, or $E > V(x)$, equation (4.95) becomes

$$\frac{\partial^2\phi(x)}{\partial x^2} + \kappa^2(x)\phi(x) = 0, \tag{4.96}$$

When $\kappa'^2(x) = \frac{2m}{\hbar^2}[V(x) - E] > 0$, or $E < V(x)$,equation (4.96) becomes

$$\frac{\partial^2\phi(x)}{\partial x^2} - \kappa^2(x)\phi(x) = 0, \tag{4.97}$$

We now assume

$$\phi(x) = \exp[ia(x)/\hbar] \tag{4.98}$$

and inserting it into Eq.(4.97) we can obtain

$$i\hbar\frac{d^2a(x)}{dx^2} - \left(\frac{da(x)}{dx}\right)^2 + \hbar^2\kappa^2 = 0, \tag{4.99}$$

Let $a(x) = a_0(x) + \hbar a_1(x) + \hbar^2 a_2(x) + \ldots$. (4.100)

Substituting Eq.(4.100) into Eq.(4.99) we can get the representations of $a_0(x)$, $a_1(x)$ and $a_2(x)$…. Thus, the solution of Eq.(4.96) can be approximately expressed as

$$\phi(x) = \frac{A}{\sqrt{\kappa(x)}}\exp[\pm i\int_a^x \kappa(x')dx'], (E > V(x)) \tag{4.101}$$

Analogously, the solution of Eq.(4.97) can also be obtained, which is

$$\phi(x) = \frac{B}{\sqrt{\kappa'(x)}} \exp[\pm\int_{a'}^{x} \kappa'(x')dx'], (E < V(x)) \tag{4.102}$$

These representations are very useful in investigations of quantum transition problems [1-3].

4.7.2. In the Nonlinear Quantum Theory

Obviously, the above methods solving the linear Schrödinger equation cannot be applied directly to solve the nonlinear Schrödinger equation (2.66) at $A(\phi)=0$ due to the nonlinear interaction. However, plenty of methods are used to solve the nonlinear Schrödinger equation at present, which are stated as follows.

(A) Function and Variable Transformations

A complicated equation can be transformed into a simpler or a standard equation, which is readily solved by various functions and variable transformations. The forms of transformation used depend on the properties of dynamic equations. The following transformations are often used.

(1). Function transformation. For example the transformation, $\phi = \sqrt{\rho}\exp[i\theta(x',t')]$, is used,where $\varphi = \sqrt{\rho}$. Using this transformation, the nonlinear Schrödinger equation(2.66) at $A(\phi)=0$ can be written as two equations (2.70) and (2.71), which can be readily solved [18-22].

(2).Galilei transformations. As an example, we consider the nonlinear Schrödinger equation (3.1) with $b = 1$.Ifit is a special solution, it is written as

$$\phi(x',t') = A\sec h(Ax')\exp\left(-i\frac{A^2t'}{2}\right)$$

where $x' = x\sqrt{2m/\hbar^2}, t' = t/\hbar$. Since this nonlinear Schrödinger equation is invariant under the Galilei transformationin Eq.(3.70), its general solution can thus be obtained, which is:

$$\phi(x',t') = A\sec h\left[A(x'-vt')\right]\exp\left(-ivx + i\frac{(v^2 - A^2)t'}{2}\right).$$

Let $v = 2\xi$, and $A = 2\eta$, then it reduces to the soliton solutionin Eq.(3.9), which wasobtained by the inverse scattering method.

(3).The variable transformation in the traveling-wave method. If we can write the solution of an equation of motion in the form of $\phi(x,t) = \phi(\xi')$, with $\xi' = x - vt$, it is called a traveling-wave solution. This method is widely used in wave mechanics and nonlinear quantum mechanics,such as the nonlinear Schrödinger equation given in Eq. (3.1), its traveling-wave solution under the transformation of $\xi' = x' - vt'$ can be written as: $\phi(x,t) = \phi(\xi')e^{i(kx'-\omega t')}$. In terms of the new variables, equation (3.1) can be written as

$$\frac{d^2\phi(\xi')}{d\xi'} + (\omega - k^2)\phi(\xi') - b\phi^3 = 0.$$

The soliton solution of this equation can be easily obtained when $b > 0$and $\gamma = \omega - k^2 > 0$,

which is

$$\phi(x,t) = \sqrt{\frac{2\gamma}{b}}\,\text{sech}\left(\sqrt{\gamma}\left(\xi' - \xi_0'\right)\right)e^{i(kx'-\omega t')}.$$

(B) The Inverse Scattering Method

The inverse scattering method was first proposed by Gardner, Green, Kruskal, and Miura (GGKM) in 1967 for solving the KdV equation [51-60], which connects a nonlinear dynamic equation to the linear Schrödinger equation;

$$\frac{\partial^2 \psi}{\partial x^2} + (\lambda - \phi)\psi = 0,$$

through the GGKM transformation:

$$\phi = \frac{1}{\psi}\frac{d^2\psi}{dx^2} + \lambda(t),$$

where ψ satisfies the nonlinear dynamic equation. They and Zakharov et al., [12-13] found out the soliton solution of KdV equation using these transitions. In this case, Lax gave a more general formulation of the inverse scattering method [11], then the soliton solution of Eq.(3.1) can be obtained, which was finished in 1967 [51]. Zakharov and Shabat [12-13] also found out the soliton solutions of the nonlinear Schrödinger equation (3.1) using the inverse scattering method in 1971-1972. In 1973, the Sine-Gordon equation was also solved by Ablowitz et al. [52-55].

According to Lax [11], given a nonlinear equation:

$$\varphi_{t'} = K(\varphi), \phi = \phi(x,t),$$

illustrating that the inverse scattering method for solving the equation consists of the following three steps [55-60].

(1).Find operator L and $\hat{B}$, which depend on the solution ϕ, such that $i\hat{L}_{t'} = \hat{B}\hat{L} - \hat{L}\hat{B}$, such as given in Eqs. (4.4)-(4.7),in the nonlinear Schrödinger equation (3.1).

(2).The scattering operator, L, satisfies its eigenvalue equation, $L\Psi = k\Psi$, where k is a spectral parameter (generally speaking, complex), and is often called a Jost function.

(3) The time dependence of the scattered wave is determined by $i\Psi_{t'} = \hat{B}\Psi$. Instead of directly solving the nonlinear equation to obtain $\varphi(x', t')$ for the given initial condition $\varphi(x', 0)$, the inverse scattering method proceeds in the above three steps and obtains a solution of the nonlinear integral-type equations. The major difficulty of the inverse scattering method is the lack of a systematic approach for finding the operators L and $\hat{B}$ even if they exist.

(C) The Perturbation Method of the Nonlinear Schrödinger Equation

Most of dynamic equations cannot be solved analytically in the nonlinear quantum theory, but can be found by the perturbative approach, when certain terms are small relative to the nonlinear energy or the kinetic energy in the nonlinearSchrödinger equation.

In reality, due to the existence of boundaries, defects, impurities, dissipation, external fields, *etc.*, a system usually experiences some perturbation. In fact, when the soliton behavior of microscopic particles in the system is probed, the measurement itself is bound to alter the state of the system to some extent. Thermal excitation at finite temperature is another form of perturbation that no system can avoid. Therefore, it is necessary to solve the influences of these perturbations on the properties of the soliton solution of the nonlinearSchrödinger equation.

In fact, when such perturbations are relatively weak, they may be treated by perturbation methods. In contrast to the linear quantum mechanics, a systematic perturbative approach and a universal formula, as mentioned above, are impossible to obtain in the nonlinear quantum theory due to the complexity of nonlinear interactions. Therefore, there are many ways to solve the perturbed dynamic-equations besides the direct perturbation theory, which is applicable to both perturbed integrable and nonintegrable systems; then the perturbative method is an important method solving nonlinearSchrödinger equation. In this section, two different kinds of perturbed methods, which are often used, are briefly described as follows.

Linear perturbation theory allows one to study the influence of small external perturbations on behaviors of microscopic particles described by the nonlinear Schrödinger equation and the Sine-Gordon equation. Many scientists, such as Pang [61-65], Fogel, Trullinger, Bishop, and Krumbansl [66-68], contributed to the development of the linear perturbation theory. The theory is based on an expansion of the soliton wave function in terms of the complete set of eigenfunctions of the self-conjugate differential operator. Such an operator has, among its eigenfunctions, a single function representing the stationary wave, propagating with the same velocity as the free microscopic particle (soliton). We here describe first the linear perturbation technique for the nonlinear Schrödinger equation (2.66) at $A(\phi) = 0$ as put forward by Pang [61-65] and by Fogel et al., [66-68], respectively.

(a) The Theory of Linear Perturbation in the Nonlinear System

Assume that $\phi_0(r,t)$ is the soliton solution of the nonlinear Schrödinger equation of an unperturbed system. The wave function of the perturbed system can be written as:

$$\phi = \phi_0 + \varepsilon\phi_1 + \varepsilon^2\phi_2 + ... \tag{4.103}$$

where εis a small quantity. Substituting Eq. (4.103) into the original nonlinear equation, then equations of ϕ_1 and ϕ_2 ...., corresponding to different orders of ε,can be respectively obtained. Thus ϕ_1 and ϕ_2 ... can be determined by solving the equations, and finally, it can also be obtained from Eq. (4.103).

For example, the nonlinear Schrödinger equation(2.66) at $A(\phi)=0$, when it is small we can derive its solution by the linear perturbed theory. In such a case, Pang [165-169] introduced a small parameter ε to denote the perturbation effect of potential $V(x)$. Thus, equation (2.66) can now be written as:

$$i\frac{\partial\phi}{\partial t'} + \frac{\partial^2\phi}{\partial x'^2} + b|\phi|^2\phi = \varepsilon V(x')\phi, \tag{4.104}$$

where $t' = t/\hbar$, $x' = \sqrt{2m/\hbar^2}x$. Pang [61-65] assumed the following form for the linear perturbed solution of Eq. (4.104)

$$\phi = \phi_0 + \phi_1 = (f + \varepsilon F(x))e^{i\theta(x',t')} \tag{4.105}$$

where $\phi_0 = f(x',t')e^{i\theta(x',t')}$, is an unperturbed solution of Eq. (4.104) with $\varepsilon = 0$. This is the same as (3.9), where 2η and $4\xi'$ are respectively related to the amplitude and group velocity of the microscopic particle. Inserting Eq. (4.105) into Eq. (4.104) and neglecting terms higher than the second order in ε, we obtain the following equation for $F(x',t')$:

$$i\frac{\partial F}{\partial t'} + \frac{\partial^2 F}{\partial X'^2} - v^2\left[1 - 4\sec h^2(vX')\right]F + 2v^2\sec h^2(vX')F^* = v\sqrt{\frac{2}{b}}\sec h(vX')V(X'+v_e t'), \tag{4.106}$$

where

$$X' = x' - v_e t', v^2 = (v_e^2 - 2v_c v_e)/4 = (v_e/2)^2\delta, \delta = 1 - 2v_c/v_e \tag{4.107}$$

From Eq.(4.106) we find out the representation of F(x', t'), then the perturbation solutions of Eq.(4.104) can be obtained from Eq.(4.106). Obviously, it is different from that in linear quantum mechanics, as mentioned above.

(b) The Structure Perturbed Method

We here briefly discuss the structural perturbed method for the perturbed nonlinear Schrödinger equation. The structure perturbed method describes solutions to the dynamic equation, denoted by a new function, which is not a combination of the unperturbed solution and corrected solutions. Thus, the new function changes the nature of solutions of dynamic equations, in this case. For example, the nonlinear Schrödinger equation:

$$i\phi_{t'} + \phi_{x'x'} + b|\phi|^2 \phi = \beta\phi^*_{x'} \tag{4.108}$$

when β is small, then $\beta\phi^*_{x'}$ in Eq. (4.108) can be treated as a perturbation. In such a case, the solution of Eq.(4.108) cannot be represented by the above method, but by the structureperturbed method. In the latter Makhankov et al., [69-72]represented the structure perturbation solution of Eq. (4.108) by

$$\phi(x',t') = D(x',t')\exp[i\theta(x',t')] \tag{4.109}$$

where $\theta(x',t')$ and $D(x',t')$ are all unknown and need to be acquired from Eq.(4.108). In practice, byinserting Eq. (4.109) into Eq. (4.108) the following equation can be obtained:

$$D_{t'} + 2\theta_{x'}D_{x'} - \beta D_{x'}\sin 2\theta - \beta\theta_{x'}D\cos 2\theta + D\theta_{x'x'} = 0 \tag{4.110}$$

$$-D\theta_{t'} - D + D_{x'x'} - D\theta^2_{x'} + D^3 + \beta D_{x'}\cos 2\theta - \beta\theta_{x'}D\sin 2\theta = 0 \tag{4.111}$$

Multiplying Eq. (4.110) by *D (x',t')*,it then becomes:

$$\partial_{t'}D^2 + \partial_{x'}[(2\theta_{x'} - \beta\sin 2\theta)D^2] = 0\,.$$

Finally, the solution of Eq. (4.108) can be approximately obtained, which is denoted

$$\phi = \mu_0 \operatorname{sec}h\{\mu_0[x' - (\beta\cos 2t')/2]/\sqrt{2}\} \times \exp\{-i[(1-\mu_0^2/2)t' + \beta\mu_0\tanh(\mu_0 x'/\sqrt{2})\sin 2t'/2\sqrt{2}]\} \tag{4.112}$$

which holds when an $\mu_0^2 << 1$.

Equation(4.112) is just an example of a structure perturbation solution of the nonlinear Schrödinger equation (4.108). It shows that microscopic particles can still maintain the soliton feature in motion under the structure perturbation.

The above result shows that the influence of the small term can lead to interesting results. This means that the structural stability of solutions in nonlinear quantum mechanical systems is a rather delicate problem, which must be solved either for a specific system or for a restricted class of systems.

(d) Hirota's D Operator Method

Hirota method changes the independent variables and rewrites the original dynamic equation into the form of $F\left(D_x^m D_t^n\right)=0$, where D is an operator. Analytic solutions of the nonlinear dynamic equations in nonlinear quantum theory are then obtained using the perturbation method and D-operator proposed by Hirota [73-77]. The D-operator is defined by

$$D_x^m D_t^n (f,g)=\left[\left(\frac{\partial}{\partial x}-\frac{\partial}{\partial x}\right)^m \left(\frac{\partial}{\partial t}-\frac{\partial}{\partial \tilde{t}}\right)^n f(x,t)g\left(x,\tilde{t}\right)\right]_{x=x,t=\tilde{t}} \cdot \quad (4.113)$$

Let's further define an operator D_z and a differential operator $\partial/\partial z$ as follows:

$$D_z=\delta D_t+\varepsilon D_x, \quad \frac{\partial}{\partial z}=\delta\frac{\partial}{\partial t}+\varepsilon\frac{\partial}{\partial x}, \quad (4.114)$$

where δ and are constants. For the nonlinear Schrödinger equation:

$$i\phi_{t'}+p\phi_{x'x'}+b|\phi|^2\phi=0, \quad (4.115)$$

Hirota assumes $\phi=g/f$, where g is complex and f is real. Using the identity

$$\exp\left(\varepsilon\frac{\partial}{\partial z}\right)\left(\frac{g}{f}\right)=\frac{g(x'+\varepsilon)f(x'-\varepsilon)}{f(x'+\varepsilon)g(x'-\varepsilon)}=\frac{\exp(\varepsilon D_{x'})g\cdot f}{\exp(\varepsilon D_{x'})f\cdot f}$$

where $t'=t/\hbar$, $x'=x\sqrt{2m/\hbar^2}$, he found that g and f must satisfy:

$$\frac{\left(iD_{t'}+D_{x'}^2\right)g\cdot f}{f^2}-\frac{g}{f}\left(\frac{pD_{x'}^2 f\cdot f-bgg^*}{f^2}\right)=0.$$

Substituting $\phi=g/f$ into Eq. (4.115) yield:

$$\left(iD_{t'}+pD_{x'}^2\right)g\cdot f=0, \quad pD_{x'}^2 f\cdot f-bgg^*=0,$$

where $g$and $f$satisfy Eq.(4.113).Then the one-envelope-soliton solution of the nonlinear Schrödinger equation (4.115) is given by

$$g=Ae^{i\theta}e^{\eta}, \quad f=1+ae^{\eta+\eta^*}, \quad (4.116)$$

If A, θ, α, η, and η' are determined, then the soliton solution of Eq.(4.115) can be obtained.

(e) The Backlund Transformation Method

From mathematics we know that, for a given equation $d\phi = Pdx + Qdt$, then the integrability of the corresponding first-order differential equations,

$$\phi_x = P, \phi_t = Q\ ,$$

requires $P_t = Q_x$ .This transformation relation of ϕ deciding by P and Q is called a Backlund transformation [78-83], which consists of the following three steps. The *first* step is to find P and Q, as functions of ϕ ; the *second* is to choose a trial function ϕ_0 , and the *third* is to use $\phi_{1x} = P(\phi_1, \phi_0)$ and $\phi_{1t} = Q(\phi_1, \phi_0)$ for finding a new solution ϕ_1 . Therefore, the Backlund transformation method enables one to find a new solution from a given one. When applied repeatedly, the Backlund transformation can be used to obtain the N-soliton solutions of the nonlinear dynamics equations. Equation (4.15) or Eq. (4.25) is simply a Backlund transformation of the nonlinear Schrödinger equation (3.1).The difficulty for the Backlund transformation method is to find the functions P and Q. However, there are many different types of Backlund transformations in nonlinear quantum mechanics. Therefore, it is important to choose an appropriate Backlund transformation in order to obtain the solutions of a given equation. In general, auto-Backlund and Hirota's transformation methods[73-77] are often used together to obtain the soliton solutions of nonlinear dynamic equations.

(f) The Numeral Simulation Method

The numeral simulation methods, such as Monte-Carlo, Runge-Kutta, and finite element methods, are often used to solve the nonlinear Schrödinger equation. In the Runge-Kutta method [84-85], we should first decompose the nonlinear Schrödinger equation as two dynamic equations, and then again become the two continuum dynamical equations, as discrete equations, to solve them as shown above.

(g) The Method Depressing Dimension for Higher-Dimensional Equations

We here introduce a method of reduction from higher to lower dimensional equations proposed by Hirota[73-77] and Tajiri [86-87] to solve nonlinear dynamics equations in high dimensional systems.

In fact, for the two-dimensional case, the solutions for some similarity variables have been studied by Nakamura, et al. [88-89].They obtained an explosion–decay mode solution by generalizing the similarity type plane wave solution. Tajiri investigated the similarity of solutions of one and many-dimensional cases using Lie's method. For 3D-nonlinear Schrödinger equation:

$$i\phi_{t'} + p\phi_{x'x'} + q'\phi_{y'y'} + r'\phi_{z'z'} + b|\phi|^2\phi = 0, \tag{4.117}$$

where p, q', r', b areconstants, $t' = t/\hbar$, $x' = x\sqrt{2m}/\hbar, y' = y\sqrt{2m}/\hbar$, $z' = z\sqrt{2m}/\hbar$. Tajiri [86-87] considered the following transformation of an infinitesimal one-parameter(ε) Lie's group in the space

$$\bar{x} = x' + \varepsilon X(x', y', z', t', \phi) + O(\varepsilon^2),\ \bar{y} = y' + \varepsilon Y(x', y', z', t', \phi) + O(\varepsilon^2),$$

$$\bar{z} = z' + \varepsilon Z(x', y', z', t', \phi) + O(\varepsilon^2),\ \tilde{t} = t' + \varepsilon T(x', y', z', t', \phi) + O(\varepsilon^2),$$

$$\bar{\phi} = \phi' + \varepsilon U(x', y', t', \phi) + O(\varepsilon^2). \tag{4.118}$$

He assumed that the coefficients, p, q' and r are not all the same sign, for example, $p, q' > 0$ and $r < 0$, and using the transformation,

$$X = \frac{1}{\sqrt{p}}(x' - \theta_1 t'), Y = \frac{1}{\sqrt{q'}}(y' - \theta_2 t'), Z = \frac{1}{\sqrt{-r}}(z' - \theta_3 t'),$$

and

$$\phi = \exp\left\{ i \left[\begin{array}{l} \frac{\theta_1}{2p}(x' - \theta_1 t') + \frac{\theta_2}{2q'}(y' - \theta_2 t') + \\ \frac{\theta_3}{2r'}(z' - \theta_3 t') + \theta_4 t' \end{array} \right] \right\} U(X, Y, Z),$$

Tajiri obtained [86-87]

$$U_{ZZ} - U_{XX} - U_{YY} + cU - b|U|^2 U = 0, \tag{4.119}$$

which is formally the 2D-nonlinear Klein-Gordon equation, where $\theta_1, \theta_2, \theta_3$ and θ_4 are arbitrary constants, and

$$c = \theta_4 - \frac{\theta_1^2}{4p} - \frac{\theta_2^2}{4q'} - \frac{\theta_3^2}{4r'}.$$

Making the transformation once more:

$$\bar{\bar{\xi}} = X + \alpha Y = \frac{x'}{\sqrt{p}} + \alpha\frac{y'}{\sqrt{q'}} - \left(\frac{\theta_1}{\sqrt{p}} + \alpha\frac{\theta_2}{\sqrt{q'}} \right) t',\ \tau = X - \frac{1}{\alpha}Y \mp \frac{\sqrt{1+\alpha^2}}{\alpha}Z$$

$$=\frac{x'}{\sqrt{p}}-\frac{y'}{\alpha\sqrt{q'}}\mp\frac{\sqrt{1+\alpha^2}}{\alpha\sqrt{-r'}}z'-\left(\frac{\theta_1}{\sqrt{p}}-\frac{\theta_2}{\alpha\sqrt{q'}}\mp\frac{\theta_3\sqrt{1+\alpha^2}}{\alpha\sqrt{-r'}}\right)t',$$

$$U=\exp\left\{i\left[\begin{array}{l}\pm\dfrac{\alpha\left(a^2-c\right)}{2a\sqrt{1+\alpha^2}}\left(X-\dfrac{1}{\alpha}Y\right)+\\ \dfrac{1}{2}\left(a+\dfrac{c}{a}\right)Z\end{array}\right]\right\}G'\left(\overline{\overline{\xi}},\tau\right), \tag{4.120}$$

Tajiri [86-87] found:

$$iG'_{\tau}\pm\frac{\alpha}{2a}\sqrt{1+\alpha^2}G'_{\overline{\overline{\xi}}\overline{\overline{\xi}}}\pm\frac{\alpha b}{2a\sqrt{1+\alpha^2}}\left|G'\right|^2G'=0, \tag{4.121}$$

which is formally the 1D-nonlinear Schrödinger equation, its solution is easily found, where $a$and a are arbitrary constants.

In the meanwhile, Tajiri [86-87]generalized the above Hirota method to two- and three-dimensional systems and solved nonlinear dynamic equations. For Eq. (4.112) with $b>0$ (for $p,q'>0$ and $r'<0$), Tajiri here introduces the variable transformation,

$$\phi=\frac{g(x',y',z',t')}{f(x',y',z',t')},$$

where $f=f^*$, and found the soliton solution of Eq.(4.117), using the above Hirota method.

Therefore, the methods solving the dynamic equations in linear and nonlinear quantum mechanics are fully different.

4.8. Distinctions of Feature of Microscopic Particles in Two Theories at MinimumNonlinear Interaction

From previous studies and the soliton theory, we know that the microscopic particles are a soliton and have a wave-corpuscle duality, when the nonlinear interaction can balance, and thus cancels the dispersion effect. In this case, the properties of microscopic particles in the systems must be described by the nonlinear Schrödinger equation. In practice,the nonlinear interactions are different in different systems; the nonlinear interactions insome systemsare stronger, but other systems are not so.In this case, it is quite necessary to know if linear quantum mechanics could be used to describe the states and properties of microscopic particles when the nonlinear interactions are so small that they approach zero. This is an interesting problem and worthy of deep research.

For a quite weak nonlinear interaction (b<<1) and small skirt $\phi(x',t')$, Pang [18-22] represents the solution of the nonlinear Schrödinger equation (3.1) at by

$$\phi = 4\sqrt{2/bk}e^{-2k\left(x'-x_0'-\upsilon_e t'\right)}e^{i\upsilon_e\left(x'-x_0'-\upsilon_c t'\right)/2} \tag{4.122}$$

where $x' = x/\sqrt{\hbar^2/2m}$, $t' = t/\hbar$. Because $b|\phi|^2\phi$ is very small and approaches zero in this case, then Eq. (3.1)can be approximated by:

$$i\phi_{t'} + \phi_{x'x''} \approx 0 \tag{4.123}$$

Substituting Eq. (4.122) into Eq. (4.123), we see that it satisfies Eq. (4.123), and we can get $\upsilon_e \approx 4k$ from this equation, which is the group speed of the microscopic particle (Near the top of the peak, we must take both the nonlinear and dispersion terms into account because their contributions are of the same order. Thus, it is the group speed). However, if Eq. (4.123) is treated as a linear Schrödinger equation, its solution is Eq. (1.8). In a one-dimensional case, it takes the form:

$$\phi'(x',t') = Ae^{i(kx'-\omega t')} \tag{4.124}$$

Thus, we now have $\omega = k^2$, which gives the phase velocityω/k as $\upsilon_c = k$ and the group speed $\partial\omega/\partial k = \upsilon_{gr} = 2k$. Apparently, this is different from $\upsilon_e = 4k$. This is because the solutionin Eq. (4.124) is considerably different from Eq. (4.122). Therefore, the solution Eq. (4.124) is not the solution of the nonlinear Schrödinger equation (3.1) in the case of a very weak nonlinear interaction.

Solution Eq. (4.122) is a "divergent solution" (i.e. $\phi(x,t) \to \infty$ at $x \to -\infty$), or a "decaying solution"(i.e., $\phi(x,t) \to 0$ at $x \to \infty$), hence it is not an "ordinary plane wave". In such a case, we can say that the soliton comes from the evolution of a divergent wave, instead of a plain wave in Eq.(4.124). Then the divergence develops by the nonlinear interaction to yield a solitary wave of finite amplitude. When the nonlinear interaction is very weak, the soliton will diverge, but suppression of the divergence will not result in a soliton [18-22].

These circumstances are clearly seen from the soliton solution in Eq. (3.8) in the case of nonlinear coefficient $b \neq 1$. If the nonlinear term approaches zero ($b \to 0$), and the solitary wave diverges $\left(\phi(x,t) \to \infty\right)$. If we want to suppress the divergence, then we have to set$k$ = 0. In such a case, we get Eq. (4.124) from Eq. (4.122). This illustrates that the nonlinear Schrödinger equation can be reduced to the linear Schrödinger equation if, and only if, the nonlinear interaction and the group speed of the particles are simultaneously zero.

Therefore, we can conclude that the microscopic particles, described by the nonlinear Schrödinger equation, in the very weak nonlinear interaction limit is also *not* the same as that in the linear Schrödinger equation in quantum mechanics. Only if the nonlinear interaction is equal to zero, can the nonlinear Schrödinger equation reduce to the linear Schrödinger equation. This means that linear quantum mechanics can only be used to describe the properties of microscopic particles in the systems without nonlinear interactions, i.e., nonlinear quantum mechanics becomes linear quantum mechanics, when the nonlinear interaction is equal to zero. Once the nonlinear interactions are considered then we must use nonlinear quantum mechanics to study the motions of microscopic particles in any physical systems, even though it is very small. Therefore, nonlinear and linear quantum mechanics are self-consistent each other, the latter is only an approximate theory and a special case of the former.

However, real atoms and molecules are made up of a great number of microscopic particles, and nonlinear interactions always exist in the systems. Hence, the physical systems, in which the nonlinear interaction is zero, are not existent in the real world. This clearly shows that nonlinear quantum mechanics are truly a correct and universal theory for describing microscopic particles in real physical systems. It can be used extensively in atoms, molecules, solids, condensed matter, polymers, and biomolecules, even though the nonlinear interaction is very weak, including the hydrogen atom. Thus, we can affirm that linear quantum mechanics is only an approximate theory, and can only give the wave feature of particles at zero nonlinear interaction.

To sum up, the above discussions and comparisons show clearly that the properties of microscopic particles described by the nonlinear Schrödinger equation are completely different from those in linear quantum mechanics, even although the nonlinear interactions are very small. Upon that, the new theoryisa universal theory. It can be used to study the states and properties of microscopic particles in all physical systems.

4.9. The Reasons of the Differences between the Two Theories As Well As the Physical Effects of the Nonlinear Interaction

Quite clearly, the reason for differences between the two theories are caused by the nonlinear interactions in the dynamic equations because nonlinear quantum mechanics becomelike linear quantum mechanics once the nonlinear interaction is zero.

In such a case, we must first know, what would be the effects of nonlinear interactions on microscopic particles? To answer this question, we should carefully observe the properties of the macroscopic motion of a water wave in the sea. When a water wave approaches the beach, its shapevaries gradually from a sinusoidal cross section to triangular, and eventually a crest, which moves faster than the rest, as shown in Figure 4.1, which changes as time goes on. This is a result of the nonlinear nature of a wave. As the wave approaches the beach, it will be broken up due to the fact that the nonlinear interaction is enhanced. Therefore, we can affirm that the nonlinear interaction destructs the wave shape and propagation of the wave. In this case, the speed of wave propagation depends on the height of the wave. If the phase velocity of the wave is denoted by v_c, since it depends weakly on the height of the wave, *h*, then we can express the relation by

$$v_c = \frac{\omega}{k} = v_{co} + \sigma_1 h, \tag{4.125}$$

where $\sigma_1 = \frac{\partial v_c}{\partial h}\Big|_{h=h_0}$, h_0 is the average height of the wave surface, v_{co} is the linear part of the phase velocity of the wave, σ_1 is a coefficient denoting the nonlinear effect.

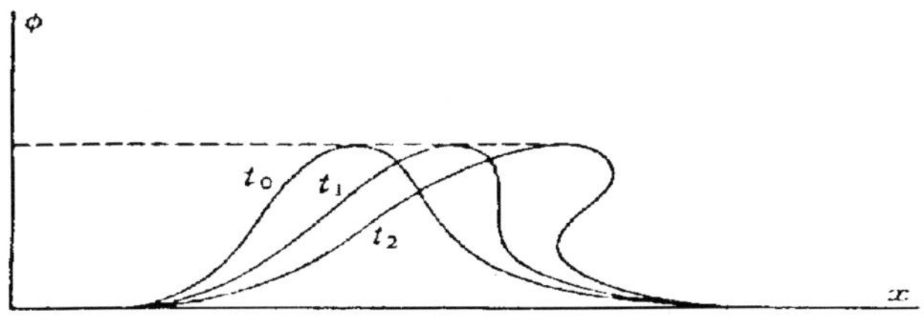

Figure 4.1. Changes in the shape of a water wave at different times.

This expresses that the nonlinear interaction results in changes in both form and velocity of waves. This is the same for the dispersion effect, but their mechanism and rules are different [57, 60, 90].

In the dispersion medium the relation between the wavelength and frequency of the light (wave) is expressed by $\omega = \omega(\vec{k})$ or $G(\omega, \vec{k}) = 0$, where $\partial^2\omega / \partial^2 k^2 \neq 0$ in a one-dimensional case. It specifies how the velocity or frequency of the wave (light) depends on its wavelength or wavevector. The equation depicts wave propagation in a dispersive median and is called as dispersion equation. The linear Schrödinger equation (1.8) in linear quantum mechanics is just a dispersion equation. If Eq. (1.9) is inserted into Eq. (1.8), we can get $\omega = \hbar k^2/2m$ at V(x)=0. From (1.10) we obtain $E = \hbar\omega, \vec{p} = \hbar\vec{k}$. The quantity $v_c = \omega/k$ is called the phase velocity of the microscopic particle (wave), but the wave vector $\vec{k}$ is a vector designating the direction of the wave propagation. Thus, the phase velocity can be denoted by $\vec{v_c} = (\omega/k^2)\vec{k}$. This is a standard dispersion relation. Therefore, the systems described by linear quantum mechanics constitute a dispersed medium, then the solution Eq. (1.9) of the linear Schrödinger equation (1.8) is affirmatively a dispersive wave as stated in Chapter 1. Thus, the microscopic particles represented by Eq. (1.9) possess only a wave feature and are unstable.

When the dispersive effect is weak, the velocity of a wave can be denoted by

$$v_c = \omega / k = v'_{co} + \sigma_2 k^2, \tag{4.126}$$

where $v_{co}^{'}$ is the dispersionless phase velocity, and $\sigma_2 = (\partial^2 v_c / \partial k^2)\big|_{k=k_0}$ is the coefficient of the dispersion feature of the wave. The lowest order dispersionoccurring in the phase velocity is proportional to k^2, and the term proportional to k gives rise to the dissipation effect.

Therefore, the dispersed effect and the nonlinear interaction can all deform the form of the wave, but their mechanisms and effects are different. If they occur simultaneously in a dynamic equation, then the nature of microscopic particles described by the dynamic equation will be changed. If the balance between two deformation effects of wave each other, namely, the dispersion effect of the wave is suppressed by the nonlinear interaction, then the localization of microscopic particles appears possible, which can be verified in the following example.

For the particle described by the equation:

$$\phi_t + \phi\phi_x = 0\,, \tag{4.127}$$

where $\phi\phi_x$ is a nonlinear interaction without the dispersive term. It is easy to verify that $\phi = \Phi^{'}(x - \phi t)$ satisfies Eq. (4.127). This means that ϕ changes with x at different t obtained from Eqs.(4.127) in the velocity ϕ [1-3]. Thus, as time elapses, the front side of the wave gets steeper and steeper, until it becomes a triple-valued function of x due to the nonlinear interaction. This is just a deformation effect of the wave resulting from the nonlinear interaction. If we let $\phi = \Phi^{'} = \cos \pi x$ at $t = 0$, then at $x = 0.5$ and $t = \pi^{-1}, \phi = 0$ and $\phi_x = \infty$. The time $t_B = \pi^{-1}$ at which the wave becomes very steep is called the destroyed period of the wave. However, the collapsing phenomenon of the nonlinear interaction can be suppressed if a dispersion term ϕ_{xxx} is added into Eq. (4.127). In fact, in such a case, Eq. (4.127) becomes a KdV equation, $\phi_t + \phi\phi_x = v'\phi_{xxx}, (v' > 0)$ [57,60,90], which has a stable soliton solution, of which the form is $\sec h^2(X)$. This indicates clearly that although the dispersed force and nonlinear interaction can all distort a wave, a stable solitary wave and localization of particle can occur through the mutual interaction of these deformation effects only if the nonlinear interaction and dispersive effect exist simultaneously, and can balance and cancel each other in the system. This is just the basic condition and mechanism of form of a soliton and localization of the microscopic particles.

4.10. TheReal Existences of the Nonlinear Interaction in Physical Systems and Its Different Mechanism of Generation

4.10.1. TheReal Existences of the Nonlinear Interaction

The results, investigated above, tell us that the deformational influence of the dispersive effect on the wave can be suppressed by a nonlinear interaction, thus, the soliton solution of dynamic equations and localization of particles can occur in this case. These conditions of localization of the microscopic particles can be satisfied and guaranteed in the nonlinear Schrödinger equation and the nonlinear Klein-Gordon equation shown in Eqs.(2.66)-(2.69) in nonlinear quantum mechanics. Thus we can affirm that the microscopic particles described by these dynamic equations in the new theory can be localized as a soliton having a wave-corpuscle duality, which was already verified in Chapter 3.

Present questions are whether the nonlinear interactions really exist in physical systems. If they are existent, then what are their mechanisms of generation? Or, how do these nonlinear interactions produce in a physical system? These problems are worth serious study. Investigations show that the nonlinear attraction-interactions can be produced by the following five mechanisms. In the first mechanism, the attractive nonlinear interaction, which results in the localization of microscopic particles, is due to the motions of the studied particle, which changes the positions of other particles or the field of environment around the studied particle, thus it accepts an action of new interaction. Thus, this is called a self-interaction mechanism. This phenomenon occurs often in the system of may-bodies and many particles. In the second mechanism, the medium, in which the microscopic particles move, has an anomalous dispersion effect (*i.e.* $k^{''} = \partial^2 k/\partial\omega^2\big|_{\omega_0} < 0$) and nonlinear features resulting from the anisotropy of structure, then the state of the microscopic particles are modulated by the anomalous effect or nonlinear interaction and are finally localized. This is called aself-focusing mechanism. The third mechanism is called self-trapping, in which the microscopic particles are acted through the nonlinear interaction provided by the interaction between the moved microscopic particles and moved background field or lattice. The fourth mechanism is self-condensed, in which there is an attractive interaction among these particles; they will be condensed to the state with lower energy due to the lowering of temperature of the system to below the critical temperature, where the particle is Boson, thus it is often called the Bose condensation, or Bose-Einstein condensation of microscopic particles due to the attraction effect among the Bose particles. The fifth mechanism is called the self-localized mechanism, which is due to the double- or many- well interaction potentials generated by special structure of the systems or materials. In this case, the microscopic particles are spontaneously localized as a soliton with wave-corpuscle duality due to the effects of the double- or many- well potentials.

We can affirm that these mechanisms of nonlinear interaction are actually existent in physical systems because they are caused by the motions and interactions of microscopic particles, and the specialty and peculiarity of structure of physical systems. Their details are described as follows.

4.10.2. The Self-Interaction Mechanism of Nonlinear Interaction and Its Features

This mechanism was proposed first by Fermi [91]. In order to describe the dynamical property of radiation due to the charge distributionin an atom, Fermi [91] proposed his first nonlinear Schrödinger equation, inwhich the nonlinear interaction is due to the moved electron in an atom, which accepted an interaction by variation of electromagnetic radiation-of the electron itself. The nonlinear interaction is denoted by a reaction principle combined with the Schrödinger interpretation of the wave function, which is denoted by $V_{RR}(\psi^*\psi)\psi$, where $V_{RR}=-(2e/3c^3)\dddot{d}.\vec{r}$, is a radiation-reaction potential of a dipole generated by theSchrödinger 's charge distribution $\rho(r)=e\psi^*(r)\psi(r)$ as mentioned in Chapter 2.

Davydov and Pestryakov [92]studied the nonlinear localization phenomenon of a complex scalar field of spinless quasi-particles with innertialess self-interaction. The Hamiltonian density of the scalar field with self-interaction in one-dimensional infinite space is given by

$$H'=\frac{\hbar^2}{2m}\left|\frac{\partial\phi}{\partial x}\right|^2\mp b|\phi|^4 \tag{4.128}$$

where H' $=\mathcal{H}$ is the Hamiltonian density, b is a nonlinear parameter independent of the velocity, the "-" and "+" signs correspond to attraction and repulsion self-interactions, respectively. Such a self-interaction can be realized by local interaction between the microscopic particles, which correspond to the field, or by local interaction of these quasiparticles with the field of inertialess displacements of the other particles, which are not considered explicitly here. This model resembles that of the ϕ^4 -type self-interaction.

If the self-interaction in the systems is absent, i.e., $b=0$, the field $\phi(x,t)$ describes the states of non-interacting quasiparticles with mass m . Their motion is described by a plane wave with wave vector k and energy $\hbar^2k^2/2m$. If $b\neq 0$, the corresponding equation of motion is

$$i\hbar\frac{\partial\phi}{\partial t}+\frac{\hbar^2}{2m}\frac{\partial^2\phi}{\partial x^2}\mp 2b|\phi|^2\phi=0 \tag{4.129}$$

This is a standard nonlinear Schrödinger equation and the same with Eq.(3.1) in nonlinear quantum mechanics, then it has a soliton solution as given by (3.8). Therefore, electrons in these organic conductors are self-localized due to the self-interaction. In other words, equation (4.129) provides stationary solutions in the form of modulated plane waves with an inhomogeneous spatial distribution of the density of the microscopic particle with a small constant velocity.

4.10.3. The Self-Focusing Mechanism of Nonlinear Interaction and Corresponding Dynamic Equations

We now elucidate the self-focusing mechanism of nonlinear interaction. As it is known as mentioned above, the nonlinear interaction in this mechanism is due to the anomalous dispersion effect or a nonlinear effect of the systems or materials. We now establish a corresponding dynamic equation of microscopic particles in the system using this mechanism.

It is well known that a linear wave equation in linear quantum mechanics is often denoted by a linear operator L' which consists of ∂_t and ∇, i.e.,

$$L'(\partial_t, \nabla)\Phi = 0. \tag{4.130}$$

Its solution can be approximately denoted by the following monochromatic wave:

$$\Phi = \varepsilon\varphi e^{i(\vec{k}\cdot\vec{x}'-\omega t')}$$

with a constant amplitude $\varepsilon\varphi$, corresponding dispersion relation of frequency ω and wave vector k can be represented as

$$L(-i\omega, ik) = 0, \text{ or } \omega = \omega(\vec{k}). \tag{4.131}$$

If comparing Eq.(4.130) with Eq.(4.131) we find that ω and k correspond to $-i\partial_{\vec{x}'}$ and $-i\partial_{t'}$, respectively [20-22]. Thus, Eq.(4.130) becomes

$$\left[i\partial_{t'} - \omega(-i\partial_{\vec{x}'})\right]\varphi e^{i(\vec{k}\cdot\vec{x}'-\omega t')} = 0 \tag{4.132}$$

where $\partial_{\vec{x}'}$ is the gradient with respect to $\vec{x}'$ and $\omega(-i\partial_{\vec{x}'})$ is the pseudo-differential operator obtained by replacing k with $-i\partial_{\vec{x}'}$ in $\omega(\vec{k})$.

In a weak nonlinear medium, which responds adiabatically (or instantaneously) to a wave of finite amplitude, the nonlinearity is expected to affect the dispersion relation of the carrier wave (in addition to the generation harmonics of smaller amplitudes), thus, the frequency of the microscopic particle relates to its intensity.

Sulem et al., [93,57,22]replaced the frequency $\omega(\vec{k})$ by a function $\Omega(\vec{k}, \varepsilon^2|\varphi|^2)$ with $\Omega(\vec{k}, 0) = \omega(\vec{k})$ due to the nonlinear feature of the medium. This effect is just called a self-focusing effect. In this case a complex wave amplitude φ is no longer a constant but is modulated in space and time, and thus becomes dependent on the slow variables $X = \varepsilon x'$ and $T = \varepsilon t'$. In Eq. (4.132), the derivatives $\partial_{t'}$ and $\partial_{\vec{x}'}$ should also be thus replaced with $\partial_{t'} + \varepsilon\partial_T$ and $\partial_{\vec{x}'} + \varepsilon\nabla$, where ∇ now denotes the gradient with respect to the slow spatial variable*X*. Then the equation (4.132) becomes

$$\left[i\partial_{t'} + i\varepsilon\partial_T - \Omega(-i\partial_{\vec{x}'} - i\varepsilon\nabla, \varepsilon^2|\varphi|^2)\right]\varphi e^{i(\vec{k}\cdot\vec{x}'-\omega t')} = 0.$$

It is natural to obtain the weakly nonlinear dispersion relation

$$\left[\omega + i\varepsilon\partial_T - \Omega(\vec{k} - i\varepsilon\nabla, \varepsilon^2|\varphi|^2)\right]\varphi = 0.$$

where ε is small. If we expanded various quantities in this equation to powers of ε and keep the terms up to the second order, the following equation is obtained [93,20-21]

$$i(\partial_T + \overrightarrow{v_g} \cdot \nabla)\varphi + \varepsilon\left[\nabla \cdot \nabla + b|\varphi|^2\right]\varphi = 0. \tag{4.133}$$

where $v_g = \nabla_k \omega$ is the group velocity, the coupling coefficient b is related to the expansion coefficient of wave function of microscopic particles (or intensity of wave) and is given by $\partial\Omega/\partial(|\varphi|^2)$ evaluated at $|\varphi|^2 = 0$ and at the carrier wave vector $\vec{k}$. According to Sulem et al.'s method [93], equation (4.133) may serve as an initial value problem in time. This equation is conveniently written in a reference frame moving at the group velocity by defining $\vec{\xi} = \overrightarrow{X} - T\overrightarrow{v_g}$. Rescaling the time in the form $\tau = \varepsilon T$, the following nonlinear Schrödinger equation can be obtained

$$i\frac{\partial\varphi}{d\tau} + \nabla^2\varphi + b|\varphi|^2\varphi = 0 \tag{4.134}$$

where the spatial derivatives are now taken with respect to ξ. This equation is the same as Eq. (3.1) in nonlinear quantum mechanics and Eq.(3.170) or Eq.(3.171) in light fiber. This affirmed that the photon soliton in the light fiber is formed by the self-focusing mechanics by means of the Kerr effect, a nonlinear effect.

The above discussion illustrates that the motion of a microscopic particle (or wave) is always described by the nonlinear Schrödinger equation under this self-focusing mechanism, which results in the nonlinear interactions in the systems.

4.10.4.The Self-Trapping Mechanism of Nonlinear Interaction and Its Properties

(1) Form of Self-Trapping Mechanism

We are quite interested in the self-trapping mechanism of the nonlinear interaction. Through this mechanism, the microscopic particles in the systems become as a soliton with wave-corpuscle duality due to the nonlinear interaction. The mechanism occurs in crystals, polymers, and biomacromolecules, for example, molecular crystal-acetanilide (ACN) and protein molecules. As it is known, Landau early proposed this concept or mechanism of self-trapping on the motion of an electron in a crystal lattice. He suggested that localized effect of electrons would be to polarize the crystal, which, in turn, would lower its energy. Pekar [94], Frohlich [95,96], and Holstein [97] discussed Landan's suggestion [98-99] in detail. Other examples of the self-trapping include electromagnetic energy in a plasma and hydrodynamic energy in a water tank. Subsequently, Davydov and co-workers [100-104] bring it in the acetanilide and protein molecules to study the form of soliton of the excitons. In these systems the peptide groups (or amino acid molecules in proteins), which consist of four atoms (N, H, O and C) bound by chemical forces, are arranged along the molecular chains in the form of...H-N-C=0...H-N-C=0...H-N-C=0..., of which ...H-X is a hydrogen bond, X is N-C=0 in which an amide group, C=0 is a carbohyl group. The neighboring peptide groups in a chain are linked by hydrogen bonds. In acetanilide, two close chains of hydrogen bonded amide-I groups run through the crystal. In alpha-helix protein molecules there are three amino acid molecular chains. The collective excitations in these systems that correspond to intramolecular excitations are very specific in nature and are different from ordinary elementary excitations in three-dimensional crystals. In protein molecules and acetanilide, the intramolecular excitations include amide-I vibrativonal quantum with energy of 0.205 eV excited by the energy released in the hydrolysis of adenosinc triphosphate (ATP) molecules, and an electric dipole moment of 0.3 Debye directed along the axis of the molecular chains. Davydov [100-104]referred to the intramolecular excitation or quantum of amide-I vibration as an extra electron or exciton. When the true vibrations of the lattices or peptide groups (or amino acid molecules) and their interaction with the excitons are considered, a nonlinear interaction will occur in these systems. Thus, these excitons could be self-trapped as a localized soliton. Davydovet.al. [38-44] studied its properties in simple one-dimensional systems.

(2)TheWave Features of Excitons in Uniform Crystals without Exciton-Phonon Interaction

In cases without exciton-phonon interaction, the excitons have only a wave feature and move freely in uniform systems, while the true vibrations of the lattices or peptide groups (or amino acid molecules) and their interaction with the excitons are not considered, i.e., the true motions of lattices or peptide groups are completely frozen, and the interaction of the exciton with lattice or peptide groups is replaced by an average field. For an infinite chain of weakly bound molecules (or groups of atoms) with mass*M*, which are separated at a distance r_0 , we assume that the excitons are characterized by energy ε_0 and an electric dipole moment $\vec{d}$ directed along the chains. When the molecule is excited the states of the exciton are changed due to its interacting with neighboring peptide groups. This results in a change in the equilibrium distances between the peptide groups or amino acid molecules. If the molecules are identical, there arise some additional resonance interactions, which cause the transfer of excitations from one molecule to another. In this case, Davydov [100-104] gave the Hamiltonian operator of the excitons with energy ε_0 by the following form

$$H_{ex} = \sum_n \left[\left(\varepsilon_0 - D \right) B_n^+ B_n - J (B_n^+ B_{n+1} + B_{n+1}^+ B_n) \right] \tag{4.135}$$

where $B_n^+ (B_n)$ is the exciton's creation (annihilation) operator at the *n*th site with an energy $\varepsilon_0 = 0.205$ eV. They satisfy the commutation relation:

$$\left[B_n, B_m^+\right] = \delta_{nm}, \left[B_n, B_m\right] = [B_n^+, B_m^+] = 0. \tag{4.136}$$

Also in Eq. (4.135), the $\varepsilon_0 B_n^+ B_n$ denotes the kinetic energy of the exciton, $J(B_n^+ B_{n+1} + B_{n+1}^+ B_n)$ represents the resonant (or dipole-dipole) interaction between neighboring excitons, $J = 2\vec{d}^2 / r_0^3$ is the resonance (or dipole-dipole) interaction that determines the transition of an exciton from one molecule to another, and $DB_n^+ B_n$ denotes the interaction of the exciton with the lattice or peptide groups. *D* is the deformation excitation energy, and is approximately a constant. This means that the nonlinear interaction is frozen, and is denoted by a constant. Thus, the true motions of lattices or peptide groups are completely neglected and the interaction of the lattice or peptide groups with the exciton is replaced by an average constant field. Thus, we can conjecture that the motion of the exciton satisfies the linear quantum mechanics, in this case.

Davydov [100-104] represented the excitations state of the excitons in the system by the wave function:

$$|\Psi\rangle = \sum_n \psi_n(t) B_n^+ |0\rangle_{ex} \tag{4.137}$$

where $|0\rangle_{ex}$ is the ground state of the exciton, the coefficient $\psi_n(t)$ satisfies the normalization condition $\sum_n |\psi_n(t)|^2 = 0$. Davydov assumed that the wave function $|\Psi\rangle$ satisfies the following the Schrödinger equation,

$$i\hbar \frac{\partial}{\partial t} |\Psi\rangle = H |\Psi\rangle. \tag{4.138}$$

Then we can get

$$i\hbar \frac{\partial \psi_n}{\partial t} = \left[(\varepsilon_0 - D)\psi_n\right] - J(\psi_{n+1} + \psi_{n-1}). \tag{4.139}$$

In the continuum approximation,

$$\psi_n(t) = \psi(x,t), \psi_{n\pm1}(t) = \psi(x,t) \pm r_0 \frac{\partial \psi}{\partial x} \mp \frac{1}{2!} r_0^2 \frac{\partial^2 \psi}{\partial x^2}, \tag{4.140}$$

We get the equation of motion of the excitons

$$i\hbar\frac{\partial\psi(x,t)}{\partial t}=(\varepsilon_0-D-2J)\psi(x,t)-\frac{\hbar^2}{2m}\frac{\partial^2\psi(x,t)}{\partial x^2} \tag{4.141}$$

where $m=\hbar^2/2Jr_0^2$ is the effective mass of the exciton. Equation (4.141) is just a linear Schrödinger equation; this shows that the exciton, an elementary excitation, satisfies the rules of the linear quantum mechanics and only has a wave feature.

Now,makingthe transformation $B_n=\frac{1}{\sqrt{N}}\sum_k A_k e^{iknr_0}, A_n=\frac{1}{\sqrt{N}}\sum_k B_k e^{-iknr_0}$ and insertingit into Eq. (4.141), we get $H_{ex}=\sum_\kappa E(k)A_k^+A_k$, where $E(k)=\varepsilon-D-2J\cos\kappa r_0$, which is the effective energy of the exciton. Then thewave function of the exciton in this case is given by

$$\psi_k(x,t)=\frac{1}{\sqrt{N}}\sum_\kappa e^{i[knr_0-\omega(k)t]}\psi'(x-nr_0),$$

where $\hbar\omega(k)=E(k)$, and $\psi'(x-nr_0)$ is the amplitude of the exciton at the nth site. This representation shows that once an exciton occurs, all components in the chains are excited. This is called a collective excitation state. This manifests that the exciton state is represented by a linear wave packet, which is a linear superposition of many plane waveswith energy $E(k)$ concerning$k$, spread over the entire chain, rather than localized at one molecule. Hence, it is an expanded elementary excitation state. It may be represented by $\psi_k(x,t)=A(k)e^{i[kx-\omega(k)t]}$ with $A(k)=\sum_n e^{ik(nr_0-x)}\psi'(x-nr_0)$. This shows that the exciton moves as a plane with a wavelength $\lambda=2\pi/k$, but its amplitude is modulated in this case.

(3)The Nonlinear Interaction and Localization of Excitons in the Self-Trapping Mechanism

However, when the exciton is excited in the peptide groups due to the vibrations of the amide-I (or the C=O bonds), the states and positions of the peptide groups will be also changed due to its coupling effect with the vibrations of the amide-I. If the interaction and the low-frequency vibration of the peptide group are considered, and the displacement of the peptide groups is denoted by u_n, Davydov et al., [100-104] gave the low-frequency vibrational Hamiltonian of the peptide groups and the interaction Hamiltonian between the exciton and the vibration of the peptide group by

$$H_{ph}=\sum_n\left[\frac{P_n^2}{2M}+\frac{1}{2}w(u_n-u_{n-1})\right] \tag{4.142}$$

and

$$H_{\text{int}} = \sum_n \chi(u_{n+1} - u_{n-1})B_n^+ B_n \tag{4.143}$$

respectively, where M is the mass of the peptide group, w is the spring constant of the molecular chain, P_n is the conjugate moment of u_n, $\chi = \partial J/\partial u_n$ is the coupling constant. In such a case, Davydov rewrote the Hamiltonian of the system, which is

$$H = H_{ex} + H_{ph} + H_{\text{int}}. \tag{4.144}$$

In this case, Davydov further used the following wave function to express the collective excitation state in the system

$$|\Phi(t)\rangle = \sum_n \phi_n(t)B_n^+ |0\rangle_{ex} \exp\left\{-\frac{i}{\hbar}\sum_n [\beta_n(t)P_n - \pi_n(t)u_n]\right\}|0\rangle_{ph} \tag{4.145}$$

where $|0\rangle_{ph}$ is the ground state of the phonon. Using the functional $\langle\Phi(t)|H|\Phi(t)\rangle$ and the variational approach, Davydov et al., [100-104]got

$$i\hbar\frac{\partial\phi_n}{\partial t} = [\varepsilon_0 + W + \chi(\beta_{n+1} - \beta_{n-1})]\phi_n - J(\phi_{n+1} + \phi_{n-1}) \tag{4.146}$$

$$\frac{\partial^2\beta^2}{\partial t^2} + \frac{w}{M}(2\beta_n - \beta_{n-1} - \beta_{n+1}) = \frac{\chi}{M}(|\phi_{n+1}|^2 - |\phi_{n-1}|^2) \tag{4.147}$$

where,

$$\beta_n(t) = \langle\Phi(t)|u_n|\Phi(t)\rangle, \pi_n = \langle\Phi(t)|P_n|\Phi(t)\rangle = M\frac{\partial\beta_n}{\partial t},$$

$$W = \frac{1}{2}\sum_n\left[M(\frac{\partial\beta_n}{\partial t})^2 + w(\beta_n - \beta_{n-1})^2\right].$$

In the continuity approximation, Eqs. (5.56) and (5.57) become

$$\left[i\hbar\frac{\partial}{\partial t} - \Lambda + \frac{\hbar^2}{2m}\frac{\partial^2}{\partial x^2} - 2\chi\frac{\partial\beta(x,t)}{\partial x}\right]\phi(x,t) = 0, \tag{4.148}$$

$$\left[\frac{\partial^2}{\partial t^2} - v_0^2\frac{\partial^2}{\partial x^2}\right]\beta(x,t) - \frac{2\chi r_0}{M}\frac{\partial}{\partial x}|\phi(x,t)|^2 = 0 \tag{4.149}$$

where $\Lambda = \varepsilon_0 - 2J + W$, $v_0 = r_0\sqrt{w/M}$ is the sound speed of the molecular chain. Now let $\xi = x - x_0 - vt$, $G = 4\chi^2/\left[w\left(1-s^2\right)\right]$, $s = v/v_0$, $\mu = G/4J$,we can get from Eqs.(4.148) and (4.149) that

$$i\hbar\frac{\partial\phi}{\partial t} - \Lambda\phi + \frac{\hbar^2}{2m}\frac{\partial^2\phi}{\partial x^2} + G|\phi|^2\phi = 0 \tag{4.150}$$

$$\frac{\partial\beta(x,t)}{\partial x} = -\frac{2\chi r_0|\phi(x,t)|^2}{w(1-s^2)} \tag{4.151}$$

Equation (4.150) is a nonlinear Schrödinger equation, which is the same Eq.(2.66) at $A(\phi) = 0$. It has a soliton solution as given by [[100-108, 57,60,20-22]

$$\phi(x,t) = \sqrt{\frac{\mu}{2}}\sec h\left[\frac{\mu}{r_0}(x - x_0 - vt)\right]\exp\left\{\frac{\hbar v}{2Jr_0^2}(x - x_0) - \frac{Et}{\hbar}\right\} \tag{4.152}$$

Thus, from Eqs.(4.151)-(4.152) we can give the solution of Eq. (4.149) as follows

$$\beta(x,t) = -\frac{\chi r_0^{\ 2}}{w(1-s^2)}\tanh\left[\frac{\mu}{r_0}(x - x_0 - vt)\right] \tag{4.153}$$

Equations (4.150)-(4.153) show clearly that the exciton now is localized to become a soliton due to the nonlinear interaction $G|\phi|^2\phi$. We can find out the energy of the soliton as

$$E = E_0 + \frac{1}{2}m_{sol}v^2 \tag{4.154}$$

where v is the soliton velocity, E_0 is the rest energy of the energy of the soliton, given by

$$E_0 = \varepsilon_0 - 2J - \frac{\chi^2}{3w^2J} < E_{ex} = \varepsilon_0 + 2J \tag{4.155}$$

and

$$m_{sol} = m \mid \frac{4\chi^2(1 + 3s^2/2 - s^4/2)}{3w^2Jv_0^2(1-s^2)^3} > m.$$

Equation (4.155) shows that the soliton energy is lower than that of the exciton given in Eq. (4.141) by $E_B = -\chi^4/3w^2J$, which is just the binding energy of the soliton. This indicates clearly that the exciton is self-trapped as a stable soliton through energy depression in this case. Upon that the localization of the exciton is called self-trapping, the nonlinear interaction $G|\phi|^2\phi$ accepted by it in the dynamic equation(4.150) is produced by virtue of the self-trapping mechanism. Concretely speaking, it is caused by the interaction between the moved exciton and the moved peptide group, which are exhibited clearly in Eqs.(4.148) and (4.149), their coupling interaction is expressed by Eq.(4.150). This is the essence of the self-trapping mechanism, and corresponding nonlinear interaction.

4.10.5. The Self-Localized Mechanism of the Nonlinear Interaction and Its Dynamic Equations

The self-localized mechanism of nonlinear interaction often occurs in systems with anisotropic structure, its familiar example is hydrogen-bonded systems.

A hydrogen-bonded system is composed of hydrogen ions and heavy ions or a heavy ion group through covalent bonds and hydrogen bonds. It is a chain structure and exists widely in many condensed matters and living systems, such as ice, solid alcohol, water, carbon hydrates, DNA and proteins [109-113], Experiments show that there is considerable electrical conductivity even though electron transfer through the systems are hardly supported in the systems, such as ice crystal. The mechanism of the electric conductivity in the hydrogen-bonded systems is a long-standing problem and was studied widely [113-128]. This investigation becomes even more important in view of the close connection with the problem of proton transfer across biological membranes, which is something that could explain some fundamental properties of life.

An ice crystal is a typical hydrogen-bonded system [110-112].Itis a system considered as one-dimensional chains, so-called Bernal- Fowler filaments [109-112]. In the normal state, the two kinds of arrangements of the type OH__H.... OH__H.... OH__H....OH__H and the type H__OH....H__OH....H__ OH....H__OH have the same energy. Thus, the potential energy of the proton should have the form of nonlinear double-well with two minima corresponding to its two equilibrium positions. The barrier of the double-well potential has the order of the binding energy of the covalent bond H__O, which is approximately 20 times larger than that in a hydrogen bond. Experimental observations found that ice crystals exhibit considerable electric conductivity; along the chains it is about 10^3-10^4 times larger than that in perpendicular direction [110-112]. We infer that when the system is perturbed by an externally applied field, the states and positions of the protons in the double-wells are changed. In this case, the protons move from one well to another by means of a translation and a jump or migration in the interbonds and intrabonds to form an ionic defect, hydroxonium, H_3O^+, and another water molecule is dissociated by losing one of its protons to its neighbors, forming another ionic defect, hydroxyl, OH^{-}, each proton (H^+) can be transferred inside the H__OH....H bridge interchanging role of the covalent and bonds hydrogen with the oxygen atom. Thus, the proton transfer along the hydrogen-bonded chain is carried out by means of the migration of hydroxonium and hydroxyl ionic defects in the intrabonds, which is shown in Figure 4.2a. However, when a proton approaches a molecule occupying a boundary of the chain, the transfer is not possible in the same direction, but can be achieved with a re-orientation of OH groups by the Bjerrum(or bonded) defect, a pair of D and L defects, which is induced by the rotations of OH groups as shown in Figure 4.2b.

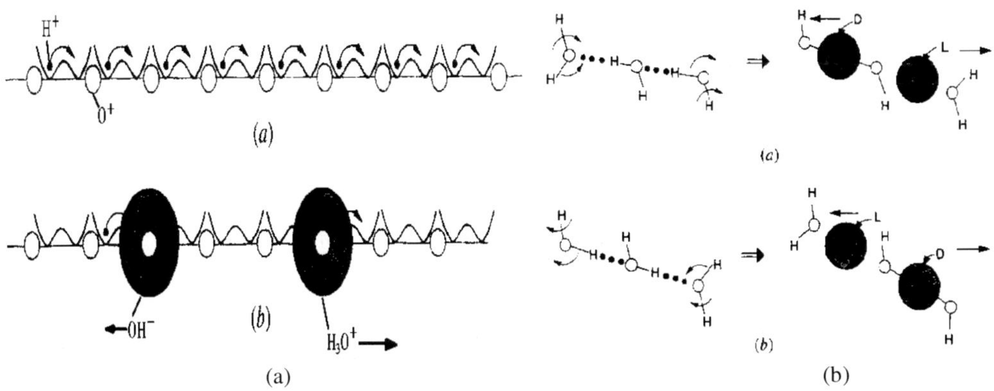

Figure 4.2. Two ionic defects (a) and two Bjerrun defects (b) in a hydrogen-bonded system (ice crystal), respectively.

Therefore the proton transfer is carried out by a combination motion of the ionic and bonded defects in the hydrogen-bonded system. Plenty of researchers [113-128] investigated the properties and rules of proton transfer in hydrogen-bonded systems using different models. In Pang's model the features of structure of the molecule and nonlinear property of the proton transfer are sufficiently considered, a new model of proton transfer with two components, shown in Figure 4.3, and the corresponding Hamiltonian of the systems are proposed. In the one-dimensional hydrogen-bonded chains, the double-well potential of the proton is represented by $U(R_n)=U_0[1-(R_n/r_0)^2]^2$.

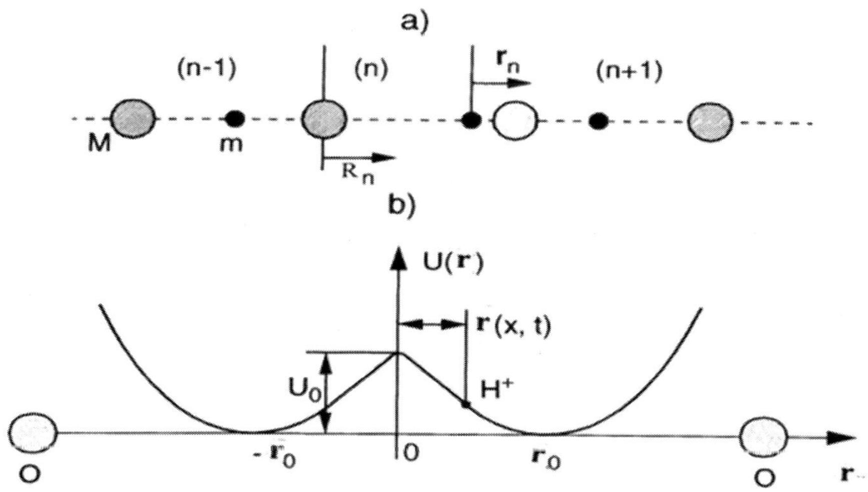

Figure 4.3.One-dimensional lattice model of double-well potential in the hydrogen-bonded systems.

If considering again the elastic interaction, which is caused by the covalent interaction and the coupled interaction between protons and heavy ions, and the resonant or dipole-dipole interaction between neighboring protons and the changes of relative positions of neighboring heavy ions, resulting from this interaction, as well as the harmonic model with acoustic vibrations of low frequency for the heavy ionic sublattice, then the Hamiltonian of the systems is expressed by [121-128]

$$H = H_p + H_{ion} + H_{int} = \sum_n [\frac{1}{2m} p_n^2 + \frac{1}{2} m\omega_0^2 R_n^2 - \frac{1}{2} m\omega_1^2 R_n R_{n+1} + U_0 [1 - (\frac{R_n}{r_0})^2]^2] + \sum_n [\frac{1}{2M} P_n^2 + \frac{1}{2} W (u_n - u_{n-1})^2] + \sum_n \left[\chi_1 m (u_{n+1} - u_n) R_n^2 + m\chi_2 (u_{n+1} - u_n) R_n R_{n+1} \right] \quad (4.156)$$

where the proton displacements and momentum are R_n and $p_n = m\dot{R}_n$, respectively, the first being the displacement of the hydrogen atom from the middle of the bond between the nth and the (n+1)th heavy ions (oxygen atoms in ice), r_0 is the distance between the central maximum and one of the minima of the double-well, U_0 is the height of the barrier of the double-well potential. Similarly, u_n, and $P_n = M\dot{u}_n$ are the displacement of the heavy ion from its equilibrium position and its conjugate momentum, respectively. $\chi_1 = \partial\omega_0^2 / \partial u_n$ and $\chi_2 = \partial\omega_1^2 / \partial u_n$ are coupling constants between the protonic and heavy ionic sublattice, which represent the changes of the energy of vibration of the protons and of the dipole-dipole energy between the neighboring protons due to a unit extension of the heavy ionic sublattice, respectively. $m\omega_1^2 R_n R_{n+1} / 2$ show the correlation interaction between neighboring protons caused by the dipole-dipole interactions, and ω_0 and ω_1 are diagonal and non-diagonal elements of the dynamic matrix of the proton, respectively. ω_0 is also the Einstein resonant frequency of the proton,W is the elastic constant of the heavy ionic sublattice. m and M are the masses of the proton and heavy ion, respectively. $C_0 = u_0 (W/M)^{1/2}$ is the sound velocity in the heavy ionic sublattice, and u_o is the lattice constant. The r_0, R_n and r_n and U_0 are shown in Figure 4.3 The part H_P of H is the Hamiltonian of the protonic sublattice with an on-site double-well potential $U(r_n)$, H_{ion} being the Hamiltonian of the heavy ionic sublattice with low-frequency harmonic vibration, and H_{int} is the interaction Hamiltonian between the protonic and heavy ionic sublattices. This model is significantly different from the ADZ model [113], and the Pnevmatikos etal. models [115-120], although it is also a coupled model of two oscillators (proton and heavy ion).

However, the proton isquantized, thus the Hamiltonian of the systems should be quantized by using the standard transformation [121-128]:

$$R_n = (2m\omega_0/\hbar)^{-1/2}(a_n + a_n^+),\ p_n = m\dot{R}_n = (\hbar m\omega_0/2)^{1/2}(-i)(a_n - a_n^+) \quad (i = \sqrt{-1}) \quad (4.157)$$

where $a_n^+(a_n)$ is the creation (annihilation) operator of the proton. Then Eq. (4.156) becomes

$$H=\sum_n[\hbar\omega_0(a_n^+a_n+1/2)-(\hbar\omega_1^2/4)(a_n^+a_{n+1}+a_{n+1}^+a_i+a_n^+a_{n+1}^++a_{n+1}a_n)-$$
$$(U_0\hbar/mr_0^2\omega_0)(a_n^+a_n+a_na_{=n}^++a_n^+a_n^++a_na_n)+(\hbar^2U_0/4m^2r_0^4\omega_0^2)(a_n^++a_n)(a_n+a_n^+)(a_n+a_n^+)(a_n+a_n^+)]$$
$$+\sum_n[\frac{M}{2}P_n^2+\frac{W}{2}(u_n-u_{n-1})^2]+(\hbar\chi_1/4\omega_0)\sum_n[(u_{n+1}-u_{n-1})(a_n^+a_n+a_na_n^++a_n^+a_n^++a_na_n)]+$$
$$(\hbar\chi_2/2\omega_0)\sum_n[(u_{n+1}-u_n)(a_{n+1}^+a_n+a_{n+1}a_n^++a_{n+1}^+a_n^++a_{n+1}a_n)]+U_0/4 \tag{4.158}$$

In this system the collective excitations of the protons and vibrations of the heavy ions, arising from the nonlinear interactions caused by the localized fluctuation of the protons and the deformation of structure of the heavy ionic sublattice have coherent features. Thus the wave function, describing the collective excitations in the systems, can be represented by [121-128]

$$|\Phi>=|\varphi>|\beta_q>=(1/\lambda')(1+\sum_n\varphi_n(t)a_n^+)|0>_{pr}\exp\{\sum_n 1/i\hbar[u_n(t)P_n-\pi_n(t)u_n]\}|0>_{ph} \tag{4.159}$$

where $|0\rangle_{pr}$ and $|0\rangle_{ph}$ are the ground states of the proton and the vibrational excitation of the heavy ionic sublattice, respectively. φ_n (t), $\pi_n(t)=\langle\Phi|P_n|\Phi\rangle$ and $u_n(t)=\langle\Phi|u_n|\Phi\rangle$ are three unknown functions, λ′is a normalization constant.

Utilizing the |Φ> in Eq.(4.159) and the Hamiltonian (4.158), from the Heisenberg equation:

$$i\hbar\frac{\partial}{\partial t}<\Phi|u_n|\Phi>=i\hbar\frac{\partial u_i(t)}{\partial t}=<\Phi|[u_n,H]|\Phi>,$$

$$i\hbar\frac{\partial}{\partial t}<\Phi|P_n|\Phi>=i\hbar\frac{\partial}{\partial t}\pi_n(t)=<\Phi|[P_n,H]|\Phi>,$$

we can get the equation of motion for the heavy ion, it is

$$M\frac{\partial^2}{\partial t^2}u(x,t)=W(u_{n+1}+u_{n-1}-2u_n)+\hbar\chi_1/2\omega_0(|\varphi_{n+1}|^2-|\varphi_{n-1}|^2)$$
$$+\hbar\chi_2/2\omega_0(\varphi_{n-1}^*\varphi_n+\varphi_n^*\varphi_{n-1}-\varphi_n\varphi_{n+1}^*-\varphi_n^*\varphi_{n+1}) \tag{4.160}$$

In the continuum approximation, Eq.(4.160)is represented by

$$M\frac{\partial^2}{\partial t^2}u(x,t)=Wu_0^2\frac{\partial^2}{\partial x^2}u(x,t)+\hbar(\chi_1+\chi_2)u_0\frac{\partial}{\partial x}|\varphi(x,t)|^2 \tag{4.161}$$

If using the time-dependent Schrödinger equation $i\hbar\frac{\partial}{\partial t}|\Phi(t)>=H|\Phi(t)>$ and in the continuum approximation we can get

$$i\hbar\frac{\partial\varphi(x,t)}{\partial t}=\varepsilon\varphi(x,t)-Ju_0^2\frac{\partial^2}{\partial x^2}\varphi(x,t)-G|\varphi(x,t)|^2\varphi(x,t) \tag{4.162}$$

where g' is an integral constant,

$$s=v/C_{0,}\varepsilon=3\hbar\omega_0/2-G(t)+U_o/4\ -\hbar U_0/mr_0^2\omega_0+g'\hbar(\chi_1+\chi_2)u_0/\omega_0$$

$$-\hbar\omega_1^2/2\omega_0+3\hbar^2U_0/(2m^2r_0^4\omega_0^2),J=\hbar\omega_1^2/4\omega_0,\varepsilon_0=\hbar\omega_0, \tag{4.163}$$

$$G=\{3\hbar^2U_0/2m^2r_0^4\omega_0^2-[\hbar^2u_0^2(\chi_1+\chi_2)^2/(\omega_0^2MC_0^2(1-s^2))]\},\mu=\frac{G}{4J} \tag{4.164}$$

We see from the above result, that the properties of the proton transfer in the hydrogen bonded systems are determined by the nonlinear Schrödinger equation (4.162), which is the same Eq.(2.66) at $A(\phi)=0$. This indicated that the proton transfer can be described by nonlinear quantum mechanics, where the nonlinear interaction $G|\varphi(x,t)|^2\varphi(x,t)$ is generated by thedouble-well potential because the height of the barrier of the double-well potential,U_0, is contained in the nonlinear interaction energy G. Thus, the nonlinear interaction is formed by the self-localized mechanism.

In one word, we here established and provided five mechanisms for the forms of the nonlinear interaction in Pang's nonlinear quantum mechanics in variously different physical-systems. This manifested that there are different nonlinear interactions in different systems, but they can all be described by the nonlinear quantum mechanics. Therefore, Pang's nonlinear quantum mechanics isa universal theory and can be used widely in all physical systems.

In Chapters 5-8, we will describe in detail the nonlinear quantum mechanical properties of microscopic particles in different physical systems using the new theory. Applications of the new theory are used to establish the concrete dynamic equations of microscopic particles and to seek the mechanism of nonlinear interaction and its representation forms, starting from features of structure of the system and features of interaction of microscopic particles and the background field, as well as to give the properties and rules of motion of microscopic particles, and to explain some physical phenomena in these systems using nonlinear quantum mechanical method and theory. These problems are described in Chapters 5-8.

REFERENCES

[1] J. Potter, *Quantum mechanics*, North-Holland publishing Co. 1973.

[2] M. Jammer, The conceptual development of quantum mechanics, Tomash, Los Angeles, 1989.

[3] Zhou Si Xun, Quantum mechanics, *Shanghai Press of Science and Technology*, Shanghai, 1960.

[4] F.Calogere, Degasperis A, *Nuovo Cimente*, B 32(1976)201.

[5] R. H. Bullongh, P. J.Caudrey, *Solitons,* Springer, Berlin, 1980, pp301-324.

[6] Konopelchenko, B. G., *Phys. Lett.* A 74 (1979) 189.111.

[7] Konopelchenko, B. G., *Phys. Lett.* A 87 (1982) 445.

[8] Konopelchenko, B. G., *Phys. Lett.* B 100 (1981) 254.

[9] B. G. Konopelchenko, *Nonlinear integrable equations*, Springer, Berlin 1987.

[10] B. G. Konopelchenko, *Solitons in multidimensions*, World Scientific, Singapore, 1995.

[11] P.L.Lax, *Comm.Pure and Appl. Math.* 21(1968)467.

[12] V. E.Zakharov, & A. B. Shabat, *Sov. Phys.-JETP*, 34 (1972) 62.

[13] V. E.Zakharov, & Shabat, A. B.. *Sov. Phys.-JETP*, 37(1973)823.

[14] F. Calogero, and A. Degasperis, *Nuovo Cimento* 32B (1976) 201.

[15] F.Calogero, *Nonlinear Evolution Equations Solvable by Spectral Transform*, London, Plenum Press, 1978.

[16] V. Zakharov, and I. Fadeleev, *Funct. Analiz* 5 (1971) 18.

[17] Pang Xiao-feng,*Physica B* 405(2009)4327.

[18] Pang Xiao-feng, *Physica B* 403 (2008) 4292.

[19] Pang Xiao-feng, *Frontiers of physics in China*, 3 (2008)413.

[20] Pang Xiao-feng, *Quantum Mechanics in Nonlinear Systems*, Singapore: World Scientific Publishing Co., 2005.

[21] Pang Xiao-feng, *nonlinear quantum mechanics*, Beijing, Chinese Electronic Industry Press, 2009.

[22] Pang Xiao-feng, *The Theory of nonlinear quantum mechanics*, Chongqing, Chinese Chongqing Press, 1994.

[23] R. J. Glanber. Coherent and incoherent states of the radiation field, *Phys. Rev.* 13(1963)2766.

[24] A. S.Davydov, *Theory of Molecular Excitons*, lnd Nauka, Moscow, 1968; 2nd New York, Plenum, 1975.

[25] Pang Xiao-feng, *Science in China*, G, 51(2008) 225.

[26] Pang, Xiao-feng, *European Physical Journal B* 10(1999)415.

[27] Pang, Xiao-feng, *European Phys. J.B* 19(2001)297.

[28] Pang, Xiao-feng, *J. Phys. Condens. Matter* 2 (1990) 9541.

[29] L. N.Bardeen, L. N. Cooper and J. R. Schrieffer, *Phys. Rev.* 108 (1957) 1175.

[30] Pang Xiao-feng and Chen Xianron, *Inter. J. Infr. Mill. Waves* 22 (2001) 291.

[31] Pang Xiao-feng and Chen Xianron, *Inter. J. Infr. Mill. Waves* 23 (2002) 375.

[32] Pang Xiao-feng and Chen Xianron, *J. Phys. Chem. Solids* 62 (2001) 793.

[33] Pang Xiao-feng and Chen Xian-ron, *Phys. Stat. Sol.* 229 (2002) 1397.

[34] Pang Xiao-feng and Chen Xian-ron, *Chin. Phys.* 9 (2000) 108.

[35] Pang Xiao-feng and Chen Xian-ron, *Int. J. Model Phys.* 20(18) (2006) 2505.

[36] Pang Xiao-feng and Zhang Huai-wu, *Model Phys. Lett*. 20(30) (2006) 1923.
[37] A. C.Scott, and J. C. Eilbeck, *Chem. Phys. Lett*. 132 (1986) 23.
[38] A. C. Scott, and J. C. Eilbeck, *Phys. Lett*. 119A (1986) 60.
[39] A. C. Scott, D. S. Lomdahl and J. C. Eilbeck, *Chem. Phys. Lett*. 113 (1985) 29.
[40] A. C. Scott, L. Bernstein and J. C. Eilbeck, *J. Biol. Phys*. 17 (1989) 1.
[41] Chen Xian-ron, Gou Qing-quan and Pang Xiao-feng, *Acta. Phys. Sin*. 7 (1998) 329.
[42] Chen Xian-ron, Gou Qing-quan and Pang Xiao-feng, *Acta. Phys. Sin*. 8 (1999) 1313.
[43] Chen Xian-ron, Gou Qing-quan and Pang Xiao-feng, *Chin. J. Atom. Mol. Phys*. 14 (1997) 393.
[44] Chen Xian-ron, Gou Qing-quan and Pang Xiao-feng, *Chin. J. Chem. Phys*. 10 (1997) 145.
[45] Chen Xian-ron, Gou Qing-quan and Pang Xiao-feng, *Chin. J. Chem. Phys*. 11 (1999) 240.
[46] Chen Xian-ron, Gou Qing-quan and Pang Xiao-feng, *Chin. J. Comput. Phys*. 16 (1999) 346.
[47] Chen Xian-ron, Gou Qing-quan and Pang Xiao-feng, *Chin. Phys. Lett*. 13 (1996) 660.
[48] Chen Xian-ron, Gou Qing-quan and Pang Xiao-feng, *Commun. Theor. Phys*. 31 (1999) 169.
[49] Chen Xian-ron, Gou Qing-quan and Pang Xiao-feng, *J. Sichuan University (nature) Sin*. 35(1998).
[50] J. Satsuma, and N. Yajima, *Prog. Theor. Phys*. (Supp) 55 (1974) 284.
[51] C. S.Gardner, J. M. Greene, M. D. Kruskal and R. M. Miure, *Phys. Rev. Lett*. 19 (1967)1095.
[52] M. J.Ablowitz, D. J. Kaup, A. C. Newell and H. Segur, *Phys. Rev. Lett*. 30 (1973) 1462.
[53] M. J.Ablowitz, D. J. Kaup, A. C. Newell and H. Segur, *Phys. Rev. Lett*. 31 (1973) 125.
[54] M. J.Ablowitz, D. J. Kaup, A. C. Newell and H. Segur, *Stud. Appl. Math*. 53 (1974)249.
[55] P. C.Sabatier, *Inverse method in action*, Springer, Berlin, 1990.
[56] M. J.Ablowitz, & H.Segur, (1981). *Solitons and the inverse scattering transform*, SIAM.
[57] Pang, Xiao-feng. Soliton physics, *Sichuan Sci. and Tech*. Press Sin., Chengdu, 2003.
[58] P. G.Drazin and R. S. Johnson, *Solitons, an Introduction*, Cambridge Univ. Press, Cambridge, 1989.
[59] P. E. Zhidkov, Korteweg-De, *Vries and nonlinear Schrödinger equation: qualitative theory*, Springer, Berlin, 2001.
[60] Guo, bai-lin. & pang, Xiao-feng, *solitons,* Chinese Science Press, Beijing, 1987
[61] Pang, Xiao-feng, J. *Low Temp. Phys*. 58 (1985) 334.
[62] Pang, Xiao-feng, *Acta Phys. Sin*. 43 (1994) 1987.
[63] Pang, Xiao-feng, *Chin. J. Phys. Chem*. 12 (1995) 1062.
[64] Pang, Xiao-feng, *J. Huanghuai Sin*. 11 (1995) 21.
[65] Pang, Xiao-feng, Properties of macroscopic quantum effects and dynamic natures of electrons in Superconductors, in *Superconductivity -Theory and Applications* edn.by In Tech-Open Access Publisher, Croatia, July, 2011, pp173-220.
[66] M. B. Fogel, S. E. Trulinger, A. R. Bishop and J. A. Krumhansl, *Phys. Rev*. B 15 (1977)1578.

[67] M. B.Fogel, S. E. Trulinger, A. R. Bishop and J. A. Krumhansl, *Phys. Rev. Lett.* 36 (1976) 1411.
[68] M. B. Fogel, S. E. Trullinger and A. R. Bishop, *Phys. Lett.* A 59 (1976) 81.
[69] V.G. Makhankovand N. Makhaidiani and O. Pashaev, *Phys. Lett.* A81 (1981) 161.
[70] Makhankov, V. & V.Fedyanin, *Phys. Scr.*, 20 (1979) 552.
[71] V. G.Makhankov, & V. K.Fedyanin,*Phys. Rep.*, 104(1984) 1.
[72] V. G. Makhankov, *Phys. Rep.*, 35 (1978) 1.
[73] R., J. Hirota, *Math. Phys.* 14 (1973) 805.
[74] R., J.Hirota, *Phys. Soc.* Japan 45 (1978) 174.
[75] R., J.Hirota, *Phys. Soc.* Japan 51 (1982) 323.
[76] R. Hirota, Phys. *Rev. Lett.* 27 (1972) 1192.
[77] R.Hirota, Prog. *Theor. Phys.* 57 (1977) 797.
[78] S S. R.Kinner, G. R. Allan, D. R.Andersen, & A. L.Smirl, *Dark spatial solitonpropagation in bulk ZnSe in Proc. Conf. on Lasers and Electro-Optics,* Baltimore (Opt. Soc. Am.),1991.
[79] R. N.Thurston, & A. M. Weiner, *J. Opt. Soc. Am.*, B8 (1991)471.
[80] V. E. Zakharov, S. V.Manako, S. P. Noviko, & I. P. Pitayersky, *Theory of Solitons*, Plenum, New York, 1984.
[81] T. A. Rao, and A. A. Rangwala, Backlund transformation and soliton wave equation, in solitons, *An introduction and application*, eds. M. Lakshwanan, Springer, Berlin, 1988, p.176.
[82] C.Rogers, and W. F. Shadwick, *Backlund transformations and their applications*, Academic Press, New York, 1982.
[83] R. M.Miura, *Backlund transformations and the inverse scattering method, solitons and their applications*, Springer, Berlin, 1976.
[84] J.Stiefel, *Einfuhrung in die Numerische Mathematik*, Teubner Verlag, Stuttgart, 36(1965)123.
[85] K. E. Atkinson, *An Introduction to Numerical Analysis*, Wiley, New York Inc. 1987.
[86] M.Tajiri, *J. Phys. Soc.* Japan 52 (1983) 1908 and 2277.
[87] M.Tajiri, *J. Phys. Soc.* Japan 53 (1984) 1634.
[88] A.Nakamura, *J. Math. Phys.* 23 (1982) 417.
[89] A.Nakamura, *J. Phys. Soc.* Japan 50 (1981) 2469.
[90] A. Hasegawa, *Optical solitons in fiber*, Berlin, Springer, 1989.
[91] E. Fermi, *Rend. Lincei.* 5(1927) 795.
[92] A. S. Davydov, and G. M. Pestryakov, Excitation states of the field with inertial self-action, preprint TTP-112R, *Inst.Theor. Phys. Kiev*, 1981 (in Russian).
[93] C.Sulem, and P. L. Sulem, *The nonlinear Schrödinger equation: self-focusing and wavecollapse,* Springer, New York, 1999.
[94] S.Pekar, *J. Phys.* USSR. 10 (1946) 341 and 347.
[95] H. Frohlich, *Adv. Phys.* 3 (1954) 325.
[96] H.Frohlich, *Proc. Roy.* Soc. London A 223 (1954) 296.
[97] T. Holstein, *Ann. Phys.* 8 (1959) 325 and 343.
[98] L. D.Landau, and E. M. Lifshitz, *Quantum mechanics*, Pergamon Press, Oxford, 1987.
[99] L. D.Landau, Phys. Z. Sowjetunion 3 (1933) 664.
[100] A. S.Davydov, and N. I. Kislukha, *Phys. Stat. Sol.* (b) 59 (1973) 465.
[101] A. S.Davydov, and N. I. Kislukha, *Phys. Stat. Sol.* (b) 75 (1976) 735.

[102] A. S.Davydov, *Phys. Scr.* 20 (1979) 387.
[103] A. S.Davydov, *Phys. Stat. Sol.* (b) 102 (1980) 275.
[104] A. S. Davydov, Solitons in molecular systems, D. Reidel Publishing, Dordrecht, 1985.
[105] Pang, Xiao-feng, *Phys. Rev.* E 62 (2000) 6898.
[106] Pang Xiao-feng, *Physics of Life Review*, 8 (2011)264.
[107] Pang Xiao-feng, *Progress in Biophysics and Molecular Biology*, 108(2012)1.
[108] Pang Xiao feng, *Biophysical Reviews and Letters*, 9(2014)1.
[109] P. Schuster, G. Zundel and C. Sandorfy, *The Hydrogen Bond, Recent Developments inTheory and Experiments*, North Holland, Amsterdam, 1976, p 47-98.
[110] P.V. Hobbs, *Ice Physics*, Oxford, Clarendon Press, 1974,p12-87.
[111] M. Eigen, L. de Maeyer and H. C. Spatz, Coll. *Physics of ice crystals*, 1962, p45-103.
[112] E.Whalley, S. J. Jones and L. W. Grold, *physics and chemistry of ice*, Ottawa, Roy. Soc Canada, 1973, P32-154.
[113] V. Ya. Antonchenko, A. S. Davydov and A. V. Zolotariuk, *Phys. Stat. Sol.* (b) 115 (1983) 631.
[114] E.W.Laedke, K.H.Spatschek, M.Wilkens and A.V.Zolotaryuk, *Phys. Rev*.A 32 (1985) 1161.
[115] J. F.Nagle and H.J.Morowitz, *Proc.Natl.Acad.Sci.USA* 75 (1978) 298.
[116] .J. F.Nagle, M.Mille and H.J.Morowitz, *J.Chem.Phys*.72 (1980) 3959.
[117] M. Peyrared, St.Pnevmatikos and N.Flyzanis, *Phys. Rev*.A36 (1987) 903.
[118] St. Pnevmatikos, *Phys.Rev.Lett.* 60 (1988) 1534.
[119] L. N. Kristoforov and A.V.Zolotaryuk, *Phys.Stat*.Sol. (b) 146 (1988) 487.
[120] A. V. Zolotaruk, R.H.Spatschek and L.E.W.Ladre, *Phys. Lett.* A101(1984)517.
[121] Pang Xiao-feng, H.J.W.Miller-kirsten, *J.Phys.Condensed matter* 12(2000)885.
[122] Pang Xiao-feng and Feng Yuan-ping, *Chem. Phys. Lett.*, 373(2003)392.
[123] Pang Xiao-feng, Yu Jin-.feng, *Chinese Physics Letters*. 24 (2007) 1452.
[124] Pang Xiao-Feng and A.F.Jalbout, *Physics Letters* A 330(2004) 245.
[125] Pang Xiao-feng, Feng Yuan Ping, *Journal of Biomolecular Structure & Dynanics* 25 (2007)435.
[126] Pang Xiao-feng, proton transfer in hydrogen-bonded systems and its applications, LAP Lambert Publishing, Germany, 2013.
[127] Pang Xiao feng, water, *molecular structure, and properties*, World Scientific Publishing Co., Singapore, 2014(472 pages).
[128] Pang Xiao-feng, *Progress in Biophysics and Molecular Biology*, 109 (2013)1.

Chapter 5

THE NONLINEAR QUANTUM PROPERTIES OF THE EXCITON AND ITS EXPERIMENTAL CONFIRMATION IN MOLECULAR CRYSTALS

We here depicted in detail the applications of new nonlinear theory and gave further the nonlinear quantum mechanical properties of the exciton in the molecular crystals by using the new nonlinear quantum mechanical theory and methods, in which we first established the Hamiltonian of system and corresponding dynamic equation of exciton and phonon based on the molecular structure of acetanilide and interaction features of exciton and phonon, or peptide group.

The results indicated the exciton satisfied a nonlinear Schrödinger equation, where the nonlinear interaction is formed in virtue of the self-trapping mechanism arising from the interaction between the moved exciton and moved phonon. In this case the exciton becomes a exciton-soliton with wave-corpuscle duality, its existences and properties are further confirmed by many experiments, such as the infrared spectrum of absorption, the Raman spectrum and transient-infrared-bleaching method and pump-probe experiment as well as linear absorption spectra and 2D-IR pump probe spectra, and so on. Thus, the correctness of the nonlinear quantum mechanics are affirmed in these experimental results. On the other hand, we used these results and nonlinear quantum mechanical theory to give and elucidate the properties of infrared absorption, Raman scattering and eigenenergy spectrum of acetanilide. In this investigation, we obtained the rule of exponential decreases of infrared absorption strength of 1650cm^{-1} band, which corresponds to the soliton, with increasing the temperature of acetanilide using this soliton theory, and obtained specific heat of the acetanilide, which is consistent with experimental results, using nonlinear quantum mechanical theory.

Finally, the true information of existence and lifetime of the exciton-soliton was captured and obtained using the technique of Mark free-electron laser, and the pump-probe spectroscopy, which is 15ps in the former and 35ps in the latter for acetanilide, respectively. These results affirmed both the true existence of the exciton-soliton and the correctness of nonlinear quantum mechanics in the experiments and theoretical deduction.

5.1. The Nonlinear Quantum Properties of the Excitons in Molecular Crystal and Its Self-Trapping Mechanism of the Nonlinear Interaction

5.1.1. TheStructure of Molecular Crystals and Nonlinear Excitation of the Exciton

We first discuss a particular molecular crystal, acetanilide $(CH_3COHNC_4H_5)_x$. Two close chains of hydrogen-bonded amide-I groups which consists of atoms of carbon, oxygen, nitrogen and hydrogen (CONH), run through the acetanilide crystal [1-6]. This crystal structure of acetanilide has been determined and a unit cell is shown in Figure 5.1. The space group is $D_{2h}^{15}(P_{bca})$ and the unit cell or factor group is D_{2h} for this crystal. The average lattice constants are a = 1.9640 nm, b = 0.9483 nm, and c = 0.7979 nm. There are eight molecules in an unit cell and at the amide-I frequency, each of these has one degree of freedom (d.f.). Thus, there are three infrared-active modes (B_{1u}, B_{2u}, and B_{3u}), four Raman-active modes (A_g, B_{1g}, B_{2g} and B_{3g}), and one inactive mode (A_u). However, at low frequency (<200 cm^{-1}), each molecule exhibits 6 d.f. (three translations and three rotations. This gives 48 low-frequency modes: 24 Raman active modes ($6A_g + 6B_{1g} + 6B_{2g} + 6B_{3g}$), 18 infrared-active modes ($6B_{1u} + 6B_{2u} + 6B_{3u}$) and six (A_u) modes corresponding to acoustic modes of translation and rotation). All of these active modes are seen in infrared absorption and Raman experiments. Meanwhile, in acetanilide, the nearly planar amide-I groups have bond lengths, which are close to those found in polypeptide (see Figure 5.2). Since the physical properties of such hydrogen bonded amide-I systems are very sensitive to bond lengths, hence, studies of acetanilide can reveal new phenomena. For example, in experiments of infrared absorption and Raman scattering, a new amide-I band red-shifted from the main peak at 1666 cm^{-1} by about 16 cm^{-1}, when the crystalline acetanilide is cooled from 320 to 10 K, but no other major changes occur from 4000 to 800 cm^{-1}. The intensity of this new band increases steadily from room temperature to 70 K. The band at 1650 cm^{-1} is not present in amorphous materials or acetanilide methylated at the position where hydrogen-bonded distances occur, but it is recovered after annealing. Similar phenomena can be observed in Raman scattering experiments [7-27]

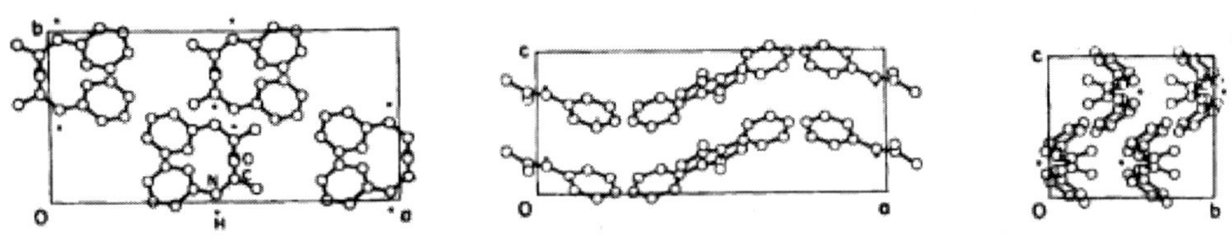

Figure 5.1. Various views of the cell unit of acetanilide, with cell parameters a = 19.640 °A, b = 9.483°A, and c = 7.979 °A [Ref.1-6].

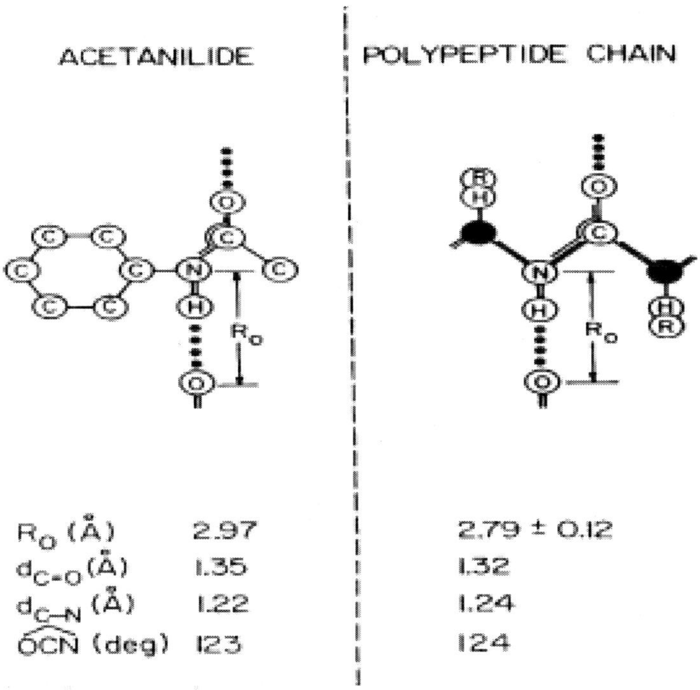

Figure 5.2. Comparisons between of acetanilide and a protein molecule [Ref.1-6].

As it is known, the characteristic feature of amide-I group, CONH, in polypeptides is amide-I mode, which mainly involves the stretching of the C=O bond. This mode is observed as an infrared absorption peak at 1666 cm^{-1} in acetanilide and near this value in a wide variety of materials, including the amide-I groups. The corresponding spectroscopic evidence of the new band at 1650 cm^{-1} has been mentioned earlier, but detailed measurements of the crystal structure and specific heat as a function of temperature of peptide groups preclude the assignment of the new band to (1)a conventional amide-I mode, (2) crystal defect states, (3) Fermi resonance or (4) frozen kinetics between two different subsystems. The correct assignment is the self-trapping of the amide-I vibrational energy. These conclusions are based on the following experimental facts: (1) the ^{15}N substitution induces a small shift to the amide-I at 1666 cm^{-1}, and the new band is also shifted by the same amount; (2) deuterium substitution at the NH position strongly affects both the amide-I and the new band in a complicated way; (3) upon cooling a decrease in the integrated absorption of the normal amide-I and a corresponding increase in the integrated absorption of the 1650 cm^{-1} band are observed; (4) the 1650 cm^{-1} band and the amide-I band show the same dichroism over the temperature range integrated; and (5) the measurements of the specific heat, dielectric constant and volume expansion as a function of temperature rule out the occurrence of rotational isomerism or polymorphic transition which would affect some other infrared and Raman absorption bands, but not the new band. The self-trapping mechanism of amide-I vibrational energy used by Scott and Eilbeck et al., [8-16] comes from the Davydov model of vibration energy transport in alpha-helix protein molecules [17-27], which was described in Chapter 4.They have given a good account of the properties of first excited state in this model. Scott and co-workers, Alexander and Krumbansl have also obtained an exponential dependence of the absorption intensity on temperature, exp ($-\beta T^2$) [15-16], and explained the experimentally observed intensity change of the new band with decreasing temperature in terms of a complementary polaron or soliton of the self-trapping state of exciton on the basis of the Davydov model (Davydov referred to the intramolecular excitation occurred in the systems as exciton in such a case). However, the red shift of a few cm^{-1} given by this model is much smaller than the experimental value of 16 cm^{-1}. This indicates that these models need further improvement and development.

5.1.2. TheNonlinear Quantum Mechanical Properties of the Excitons and the Self-Trapping Mechanism of the Nonlinear Interactions

Pang and co-workers [28-40] also proposed that the new band of amide-I is caused by the self-trapping of amide-I vibrational quanta (exciton) in their soliton model (the vibrational quantum is referred to as exciton, in accordance with the exciton theory by Davydov). But this model is different from Davydov's soliton model or Alexander's complementary polaron model.

Here it is a nonlinear interaction between the localized amide-I vibrational-quantum and low frequency vibrations of the lattice (phonons) that result in self-trapping of the exciton as a soliton. The mechanism of this soliton model can be described as follows: Through their intrinsic nonlinear interaction, an amide-I vibrational quantum acts as a source of low-frequency phonons and causes shifts in the average positions of the lattice molecules in ground state. These shifts, or lattice distortions, in turn, react through the same nonlinear interaction, as a potential well to trap the amide-I vibrational quantum, resulting in the soliton, and preventing the energy dispersion of the amide-I vibrational quantum via the dipole-dipole interaction that occurs in neighboring peptides with certain electric moments. Evidently, the soliton is formed by the self-trapping of excitons interacting with low frequency lattice phonons, and is a dynamic self-sustaining entity. This entity propagates, together with the lattice deformation along the molecular chains. The main property of the soliton is that it can move over macroscopic distances at velocity$v$ and retain its wave form, energy, momentum, and other quasi-particle properties. Using this model, Pang and coworkers also assigned an exponential dependence of the infrared-absorption intensity to the temperature, which is consistent with the experimental result [28-40].

From these studies, we know that the collective excitations can result from a localized fluctuation of the amide-I vibrational quanta and structural deformation of peptide groups due to the phonon-excition interaction in molecular crystals. In this case, the Hamiltonian of the systems given by Pang [28-40] can be presented by:

$$H = H_{ex} + H_{ph} + H_{\text{int}} = \frac{1}{2m}\sum_n p_n^2 + \frac{m\omega_0^2}{2}\sum_n r_n^2 + \frac{m\omega_1^2}{2}\sum_n r_n r_{n+1} + \frac{1}{2M}\sum_n P_n^2 + \tag{5.1}$$

$$\frac{\beta}{2}\sum_n (R_n - R_{n-1})^2 + m\chi_1 \sum_n (R_{n+1} - R_{n-1}) r_n^2 + m\chi_2 \sum_n (R_{n+1} - R_n) r_n r_{n+1} - m\chi_2 \sum_n (R_{n-1} - R_n) r_n r_{n-1} \quad (5.1)$$

This Hamiltonian includes the vibrational excitation of amide-I caused by localized fluctuation and vibration of the peptide groupscaused by structural deformation of molecular chains as well as the interaction between the two modes of motion in the molecular crystal. Here m is the mass of amide-I vibrational quantum (exciton), $\omega_0$andω_1 are diagonal and off-diagonal elements of the dynamic matrix of vibrational quantum, ω_0 is also a Einstein vibrational frequency of the exciton, and $m\omega_1^2 r_n r_{n+1}/2$ is the interaction between the nearest neighboring excitons caused by the dipole-dipole interaction in molecular chains. Likewise, r_n and $p_n = m(\dot{r}_n)$ *are* the normal coordinates of the nth excitons and its canonical conjugate momentum, respectively, M the mass of a peptide group in the unit cell, $\chi_1=\partial\omega_0^2/\partial R_n$ and $\chi_2=\partial\omega_1^2/\partial R_n$ represent the changes of energy of exciton and of coupling interaction energy between the excitons due to an unit extension of the molecular chain, respectively. R_n and $P_n = M\dot{R}_n$ arethe canonically conjugate operators of displacement and momentum of the peptide group, β the elastic constant of molecular chains. H_{ex}in H is the Hamiltonian of harmonic vibration of amide-I including the off-diagonal factor, $H_{ph}$is the Hamiltonian of harmonic vibration of the peptide group, and H_{int} is the interaction Hamiltonian between the two modes of motion.

This Hamiltonian is significantly different from that of the Davydov model [17-22],which is denoted in Section4.10 in Chapter4used by Scott and co-workers[7-14]. As far as the vibration of amide-I is concerned, we adopt a harmonic oscillator model with optical vibration that includes an off-diagonal factor, which comes from the interaction between excitons, and interaction with displacement of peptide groups. Thus, vibrational frequencies of amide-I are related to the displacements of peptide groups, which shows the occurrence of the interaction between amide-I vibrations and displacement of peptide groups. Their relationship is described by:

$$\omega_0^2(R_n)\approx\omega_0^2-\frac{\partial\omega_0^2}{\partial R_n}(R_n-R_{n-1})=\omega_0^2+\chi_1(R_n-R_{n-1}),\ \omega_1^2(R_n)\approx\omega_1^2+\chi_2(R_n-R_{n-1}).$$

Inserting the above into the vibrational Hamiltonianbetween the amide-I and peptide group, and taking further into account the effect of the neighboring peptide groups on the excitons, thenHamiltonian shown in Eq. (5.1) can be easily obtained. Therefore, the Hamiltonian in Eq.(5.1) has a correct description of the interactions in the molecular crystal. Thus,Pang's theory can give a complete description on the dynamic properties of the systems.

However, the Hamiltonian in the Davydov model did not give such a one-to-one description of interactions in this system, namely, the Davydov model [17-22] did not condense the interaction between the amide-I vibrational quantum and displacement of the peptide groups resulting from the interaction between the neighboring excitons, so Davydov Hamiltonian possesses not in the one-to-one correspondence for the interactions between $H_{ex}$and H_{int}. Thus, the Davydov model has encountered many difficulties in the study of the properties of nonlinear excitations in the systems.

Since amide-I vibration and vibration of peptide groups are all quantized, thus we here should introduce the following canonical second quantization transformation,

$$r_n=\sqrt{\frac{\hbar}{2m\omega_0}}\left(b_n^+ + b_n\right), p_n=-i\sqrt{\frac{2m\omega_0}{\hbar}}\left(b_n - b_n^+\right) \tag{5.2}$$

$$R_n=\sum_q\left[\frac{\hbar}{2NM\omega_q}\right]^{1/2} e^{iqnr_0}\left(a_{-q}^+ + a_q\right), P_n=i\sum_q\left[\frac{M\hbar\omega_q}{2N}\right]^{1/2} e^{iqnr_0}\left(a_{-q}^+ - a_q\right) \tag{5.3}$$

to quantize the Hamiltonian bin Eq.(5.1), where $i=\sqrt{-1}$, $\omega_q=2(\beta/M)^{1/2}\sin(r_0q/2)$ is the frequency of the phonon with wave vector$q$, N the number of unit cells in the molecular chain, r_0 the distance between the molecules, $b_n^+(b_n)$ and $a_q^+(a_q)$ are the creation (annihilation) operator of the exciton and phonon, respectively.

Using Eqs. (5.2) and (5.3), the equation (5.1) becomes

$$H=\sum_n\varepsilon_0\left(b_n^+b_n+\frac{1}{2}\right)-J\sum_n\left(b_n^+b_{n+1}+b_nb_{n+1}^+\right)+\sum_q\hbar\omega_q\left(a_q^+a_q+\frac{1}{2}\right)+$$
$$\sum_{n,q}\left[g(q)\left(b_n^+b_n+b_nb_n^+\right)+g_1(q)\left(b_n^+b_{n+1}+b_nb_{n+1}^+\right)-g_2(q)\left(b_n^+b_{n-1}+b_nb_{n-1}^+\right)\right]\left(a_q+a_{-q}^+\right)e^{inr_0q} \tag{5.4}$$

where

$$\varepsilon_0=\hbar\omega_0, J=\hbar\omega_1^2/4\omega_0, g(q)=\left(\frac{\hbar}{2NM\omega_q}\right)^{1/2}\left(\frac{\hbar\chi_1}{2\omega_0}\right)\left[e^{ir_0q}-e^{-ir_0q}\right],$$

$$g_1(q)=\left(\frac{\hbar}{2NM\omega_q}\right)^{1/2}\left(\frac{\hbar\chi_2}{2\omega_0}\right)\left[e^{ir_0q}-1\right], g_2(q)=\left(\frac{\hbar}{2NM\omega_q}\right)^{1/2}\left(\frac{\hbar\chi_2}{2\omega_0}\right)\left[e^{-ir_0q}-1\right].$$

Because the collective excitations generated by the localized fluctuation of the excitons and the structural deformation of the peptide groups have a coherent feature, thus Pang [28-35] represented the excitation states of the excitons and vibrations of peptide group (phonons) by the following wave function:

$$|\Phi\rangle=|\varphi(t)\rangle U(t)|0\rangle_{ph}=\frac{1}{\rho}\left(1+\sum_n\varphi_n(t)b_n^+\right)|0\rangle_{ex}\cdot U(t)|0\rangle_{ph} \tag{5.5}$$

with

$$U(t) = \exp\left\{\sum_n \frac{1}{i\hbar}\left[u_n(t)P_n - \pi_n(t)R_n\right]\right\}$$

$$\text{or } U(t) = \exp\left\{\sum_q\left[\alpha(t)a_q^+ - \alpha_q^*(t)a_q\right]\right\} \tag{5.6}$$

where $|0\rangle_{ex}$ and $|0\rangle_{ph}$ are the excitonic and phononic ground-states, respectively. The ρ is the normalization factor. We assume hereafter that $\rho = 1$ for convenience of calculation unless it is explicitly noted otherwise. $\phi_n(t)$, $u_n(t) = \langle\Phi|R_n|\Phi\rangle$ and $\pi_n(t) = \langle\Phi|P_n|\Phi\rangle$ are three unknown functions.

In Eq. (5.5) the wave function of the excitons is

$$|\varphi\rangle = \frac{1}{\rho}\left(1 + \sum\nolimits_n \varphi_n(t)b_n^+|0\rangle_{ex}\right),$$

which is different from $|\varphi_D\rangle = \sum_n \varphi_n(t)B_n^+|0\rangle_{ex}$ in the Davydov wave function $|D_2\rangle$ [17-22], and it is not an excitation state of a single particle, but rather a coherent state, or more accurately, a quasi-coherent state. In the case of small $\varphi_n(t)$ (i.e., $|\phi_n(t)| << 1$) and it can be approximately written as

$$|\varphi\rangle = \frac{1}{\rho}\exp\left[-\frac{1}{2}\sum_n|\varphi_n(t)|^2\right]\exp\left[\sum_n \varphi_n(t)b_n^+\right]|0\rangle_{ex} = \frac{1}{\rho}\exp\left[\sum_n \varphi_n(t)b_n^+ - \varphi_n^*(t)b_n\right]|0\rangle_{ex}$$

in which the final representation is a standard coherent state, which thus is normalized. Since the condition $\sum_n|\varphi_n(t)|^2 = 1$, is required in this calculation, then the condition, $|\phi_n(t)| << 1$, is naturally satisfied for the acetanilide consisting of several hundreds of peptide groups. Therefore, the above representation is correct. Thus, the wave function in Eq. (5.5) is also normalized at $\rho = 1$.

In fact, the wave function of the exciton state in Eq. (5.5) should be viewed as an effective truncation of a standard coherent state, in which the aforementioned two terms are retained. Thus, it is referred to as a quasi-coherent state. However, it is not an eigenstate of the number operator $\hat{N} = \sum_n b_n^+ b_n$ since

$$\hat{N}|\varphi\rangle = \sum\nolimits_n b_n^+ b_n|\varphi\rangle = |\varphi\rangle - |0\rangle_{ex} \neq |\varphi\rangle.$$

Hence, $|\varphi\rangle$ represents a coherent superposition of the one-quantum amide-I vibrational state and the ground state of the exciton. But, the number of quantums are determinate in this state and it contains only one exciton because:

$$N=\langle\varphi|\hat{N}|\varphi\rangle=\langle\varphi|\sum_n b_n^+ b_n|\varphi\rangle=\sum_n|\varphi_n(t)|^2=1.$$

Therefore, $|\varphi\rangle$ given in Eq. (5.5) not only exhibits coherent feature of collective excitations of the excitons caused by the nonlinear interaction generated by means of the exciton-phonon interaction, which makes the wave functions of the states of the exciton and phonon symmetrical, or speaking, the wave functions of the exciton and phonon have simultaneously the coherent feature(because the wave function of the phonon was already a standard coherent sate), but also it can also make the number of the excitons conserved for the Hamiltonian in Eq. (5.4) as wellin this case.

On the other hand, the wave functionin Eq.(5.5) has the advantage, i.e., the equation of motion of the exciton can be obtained from the following Heisenberg equations for the creation and annihilation operators in Eqs. (5.4) – (5.6)

$$i\hbar\frac{\partial}{\partial t}\langle\Phi|b_n|\Phi\rangle=|\Phi\rangle[b_n,H]|\Phi\rangle, i\hbar\frac{\partial}{\partial t}\langle\Phi|a_q|\Phi\rangle=|\Phi\rangle[a_q,H]|\Phi\rangle \quad (5.7)$$

From these equations we can obtain the following equations of $\varphi_n(t)$, $\alpha_q(t)$ and $\alpha_{-q}^*(t)$

$$i\hbar\dot{\varphi}_n=\hbar\omega_0\varphi_n-J(\varphi_{n+1}+\varphi_{n-1})+\sum_q\{2g(q)(\alpha_q+\alpha_{-q}^*)\varphi_n+$$

$$(g_1(q)-g_2(q))[(\alpha_q+\alpha_{-q}^*)\varphi_{n+1}+(\alpha_q+\alpha_{-q}^*)\varphi_{n-1}]\}e^{ir_0qn}, \quad (5.8)$$

$$i\hbar\dot{\alpha}_q=\hbar\omega_q\alpha_q+\sum_n\{2g(q)|\varphi_n(t)|^2+[g_1(q)-g_2(q)](\varphi_n^*\varphi_{n+1}+\varphi_n^*\varphi_{n-1})\}e^{-ir_0qn} \quad (5.9)$$

$$i\hbar\dot{\alpha}_{-q}^*=-\hbar\omega_q\alpha_{-q}^*-\sum_n\{2g(q)|\varphi_n(t)|^2+[g_1(q)-g_2(q)](\varphi_n^*\varphi_{n+1}+\varphi_n^*\varphi_{n-1})\}e^{-ir_0qn} \quad (5.10)$$

From Eqs. (5.9) and (5.10), we can get:

$$\ddot{\alpha}_{-q}^*+\ddot{\alpha}_q=-\omega_q^2(\alpha_q+\alpha_{-q}^*)-\sum_n\left\{\frac{4g(q)\omega_q}{\hbar}|\varphi_n|^2+\frac{2g_1(q)-2g_2(q)}{\hbar}\omega_q(\varphi_n^*\varphi_{n+1}+\varphi_n^*\varphi_{n-1})\right\}e^{-ir_0qn}.$$

Also, from Eq. (5.3) we get the Fourier transformation of the variable, $u_n=\langle\Phi|R_n|\Phi\rangle$, as:

$$u_n = \frac{1}{\sqrt{N}} \sum_q u_q(t) e^{iqx}$$, where x=n r_0 and $u_q(t) = \sqrt{\frac{\hbar}{2M\omega_q}} (\alpha_q + \alpha_q^*)$.

Thus, we get in the long-wave length approximation

$$\ddot{u}_q + \upsilon_0^2 q^2 u_q = \frac{1}{\sqrt{N}} \sum_n \left\{ \frac{2iq\hbar r_0 \chi_1}{M\omega_0} |\varphi_n|^2 + \frac{iq\hbar r_0 \chi_2}{M\omega_0} \left(\varphi_n^* \varphi_{n+1} + \varphi_n^* \varphi_{n-1} \right) \right\} e^{-ir_0 qn}$$

where $\omega_q = v_0 q, v_0 = r_0 (\beta/M)^{1/2}$.

Now, multiplying the above equation by $\exp(ir_0 qn)/\sqrt{N}$, summing over the wave number, q, and making again the continuum approximation, then we [28-34] can get:

$$\frac{\partial^2 u(x,t)}{\partial t^2} - \upsilon_0^2 \frac{\partial^2 u(x,t)}{\partial x^2} = \frac{2\hbar r_0 (\chi_1 + \chi_2)}{M\omega_0} \frac{\partial |\varphi(x,t)|^2}{\partial x}. \tag{5.11}$$

At the same time, equation (5.7) becomes also:

$$i\hbar \frac{\partial \varphi(x,t)}{\partial t} = (\hbar\omega_0 - 2J)\varphi(x,t) - Jr_0^2 \frac{\partial^2 \varphi(x,t)}{\partial x^2} + \frac{2\hbar r_0 (\chi_1 + \chi_2)}{\omega_0} \varphi(x,t) \frac{\partial u(x,t)}{\partial x} \tag{5.12}$$

Inserting Eq. (5.11) into Eq. (5.12), we can get the following nonlinear Schrödinger equation for motion of the exciton:

$$i\hbar \frac{\partial \varphi(x,t)}{\partial t} = (\hbar\omega_0 - 2J)\varphi(x,t) - Jr_0^2 \frac{\partial^2 \varphi(x,t)}{\partial x^2} + \frac{4\hbar^2 r_0^2 (\chi_1 + \chi_2)^2}{M\omega_0 (\upsilon^2 - \upsilon_0^2)} |\varphi(x,t)|^2 \varphi(x,t) \tag{5.13}$$

where $\hbar^2 / 2m = Jr_0^2$.Here m is the effective mass of the exciton.

Clearly, equation(5.13) is same with dynamic equation (2.66) at $A(\phi) = 0$ in nonlinear quantum mechanics. This shows that the properties and rules of motion of the exciton can be described by nonlinear quantum mechanics, or speaking, the exciton in the molecular crystal has the wave-corpuscle duality presented in Chapter 3. Very evidently, where the nonlinear interactions, $\frac{4\hbar^2 r_0^2 (\chi_1 + \chi_2)^2}{M\omega_0 (\upsilon^2 - \upsilon_0^2)} |\varphi(x,t)|^2 \varphi(x,t)$ is generated by the interaction between the moved exciton and moved peptide group, which is affirmed by

$$\frac{\partial u(x,t)}{\partial x}=-\frac{2hr_0\left(\chi_1+\chi_1\right)}{M\omega_o\left(1-s^2\right)}|\varphi(x,t)|^2,$$

their states of movement were described by Eqs.(5.11)-(5.13). Therefore, we can confirm that the nonlinear interaction in Eq.(5.13) is due to the interaction between the moved exciton and moved phonon, or displacement of the peptide ground. Then this is naturally called the self-trapping mechanism according to its define shown in Chapter4.Its detailed description is as follows.

The excitation of the exciton or C=O bond stretchingvibration leads to distorting of the structure of the molecular chains or vibration of the peptide groups due to the coupling action between them, but the chain deformation provides a double-well potential to react in turn on the excitons, thus the exciton is self-trapped as a soliton through the coupling interaction to prevent its energy dispersion. In such a case, the exciton is localized as a soliton on the chains.

5.1.3. TheProperties of the Exciton–Soliton in Acetanilide

In accordance with the nonlinear quantum mechanics in Chapter 3, we easily give the soliton solutionof Eq.(5.13), which is represented by

$$\varphi(x,t)=\sqrt{\frac{\mu}{2}}\operatorname{sec}h\left[\mu'\left(x-x_0-\upsilon t\right)\right]\exp\left\{\frac{i}{\hbar}\left[\frac{\hbar^2\upsilon\left(x-x_0\right)}{2Jr_0^2}-E_{sol}t\right]\right\}, \tag{5.14}$$

$$u(x,t)=-\frac{\hbar\left(x_1+x_2\right)}{2\omega_0\beta\left(1-s^2\right)}\tanh\left[\mu'\left(x-x_0-\upsilon t\right)\right] \tag{5.15}$$

where

$$\mu=\frac{G}{4J},G=\frac{4\hbar^2r_0^2\left(\chi_1+\chi_2\right)^2}{M\upsilon_0^2\omega_0^2\left(1-s^2\right)},\mu'=\frac{\mu}{r_0},s=\frac{\upsilon}{\upsilon_0},\varepsilon=\hbar\omega_0-2J,\varepsilon_0=\hbar\omega_0, \tag{5.16}$$

Using

$$\alpha_q=\sqrt{\frac{M\omega_q}{2\hbar}}u_q+i\sqrt{\frac{1}{2M\hbar\omega_q}}\pi_q,$$

and the results given above, we found that $\alpha_q(t)$ in Eq. (5.3) is given by:

$$\alpha_q(t)=\frac{1}{\sqrt{N}}\sum_n \alpha_n(t)e^{-iqx}=\frac{i\hbar(\chi_1+\chi_2)\pi M(\omega_q+\upsilon q)}{2M\upsilon_0^2\omega_0(1-s^2)\hbar N\omega_q}\sinh\left(\frac{\pi r_0 q}{2\mu}\right)e^{iq\upsilon t}=\alpha_q e^{iq\upsilon t}. \quad (5.17)$$

These results tell clearly us that the motions of the excitons in the acetanilide satisfy the nonlinear Schrödinger equation, thus obey the laws of nonlinear quantum mechanics.In this case, the exciton is as self-trapped as a soliton by means of the self-trapping mechanism. Then we can confirm that the exciton-soliton has a wave-corpuscle duality [38-40].

In such a case, the energy of the soliton can be found by using Eqs. (5.14) and (5.15), which is expressed by:

$$E_{sol}=\int_{-\infty}^{\infty}Hd\xi=\varepsilon_0-2J+\frac{\hbar^2\upsilon^2}{4Jr_0^2}-\frac{\mu^2 J}{3}=E_0+\frac{M_{sol}\upsilon^2}{2},$$

where

$$E_0=\varepsilon_0-2J-\frac{\hbar^4(\chi_1+\chi_2)^4}{3\beta^2 J\omega_0^4}$$

is the rest energy of the soliton. The effective mass of the soliton is given by:

$$M_{sol}=m+\frac{4\hbar^4(\chi_1+\chi_2)^4\left(1-3s^2/2-s^4/2\right)}{3\beta^2 J\omega_0^4\upsilon_0^2\left(1-s^2\right)^2}>m$$

We see from the above results that the energy and the rest energy of the soliton are lower about $\hbar^4(\chi_1+\chi_2)^4/3\beta^2 J\omega_0^4(1-s^2)$ and $\hbar^4(\chi_1+\chi_2)^4/3\beta^2 J\omega_0^4$ than those of the exciton, $E'=\varepsilon_0-2J+\mathrm{mv}^2/2$ and $E_0'=\varepsilon_0-2J$, respectively. This means that the exciton-soliton is stable. But the mass of the soliton is greater than that of the exciton, *m*.

5.1.4. The Features of Nonlinear Excitations of Phonons in Molecular Crystal

From the above studies, we know that exciton-soliton or localization of exciton occurs due to nonlinear exciton-phonon interaction. In such a case, the phonon is also localized, and moves in a soliton in the systems. This can be verified from Eqs. (5.11) – (5.12). In this case, from Eqs. (5.14) –(5.15) and

$$\mu=(\chi_1+\chi_2)^2\hbar^2/JM\upsilon_0^2\omega_0^2\left(1-\mathrm{s}^2\right)^2,$$

where $s = v / v_0$, we can get

$$\frac{\partial}{\partial x}\left|\varphi(\chi,\ t)\right|^2 = \frac{2(\chi_1+\chi_2)^3\hbar^3}{r_0^3\beta J^2(1-s^2)\omega_0^3}\left[u - \frac{4\omega_0^2(1-s^2)^2\beta^2}{\hbar^2(\chi_1+\chi_2)^2}u^3\right] \tag{5.18}$$

Substituting Eq. (5.18) into Eq. (5.11), we obtain

$$u_{tt} - \upsilon_0^2 u_{xx} = \frac{4(\chi_1+\chi_2)^4\hbar^4}{Mr_0^2\beta J^2(1-s^2)\omega_0^4}\left[u - \frac{4\omega_0^2(1-s^2)^2\beta^2}{\hbar^2(\chi_1+\chi_2)^2}u^3\right] \tag{5.19}$$

This is a ϕ^4 -equation for $u(x, t)$. It has a soliton solution given in Eq. (5.15). Thus, phonon has been localized in this case. Obviously, the localization of phonon is due to the nonlinear exciton-phonon interaction in acetanilide. However, this nonlinear interaction is different from that of the exciton because the natures and moved feature of the two particles are different. If $\chi_1 = \chi_2 = 0$, then Eq. (5.19) reduces to a linear wave equation, $u_{tt} - v_0^2 u_{xx} = 0$, which has a solution of plane wave, $u = A\sin(\omega t - \theta_0)$,with a certain dispersive effect. This manifests that the peptide groups performs harmonic vibration in this case. This denotes clearly that the localization of phonons is carried out by virtue of the exciton-phonon interaction, in which the nonlinear interaction of B'u^3 occurs in Eq.(5.19), where $B' = 16(\chi_1+\chi_2)^2\hbar^2\beta(1-s^2)/M\upsilon_0^2\omega_0^2 J$. This indicated that the motion of the phonon should be described by nonlinear quantum mechanics, instead of linear quantum mechanics, in the system.

These results show clearly that both microscopic particles (or excitons) and background field (or phonons) are simultaneously acted by the nonlinear interactions. In this case, the microscopic particles (excitons) and vibration of background field (or phonons) simultaneously are localized and possess the wave-corpuscle duality under action of the different nonlinear interactions, respectively. This indicated also that there are also the action and counteraction of microscopic particles when they interact with each other in nonlinear quantum mechanics [38-39], but the sizes of the action and counteraction accepted by different microscopic particles are different because the natures and moved feature of the two particles are different. However, we should note that this feature of interaction between two microscopic particles in nonlinear quantum mechanics is different from that between two classical particles in classical physics. This is an interesting result.

5.2. The Behaviors of the Nonlinear Excitation of the Exciton in the Case of Anharmonic Vibration of Lattices in Acetanilide

Pang etal. [28-40] also studied the influence of anharmonic vibration of the peptide groups on the exciton-soliton. Then the Hamiltonian of peptide group in Eq. (5.1) is now replaced by

$$H'_{ph} = \frac{1}{2M}\sum_n P_n^2 + \sum_n [\frac{\beta}{2}(R_n - R_{n-1})^2 + \frac{1}{3}\lambda_1 (R_n - R_{n-1})^3] \tag{5.20}$$

It was found that term of $(\hbar\omega_0 - 2J)\varphi(x,t)$ in Eq. (5.13) is only changed in this case, thus the phase, velocity, energy, amplitude and width of the exciton-soliton are only varied.

Inserting Eq.(5.3) into Eq.(5.20) we obtain

$$H_{ph} = \sum_q \hbar\omega_q \left(a_q^+ a_q + \frac{1}{2} \right) + \sum_{q_1 q_2} F(q_1, q_2)\left(a_{-q_1}^+ + a_{q_1}\right)\left(a_{-q_2}^+ + a_{q_2}\right)\left(a_{-(q_1+q_2)}^+ + a_{(q_1+q_2)}\right) \tag{5.21}$$

where

$$F(q_1, q_2) = \frac{8}{3} i\lambda_1 \left(\frac{\hbar}{2M} \omega_{q_1}\omega_{q_2}\omega_{(q_1+q_2)} \right)^{1/2} \sin(r_0 q_1/2)\sin(r_0 q_2/2)\sin[r_0(q_1+q_2)/2]$$

From Eq.(5.21) and the phonon part, $U|0\rangle_{ph}$,in Eq.(5.5) and the Heisenberg equation (5.7), we can get

$$i\hbar\dot{\alpha}_q = \hbar\omega_q\alpha_q - \sum_k \left\{ F(k-q)\left(\alpha_k + \alpha_{-k}^*\right)\left(\alpha_{k-q} + \alpha_{k-q}^*\right) \right\} \tag{5.22}$$

$$i\hbar\dot{\alpha}_{-q}^* = -\hbar\omega_q\alpha_{-q}^* + \sum_k \left\{ F(k-q)\left(\alpha_k + \alpha_{-k}^*\right)\left(\alpha_{k-q} + \alpha_{k-q}^*\right) \right\} \tag{5.23}$$

From Eqs. (5.22) and (5.23), we get:

$$\ddot{\alpha}_{-q}^* + \ddot{\alpha}_q = -\omega_q^2\left(\alpha_q + \alpha_{-q}^*\right) - \sum_n \left\{ \frac{2F(k-q)\omega_q}{\hbar}\left(\alpha_k + \alpha_{-k}^*\right)\left(\alpha_{k-q} + \alpha_{k-q}^*\right) \right\} \tag{5.24}$$

Because

$$u_n = {}_{ph}\langle 0|U|R_n|U|0\rangle_{ph} = \frac{1}{\sqrt{N}}\sum_q u_q(t)e^{iqx},$$

where x=n r_0 and $u_q(t) = \sqrt{\frac{\hbar}{2M\omega_q}}\left(\alpha_q + \alpha_q^*\right)$,then, equation(5.24) becomes

$$\ddot{u}_q + \omega_q^2 u_q = \frac{-8}{3M} i\lambda_1 \sin(r_0 q_1/2)\sum_k \sin(r_0 k/2)\sin[r_0(q-k)/2]u_k u_{q-k} e^{-ir_0(k-q)} \quad (5.25)$$

Taking into account the dispersion relation,

$$\omega_q \approx \frac{1}{2}\sqrt{\frac{\beta}{M}}\left[\frac{1}{2}v_0 q - \frac{1}{3!}\left(\frac{1}{2}r_0 q\right)^3\right] = v_0 q\left(1 - \frac{1}{24}(r_0 q)^2\right),$$

equation(5.25) in long wavelength and the second order approximation can be written as

$$\ddot{u}_q + v_0^2 q^2 u_q - \frac{1}{12} v_0^2 r_0^2 q^4 u_q = \frac{-i}{3M}\lambda_1 r_0^3 q \sum_k (q-k)k u_k u_{q-k}$$

In the continuum approximation, applying the following relation,

$$\frac{\partial u(x,t)}{\partial x} = \frac{1}{\sqrt{N}}\sum_q (iq)u_q(t)e^{iqx}$$

The above equation can be represented by

$$\frac{\partial^2 u(x,t)}{\partial t^2} - \upsilon_0^2 \frac{\partial^2 u(x,t)}{\partial x^2} - \frac{v_0^2 r_0^2}{12}\frac{\partial^4 u(x,t)}{\partial x^4} = \frac{\lambda_1 r_0^3}{3M}\left|\frac{\partial u(x,t)}{\partial x}\right|^2 \quad (5.26)$$

Letting $Q(x,t) = -\partial u(x,t)/\partial x$, then the above equation can be rewritten as

$$\frac{\partial^2 Q(x,t)}{\partial t^2} - \upsilon_0^2 \frac{\partial^2 Q(x,t)}{\partial x^2} - \frac{v_0^2 r_0^2}{12}\frac{\partial^4 Q(x,t)}{\partial x^4} = \frac{\lambda_1 r_0^3}{3M}|Q(x,t)|^2 \quad (5.27)$$

This is a nonlinear equation of the phonon. It shows again that the phonon is a nonlinear microscopic particle in such a case. Applying the boundary conditions, $Q(\pm\infty) = u(\pm\infty) = 0$, it has the following soliton solution

$$Q(x,t)=\frac{3M}{4\sqrt{2}\lambda_1 r_0^3}(v^2-v_0^2)\operatorname{sec}h^2\left[\sqrt{2}\left(x-x_0-\upsilon t\right)/(r_0\sqrt{(v^2/v_0^2)-1})\right],(v>v_0) \quad (5.28)$$

Thus

$$u(x,t)=\frac{-3M}{4\sqrt{2}\lambda_1 r_0^2}(v^2-v_0^2)^{3/2}\tanh\left[\sqrt{2}\left(x-x_0-\upsilon t\right)/(r_0\sqrt{(v^2/v_0^2)-1})\right],(v>v_0) \quad (5.29)$$

where v is the velocity of the phonon-soliton. The shape of the soliton is similar to the soliton solution of Eq.(5.15). But Eq.(5.29) is a supersonic soliton (v> v_0), which satisfies the nonlinear equation (5.26), while Eq.(5.15) is a subsonic soliton (v< v_0), which satisfies the ϕ^4 – equation (5.19). Therefore, the mechanism and properties of localization of the phonons in Eqs.(5.19) and (5.27) are obviously different in the two cases.

These results show clearly that both exciton and phonon become solitons with wave-corpuscle duality under action of the nonlinear interaction arising from their motion and interaction between them. Thus, their rules and properties of motion can be described using nonlinear quantum mechanics [38-40].

5.3. The Raman Scattering Arising from the Motions of the Exciton-Solitons in Molecular Crystal

5.3.1. Theoretical Description of Raman Scattering Effect

At present, we study the physical phenomena and effects arising from the nonlinear excitations and motions of the excitons, such as, Raman scattering and infrared absorption, in the following three sections.

In order to study the Raman scattering effect we first should diagonalize the soliton Hamiltonian equation in nonlinear quantum mechanics [38-40]. Then we should transform it to the intrinsic reference frame moving with the soliton at velocityv[28-35]. In this case, the equation(5.4) must be replaced by

$$\overline{H}=H-\sum\nolimits_{\kappa}\hbar k v(b_{\kappa}^{+}b_{\kappa}+a_{\kappa}^{+}a_{\kappa}) \quad (5.30)$$

where

$$b_k^+=N^{-1/2}\sum_n \exp(iknr_0)b_k^+,b_k=(b_k^+)^+$$

Since the soliton excitation is always connected with the deformation of the molecular chain or intramolecular excitations, then it is very necessary to pass in Eq. (5.30) to new phonons taking this deformation into account. In the light of Eremko, et al.'s method [23-27], such a transition is realized by the replacement $(R_i, P_i) \to (a_q^+, a_q) \to (A_q^+, A_q)$ according to the relations: $A_q^+ = a_q^+ - N^{-1/2}\alpha_q^*, A_q = a_q - N^{-1/2}\alpha_q$,where A_q^+ (A_q) is the creation (annihilation) operator of the new phonon. The coherent phonon state (lattice distortion) in Eq.(5.6), thus, becomes as the vacuum state of new phonons in this case

$$\left|\hat{0}\right\rangle_{ph} = \exp[(\alpha_q(t)a_q^+ - \alpha_q^*(t)a_q^+]\left|0\right\rangle_{ph}$$

where $A_q\left|\tilde{0}\right\rangle_{ph} = 0$. We now carry out the canonical transformation for the partial diagonalization of the soliton Hamiltonian by the following transformation:

$$B_\lambda^+ = \sum\nolimits_n \varphi_\lambda(t) b_n^+, B_\lambda = (B_\lambda^+)^+, \text{ with } \sum\nolimits_\lambda \varphi_\lambda^*(n)\varphi_\lambda(n') = \delta_{nn'}, \sum\nolimits_n \varphi_\lambda^*(n)\varphi_{\lambda'}(n) = \delta_{\lambda\lambda'}$$

The so-called partial diagonalization of the Hamiltonian implies the diagonalization of that part of the Hamiltonian, which does not contain the creation and annihilation operators of the new phonons. The conditions imposed onto the function $\varphi_\lambda(n)$ to realize such a diagonalization are equivalent in the continuum approximation to the following problem on eigenfunction $\varphi(n\lambda)$ and the eigenvalues E_λ:

$$\left[-J\frac{\partial^2}{\partial n^2} - (\frac{i\hbar v}{r_0})\frac{\partial}{\partial n} - (2\mu^2 J r_0^2)\operatorname{sec}h^2(\mu n) + \varepsilon_0 - 2J\right]\varphi\begin{pmatrix} n \\ \lambda \end{pmatrix} = E_{\lambda\varphi}\varphi\begin{pmatrix} n \\ \lambda \end{pmatrix}.$$

It has the unique bound state solution:

$$\varphi\begin{pmatrix} n \\ s \end{pmatrix} = (\frac{\mu}{2})^{1/2} \operatorname{sec}h(\mu n)\exp\left[\frac{i\hbar v n}{2Jr_0^2}\right],$$

with

$$E_s = \varepsilon_0 - 2J - [(\hbar^2 v^2)/(4Jr_0^2)] - \mu^2 J \tag{5.31}$$

and an unbound state solution given by

$$\varphi\begin{pmatrix} n \\ k \end{pmatrix} = \frac{\mu.\tanh(\mu n) - ikr_0}{\sqrt{2}\pi(\mu - ikr_0)}\exp\left[\frac{i\hbar v n}{2Jr_0^2} + ikr_0 n\right],$$

with

$$E_\kappa = \varepsilon_0 - 2J - [(\hbar^2 v^2)/(4Jr_0^2)] + (kr_0)^2 J \tag{5.32}$$

$E_s$is less than that of lowest unbound state $E_\lambda^0 = \varepsilon_0 - 2J$ by $\mu^2 J$ in the statistical state. Thus, the partially diagonalized Hamiltonian in Eq.(5.31) becomes

$$\overline{H} = E_s B_s^+ B_s + \sum_k E_k B_k^+ B_k + \sum_q \hbar(\omega_q - vq) A_q^+ A_q + \frac{1}{\sqrt{N}} \sum_q \hbar(\omega_q - vq)(A_q^+ \alpha_q + A_q^+ \alpha_q^*)(1 - B_s^+ B_s) + \\ \frac{1}{N} \sum_{qk} \overline{F}(k,q)(B_s^+ B_{-k} + B_k^+ B_s)(A_{-q}^+ + A_q) + \frac{1}{\sqrt{N}} \sum_{qk} F(k,q)(B_{k+q}^+ B_k)(A_q + A_{-q}^+) + \frac{G^2}{48J} \tag{5.33}$$

where

$$\overline{F}(k,q) = \left[\frac{2ir_0 qJ\pi\hbar}{\omega_0(G + 4Jir_0 k)}\right] \left(\frac{\hbar J}{GM\omega_q}\right)^{1/2} [x_1(e^{iqr_0} - e^{-iqr_0}) + x_2(e^{iqr_0} - e^{-iqr_0})] \operatorname{sec} h \left[\frac{2\pi r_0 J(q-k)}{G}\right],$$

and

$$F(k,q) = \left(\frac{\hbar^3}{2M\omega_0^2\omega_q}\right)^{1/2} [x_1(e^{iqr_0} - e^{-iqr_0}) + x_2(e^{iqr_0} - e^{-iqr_0})] \left\{1 - \frac{4iGqr_0}{(G - 4ikr_0 J)[G - 4o(l+q)r_0 J]}\right\}$$

are the coupled constants of the bound state and the unbound state with the new phonons, respectively. Obviously, the distribution amplitude of the amide-I vibrational excitations, Eqs.(5.31) and (5.32), can be found if an exactly specified deformation potential is assumed to be given a priori. Thus this treatment, mentioned above, makes us see that the Hamiltonian in in Eqs.(5.33) or (5.30) describing the molecular chain in the reference frame moving with velocity$v$possesses both the soliton state, equation (5.31), with the energy

$$E_{sol} = E_s + W = \varepsilon_0 - 2J - \left(\frac{\hbar^2 v^2}{4Jr_0^2}\right) - \frac{G^2}{48J},$$

and also the delocalized state corresponding to the exciton in the undeformed chain with the energy E_k in Eq.(5.32). The former is a localized coherent structure with a size of the order of $2\pi\ r_0/\mu$ that propagates with velocity$v$and can transfer energy $E_{sol} < \varepsilon_0$. Unlike a bare exciton that can be scattered by interactions with the phonons, the soliton state describes a quasi-particle consisting of the exciton plus the lattice deformation (for this point we can see from E_{sol}) and hence it already includes the interaction with the acoustic phonons. Therefore, it is not scattered and speeded by this interaction, and can maintain its form, energy, momentum and other quasi-particle properties over a macroscopic distance in the absence of external fields and /or impurities.

The above representation shows sufficiently that the soliton energy is lower than that of the lowest exciton state(k =0) by $G^2/48J$. Thus, an energy gap between the soliton state and the exciton state occurs, which corresponds to the binding energy of the soliton. This is due to the fact that the soliton, unlike the exciton, is localized, and is a dynamically self-sustaining entity resulting from a self-trapping of the exciton through interaction with the deformation of the molecular chain. Because of the self-trapping, the energy of the exciton drops about $G^2/24J$,which is the same as the deformation energy of the lattice,

$$W = \frac{1}{N}\sum_q \hbar(\omega_q - vq)|\alpha_q|^2 = \frac{G^2}{24J}$$

The energy of the soliton formed by this mechanism is about $G^2/48J$ less than that of the exciton. Thus, the energy gap between them is $G^2/48J$. The energy gap can result in the red shift from the main peak at 1666 cm^{-1} in the infrared absorption and Raman spectra. The red shift of about 16 cm^{-1} from the main peak of 1666 cm^{-1} was observed experimentally in infrared and Raman spectra in the acetanilide as mentioned earlier.

Utilizing the following values of the physical parameters of acetanilide [1-6, 28-35],

$$\beta=(9.6\sim26)\text{N/m};\ M=(3.42\sim4.50)\times10^{-25}\text{kg},\ \chi=\hbar\chi_1/2\omega_0=(56-62)PN,$$
$$\chi'=\hbar\chi_2/2\omega_0=(6-9)PN, \omega_0=(2-4)\times10^{14}s^{-1}, \omega_1=(3-6)\times10^{13}s^{-1}$$
$$\varepsilon_0=(0.205\sim0.210)\text{eV},\ J=3.4\text{-}4\text{cm}^{-1},\ r_0=4.0\ \overset{0}{A} \tag{5.34}$$

Pang obtained the binding energy of the soliton or the energy gap between the soliton and exciton to be 18.1–33 cm^{-1}, which is basically consistent with the experimental value of the red shift, 16 cm^{-1}. However, the binding energy of the soliton is about 3–7 cm^{-1} for the Davydov model, which is too small. The soliton or the localization of the exciton is formed due to the fact that the nonlinear interaction energy $G=(1.5\text{-}3.2)\times10^{-21}$J balances and suppresses the dispersive energy $J=0.795\times10^{-22}$J. This indicates clearly that the observed red shift is caused by the soliton. Thus, it is experimentally confirmed that the soliton is actually localized in the acetanilide.

Identification of the corresponding Stokes component in the Raman scattering spectrum is one of the ways of experimentally determining the energy gap between the soliton and the exciton and, hence, the solitons themselves. Therefore, it is necessary to calculate the differential cross-section of the Raman scattering due to the solitons. We assume here that the Raman scattering process is activated by some intermediate states of molecular chains associated with the soliton, e.g., the electronic excitation with Hamiltonian is

$$H_e = \sum_{kn} \varepsilon_n(k) D_{kn}^+ D_{kn}$$

where $D_{kn}^+(D_{kn})$ is the creation (annihilation) operator of the electronic excitation of the nth band with wave vector$k$and energy $\varepsilon_n(k)$. At the same time, let the incident light wave be quantized in the volume $V'=Nr_0S'$, then its Hamiltonian can be denoted by

$$H_Q = \sum_{Q\sigma} \hbar\omega_Q C^+_{Q\sigma} C_{Q\sigma}$$

where $C^+_{Q\sigma}(C_{Q\sigma})$ is the creation (annihilation) operator of the photon with the wave vector Q, the energy $\hbar\omega_Q$ and unit polarization vector $\vec{e}_\sigma(Q)$. In such a case, the interaction Hamiltonian among the soliton, electronic excitation states and light wave leading to the Raman scattering can be denoted by

$$H_{\text{int}} = \frac{1}{N}\sum_{kk'}\sum_{q} X_{nn'}(q) D^+_{k+qn'} D_{kn} B^+_{k'-q} B_{k'} + \frac{1}{N}\sum_{kk'}\sum_{q} \overline{X}_{nn'}(q) D^+_{kn} D_{kn'} B^+_{k'-q} B_{k'} (B^+_k B_{k'} - B^+_k B_k) + \sqrt{N}\sum_{Q\sigma n} U_{n\sigma}(Q)(C^+_{Q\sigma} + C_{-Q\sigma})(D^+_{Qn} - D_{-Qn}) \tag{5.35}$$

where $X_{nn'}$ and $\overline{X}_{nn'}$ are the interaction coefficients among the excitons and solitons and electronic excitations with different wave vectors, respectively, $U_{n\sigma}$ (Q) is the coupled constant between the electronic excitations and the light wave, i.e.,

$$U_{n\sigma}(Q) = -i\varepsilon_n(0)\sqrt{2\pi / h\omega_Q V'}(\vec{e}_\sigma(Q).\vec{d}_n)$$

where $\vec{d}_n$ is the dipole moment of the transition from the ground state of the electron into the state of the *n*th band, the Z –axis is directed along the molecular chain.

Pang calculated the transition probability per unit time of the “electron-light-soliton (exciton) system” from the initial state:

$$|m\rangle = C^+_{Q_0\sigma_0}|0\rangle_Q B^+_s|0\rangle_{ex}\prod_q \frac{1}{\sqrt{\bar{n}_q}}(A^+_q)^{\bar{n}_q}|\bar{0}\rangle_{ph}|0\rangle_e ,$$

to the final state:

$$|fk\rangle = C^+_{Q\sigma}|0\rangle_Q B^+_k|0\rangle_{ex}\prod_q \frac{1}{\sqrt{n_q!}}(a^+_q)^{n_q}|0\rangle_{ph}|0\rangle_e$$

caused by the perturbation potential, Eq.(5.35), where the initial state consists of the photons with the wave vector $\vec{Q}_0$ and polarization $\vec{e}_{\sigma_0}(\vec{Q}_0)$ and the soliton moving with the velocity *v* and the new phonons and the electronic excitation, while the final state consists of the photons with the wave vector $\vec{Q}$ and polarization $\vec{e}_\sigma(\vec{Q})$ and the exciton and a certain number of ordinary phonons and electronic excitations, where $|0\rangle_Q$ and $|0\rangle_e$ are the vacuum states of the photons and the electronic excitation states, respectively. According to the perturbation theory in quantum theory, in the case of interest, the probability of transition from the initial state to the final state can be determined by

$$\frac{d}{dt}\overline{W}[(Q_0,\sigma_0)\to(Q,\sigma)]=\frac{d}{dt}[\frac{1}{\hbar^6}\sum_{k,j}P^{(ph)}\left|\int_{-\infty}^{t}dt_1\int_{-\infty}^{t_1}dt_2\int_{-\infty}^{t_2}dt_3\langle fk|\overline{H}_{\text{int}}(t_1)\overline{H}_{\text{int}}(t_2)\overline{H}_{\text{int}}(t_2)|m\rangle\right|^2]$$

where

$$\overline{H}_{\text{int}}(t_1)=\exp[iH_0t]H_{\text{int}}(t)\exp[iH_0t]$$
$$H_0=H_0+H_Q+H_e-\hbar\sum\nolimits_{kn}kvD_{kn}^{+}D_{kn}-\hbar\sum\nolimits_{Q\sigma}(\vec{Q}.\vec{v})C_{Q\sigma}^{+}C_{Q\sigma},$$

However, we are more interested in the long-time behavior of d $\overline{W}$ /dt for estimating this probability. By straightforward calculation the final transition probability can approximately be represented by

$$\lim_{t\to\infty}\frac{d\overline{W}}{dt}\approx\frac{\pi^2r_0^2}{4V'^2}\frac{\gamma\sqrt{J}}{\mu^2J+\Delta_{ks}}(\frac{\gamma}{2.42\omega_0})^{g_0}\{[(\Delta_{ks}(0)-W)^2+(\hbar\gamma)^2][\sqrt{(\Delta_{ks}(0)-W)^2+(\hbar\gamma)^2}-$$
$$\sqrt{(\Delta_{ks}(0)-W)^2}]\}^{-1/2}\left|\sum_{nn_o}\frac{(Q_{0z}-Q_z)U_{\sigma_0n_0}(Q_o)[X_{nn_0}(Q_{0z}-Q_z)+\overline{X}_{nn_0}(Q_{0z}-Q_z)]U_{\sigma n}^{*}(Q)}{[(\varepsilon_{n_0}(Q_{0z})-\hbar\omega_{Q_{0z}})(\varepsilon_n(Q_z)-\hbar\omega_{Q_z})]}+\right.$$
$$\left.\frac{(Q_z-Q_{0z})U_{\sigma_0n_0}(Q_o)[X_{nn_0}(Q_z-Q_{0z})+\overline{X}_{nn_0}(Q_z-Q_{0z})]U_{\sigma n}^{*}(Q)}{[(\varepsilon_{n_0}(Q_{0z})-\hbar\omega_{Q_{0z}})(\varepsilon_n(Q_z)-\hbar\omega_{Q_z})]}\right|^2\times$$
$$\{2+\tanh^2[\frac{\pi}{2\mu}(\sqrt{\frac{\Delta_{ks}(0)}{J}}-(Q_{0z}-Q_z)r_0)]+\tanh^2[\frac{\pi}{2\mu}(\sqrt{\frac{\Delta_{ks}}{J}}-(Q_z-Q_{0z})r_0)]\text{............}(5.36)$$

where

$$\gamma=\frac{2J\mu r_0K_BT}{\hbar^2v_o},\Delta_{ks}=E_k-E_s-\mu^2J/3+\hbar(\omega_Q-\omega_{Q_0})$$

5.3.2. The Features of Raman Scattering of Acetanilide

According to the quantum theory, we find that the cross-section of the Raman scattering is closely associated with the transition probability, Eq. (5.36). Due to the fact that this transition probability has a sharp peak, when the frequency of scattering light is in the vicinity of the frequency of the incident light, ω_{Q_0} , we are interested only in the cross-section of the Raman scattering,in thiscase, which is

$$\frac{dN}{NdS}=\frac{d^2\delta}{d\omega_Q d\Omega}=\frac{V'^2\omega_{Q_0}^2}{4\pi^2}\frac{d\overline{W}}{dt}\approx\frac{V'^2\omega_{Q_0}^2 r_0^2}{16V'^2\hbar^2}\frac{\hbar\gamma\sqrt{J}}{\mu^2 J+\Delta_{ks}}\left(\frac{\gamma}{2.42\omega_0}\right)^{g_0}\{[(\Delta_{ks}(0)-W)^2+(\hbar\gamma)^2]\times$$
$$[\sqrt{(\Delta_{ks}(0)-W)^2+(\hbar\gamma)^2}-\sqrt{(\Delta_{ks}(0)-W)^2}]\}^{-1/2}\times$$
$$\left|\sum_{nn_o}\frac{(Q_{0z}-Q_z)U_{\sigma_0 n_0}(Q_o)[X_{nn_0}(Q_{0z}-Q_z)+\overline{X}_{nn_0}(Q_{0z}-Q_z)]U^*_{\sigma n}(Q)}{[(\varepsilon_{n_0}(Q_{0z})-\hbar\omega_{Q_{0z}})(\varepsilon_n(Q_z)-\hbar\omega_{Q_z})]}+\right.$$
$$\left.\frac{(Q_z-Q_{0z})U_{\sigma_0 n_0}(Q_o)[X_{nn_0}(Q_z-Q_{0z})+\overline{X}_{nn_0}(Q_z-Q_{0z})]U^*_{\sigma n}(Q)}{[(\varepsilon_{n_0}(Q_{0z})-\hbar\omega_{Q_{0z}})(\varepsilon_n(Q_z)-\hbar\omega_{Q_z})]}\right|^2\times$$
$$\{2+\tanh^2[\frac{\pi}{2\mu}(\sqrt{\frac{\Delta_{ks}(0)}{J}}-(Q_{0z}-Q_z)r_0)]+\tanh^2[\frac{\pi}{2\mu}(\sqrt{\frac{\Delta_{ks}}{J}}-(Q_z-Q_{0z})r_0)]$$

(5.37)

where dΩ=dθ dφ sinθ is the element of solid angle in the direction of propagationof the scattering light, $\theta_0$is an angle between the incident light $\vec{Q}_0$ and the z-axis, θ is an angle between the $\vec{Q}_0$ and the scattering light $\vec{Q}$, φ is a angle between the projection of the vector $\vec{Q}$ on the xy-plane and x-axis, as is shown in Figure 5. 3.

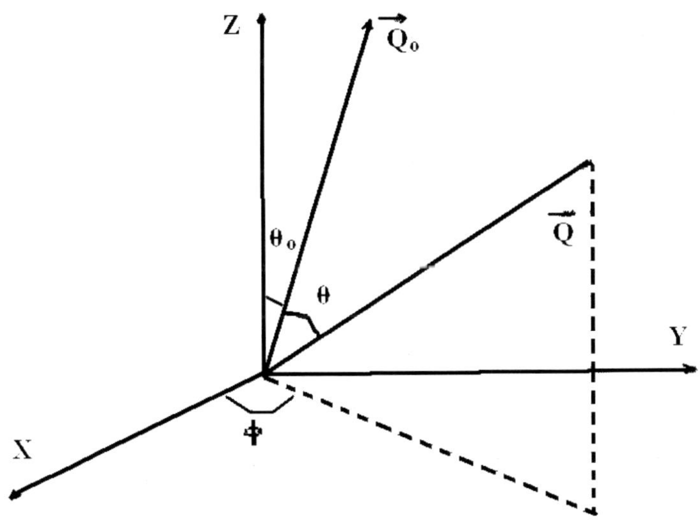

Figure 5.3. The schematic figure of directions for the incident light $\vec{Q}$ and scattering light $\vec{Q}$, here the molecular chain is in the z axis.

Obviously, the size of the Raman scattering depends closely on the size of the energy gap and the direction and intensity of the incident light and the size and orientation of the dipole moment of transition from the ground state to the electronic resonance level. This shows that the properties of the scattering light are different for different incident lights. If the frequency of the incident light is in the following region,

$$0 < \hbar\omega_{Q_0} < \mu^2 J - \varepsilon_n(0),$$

or

$$\mu^2 J + \varepsilon_n(0) < \hbar\omega_{Q_0} < 2\varepsilon_n(0) ,$$

then Eq.(5.37) represents a resonant Raman scattering. The inequalities ensure not only the smoothness of the cross-section as a function of ω_{Q_0} but also the applicability of the above perturbation theory to the calculation of the cross-section. However, when the frequency of the incident light satisfies $\mu^2 J << \hbar\omega_{Q_0} << \varepsilon_n(0)$ for all$n$, then Eq.(5.37) represents a nonresonant Raman scattering. In such case, a part of energy of incident light will be expanded and converted into the excitation energy of the soliton. Thus, the angular-dependent part in Eq.(5.37) may approximately be treated by a constant.

If studying in detail the significances of Eq.(5.37), the Raman scattering has the following properties: (1)When the frequencies of the scattering light is close to that of the incident light, then the scattering cross-section is the largest.

(2) For a fixed direction of the scattered light, the cross-section and the shapes of the spectral line are mainly determined by the factor,

$$\frac{\hbar\gamma\sqrt{J}}{\mu^2 J+\Delta_{ks}} \frac{\omega_{Q_0}^2 r_0^2 (16\hbar^2)^{-1}(\gamma/2.42\omega_0)^{g_0}}{\{[(\Delta_{ks}(0)-W)^2+(\hbar\gamma)^2][\sqrt{(\Delta_{ks}(0)-W)^2+(\hbar\gamma)^2}-\sqrt{(\Delta_{ks}(0)-W)^2}]\}^{1/2}}$$

whether it is resonant Raman scattering or nonresonant Raman scattering.

(3) In general, the scattering indicatrices (angular dependence of the scattering cross-section) are different for the resonant scattering and nonresonant scattering. For the former, the scattering indicatrix depends strongly on the orientation of the dipole moment of the transition from the ground state to the resonance electronic levels. For the latter when the frequencies of the scattered light, ω_Q, are close to that of the Raman peak, its scattering indicatrix is not related with the coupling strength, and has a simple form.

(4) The intensity of the scattered light reaches a maximum when the frequency $\omega_Q$approaches $\omega_{Q_0} + 2J - G^2/48J - \varepsilon_0$.

This maximum scattering intensity is associated with the transition of the system from the soliton state, which is accompanied by lattice deformation, to the delocalized exciton state without lattice deformation. Therefore, the soliton becomes unstable as the incident light gets scattered.

These characteristics of Raman scattering and the remarkable asymmetry exhibited in the spectral line shape (sharper on the short wave side and smoother on the long wave side),which is connected to the corresponding energy dependence of the density of the delocalized states, allows us to confirm the existence of the soliton state of excitons in the organic molecular crystalsACNthrough the experimental Raman results.

5.4. The Characters of Infrared Absorption of Acetanilide in the Nonlinear Quantum Excitation Case in Acetanilide

5.4.1. The Coefficient of Infrared Absorption Arising from the Nonlinear Excitation

In the light of Hamilton theory in the nonlinear quantum mechanics,Pang expressed the effective Hamiltonian [38-40] corresponding to Eq.(5.13) as

$$H_{eff} = \frac{1}{r_0}\int(\frac{1}{2}Jr_0^2 |\varphi_x|^2 + \frac{1}{2}\varepsilon|\varphi|^2 - \frac{1}{4}G|\varphi|^4)dx. \tag{5.38}$$

Its corresponding discrete form is

$$H_{eff} = \frac{J}{2}\sum_n(\varphi_n\varphi_{n+1}^* + \varphi_{n-1}\varphi_n^*) + \sum_n(\frac{1}{2}\varepsilon|\varphi_n|^2 - \frac{1}{4}G|\varphi_n|^4) \triangleright \tag{5.39}$$

In earlier discussion, we considered only properties of exciton-soliton, which is formed through coupling of amide-I oscillators (excitons) with longitudinal acoustic phonons. To explain the temperature dependence of intensity of the infrared absorption of the exciton-soliton, one has to take the interaction between the soliton and optical phonons in acetanilide into account. It is believed that this phenomenon is related to the zero-phonon line spectrum arising from random thermal modulations of the exciton-solitons by the optical phonons. In this case, the Hamiltonian of this system can be written as

$$H = H_{eff} + H_{op} + H' \tag{5.40}$$

where H_{eff} is the Hamiltonian of the effective amide-I oscillator system given by Eq.(5.39), which satisfies $H_{eff}|m\rangle = \hbar\omega(m)|m\rangle$. Here the egenstate and the energy eigenvalue are denoted by $|m\rangle$ and $\hbar\omega(m)$, respectively, in terms of a quantum number m. H_{op} is the Hamiltonian for the optical-phonon system, which is taken to be a set of harmonic oscillators,

$$H_{op} = \sum_n h_n(Q_n) = \sum_n[-\frac{\hbar^2}{2M^*}(\frac{\partial^2}{\partial Q_n^2}) + \frac{1}{2}M^*\omega_n^2 Q_n^2] \tag{5.41}$$

with

$$h_n(Q_n)\Phi_{\kappa n}(Q_n)=\hbar\omega_n(\kappa n+\frac{1}{2})\Phi_{\kappa n}(Q_n)=\hbar\omega_n(\kappa n)\Phi_{\kappa n}(Q_n),\ \kappa_n=0,1,2,...,$$

$$\omega\ (N)=\sum_n \omega_n(\kappa_n),$$

where h_n, Q_n and ω_n are the Hamiltonian, the normal coordinate and the eigenfrequency of the nth harmonic oscillator, respectively, in which M^* is its effective mass, and $\Phi_{k_n}(Q_{n)})$ is an eigenfunction of$h_n$characterized by the integer k_n.H' in Eq.(5.40) describes the interaction between the amide-I oscillator system and the optical phonon system, which we take to be linear in the $Q_n^{'}s$, i.e.,

$$H^{'}=\sum_{n^{'}}V_{n^{'}}(\varphi_{n^{'}})Q_{n^{'}} \tag{5.42}$$

We can find that

$$V_{n^{'}}(\varphi_{n^{'}})=\frac{1}{\sqrt{n}}\sum_{n^{"}}\lambda_{n^{"}}\varphi_{n^{'}}^{2}\vec{e}(\vec{k},\eta)e^{i\vec{k}\bullet\overline{u_{n^{"}}}}\sin(\vec{k}\bullet\overrightarrow{u_{n^{'}}}) \tag{5.43}$$

where $\vec{e}(\vec{k},\eta)$ is a polarization vector associated with the wave vector $\vec{k}$ and branch $\eta(\eta=1,2,3)$ of two transverse and one longitudinal, $\lambda_{n"}$ is an interaction constant in which the index n'' indicates the range of the oscillator-phonon interaction.

In order to study the properties of the entire system given by Eq.(5.40), we use an adiabatic approximation, i.e., the phonon state depends adiabatically on the amide-I oscillator state. This can be justified on the ground that the characteristic frequency of about 1650 cm^{-1} of the amide-I oscillator system is much higher than the maximum frequency of the phonon system, which is about 200 cm^{-1}. In order to obtain the adiabatic potential $U_m$associated with the state $|m\rangle$ of the amide-I oscillator system, we neglect the change in wave function of the amide-I oscillator system caused by the perturbation H'. Thus, the effective HamiltonianH_{eff} of this system for the state $|m\rangle$ can be written in terms of a set of the Hamiltonians of the displaced harmonic oscillators of the phonon system

$$H_{eff}(m)=\hbar\omega(m)+\sum_{n^{'}}\frac{1}{2}M^*\omega_{n^{'}}^2Q_{n^{'}}^2-\sum_{n^{'}}\frac{\hbar^2}{2M^*}\frac{\partial^2}{\partial Q_{n^{'}}^2}+\sum_{n^{'}}q_{n^{'}}(m)Q_{n^{'}}$$

$$=U_m-\sum_{n^{'}}\frac{\hbar^2}{2M^*}\frac{\partial^2}{\partial Q_{n^{'}}^2}=\sum_{n^{'}}h_{n^{'}}[Q_{n^{'}}-Q_{0n^{'}}(m)]+const., \tag{5.44}$$

where

$$q_{n'}(m) = \langle m | V_{n'}(m) | (m) \rangle, Q_{0n'}(m) = -(q_{n'}(m) / M\omega_{n'}^2)$$

Using this Hamiltonian, we can find the infrared absorption coefficient arising from the motion of the soliton. In accordance with the first order Born approximation in quantum radiation and theory of absorption of light, the absorption coefficient per unit volume of the amide-I oscillator system described by Eq. (5.44) can be expressed as

$$B_{fn'(\omega)} = \frac{4\pi\omega}{\varepsilon' chV_o} \int dte^{-i\omega t} \langle \sum_{m_f} \langle m_{n'} | P(t) | m_f \rangle \langle m_f | P | m_{n'} \rangle \times$$

$$\prod_{n'} \sum_{k_n'} \Gamma(\kappa_{n'}, \kappa_n', m_n, m_f)^2 e^{i[\omega(\kappa_n') - \omega(\kappa_{n'})]t} \rangle_T \tag{5.45}$$

where

$$P_{(t)} = e^{iH_{eff}t/\hbar} P e^{-iH_{eff}t/\hbar}$$

$$\Gamma(\kappa_{n'}, \kappa_{n'}', m_n, m_f) = \int_{-\infty}^{\infty} d\zeta_{n'} \Phi_{\kappa_{n'}} [\zeta_{n'} - \zeta_{0n'}(m_n)] \Phi_{\kappa_{n'}'} [\zeta_{n'} - \zeta_{0n'}(m_f)],$$

in which

$$\zeta_{n'} = Q_{n'} \sqrt{\frac{M^*\omega_{n'}}{\hbar}}, \; Q_{0n'}(m) \sqrt{\frac{M^*\omega_{n'}}{\hbar}} = -q_{n'}(m) \sqrt{\frac{\hbar}{M^*\omega_{n'}}} \frac{1}{\hbar\omega_{n'}} << 1$$

In the above equations, $m_{n'}(k_n)$ and $m_f(k_n')$ with $\omega(N_n) = \sum_{n'} \omega_{n'}(k_n')$ and $\omega(N_f) = \sum_{n'} \omega_{n'}(k_n')$ are the quantum numbers denoting the initial and the final states of the amide-I oscillator system (the *n*th harmonic oscillator), respectively. The quantities, *c*, *P*, ε' are the speed of light, the effective dipole moment of the amide-I oscillator and the refractive index, respectively, V_0 is the volume of the total system, $\langle .... \rangle_T$ denotes the average over all initial states in the canonical equilibrium.

Following procedure of Takeno [41-42], we get

$$\Gamma(\kappa_{n'},\kappa'_{n'},m_n,m_f)=\begin{cases}1-\frac{1}{2}(\kappa_{n'}+\frac{1}{2})\zeta^2_{0n'}(m_n,m_f), \; for\;\kappa'_{n'}=\kappa_{n'} \\ \sqrt{\frac{\kappa_{n'}+1}{2}}\zeta_{0n'}(m_n,m_f), \quad for\;\kappa'_{n'}=\kappa_{n'}+1 \\ -\sqrt{\frac{\kappa_{n'}}{2}}\zeta_{0n'}(m_n,m_f), \quad for\;\kappa'_{n'}=\kappa_{n'}-1 \\ 0, \quad for\;\kappa'_{n'}\neq\kappa_{n'},\kappa_{n'}\pm 1\end{cases}$$

and

$$\prod_{n'}\sum_{k'_{n'}}\Gamma(\kappa_{n'},\kappa'_{n'},m_n,m_f)e^{i[\omega(k'_{n'})-\omega(k_{n'})]t}\cong\prod_{n'}[1+Z_{n'}(t,m_n,m_f)]$$

where

$$Z_{n'}(t,m_n,m_f)=\frac{1}{2}\zeta^2_{0n'}(m_n,m_f)[(\kappa_{n'}+1)e^{-i\omega_{n'}t}+\kappa_{n'}e^{-i\omega_{n'}t}-2\kappa_{n'}+1],$$

$$\zeta_{0n'}(m_n,m_f)=\zeta_{0n'}(m_n)-\zeta_{0n'}(m_f)<<1$$

In accordance with the above approximation procedure, the thermal average in the integrand of Eq.(5.45) can be expressed separately for the amide-I oscillator state and the phonon state, which are denoted by the angular bracket with subscript sT and aT, respectively. Taking first the average over the phonon state, we get

$$\left\langle\prod_{n'}[1+Z_{n'}(t,m_n,m_f)]\right\rangle_{aT}=\left\langle\exp_A[\sum_{n'}Z_{n'}(t,m_n,m_f)]\right\rangle_{aT}$$
$$=\exp\{\left\langle\exp_A[\sum_{n'}Z_{n'}(t,m_n,m_f)]-1\right\rangle^D_{aT}\}\approx\exp[\left\langle\sum_{n'}Z_{n'}(t,m_n,m_f)\right\rangle_{aT}] \tag{5.46}$$

where the symbol $\exp_A(\sum_{n'}Z_{n'})$ is a leveled exponential function that levels off the product $Z_{n'}$ s by neglecting terms, which contain higher orders than unity in any of the $Z_{n'}$ s, and the superscript (*D*) attached to the bracket symbol indicates a cumulative average, as proposed by Kabo [43]. Using Eq.(5.45), equation (5.46) can be written as,

$$B_{fn^{'}(\omega)} = \frac{4\pi\omega}{\varepsilon^{'} chV_o} \int dte^{-i\omega t} \langle \sum_{m_f} \exp[-z_0(T, m_n, m_f] \exp[-z_1(T, t, m_n, m_f] > \times$$

$$\langle m_{n^{'}} \mid P(t) \mid m_f \rangle \langle m_f \mid P \mid m_{n^{'}} \rangle_{aT}$$

,

$$z_0(T, m_n, m_f) = \sum_{n'} \left\langle \varsigma_{on'}^2 (m_n, m_f)(k_{n'} + 1/2) \right\rangle,$$

$$z_1(T, t, m_n, m_f) = \frac{1}{2} \sum_{n'} \left\langle \varsigma_{on'}^2 (m_{n'}, m_f)[(k_{n'} + 1) \exp(i\omega_{n'} t) + (k_{n'} \exp(-i\omega_{n'} t) \right\rangle_{aT}$$

The function $\exp[z_1(T, t, m_n, m_f]$ can be expanded by

$$\exp[z_1(T, t, m_n, m_f)] = 1 + z_1(T, t, m_n, m_f) + \frac{1}{2!} z_1^2 (T, t, m_n, m_f) + \ldots\ldots,$$

where the first, the second and the third terms on the right-hand side are the contributions to the zero, one and two-phonon processes. In the general case, one and multi-phonon processes give only weak side-band effects [41-42]. Therefore, we may only pay attention exclusively to the zero-phonon process and the main absorption line. Then we get

$$B_{fn^{'}(\omega)} = \frac{4\pi\omega}{\varepsilon^{'} chV_o} \int dte^{-i\omega t} \langle \sum_{m_f} \exp[-z_0(T, m_n, m_f] \langle m_{n^{'}} \mid P(t) \mid m_f \rangle \langle m_f \mid P \mid m_{n^{'}} > \rangle_{aT} \tag{5.47}$$

In order to determine $Z_0(T,m_n,m_f)$, we have to find out $\varsigma_{0n}(m_n, m_f)$. Then evaluating the matrix element of H_0 in Eq.(5.42) with respect to $|m\rangle$ for low frequency phonons and in the adiabatic approximation, $\varsigma_{0n'}(m_n, m_f)$, which can be separated into two parts, phonon and soliton, is

$$\varsigma_{0n^{'}}(m_n, m_f) \cong \varsigma_{0n^{'}}^{(1)} \varsigma_s^{(2)} (m_n, m_f), \tag{5.48}$$

where

$$\varsigma_{0n^{'}}^{(1)} \cong \sqrt{2} \sum_l \lambda_l \vec{e}(n^{'}) \frac{\mu_{n^{'}}}{\upsilon_{0n^{'}}} \sqrt{\frac{1}{\hbar M^* \omega_{n^{'}}}} = \frac{F_{n^{'}}}{\sqrt{\omega_{n^{'}}}},$$

$$\varsigma_s^{(2)}(m_n, m_f) \cong \frac{1}{\sqrt{N}} \sum_i (\langle m_n \mid Q_i Q_i^* + Q_i^* Q_i \mid m_n \rangle - \langle m_f \mid Q_i Q_i^* + Q_i^* Q_i \mid m_f \rangle).$$

with

$$F_{n^{'}} = \sum_{l} \sqrt{2}\lambda_l \vec{e}(n^{'})(\frac{\mu_{n^{'}}}{\upsilon_{on^{'}}})\frac{1}{\sqrt{\hbar M^*}}, \quad \omega_{n^{'}} \cong \upsilon_{on^{'}} k,$$

v_{0n} is the speed of the nth phonon mode. Since the n'-dependence of $\vec{e}(n')$ and $v_{0n'}$ comes mainly from $\omega_{n'}$, we can neglect the n_0-dependence of $F_{n'}$ and rewrite it as F. Inserting the above formula into Eq.(5.47), and using the Debye's approximation for the frequency spectrum of phonons, Pang finally obtained [44-47]

$$Z_o(T, m_n, m_f) = Z_o(T) = [\zeta_s^{(2)}]^2 \sum_{n^{'}} \frac{1}{\omega_{n^{'}}} \left\langle F_{n^{'}}^2 \left(k_{n^{'}} + \frac{1}{2} \right) \right\rangle_{aT}$$

$$\approx \left[\zeta_s^{(2)} \right]^2 \mathrm{F}^2 \sum_{n^{'}} \frac{1}{\omega_{n^{'}}} \left(\left\langle k_{n^{'}} \right\rangle_{aT} + \frac{1}{2} \right) = C + \lambda T^2, \text{ for } T << \theta. \tag{5.49}$$

where C and , λ of which the explicit expressions are omitted here, are constants, and θ is the Debye temperature of the acetanilide. If $T << \theta$, we have $z_0(T) = C + \lambda T^2$.

5.4.2. The Features of Infrared Absorption of Acetanilide

Until now, no specification has been given for the quantum number m. It may be divided into two parts, one being associated with the bound states or the soliton states of the microscopic particle, the other with continuum or scattering states of excitons, denoted by $m_s$and m_e, respectively. Correspondingly, the absorption coefficient can be divided into the soliton absorption line $B_{sf(\omega)}$ and the exciton absorption line $B_{ex(\omega)}$, appearing at about 1650 cm^{-1} and 1665 cm^{-1}, respectively,
i.e.,

$$B_{n^{'} f(\omega)} = B_{sf(\omega)} + B_{ex(\omega)}.$$

From Eqs.(5.47) and (5.48), the following relation can be derived

$$\left\langle m_s \,|\, Q_{n^{'}} Q_{n^{'}}^* + Q_{n^{'}}^* Q_{n^{'}} \,|\, m_s \right\rangle >> \left\langle m_e \,|\, Q_{n^{'}} Q_{n^{'}}^* + Q_{n^{'}}^* Q_{n^{'}} \,|\, m_e \right\rangle$$

Thus, we have,

$$\zeta_{0n}(m_{sn}, m_{sf}) >> \zeta_{0n}(m_{en}, m_{ef})$$

or

$$Z_o(T, m_{sn}, m_{sf}) >> Z_o(T, m_{en}, m_{ef})$$

Since the mean square displacement of the exciton system associated with the state $|m_e\rangle$ is a sum of a large number of oscillatory functions with respect to the space variable *n*', the temperature dependence of $B_{ex}(\omega)$ is very weak compared to that of $B_{sf}(\omega)$ .When the soliton density is so small that their mutual interactions are negligible, we can consider only a single excited state corresponding to the first excited state of single-soliton states. Pang then approximately obtained

$$B_{ex}(\omega) = \frac{4\pi\omega}{\varepsilon' chV_o} \int_{-\infty}^{\infty} dte^{-i\omega t} \langle P_1(t)P_1\rangle_{eT} \tag{5.50}$$

$$B_{sf}(\omega) = \frac{4\pi N_s\omega}{\varepsilon' chV_o} e^{-Z_o(T)} \int_{-\infty}^{\infty} dte^{-i\omega t} \langle P_1(t)P_1\rangle_{m_sT} \approx e^{-(C+\lambda T^2)} B_s^{(0)}(\omega)$$

where

$$B_s^{(0)}(\omega) = \frac{AP_1^2}{\pi} \frac{W_s\omega}{(\omega^2 - \omega_s^2)^2 + W_s^2\omega^2}$$

and $N_s$and P_1 are the soliton number and an effective dipole moment associated with a single soliton state, respectively, ω_s is the eigenfrequency of the soliton, W_s and A are, respectively, a natural width of the single-soliton absorption line and a constant. Since the soliton energy is much higher than that of phonons, $B_s^{(0)}(\omega)$ does not depend on temperature in the above approximation.

From Eq.(5.49), we see that the soliton absorption line spectrum $B_{sf}(\omega)$ is a product of the phonon-free soliton absorption line spectrum $B_s^{(0)}(\omega)$ and the Debye-Waller factor $\exp[-(C+\alpha T^2)]$. This indicates that the intensity of infrared absorption due to the nonlinear excitation of the excitons in acetanilide depends on the temperature exponentially, which is consistent with results obtained by Alexander and Krumbansl,i.e., $\exp(-\beta T^2)$ [15-16], using a polaron model. This temperature dependence of the infrared absorption line at about 1650 cm^{-1} is analogous to that of localized or resonant harmonic phonon modes due to impurities in alkali halide crystals, where the impurity-induced infrared absorption spectra has a characteristic temperature dependence in the form of $\exp(-\alpha T^2)$ (α = constant) at low temperatures. Although the physical origins of the localized modes are different in these two cases, the Debye-Waller factor results from the same origin, i.e., random thermal modulations of low frequency optical phonons. Furthermore, the temperature dependence of the excitonabsorption line, if any, is quite different from that of the soliton in the absence of the Debye-Waller factor in $B_{ex}(\omega)$. If we identify $B_{sf}(\omega)$ and $B_{ex}(\omega)$ with the absorption spectra of the acetanilide at about 1650 cm^{-1} and 1665 cm^{-1}, respectively, which were observed experimentally, the experimental results in acetanilide can be explained. Therefore, it is firmly believed that the soliton can exist in acetanilide and the nonlinear quantum theory gives a correct description of it.

Pang et al.[44-46] used the exciton-soliton model again, an improved theory for color centers in solid state and the nonlinear quantum mechanics for the interaction of the amide-I vibrational quanta (exciton) with optical phonons to study infrared absorption at finite temperature. Its numerical results for the exponential temperature dependence of the intensity of the anomalous band at 1650 $cm^{-1}$are consistent with the experimental data [44-46].

5.5. Properties of Nonlinear Eigenenergy- Spectrum in Molecular Crystals

5.5.1. The Calculation of Nonlinear Eigenenergy-Spectrum

We now calculate the properties of energy spectra of the nonlinear excitations of the excitons, or quantum energy levels of vibrational amide-I in the molecular crystal with two channels in virtue of the Hamiltonianand wave functions , which are given in Eqs. (5.4) -(5.5), using the eigenvalue method in nonlinear quantum mechanics in Chapter 4 [28-40]. Thus, we now have to discretize Eqs.(5.11) and (5.13), which can be denoted as

$$i\hbar \dot{\varphi}_n(t) = \varepsilon_0' \varphi_n(t) - J[\varphi_{n+1}(t) + \varphi_{n-1}(t)] + \lambda'[u_{n+1}(t) - u_n(t)]\varphi_n(t) \tag{5.51}$$

and

$$\ddot{u}_n(t) = (\beta / M)[u_{n+1}(t) - 2u_n(t) + u_{n-1}(t)] + \lambda' / M[|\varphi_{n+1}(t)|^2 - |\varphi_n(t)|^2] \tag{5.52}$$

where $\varepsilon_0 = \hbar\omega_0, \beta = Mv_0^2 / r_0^2$, $\lambda' = 2\hbar(x_1 + x_2)/\omega_0$ and the following transformation relation between continuous and discrete functions are used

$$\varphi(x,t) \to \varphi_n(t) \text{ and } u(x,t) \to u_n(t),$$

$$\varphi_{n\pm1}(t) = \varphi_n(t) \pm r_0 \frac{\partial \varphi_n(t)}{\partial x} + \frac{1}{2!} r_0^2 \frac{\partial^2 \varphi_n(t)}{\partial x^2} \pm \ldots$$

$$u_{n\pm1}(t) = u_n(t) \pm r_0 \frac{\partial u_n(t)}{\partial x} + \frac{1}{2!} r_0^2 \frac{\partial^2 u_n(t)}{\partial x^2} \pm \ldots$$

From Eqs.(5.51)-(5.52)we can get the equation of motion for $\varphi_n(t)$, which is as follows

$$(i\frac{d}{dt}-\omega_0)\varphi_n + J(\varphi_{n+1}+\varphi_{n-1})+\frac{1}{2}G(|\varphi_n|^2)\varphi_n = 0 \tag{5.53}$$

where G is $G = 4\hbar^2 r_0^2 (x_1 + x_2)^2 / M\omega_0 (v^2 - v_0^2)$ for this model $\hbar = 1$, s=v/v$_0$, v is velocity of exciton, v$_0$ is the sound velocity in peptide lattice. The effective Hamiltonian of excitons corresponding to Eq.(5.53) is

$$H_{eff} = \sum_n [\omega_0 |\phi_n|^2 - J(\phi_{n+1}^* \phi_n + \phi_n^* \phi_{n+1}) - \frac{1}{2} G |\phi_n|^4] \tag{5.54}$$

In addition to the energy H$_{eff}$, Eq.(5.53) has another constant of motion, i.e. the particle number:

$$N = \sum_n |\varphi_n|^2 . \tag{5.55}$$

For discretely nonlinear Schrödinger equation (5.53), each chain has three modes of motion, $\varphi_n(t)$, $\varphi_{n-1}(t)$, $\varphi_{n+1}(t)$ in each chain, which denotes the three modes of movement. Then the dynamic equation for the acetanilide with two chains can be written in the following matrix form

$$(i\frac{d}{dt}-\omega_0)\varphi + \overline{M}\varphi + GD(|\varphi|^2)\varphi = 0 \tag{5.56}$$

where φ =Col.($\varphi_1, \varphi_2, \varphi_3, \varphi_4, \varphi_5, \varphi_6,$) is a complex 6-vectors, D($|\varphi|^2$) denotes the diagonal matrix diag.($|\varphi_1|^2, |\varphi_2|^2, |\varphi_3|^2, |\varphi_4|^2, |\varphi_5|^2, |\varphi_6|^2$), and $\overline{\mathrm{M}}$ is a real symmetric matrix with six orders, which represents various neighboring interactions between amide-Is, i.e.,

$$\overline{M} = \begin{bmatrix} 0 & J & \varepsilon & \varepsilon & \varepsilon & J \\ J & 0 & \mathrm{J} & \varepsilon & \varepsilon & \varepsilon \\ \varepsilon & J & 0 & \mathrm{J} & \varepsilon & \varepsilon \\ \varepsilon & \varepsilon & \mathrm{J} & 0 & J & \varepsilon \\ \varepsilon & \varepsilon & \varepsilon & \mathrm{J} & 0 & J \\ J & \varepsilon & \varepsilon & \varepsilon & J & 0 \end{bmatrix} \tag{5.57}$$

Upon quantization [51], the complex mode amplitudes are replaced by harmonic oscillator creation and annihilation operators $\hat{B}_j^+$ and $\hat{B}_j$ with the following properties: $\hat{B}_j^+ | m_j >= \sqrt{(m_j+1)} | m_j+1>, \hat{B}_j | m_j >= \sqrt{m_j} | m_j -1>,$ and $[B_j, \hat{B}_l^+] = \delta_{jl}$ .as mentioned in Chapter 4. Then the particle number and Hamiltonian of the systems corresponding to Eq. (5.56) become the following operators, respectively

$$\hat{N} = \sum_j^6 (\hat{B}_j^+ \hat{B}_j + \frac{1}{2}), \tag{5.58}$$

$$\hat{H} = (\omega_0 - \frac{1}{2}\gamma)\hat{N} - J\sum_j^6 \hat{B}_j^+ \hat{B}_j - \frac{1}{2}\gamma \sum_j^6 \hat{B}_j^+ \hat{B}_j \hat{B}_j^+ \hat{B}_j \tag{5.59}$$

where we take $\hbar$ =1 and assume that both frequencies and energies are measured in the same unit (cm^{-1}).

We here chose the basic function set, |m,k>=|m_1>|m_2>|m_3>|m_4>|m_5>|m_6> for Eqs.(5.58) - (5.59), where m(=m_1+m_2+m_3 m_4+m_5+m_6) is the total quantum number, k=1,2,…, d(m), and d(m) is equal to the number of ways that m particles are distributed in three states. We now consider the common eigenfunction, |$\psi_{m>}$, for two operators mentioned above, we now take

$$|\psi_m>=C_1|m,1>+C_2|m,2>+\ldots+C_d(m)|m, d(m)> \tag{5.60}$$

then the Hamiltonian equation $\hat{H} | \psi_m >= E_m | \psi_m >$ can reduce to an algebra equation on the column vector $\overline{C}_m = col$ (C_1, C_2, …, $C_{d(m)}$)□

$$H_m\overline{C}_m = E_m\overline{C}_m, \tag{5.61}$$

here H_m is a d(m)×d(m), real, symmetrical matrix with diagonal element H_{mr}=<m,r | $\hat{H}$ |m, r> and off-diagonal elements H_{rs}=<m, r | $\hat{H}$ | m, s>, 1 ≤ r ≤ d(m),1≤ s ≤d(m), and r≠s.

5.5.2. The Properties of Nonlinear Energy Spectrum of the Acetanilide

Applying the above formulae and the values of parameters in Eq.(5.34) for acetanilide, we can find the parameter values in Eq.(5.61) in these calculations, which are ω_0^1=1687 cm^{-1}, J^1=3.8 cm^{-1}, γ^1=32.48 cm^{-1}, for *Davydov* 's model [17-22] and ω_0^2 =1687.4 cm^{-1}, J^2=3.8 cm^{-1}, γ^2 =37.29 cm^{-1} for *Pang* 's model [28-40], respectively. Thus, we easily calculate numerically the quantum vibrational energy levels of the acetanilide from Eq.(5.61),the calculated results of the eigenenergy spectrum of the nonlinear excitations, or quantum vibrational energy levels. For the Davydov and Pang models, their values with the experimental data [2-12,43-44,53-54] are together presented in Table 5.1 for the acetanilide.

Table 5.1. Quantum vibrational energy levels (in cm^{-1}) of acetanilide

m	Species	Exp.[(a)]	Cal.[(b)]	Cal.(c)	m	Species	Exp.[(a)]	Cal.[(b)]	Cal.[(c)]
1	A_1	1666	1670.73	1670.86	4	A1	6225	6231.24	6243.17
	E	1662	1662.23	1662.23		A1		6231.49	6243.98
	A_1	1659	1659.43	1602.11		A2	6245	6257.19	6278.12
	A_1	1650	1649.71	1654.52		A2		6257.87	6278.12
						E	6271	6287.17	6322.17
2	A_1	3150.3	3141.29	3182.14		E		6287.96	6320.17
	E	3205.1	3187.59	3242.57		A_2	6291	6296.92	6361.22
	A_1	3216	3241.96	3245.25		A_1	6301	6316.21	6367.45
	E	3250	3259.13	3254.21		E		6347.92	6370.05
	A_1	3267	3271.92	3274.67		A_2	6354	6371.67	6375.92
	A_1	3279	3282.61	3286.76		A1		6409.21	6384.24
	E	3281	3287.67	3290.13		E	6409	6414.97	6425.71
	A_1		3298.61	3300.59		A2		6425.23	6436.64
3	A_1	4752	4763.41	4774.21		A1		6430.16	6447.09
	A_1	4762	4763.66	4788.94		A2		6433.29	6452.91
	E	4793	4793.19	4801.37		E		6468.31	6472.80
	A_2	4803	4803.43	4822.61		A2		6469.96	6483.95
	E	4813	4839.87	4845.52		E		6472.77	6486.26
	A_2	4841	4856.15	4861.37		A1		6474.87	6527.74
	A_2	4864	4863.64	4874.92		A1		6513.54	6541.05
	A_1		4872.71	4879.26		E	6515	6532.46	6564.13
	A_1		4876.41	4886.27		E		6532.86	6564.24
	E		4880.13	4893.49		A2		6546.21	6571.37
	A_2		4896.23	4897.67		A1		6568.21	6684.49
	A_2		4909.76	4913.47					
	E		4936.89	4941.15					
	A_1		4956.87	4961.28					
	A_1		4981.19	4990.20					
	A_2		4992.27	5601.75					

(a)From the compilation of Ref [1-6,23-25. (b) the results for our model, (c), the results for the D avydov model.

From Table 5.1 we see that the energy spectra for the acetanilide have the following features: (1).They consist of a series of manifolds, i.e., there are several energy levels corresponding to each vibrational quantum-number m. For example, the first excited state (from 1649.71-1670.73 cm^{-1}) contains 4 levels, there are 8, 16 and 24 levels at m=2 (from 3141.29-3298.61 cm^{-1}), m=3 (from 4763.41-4992.27 cm^{-1}) and m=4 (from 6231.24-6568.21 cm^{-1}), respectively.(2).The vibrational spectra have a strongly local-mode pattern, or otherwise speaking, the distance between energy levels depends strongly on the nonlinear interaction, γ as we see from Eps. (5.58) and (5.61). This is due to the fact that γ is greater than J. (3).The Local-mode degeneracies of energy levels appear for higher-lying vibrational states starting from m >2. Precisely speaking, there are the degeneracies at 4763cm^{-1} for m=3, and at 6231 cm^{-1},6257cm^{-1},6287cm^{-1} and 6532$cm^{-1}$for m=4, respectively. This shows that degeneracy of energy levels increases with increasing m.

The major advantage of the above method is that it allows treatment of more than one amide-I quantum; the validity (in zero dispersion limit) is directly testable through Eq.(5.61).

It is clear from the table that the theoretical values are in good agreement with the experimental data obtained from infrared absorption and Raman scatteringexperiments in the acetanilideby Careri et al. [5-14]. Obviously, we see from Table5.1 that the theoretical results obtained by Pang′s modelare a better approach than the experimental data in the values of Davydov′s model. This shows that Pang's model or the nonlinear quantum mechanical method is suitable for studying the vibrational spectra of the acetanilide.

Utilizing the energy spectra shown in Table5.1, we can get the Raman lines in the acetanilide, which are shown in Figure 5.4 for m=2. The nine lines occurring in this figure, 157 cm^{-1},145 cm^{-1},130 cm^{-1},100 cm^{-1},95 cm^{-1},84 cm^{-1},72 cm^{-1},57 cm^{-1},39 cm^{-1} are quite consistent with the experimental data of nine peaks in the Raman spectrum of acetanilide at T=50K by G. Careri et al. [1-7], which are 154 cm^{-1},148 cm^{-1},127 cm^{-1},104 cm^{-1},92 cm^{-1},84 cm^{-1},75 cm^{-1},60 cm^{-1},38 cm^{-1}, respectively, as is shown in Figure 5.8.

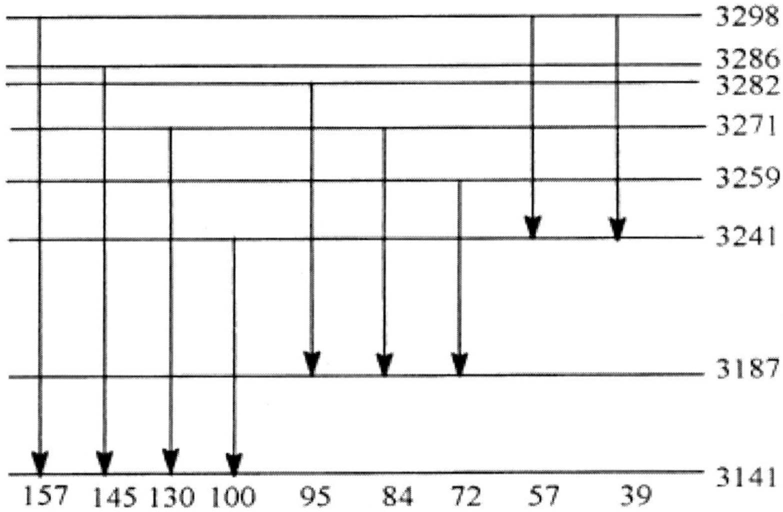

Figure 5.4. Distribution of energy levels at m=2 and corresponding Raman lines of acetanilide.

5.6. The Features of Specific Heat of Acetanilide

Careri et al., [3-6] measured the specific heat of powdered crystalline acetanilide from liquid nitrogen to room temperature. The measured data were then fitted to the formula

$$C(T)=4.59\times10^{-3}T+1.505 \tag{5.62}$$

in units of $Jg^{-1}K^{-1}$, where T is temperature in °C.

In order to find, theoretically, the specific heat of acetanilide we should first obtain the dynamic equation of the exciton-soliton at temperature T≠0, in this case using a new way [35-37].

For the system with finite temperature, the collective excitations and motions of the excitons and phonons will be changed due to the thermal effect. A straightforward result is that thermal excitation of the phonons occurs. Thus the part, U(t)10>$_{ph,}$ in Eq.(5.5) is replaced by the following representation:

$$|\alpha_v>=U_n^+|v>=\exp\left\{\sum_{qn}\left[\alpha_{nq}(t)a_q^+-a_{nq}^*(t)a_q\right]\right\}\prod_q(v_q'!)^{-1/2}(a_q^+)^{v_q'}|0>_{ph} \qquad (5.63)$$

where is the phononic complete set, which represents

$$|v>=\prod_q(v_q'!)^{-1/2}(a_q^+)^{v_q'}|0>_{ph}$$

the elementary excitation state of the single phonon due to the finite temperature T≠0K.

Utilizing the Eqs.(5.4)-(5.6) and Eq.(5.63), we can obtain the equations of motion for the exciton and phonon in the system. However, because the molecular crystal we study here is in contact with a thermostat at finite temperature T≠0K, therefore, after finishing this calculation of the expectation value in quantum theory for Eq.(5.6), we should calculate the thermal mean values of these quantities by means of the formula:

$$\overline{Y}=<Y>=Tr(\rho_{vv})_{ph}Y=\sum_v<v|\rho|v>_{ph}<\Phi_v|Y|\Phi_v> \qquad (5.64)$$

where the density matrix, $\sum_v<v|\rho|v>$, is given in the form:

$$<v|\rho|v>_{ph}=(\rho_{vv})_{ph}=<v|\exp[-\frac{H_{ph}}{K_BT}]|v>/\sum_v<v|\exp[-\frac{H_{ph}}{K_BT}]|v>$$
$$=\frac{<v|\exp[-\sum_q\hbar\omega_q(a_q^+a_q)/K_BT]|v>}{\sum_v<v|\exp[-\sum_q\hbar\omega_q a_q^+a_q/K_BT]|v>} \qquad (5.65)$$

where the diagonal matrix elements of the Hamiltonian are:

$$<\Phi\ |H|\Phi=<\Phi|H_{ex}+H_{\text{int}}|\Phi>+<\alpha_v|H_{ph}|\alpha_v>$$

Making use of previous method and the following relationships

$$<\alpha_v|(a_q+a_{-q}^+)|\alpha_v>=-[\alpha_{nq}(t)+\alpha_{n-q}^*(t)];\quad <\alpha_v|a_q^+a_q|\alpha_v>=(v_q+|\alpha_{n-q}|^2);$$

$$\sum_v\exp[-\hbar\omega_q v_q'/K_BT]<v|\exp(\alpha_{nq}^*a_q)\exp(-\alpha_{nq}a_q^+)|v>\ =(v_q+1)e^{-|\alpha_{nq}|^2(v_q+1)};$$

$$v_q=[\exp(\frac{\hbar\omega_q}{K_BT})-1]^{-1};\quad \exp[-\overline{W}_{n,n\pm1}]=<\alpha_{nq}|\alpha_{n\pm1q}>;$$

$$\overline{W}_{nn\pm1}=\exp\left\{\sum_q\left[\alpha_{n\pm q}^*\alpha_{nq}-\frac{1}{2}\left(|a_{n+1q}|^2+|\alpha_{nq}|^2\right)\right]\right\}\prod_q\left\{\sum_{m=1}^{n}\frac{(-1)^m|\alpha_{nq}+\alpha_{n\pm1q}|^2\,v_q'!}{(m\,!)(v_q'-m)!}\right\} \qquad (5.66)$$

we get, after some tedious calculations, the following equations of motion for the exciton and phonon in the molecular crystals at finite temperature.

$$\frac{\partial^2 u}{\partial t^2} - v_0^2 \frac{\partial^2 u}{\partial x^2} = \frac{\hbar r_0(\chi_1+\chi_2)}{2M\omega_0}\frac{\partial}{\partial x}|\phi|^2 \tag{5.67}$$

$$i\hbar\frac{\partial}{\partial t}\phi(x,t) = (\varepsilon_0 - 2JB(T,q))\phi(x,t) - Jr_0^2 B(T,q)\frac{\partial^2}{\partial x^2}\phi(x,t) + \frac{\hbar(\chi_1+\chi_2)r_0}{\omega_0}\frac{\partial u(x,t)}{\partial x}\phi(x,t) \tag{5.68}$$

From Eqs.(5.67)-(5.68) we can get that the motion of the exciton in the systems with finite temperature now satisfies the following nonlinear Schrödinger equation

$$i\hbar\frac{\partial\varphi(x,t)}{\partial t} = \left[\varepsilon_0 - 2JB(T,q)\right]\varphi(x,t) - Jr_0^2 B(T,q)\frac{\partial^2\varphi(x,t)}{\partial x^2} + G|\varphi|^2\varphi \tag{5.69}$$

in the first order approximation, where $JB(T,q) = (\hbar^2/2mr_0^2)\exp[-\overline{W}_n]$ with

$$\overline{W}_n \approx \overline{W}_{nn+1} = \sum_q (1+v_q)\beta_{qn\pm1}\beta_{qn}^* - v_q\beta_{qn\pm1}^*\beta_{qn} + \left(v_q+\tfrac{1}{2}\right)\left(|\beta_{qn}|^2 + |\beta_{qn\pm1}|^2\right) \approx \tfrac{1}{2}r_0\delta_0|\varphi_n|^2 B'F(q,r)$$

$$\delta_0 = \frac{2(x_1+x_2)^2}{8r_0\beta(1-s^2)J\omega_0^2},\ B' = \frac{\pi r_0(x_1+x_2)^2}{4\omega_0^2\beta v_0^2\hbar(1-s^2)},\ F(q,r) = \frac{r_0}{\pi N}\sum_q(1+r_q)|q|$$

Equation(5.69) is same with the dynamic equation (2.66) at $A(\phi)=0$ in nonlinear quantum mechanics. This shows that the properties and rules of motion of the exciton-soliton can still be described by nonlinear quantum mechanics although the system is in finite temperature, or speaking, the exciton-soliton in the molecular crystal with finite temperature has the wave-corpuscle duality. We now use the nonlinear quantum mechanics to find out the specific heat of acetanilide.

In this case, we give the soliton solution of Eq.(5.69), which is the form of

$$\varphi(x,t) = \sqrt{\frac{G'(T)}{8J}}\operatorname{sec}h\left[\frac{G'(T)}{2Jr_0}(x-x_0-vt)\right]\times\exp\left[\frac{i\hbar v}{2Jr_0^2}(x-x_0) - \frac{iE_{sol}t}{\hbar}\right]$$

where

$$G'(T) = \frac{\hbar^2(\chi_1+\chi_2)^2 r_0^2}{M\omega_0^2 v_0^2(1-s^2)}\left[1-\tfrac{1}{4}B'F(q,T)\right]$$

Clearly, the exciton-soliton has the wave-corpuscle duality; its energy can be found in the method in nonlinear quantum mechanics in Chapter 3 and is represented by

$$E_{sol} = \left(\varepsilon_0 - 2J\right) + \tfrac{1}{2}mv^2 - \frac{\hbar^4\left(x_1 + x_2\right)^4\left(1-5s^2\right)}{24\omega_0^4\beta^2\left(1-s^2\right)^3 J}\left[1 - \tfrac{1}{4}B'F\left(q,T\right)\right]^2 \quad (5.70)$$

According to the definition of the specific heat, then we obtain the specific heat arising from the thermal motion of the exciton-soliton in the acetanilide, which is

$$C_v = \frac{dE}{dT} = \frac{\hbar^4\left(x_1 + x_2\right)^4 B'\left(1-5s^2\right)}{48\omega_0^4\beta^2\left(1-s^2\right)^3 J} \times \left[1 - \tfrac{1}{4}B'F\left(q,T\right)\right]\frac{dF\left(q,T\right)}{dT}$$

At high temperature, i.e., $K_B T >> \hbar\omega_q$ we have

$$F\left(q,T\right) = \frac{2K_B T}{\hbar v_0 \pi}$$

Thus

$$C_v = K_B\left(\alpha + bT\right). \quad (5.71)$$

where T is the absolute temperature. Using the accepted values of parameters for acetanilide mentioned above in Eq.(5.34), we could obtain $\alpha = 5.15, b = 0.0199K^{-1}$. Thus, equation (5.71) is consistent with experimental data given in Eq. (5.62). This is consistent with Mrevlishvil [55] and Goldanskii and colleagues' result [58] for the linear specific heat of various biopolymers including proteins, which is $C_v = K_B(\alpha' + b'T)$.

In short, from the above result of specific heat in the acetanilide, which is in agreement with experimental datum, we can conclude that the exciton-soliton truly exists and can be described by the nonlinear Schrödinger equation in nonlinear quantum mechanics. Thus, the correctness of Pang's nonlinear quantum mechanics in Chapter 2 can be verified experimentally.

5.7. Experimental Verifications of Nonlinear Quantum Mechanical Property of the Soliton in Acetanilide

In previous sections, we studied the nonlinear quantum mechanical properties of excitons and phonons as well as some phenomena arising from their motions in acetanilide, and gave the mechanism of generation of nonlinear interaction in the system. Because the form of the exciton-soliton is associated with the vibration of the C=O stretching and deformation of the peptide groups, thus, the existence and motion of the exciton-soliton can be observed and determined by infrared absorption and Raman spectra, and so on, except for the specific heat mentioned above, in acetanilide. In this case, it is quite necessary to investigate and collect the experimental results of infrared absorption and Raman scattering and of the lifetime of the soliton or ground-state recovery time in acetanilide. In this section, we focus on these problems.

5.7.1. Experimental Results of Infrared Absorption and Raman Scattering in Acetanilide

In the early 1970s, Careri et al., [3-14] carried out a novel experimental investigation on protein dynamics. In the measurement of Careri et al., [3-15], the infrared spectra were obtained using three different infrared spectro-photometers. For the study of temperature dependence of the amide-I region, a Nicolet model 7000 Fourier-transform infrared spectrophotometer was used. Spectra were collected for 100 scans using a bandwidth of 0.5 cm^{-1}. The sample was thermostated using a closed-cycle helium refrigerator. Far-infrared absorption spectra were measured using a Michelson interferometer (model 720) equipped with a Golay cell. Samples consist of pellets obtained from a mixture of grounded acetanilide and polyethylene power. Pure polyethylene pellets were used to measure background transmission. Raman spectra were excited by a coherent radiation model 52 argon ion laser operating at 488nm or 514.5 nm, with stabilized output power of 20-200 mW. Incident light was filtered by a proper choice of interference filter and its intensity was monitored using a beam-splitter and a silicon photocell. Scattered light was analyzed by a Jarrel-Ash model 25-300 Raman spectrometer and detected by an ITT model Fw-130 cooled photomultiplier using photon counting electronics. These measured the infrared spectra of amide-I (1600-1700 cm^{-1}), amide-II (1500-1600 cm^{-1}), amide-III (1300-1500 cm^{-1}), amide-IV or VI (500-700 cm^{-1}) and amide-V (700-800 cm^{-1}). The infrared spectra of amide-I in 1600-1700 cm^{-1} and 2500-3500 cm^{-1}, 4750-4850 cm^{-1}, and 6200-6400 cm^{-1} are shown in Figure 5.5. The absorption intensity vs. temperature is shown in Figure 5.6. The Raman spectra of amide-I in 1630-1700 cm^{-1} and low-frequency modes at 300K and 50 K are shown in Figure 5.7 and Figure 5.8, respectively.

As it is known in Sect.1in this Chapter, the eigenenergies of amide-I are 1665 cm^{-1}, 1662 cm^{-1}, and 1659 cm^{-1}, which are assigned to the B_{2u}, B_{1u}, B_{3u} modes, respectively. The 1650 cm^{-1} mode in Figure 5.5(a) should be assigned to the exciton-soliton. Thus, we can affirmed that the soliton really exists in these systems.

Figure 5.6 shows that the absorption intensity decreases in the form of $\exp[-\beta T^2]$ with increasing temperature. This is consistent with the theoretical result of $\exp[-(c+\gamma' T^2)]$ given by Alexander and Krumbansl [15-16] and Pang [52-54] in Section 5.4. The comparison between experimental (denoted by "*") and the theoretical values (solid curve) is shown in Figure 5.6(b).

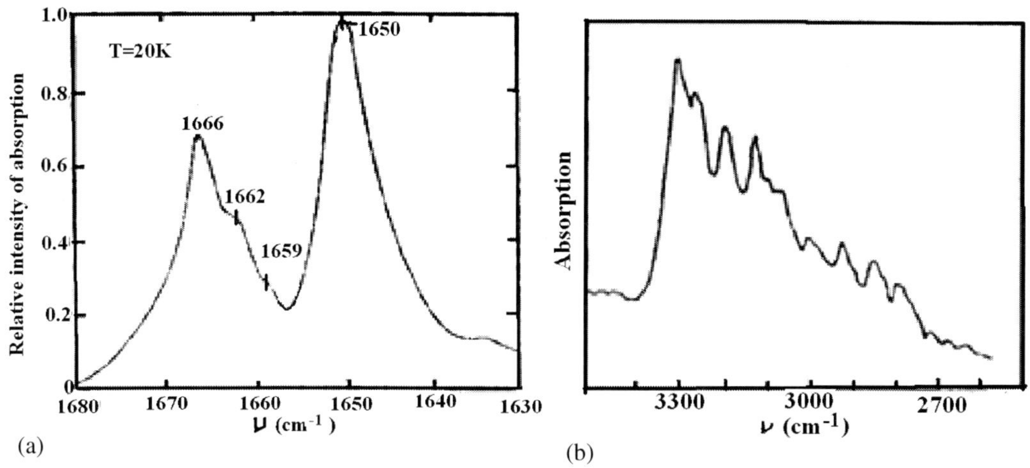

Figure 5.5. Infrared absorption spectra of acetanilide in (a) 1600-1700 cm^{-1} and (b) 2500-3500 cm^{-1} (see Ref. [3-15]).

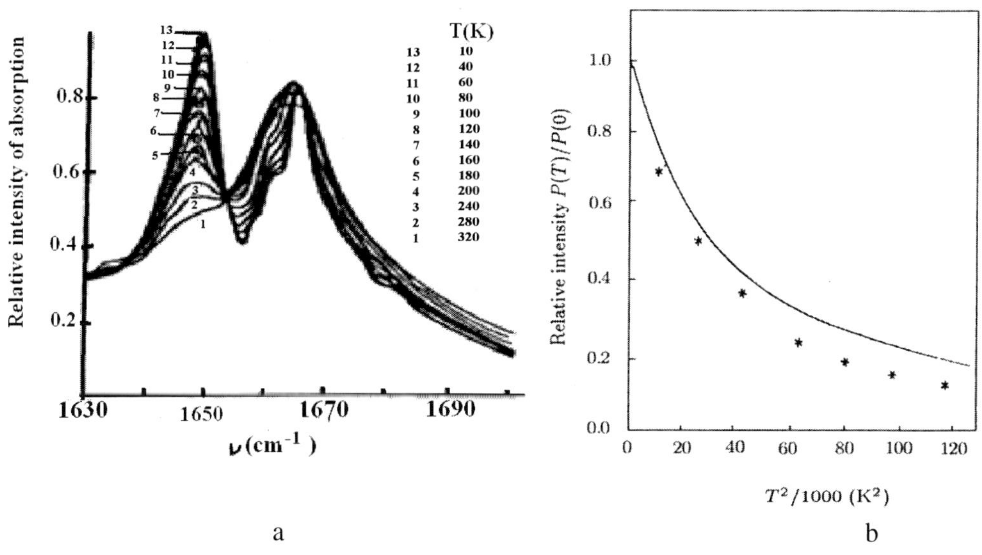

Figure 5.6. (a).The absorption intensity vs. temperature for acetanilide, and (b) comparison of experimental and theoretical values (see Ref. [3-6]).

From Figs. 5.5– 5.7, we see that the peak values of acetanilide infrared absorption agree with nonlinear eigenenergy spectra of the exciton-soliton given in Table 5.1, which were obtained from the above theory proposed by Pang. These interesting results provide experimental evidence to the existence of exciton-soliton in acetanilide and confirm the validity of the above theory.

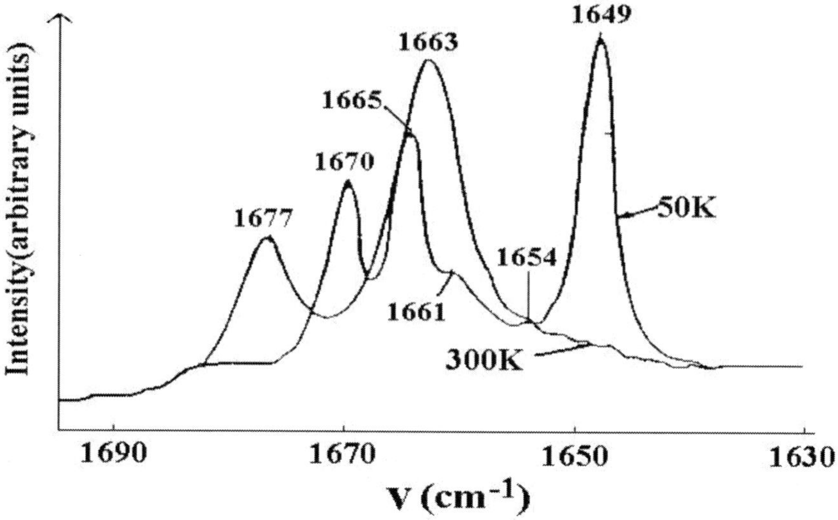

Figure 5.7. Raman spectra of acetanilide in 1630-1700cm^{-1} at 300 K and 50 K, respectively [Ref.3-10].

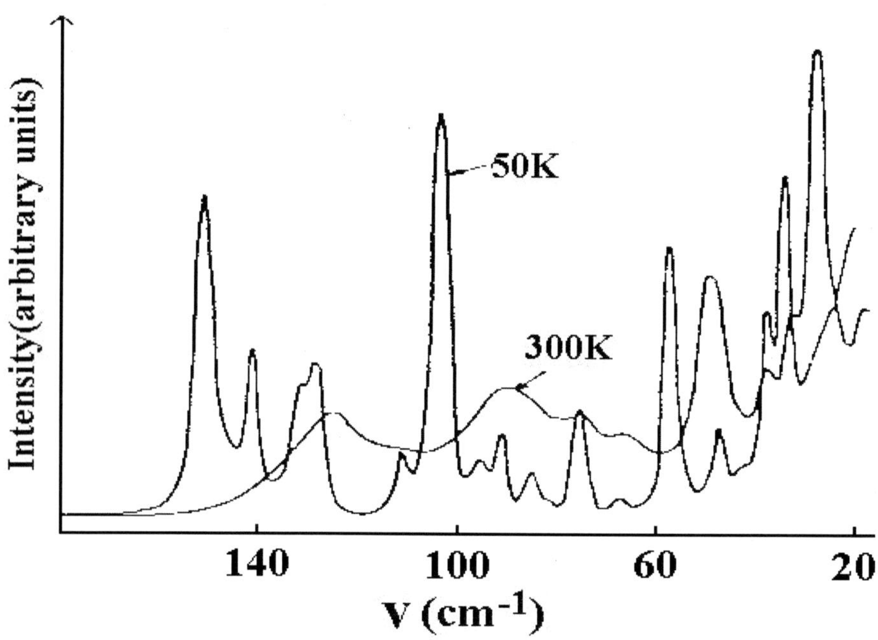

Figure 5.8. Low frequency Raman spectra of acetanilide at 300K and 50K, respectively [Ref. 3-10].

From Figure 5.8 we know that there are nine peaks at 154 cm^{-1},148 cm^{-1},127 cm^{-1},104 cm^{-1},92 cm^{-1},84 cm^{-1},75 cm^{-1},60 cm^{-1},38 cm^{-1}, in the Raman spectrum of acetanilide at T=50K by G. Careri et al. [3-7], which are quite consistent with the nine lines occurring in this figure, 157 cm^{-1},145 cm^{-1},130 cm^{-1},100 cm^{-1},95 cm^{-1},84 cm^{-1},72 cm^{-1},57 cm^{-1}, and 39 cm^{-1} in Figure 5.4, respectively, which was obtained by the nonlinear quantum mechanical theory. This result is very interesting because the Raman spectrum shown in Figure 5.8 has not been clearly explained theoretically up to now. The above agreement is not accidental, it shows clearly that the energy spectra in Table 5.1 are relevant and reliable for acetanilide, and the above method and theory we used is correct.

5.7.2. Experimental Tests of Dynamic Feature of Exciton-Soliton in Acetanilide

The experimental measurement of dynamical feature of the exciton-soliton in acetanilide is of interest. It is useful to resolve the lifetime of the soliton since the lifetime and mobility of the soliton are important criteria for its usefulness as a means of soliton features of microscopic particles and bio-energy transport in protein molecules [39-40].

Fann et al., [57] measured the relaxation time of the vibrational excitation of the 1650 cm^{-1} band by transient-infrared-bleaching experiments, which can set limits on the excitation lifetime. The source required for such measurements must be tunable around 6 μm (1650 cm^{-1}) with adequate spectral resolution to selectively excite the band (10 cm^{-1}). In addition, pulses with picosecond duration and sufficient intensity to bleach the transition are necessary to observe the bleaching recovery with adequate temporal resolution and signal-to-noise ratio. The source Fann et al., [57] chose to meet these criteria are the Mark free-electron laser (FEL), which produces pulses in bursts (macropulses) of about 1.3 μs in duration.

The Mark III laser output is divided into pump (95%) and probe (5%) beams by a CaF_2 wedge plate. Both pulses are focused by a single lens to a spot of 300-μm in diameter on the sample, which is held in a closed-cycle refrigerator with CaF_2 windows. The acetanilide was purified by repeated sublimation; crystalline domains were grown by slow cooling from an acetanilide melt between CaF_2 disks of 2 mm thickness. A substantial increase in the probe-pulse transmission at the 1650 cm^{-1} band was observed when the pump pulse was simultaneous with the probe pulse. The transmission increases due to the pump roughly follows the 1650 cm^{-1} band shape at 0 ps delay (pump and probe temporally overlapped), indicating that the effect is associated with the 1650 cm^{-1} band. This shows clearly that the exciton-soliton excitation occurs, in such case, in the acetanilide.

Figure5.9 shows the bleaching recovery dynamics at the band center as assembled from many scans. The fast component in Figure 5.9 is limited by pulse-duration, while the slow component has a decay time of 15 ± 5 ps, which can be associated with saturation recovery due to repopulation of the ground state of the 1650 cm^{-1} mode. The relaxation time or lifetime of 15 ± 5 ps indicates that the 1650 cm^{-1} band is strongly coupled to the lattice, and the 15 ± 5 ps is a typical relaxation value for a vibration line in the acetanilide. These experimental data in acetanilide support the theoretical results obtained by Pang's theory [28-40] mentioned above. Therefore, Pang's theory is appropriate to the acetanilide.

Fann et al., [57] measured further the transmission recovery of the sample at 100 ps in which the probe decays to the "cold" base line. This is an important fact since it was thought that the 1650 cm^{-1} band should disappear on deposition of energy into the 130 cm^{-1} optical phonons. Loss of the 1650 cm^{-1} band due to such a heating should result in an increase in transmission approximately by a factor of 2 at 100 ps, which, however, was not observed. Therefore, the time scale of such vibrational relaxations observed is inconsistent with the lifetime of 10^{-10} s, predicted by the theory of Pang et al. [52] in protein molecules. This manifests that Fann et al.'s measurement of the lifetime of the soliton in acetanilide cannot serve as the experimental evidence of existence of the soliton in protein molecules [39-40,52] because the molecular structure and functions of acetanilide are completely different from those of alpha-helix protein molecules. In fact, the alpha-helix proteins are biomacromolecules of helical structure of three channels, which have certain biological functions.

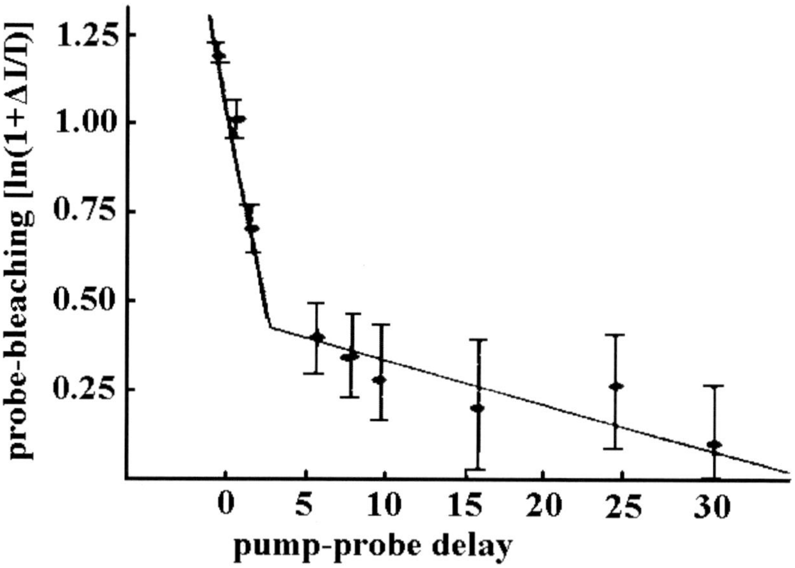

Figure 5.9. Probe-bleaching vs. pump-probe delay at 1650 cm^{-1} and 80 K [Ref.57].

5.7.3. The Lifetime of the Soliton in the Acetanilide by Pump-Probe Spectroscopy

Hamm et al., [58-64] measured the properties of infrared spectra of absorption in acetanilide by pump-probe spectroscopy. In the pump-probe experiment, the various states of the system are first excited with an intense short IR laser pump pulse, and the response upon that excitation is then probed by a second short, but much weaker probe pulse. Most of the spectroscopy performed so far on the ACN problem was stationary IR absorption spectroscopy and its temperature dependence is shown in Figure 5.10. The information gained from the pump-probe experiments, as compared to stationary IR absorption spectroscopy, is manifold. At the same time, it can measure the anharmonicities of the individual states, in which its first excited state is populated when pumping a state, and the probe pulse is probing both upward to transition to the second excited state, and then downward transition back to the ground state, through stimulated emission. If the state is harmonic, both upward and downward transitions appear at the same frequency, but with opposite signs, and cancel each other (if taking into account the effect of bleaching the ground state). On the contrary, when the state is anharmonic, they lead to two bands with opposite signs separated by the anharmonicity of the transition. Therefore, the pump-probe spectroscopy is particularly sensitive to the anharmonic (nonlinear) part of the molecular potentials, which is impossible for the stationary IR absorption spectroscopy. If two states are coupled, excitation of one state will result in a response, such as a frequency shift, of the other. This means that there are entangled transitions between them. If pumping two or more coupled states with an ultrashort pulse, whose bandwidth covers these states simultaneously, then quantum beats can be observed. Hence, we can use it to measure the lifetimes of the excited particle.

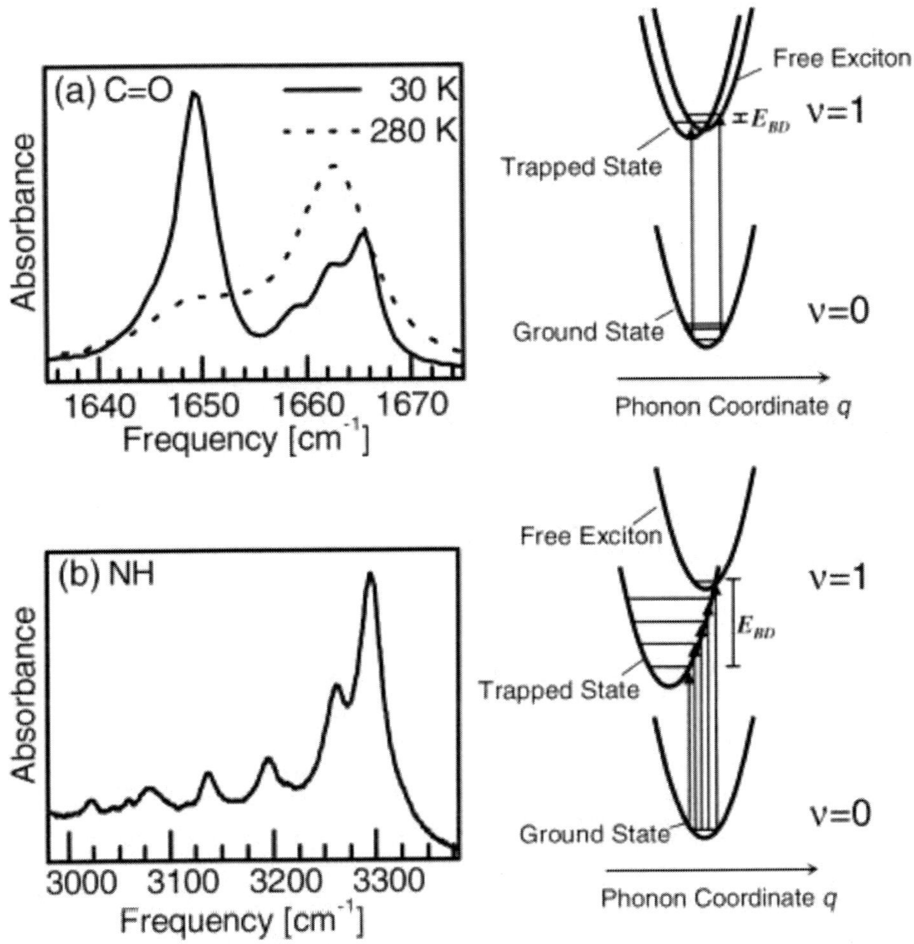

Figure 5.10. Infrared spectra of absorption of the C=O (**a**)and the N–H (**b**) band by pump-probe experiment in ACN and corresponding potential energy curves causing these spectra of the pumped states or soliton excited state by the fem-to-second time resolution(Ref. [58]).

Figure 5.11 shows two pump-probe spectra obtained by excitation of either of the two absorption lines of ACN at 90 K, where the *arrows* mark the center frequency of the pump pulse. The pump pulses are spectrally narrow in this experiment (spectral width ≈14 cm^{-1}) and excited only one band in them. When resonantly pumping the 1650 cm^{-1} band, it is bleached, but 1644 cm^{-1} band emerges. When resonantly pumping the 1666 cm^{-1} band, hardly any bleach of the band is observed. This indicates that the 1650 cm^{-1} band responds only with a signal.

In the meanwhile, they also discussed the different response of both bands of ACN [59]. They started with the statement that the involved C=O vibrators are slightly anharmonic. In this case the "anomalous" band with a 0–1 frequency of 1650 cm^{-1} and a 1–2 frequency of

1644 cm^{-1} in Figure 5.11B displays the expected anharmonic response, but the 1666 cm^{-1} band does not show the pump-probe response at all. These results manifest that the anomalous 1650cm^{-1} band shows an anharmonic response, and hence, is a localized state, but the normal 1666 cm^{-1} band shows no nonlinear response, which is only the case when it is effectively a harmonic state.(It verified that the lifetime of the normal band is long enough to potentially observe a bleach [61]). Hence, the normal band is largely delocalized, i.e., a free exciton, in agreement with the original assignments.

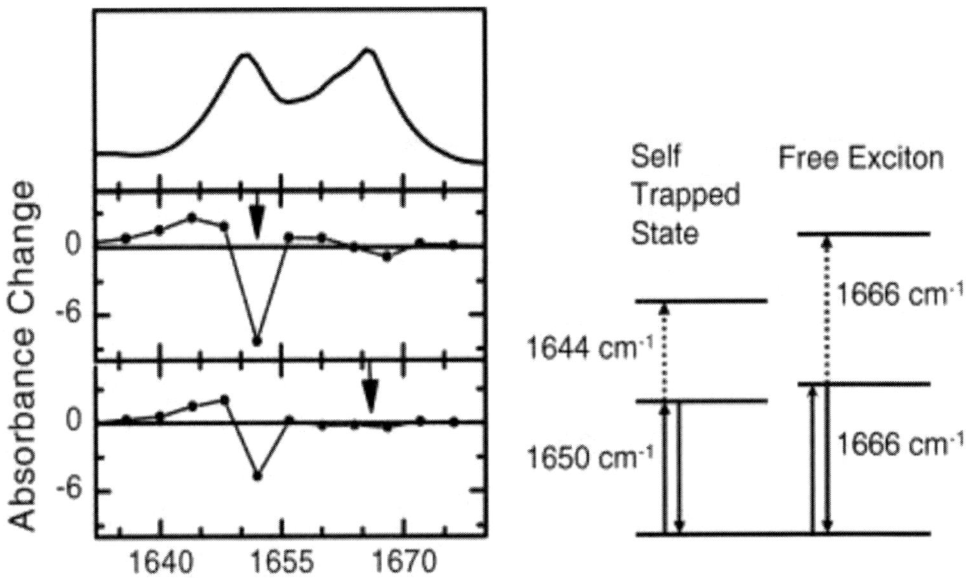

Figure 5.11. A is linear absorption spectrum (*top*) and pump probe spectra of the C=O mode of ACN at 93 K for two different narrow band pump pulses chosen to be resonant with each of the absorption bands. B is the level scheme of the system, explaining the distinctively different response of both modes (see Ref. [61]).

On the other hand, they utilized 2D-IR spectroscopy to clarify the origin of the anomalous amide I band [62]. 2D-IR spectroscopy can distinguish between various kinds of nonlinearities of the molecular Hamiltonian. In this experiment, the pump frequency they used is continuously scanned through the spectrum, and the response is plotted as a function of both pump and probe frequencies. Figure 5.10 shows the comparison of the 2D-IR spectrum of crystalline ACN with those of benzoylchloride and N-methylacetamide(NMA) dissolved in methanol, respectively, where the arrows mark the position of the pump pulse. These infrared spectra of absorption display clearly the same doublet structure in all three cases, but the 2D-IR spectra obviously differ. Benzoylchloride has a prototype example of a molecule with a strong symmetric Fermi resonance [65], whereas N-methylacetamide dissolved in methanol exists in two spectroscopically distinguishable conformations, one fraction of hydrogen bonded to the solvent and the other fraction not [66]. Thus, their molecular structures are different from that of ACN. If the two lines in ACN come from molecules in two different conformations or surroundings (topological defects), as has been suggested various times [67], pumping of one state would not cause any response of the other state (in contrast to what is seen in the experiment). The 2D-IR spectrum should in essence be diagonal, as it is for N-methylacetamide dissolved in methanol (Figure 5. 12). Otherwise, in the case of a Fermi resonance, the original explanation of the doublet in the ACN spectrum [66], a very characteristic coupling pattern, is obtained (Figure 5. 12). Thus, investigating these coupling patterns in detail, the compelling evidence against this possibility can be provided as well, and the self-trapping state or the soliton state should be substantiated [62].

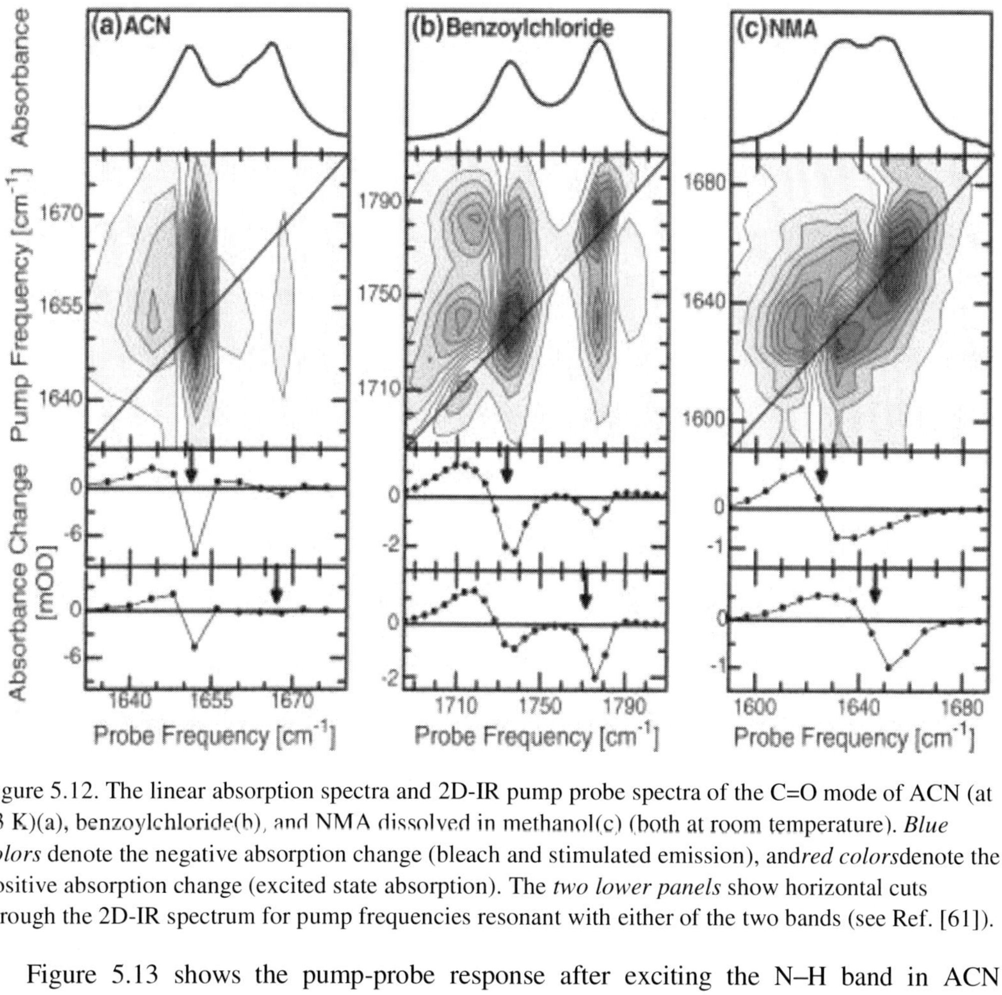

Figure 5.12. The linear absorption spectra and 2D-IR pump probe spectra of the C=O mode of ACN (at 93 K)(a), benzoylchloride(b), and NMA dissolved in methanol(c) (both at room temperature). *Blue colors* denote the negative absorption change (bleach and stimulated emission), and*red colors*denote the positive absorption change (excited state absorption). The *two lower panels* show horizontal cuts through the 2D-IR spectrum for pump frequencies resonant with either of the two bands (see Ref. [61]).

Figure 5.13 shows the pump-probe response after exciting the N–H band in ACN obtained by Edler, et al.[62]. In contrast to the C=O work, here they used a pump pulse so short that its bandwidth covers many of the side peaks of the N–H band. They were all excited at the same time, leading to pronounced quantum beats in the response. The Fourier transform of the signal in Figure 5.13 gave two peaks: a dominant peak at 48 cm^{-1} and a weaker peak at 76 cm^{-1}. The temperature dependence of the two peaks was obtained [62], which are completely consistent with the two lines in the conventional Raman spectrum (Figure 5. 13) [68]. This demonstrated that the coherences indeed represent lattice phonons; in fact, this experiment identified the two phonons that mediate vibrational self-trapping in ACN. These experimental data are interesting and worthy of further study[69-71].

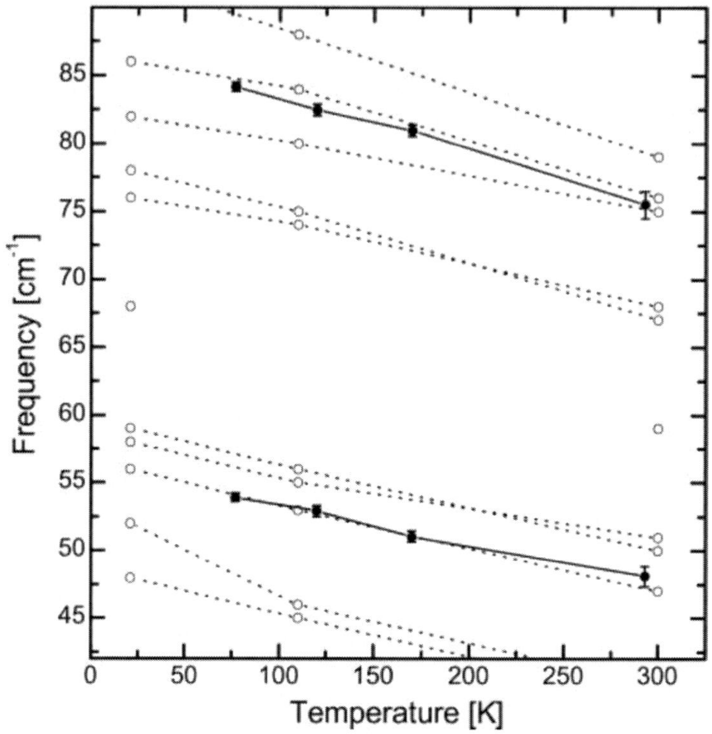

Figure 5.13. The temperature dependence of the low frequency Raman modes (*open circles*) in ACN (see Ref. [68]) and of the beating frequencies observed in the pump-probe experiment (filled circles). The temperature dependence of the beating frequencies perfectly matches that of the Raman modes(see Ref.63]).

Therefore, Hamm and co-workers have identified two phonon modes that mediate the self-trapping and found further evidence for self-trapped states in ACN and N-methylacetamide (NMA) and even in a small polypeptide [64] and speculated that such states may arise in all hydrogen-bonded systems. These experimental results are very important and interesting.

Hamm et al., [58-64] measured the lifetimes of the solitons by pump-probe spectroscopy in acetanilide and proteins. Figure 5.14 shows the pump-probe response of both the anomalous (1650 cm^{-1}, filled circles) and the "normal" band (1666 cm^{-1}, open circles) of ACN after selectively exciting the former [58]. At early times, a bleach of only the anomalous band occurs, which recovers on a fast 2-ps timescale. However, this relaxation is not complete, and a small negative signal remains. This indicates that the system does not relax back into the initial ground state, but into a state that is either spectroscopically dark or outside of spectral window. On a somewhat longer timescale (35 ps), the energy still present in the crystal thermalizes. That is, the anomalous band loses intensity (a negative signal in the difference spectroscopy of Figure 5.14) and the 1666 cm^{-1} band gains intensity, exactly as in the stationary spectra of Figure 10a when we increase the temperature. Hence, after vibrational relaxation of the initially pumped state, energy relaxes through an unknown pathway, but then reappears as heat after 35 ps. A very similar relaxation behavior was also found for the N–H band [62]. Therefore, we can determine from this experiment that the lifetime of the soliton excitation in ACN is 35ps.

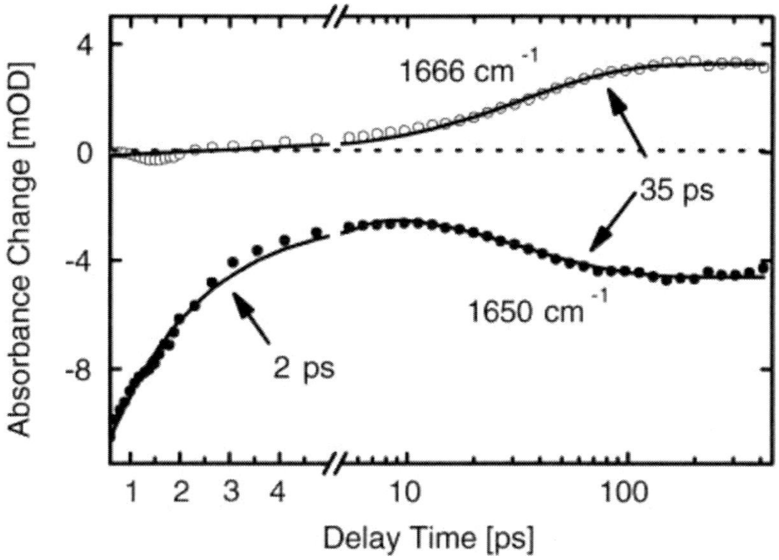

Figure 5.14. The pump-probe response of the anomalous (1650 cm^{-1}, *filled circles*) as well as of the normal band (1666 cm^{-1}, *open circles*) of ACN after selectively exciting the former (see Re. [58]).

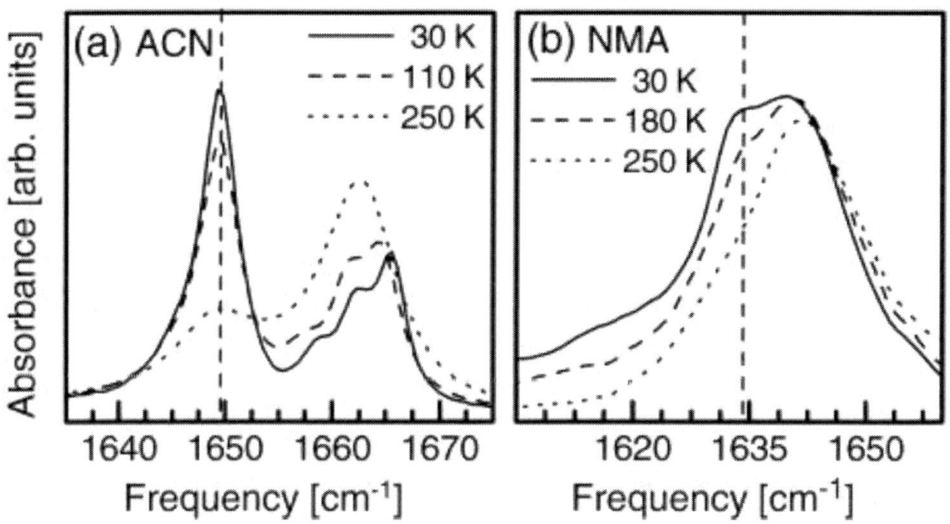

Figure 5.15.IR spectrum of the amide-I band in ACN (a) and NMA (b) at three different temperatures. The *vertical line* marks the position of the temperature-dependent sideband (see Re. [61]).

Hamm et al. [58-64] also provided compelling evidence for vibrational self-trapping in NMA, which is similar to that in crystallineACN because the former's molecular structure resembles ACN's. In particular, both crystals, ACN and NMA, have an orthorhombic structure and consist of quasi-1D chains of hydrogen-bonded peptide units (–CO–NH–) with structural properties that are similar to those of α-helices. Thus, they are often regarded to be *the* model compound for peptides and proteins. However, the mechanism is expected to be generic and should occur in this crystal. Nevertheless, no convincing experimental evidence for self-trapping in NMA had been found so far. However, they carefully measured infrared spectra of NMA [61,72]by pump-probe experiment and compared the infrared absorption spectra and the pump-probe spectra of the amide-I and NH modes of acetanilide (ACN) and *N*-methylacetamide as well as their deuteratedderivatives. The amide-I bands of NMA show a temperature-dependent sideband, but it is less distinct than in ACN (Figure 5.15). On the other hand, the N–H band of NMA is accompanied by a sequence of satellite peaks, the spacings between which are larger than in ACN (which could be since NMA is lighter, and hence these lattice phonons are tentative at higher frequencies). Comparing the pump-probe responses of these spectral anomalies in both crystals gave very similar results, providing evidence that vibrational self-trapping is indeed a common effect in hydrogen-bonded crystals.

In the meanwhile, Hamm et al., [58-64] addressed further the question of whether vibrational self-trapping can also occur in α-helices, one of the most important secondary structure motifs in proteins. Thus, they chose poly-γ-benzyl-L-glutamate (PBLG) for this investigation because it forms extremely stable and long α-helices. The monomeric unit of PBLG is a non-natural amino acid with a long side chain that stabilizes the helix; however, the helix backbone is identical to that of natural α-helices. Because of its stability, PBLG has served as a standard model helix [62].

REFERENCES

[1] M. Pope, C. E. Swenberg, *Electronic Processes in Organic Crystals*, Clarendon Press, Oxford University Press, Oxford/New York, 1982.
[2] E. A. Silinsh, V. Capek, *Organic Molecular Crystals*, AIP Press, New York, 1994.
[3] G. A. Careri, A. Gransanti and J. A. Rupley, *Phys. Rev*. A 37 (1998) 2703.
[4] G. A. Careri, E. Gratton and E. Shyamsunder, *Phys. Rev*. A 37 (1998) 4048.
[5] G. A. Careri, U. Buontempo, F. Galluzzi, A. C. Scott, E. Gratton and E. Shyamsunder, *Phys. Rev*. B 30 (1984) 4689.
[6] G. A. Careri, U. Buonttempo, F. Caeta, and A. C. Scott, Phys. Rev. Lett. 51 (1983) 304.
[7] A. C. Scott, *Physica D* 51 (1990) 333.
[8] A. C. Scott, A. C., *Phys. Lett*. A86 (1998) 603.
[9] A. C. Scott, A. C., Phys. Rep. 217 (1992) 67.
[10] J. C. Eilbeck, P.S. Lomdahl, A.C. Scott, *Phys. Rev*. B 30 (1984) 4703.
[11] A. C. Scott, E.Gratton, E.Shyamsunder and G. Careri, *Phys. Rev*.1332(1985)5551.
[12] A. C. Scott, I. J.Bigio, C. T. Johnston, *Phys. Rev*. B39(1989)12883.
[13] A. C. Careri, J.C.Eilbeck,Phys.lett.A109(1985)201.
[14] A. C. Careri, U. Buontempo, F. Galluzzi, A.C. Scott, E. Gratton, E. Shyamsunder, *Phys. Rev. B* 30 (1984) 4689.
[15] D. M. Alexander, and J. A. Krumbansl, *Phys. Rev*. B 33 (1986) 7172.
[16] D. M. Alexander, Phys. *Rev. Lett*. 60 (1985) 138.
[17] A. S. Davydov, *J. Theor. Biol*. 38 (1973) 559.
[18] A. S. Davydov, *J. Theor. Biol*. 66 (1977) 379.
[19] A. S. Davydov, *Phys. Scr*. 20(1979)387.
[20] A. S. Davydov, *The Solitons in Molecular Systems*, Dordrecht, Reidel, 1985 p17-87.

[21] A. S.Davydov, *Theory of Molecular Excitons*, lnd *Nauka*, Moscow, 1968; New York, Plenum, 1975.
[22] A. S. Davydov, *Biology and Quantum Mechanics*, Pergamon, New York,1982.
[23] T. Benkui, P. B. John, *Phys. Lett.* A 240 (1998) 282.
[24] Y. Xiao, *Phys. Lett.* A 243 (1998) 174.
[25] J. Tekec, Z. Ivic, Z. Przulj, *J. Phys.:Condens. Matter* 10 (1998) 1487.
[26] V. M. Kenkre, S. Raghvan, L. Cruzeiro-Hansson, *Phys. Rev.* B 49 (1994) 9511.
[27] A. A. Eremko, B. Yu Gaididel, A.A. Vaknenko, *Phys. Stat. Sol.* b 127 (1985) 703.
[28] Pang Xiao-feng, J. *Phys. Condens. Matter* 2 (1990) 9541.
[29] Pang Xiao-feng, *Eur. Phys. J.* B 10 (1999) 415.
[30] Pang Xiao-feng and Zhang Huai-wu, *J. Phys.Chem. Solids.* 66 （2005） 963.
[31] Pang Xiao-feng, *Chinese J. Atom. Mol. Phys.* 7 (1989) 1235.
[32] Pang Xiao-feng, *Physics, Sin.* 22 (1993) 612.
[33] Pang. Xiao-feng, *Physica* D154(2001)138.
[34] Pang Xiao-feng, *J. Phys. Chem. Solid.* 62 (2001) 793.
[35] Pang Xiao-feng, *J. Nature Sin.* 15 (1992) 915.
[36] Pang Xiao-feng, *Chinese. Phys. Lett.*10 (6) (1993) 381.
[37] Pang Xiao-feng, *Acta Phys. Sin* 42 (1993)1856.
[38] Pang Xiao-feng, Zhang Huai-wu, *Model Phys. Lett.* 20(30)(2006) 1923.
[39] Pang Xiao-feng,Feng Yaun ping, *Quantum mechanics in nonlinear system*, World Scientific Publishing Co. Singapore, 2005.
[40] Pang Xiao-feng, *The theory of nonlinear quantum mechanics*, Chongqing, Chinese Chongqing Press, 1994.
[41] S. Takeno, *J. Phys. Soc.* Japan 59 (1991) 3127.
[42] S. Takeno, *Prog. Theor. Phys.* 73 (1985) 853.
[43] R. Kabo, *J. Phys. Soc.* Japan 17(1962)206.
[44] Pang, Xiao-feng, *Chinese J. Infrared Millimeter Wave* 12 (1993) 377.
[45] Pang, Xiao-feng, *Chinese J. Infrared Millimeter Wave* 16 (1997) 288.
[46] Pang Xiao-feng, Chen Xianron, *Chin. Phys. Lett.* 19 (2000) 1091.
[47] Pang, Xiao-feng, *Acta. Phys. Sin.* 43 (1994) 1987.
[48] Pang Xiao-feng, Chen Xianron, *Int. J. Model Phys.* 20(18)(2006) 2505.
[49] Pang Xiao-feng, Chen Xianron, *Commun. Theor. Phys.* 32 (2000) 437.
[50] Pang Xiao-feng, Chen Xianron, *Phys. Stat. Sol.*(b) 229 (2002) 1397.
[51] Pang Xiao-feng, Chen Xianron, *Chin. J. Atom. Mol.* 19 (2002) 319.
[52] Pang Xiao-feng, *Nonlinear Quantum Mechanics*, Beijing, Chinese Electronic Industry Press, 2009.
[53] Xiao He-Lan, Cai Guo-ping, Sun Su-qin and Pang Xiao-feng, Chinese. *Atom. Mol. Phys.* 20 (2003) 211.
[54] Pang Xiao-feng, Xiao He-lain, Cue Guo-ping Zhang Huai-wu, Dong Bo, *Int. J.Infraredand Millimeter Waves*, 31(2010)521.
[55] G. M. Mrevlishvil, Usp. Fiz. Nauk 128 (1979) 273 [*Sov. Phys-Usp.* 22 (1979) 433.]
[56] V. I. Goldanskii, F. Yu. Krupyanskii and V. N. Flerov, Dokl. Akad. Nauk (SSSR) 272(1983)978.
[57] W. Fann, L. Rothberg, M. Roberso, S. Benson, J. Madey, S. Etemad, R. Austin, *Phys.Rev. Lett.* 64 (1990) 607.

[58] P. Hamm, *J. Biol.Phys.,* doi:10.1007/s10867-009-9126-3.
[59] J. Edler, P. Hamm, *J. Chem. Phys.* 117 (2002) 2415.
[60] J. Edler, P. Hamm, *J. Chem. Phys.* 119 (2003) 2709.
[61] J. Edler, P. Hamm, *Phys. Rev. Lett.* 88,(2002) 067405.
[62] J. Edler, P. Hamm, *Phys. Rev.* B 69 (2004) 214301.
[63] J. Edler, R. Pfister, V. Pouthier, C. Falvo, P. Hamm, *Phys.Rev. Lett.* 93 (2004) 106405.
[64] J. F.Bertran, L.Ballester, L. Dobrihalova, N. Sanchez, R.. Arrieta, *Spectrochim. Acta,Part A*, 24 (1968) 1765.
[65] S. Woutersen, Y. Mu, G. Stock, P. Hamm, *Chem. Phys.,* 266 (2001) 137.
[66] G. B. Blanchet, C. R. Fincher, *Phys. Rev. Lett.* 54 (1985) 1310.
[67] C. T. Johnston, B. I. Swanson, *Chem. Phys. Lett.* 114(1985) 547.
[68] C. T. Johnston, S. F. Agnew, J. Eckert, L. H. Jones, B. I.Swanson, C. J. Unkefer, *J. Chem.Phys.* 95 (1991) 5281.
[69] Pang Xiao-feng, Int. J. Mod. Phys. B, 29(2014)1550041.
[70] Pang Xiao-feng, *Physics of life review*, 8 (2011) 302.
[71] PangXiao-feng, *J. Appl. Phys.* 111 (2012) 093519.
[72] P. Doty, J. H. Bradbury, A .M. Holtzer, *J. Am. Chem. Soc.* 78 (1956) 947.

Chapter 6

THE NONLINEAR QUANTUM MECHANICAL FEATURES OF BOSON-EINSTEIN CONDENSED STATES

The nonlinear properties of microscopic particles, such as the Boson and helium atom in Boson-Einstein condensed states, are studied by nonlinear quantum mechanical theory and method. The investigations show that the states of the microscopic particles are described by nonlinear Schrödinger equation, they have a wave-corpuscle duality. In this case, the nonlinear interaction is due to the attractive interactions among these Boson particles in a lower temperature case through spontaneous breakdownof symmetry of the systems, in which these particles are all in a lower-energy state. We further found properties of the Boson-Einstein condensation of particles, and used experiments to verify their correctness.

6.1. BOSON-EINSTEINCONDENSATION PHENOMENON OF MICROSCOPIC PARTICLES

The condensation of matter is a change of state, for example, steam water becomes liquid water at 273K; this is the condensation of water, which is due to collection of water molecules. In the process, the energy of water molecules is depressed. Therefore, the condensation of water is related to its changes of temperature and energy. A lot of matter can produce this phenomenon. Thus, the condensation of matter is a universally thermodynamic effect and changes the state of matter. However, the Boson-Einstein condensation is different from the general condensation of matter; it is a special condensation. In the 1920s, Satyendra Boson [1] and Albert Einstein [2-3] had predicted that the Boson particles can occur with condensation, or Boson-condensation, in which the Boson particles collected to the state of lower energy due to attraction among these particles at very low temperature, which approach absolute zero OK. Therefore, this phenomenon, related to the depressions of temperature and energy of the systems, are shown in Figures 6.1 and 6.2, respectively, but their depressions for different systems are different. Therefore, the features of Boson-Einstein condensation (BEC) differ from those of the condensation of gas state forming the liquid state because the latter occurs in the coordinate real-space, but the Boson condensation exists only in momentum space due to **a** loss of energy and depression of temperature, which, in fact, approaches several Kelvin. As it is known, at zero Kelvin (absolute zero) all molecular motions stop. Scientists have figured out a way to obtain a temperature only a few billionths of a degree above absolute zero. When temperatures get that low, a BEC with a few special elements can be created. Cornell and Weiman did it with rubidium (Rb). As it is known, when the temperature becomes that low, the atomic parts can't move at all. They lose almost all of their energy. Since there is no more energy to transfer (as in solids or liquids), all of the atoms have exactly the same levels, like twins. The result of this clumping is just the BEC. The group of rubidium atoms sits in the same place, creating a"super atom".There are no longer thousands of separate atoms. They all take on the same qualities and become blobs, or bubbles, as shown in Figure 6.3 [4].

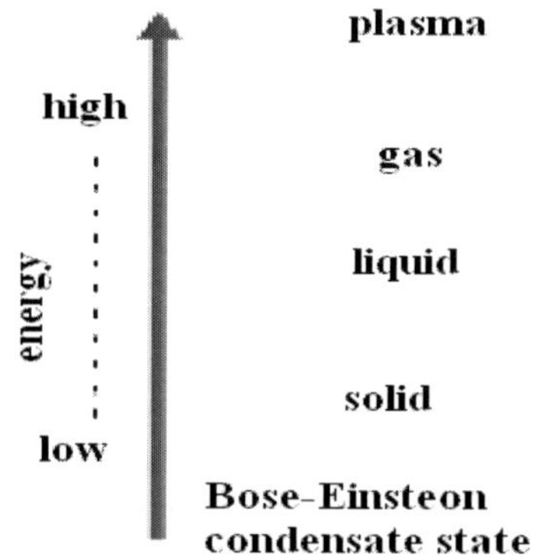

Figure 6.1.The changes of state of matter with decreasing energy [4].

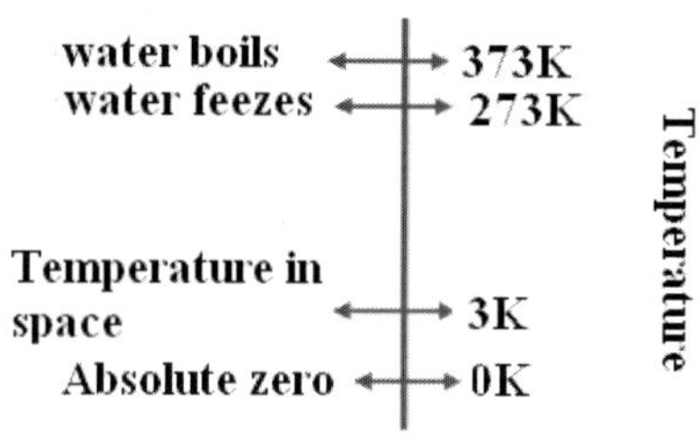

Figure 6.2. The changes of state of matter with decreasing temperature[4].

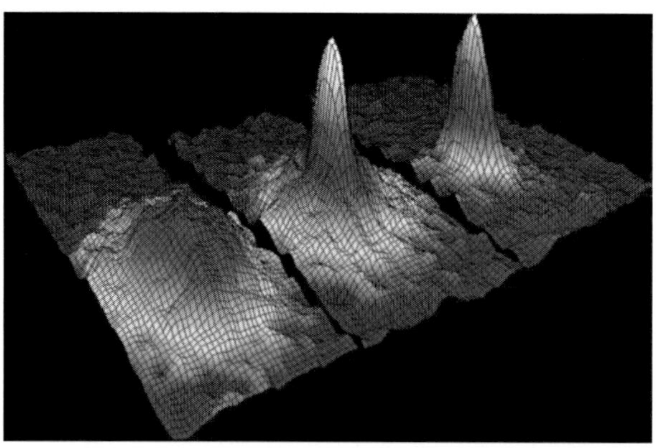

Figure 6.3. The blobs and bubbles of atoms in the case of Boson–Einstein condensation [4].

In July 14, 1999, the researchers from JILA reported in <<Science>>, their achievement of a temperature far lower than had ever been produced before, creating an entirely new state of matter predicted decades ago by Albert Einstein and Indian physicist Satyendra Nath Boson. Cooling rubidium atoms to less than 170 billionths of a degree above absolute zero caused the individual atoms to condense into a "superatom" behaving as a single entity. They showed three-dimensional successive snap shots in time in which the atoms condensed from less dense red, yellow and green areas into very dense blue to white areas as in Figure 6.3 [4]. JILA was jointly operated by NIST and the University of Colorado at Boulder. This transition to BEC occurs below a critical temperature, which for a uniform three-dimensional gas consisting of non-interacting particles with no apparent internal degrees of freedom, is given by:

$$T_c = \frac{2\pi\hbar^2}{mK_B}\left(\frac{n}{\varsigma(3/2)}\right)^{2/3} \approx 3.3125\frac{\hbar^2 n^{2/3}}{mK_B} \tag{6.1}$$

where T_c is the critical temperature, n is the density of particle, m is the mass per boson, $\hbar$ is the reduced Planck constent, K_B is the Boltzmann constent, and ς is the Riemann zeta function. Its value is $\varsigma(3/2) \approx 2.6124$.

6.2. The Theoretical Model Describing the Boson-EinsteinCondensation

6.2.1. Einstein's Non-Interacting Model

Einstein considered [2-3] a system composed of *N* noninteracting particles, in which each particle is in one of two *quantum states,* $|0\rangle$ and $|1\rangle$. If the energies of the two states are equal, then each different configuration is equally likely. If we can tell which particle is which, there are 2^N different configurations, since each particle can be in $|0\rangle$ or $|1\rangle$ independently. In almost all of the configurations, about half of the particles are in $|0\rangle$ and the other half in $|1\rangle$. The balance is in essence, a statistical effect: the number of configurations is largest when the particles are divided equally. If the particles are indistinguishable, however, there are only *N*+1 different configurations. If there are *K* particles in state $|1\rangle$, there are $N - K$ particles in state $|0\rangle$. Very clearly, whether any particular particle is in state $|0\rangle$ or in state $|1\rangle$ cannot be determined, so each value of *K* determines a unique quantum state for the whole system. If these states are all equally likely, then there is no statistical spreading out; it is just as likely for all the particles to sit in $|0\rangle$ as for the particles to be split half and half. We now assume that the energy of state $|1\rangle$ is slightly greater than the energy of state $|0\rangle$ by an amount *E*.

Then at temperature$T$, a particle will have a lesser probability to be in state $|1\rangle$ by exp($-E/kT$). In the distinguishable case, the particle distribution will be biased slightly towards state $|0\rangle$. This means that the distribution will be slightly different from half-and-half, mentioned above. However, in the indistinguishable case, since there is no statistical pressure toward equal numbers, then the most-likely outcome is that most of the particles will collapse into state $|0\rangle$ [4-10].

On the other hand, in the distinguishable case, for large N, the fraction in state $|0\rangle$ can be computed. It is the same as flipping a coin with probability proportional to $p = \exp(-E/T)$ to land tails. The probability to land heads is $1/(1 + p)$, which is a smooth function of p, and thus of the energy. But in the indistinguishable case, each value of K is a single state, which has its own separate Boltzmann probability. So the probability distribution is exponential, which is the form of

$$P(k) = Ce^{-kE/T} = CP^k$$

where the normalization constant$C$ is $(1 - p)$for large N. In this case, the expected total number of particles, which is not in the lowest energy state, butin the limit that $N \to \infty$, is equal to $\sum_{n>0} CnP^n = P/(1-P)$. Obviously, it does not grow, and approaches a constant for large$N$. This is negligible for the total number of particles. This implies that a collection of enough Boson particles in thermal equilibrium will mostly be in the ground state, with only a few in any excited state, no matter how small the energy difference [5-10].

For a gas of particles, which is in different momentum states labeled $|k\rangle$, the particles will all be in different states, if the number of particles is less than the number of thermally accessible states in the case of high temperatures and low densities. In this limit, the gas is classical. However, the density of particles increase or the temperature of the system decreases, then the number of accessible states per particle becomes smaller, and more particles will be forced into a single state, more than the maximum allowed for that state by statistical weighting. Thus, any extra particles added will go into the ground state.

In such a case, they can calculate the transition temperature at any density, and integrate, over all momentum states. The maximum number of excited particles, $p/(1 - p)$ can be obtained, and expressed[3-4]by

$$N = \frac{V}{(2\pi)^3}\int d^3k \frac{P(k)}{1-P(k)} = V\int d^3k[\exp(k^2/2mT)-1]^{-1} \text{ with } P(k) = \exp(-k^2/2mT) \tag{6.2}$$

When the integral is evaluated with the factors of k_B and $\hbar$ restored by dimensional analysis, then it gives the above critical temperature formula. Therefore, this integral defines the critical temperature and particle number corresponding to the conditions of negligible chemical potential[3-4].

6.2.2. Bogoliubov's Interacting Models in Weakly Interacting Boson Gas

As it is known, in the simplest cases, the statesof condensed particles can be described with a nonlinear Schrödinger equation [11-12], also known as Gross-Pitaevskii [13-16] or the Ginzburg-Landau equation [5-6]. The validity of this approach is actually limited to the case of ultracold temperatures and some specific atoms, such as alkali atoms. In this approach, it is assumed that the state of the BEC is described by the unique wavefunction of the condensate $\phi(\vec{r},t)$. In this case $\left|\phi(\vec{r},t)\right|^2$ is interpreted as the particle density, and then the total number of atoms is $N=\int\left|\phi(\vec{r},t)\right|^2 d\tau$ because all particles have been condensed into the ground state in BEC. Thus the energy (E) associated with the state $\phi(\vec{r},t)$ is:

$$E=\int d\tau[\frac{\hbar^2}{2m}|\nabla\phi|^2+V(\vec{r})|\phi|^2-\frac{1}{2}U_0|\phi|^4] \tag{6.3}$$

Minimizing this energy with respect to infinitesimal variations in $\phi(\vec{r},t)$, and holding the number of atoms constant, yields the Gross–Pitaevski equation (GPE) [13-16], which is a nonlinear Schrödinger equation. It may be represented by

$$i\hbar\frac{\partial\phi}{\partial t}=-\frac{\hbar^2}{2m}\nabla^2\phi-U_o|\phi|^2\phi+V\left(\vec{r}\right)\phi \tag{6.4}$$

where m is the mass of the boson, $V(\vec{r})$ is the external potential, and U_0 is representative of the inter-particle interaction.

The above representations, including the meaning of the wave function $\phi(\vec{r},t)$, are all the same as those in Pang's nonlinear quantum mechanics. This shows that the properties of Boson-Enistein condensation can be described by nonlinear quantum mechanics.

In the case of zero external potential, they gave the dispersion law of interacting Boson-Einstein-condensed particles by the Bogoliubov spectrum (for $T=0$), which is

$$\omega_P=[(p^2/2m)(p^2/2m+2U_0n_0)]^{1/2} \tag{6.5}$$

The Gross-Pitaevskii equation (GPE)(6.4) [13-16] provides a relatively good description of the behavior of atomic BEC's. However, the Gross-Pitaevskii modeldoes not take into account the temperature dependence of dynamical variables, and is therefore valid only for $T=0$. Thus, it is not applicable, for example, for the condensates of excitons, magnons and photons, where the critical temperature is up to room one. Therefore, the Gross–Pitaevskii model of BEC is, inessence, a physical approximation valid only for certain classes of BECs. By reconstruction, GPE uses the following simplifications: it assumes that interactions between condensate particles are of the contact two-body type and it neglects anomalous contributions to self-energy [17]. These assumptions are suitable mostly for the dilute three-dimensional condensates. If one relaxes any of these assumptions, the equation for the condensate wavefunction acquires the terms containing higher-order powers of the wavefunction. Moreover, for some physical systems the amount of such terms turns out to be infinite, therefore, the equation becomes essentially non-polynomial. The examples where this could happen are also the Boson–Fermicomposite condensates [18-21],effectively lower-dimensional condensates, [22] and dense condensates and superfluid clusters and droplets [23].

However, it is clear that in a general case, the behavior of the Boson-Einstein condensate can be described by coupled evolution equations for condensate density, superfluid velocity and distribution function of elementary excitations. This problem was studied by Peletminskii et al. in a microscopic approach in 1977. The Peletminskii equations are valid for any finite temperatures below the critical point. Kirkpatrick and Dorfman obtained similar equations using another microscopical approach in 1985. The Peletminskii equations also reproduce Khalatnikov hydrodynamical equations for superfluids, as a limiting case.

6.3. ARGUMENT OF CONDENSATION FEATURE OF THE PARTICLES IN BEC SYSTEM

In the last few decades, the Boson-Einstein condensation has been observed in a series of remarkable experiments using weakly interacting atomic gases, such as vapors of rubidium, sodium lithium, or hydrogen. Its formation and properties have been extensively studied. These studies show that the Boson-Einstein condensation is a nonlinear phenomenon, analogous to nonlinear optics. These microscopic parties are coherent and can be described by anonlinear Schrödinger equation in Eq.(6.4) as the Gross-Pitaerskii equation [13-16]. In fact, H. K. Bullough etal. [24-26] and Elyutin et al. [27-28] used the following nonlinear Schrödinger equation:

$$i\frac{\partial\phi}{\partial t'}=-\frac{\partial^2\phi}{\partial x'^2}-\lambda|\phi|^2\phi+V(x')\phi \tag{6.6}$$

in a one dimensional case and corresponding Hamiltonian density :

$$H'=\left|\frac{\partial\phi}{\partial x'}\right|^2+V(x')|\phi|^2-\frac{1}{2}\lambda|\phi|^4 \tag{6.7}$$

to discuss the realization of Boson-Einstein condensation in one dimension in a condensate system, where H'= $\mathcal{H}$is the Hamiltonian density of the system, $t'=t/\hbar$, $x'=x\sqrt{2m}/\hbar$, the nonlinear parameters of λ are defined as $\lambda=U_0=-2Naa_1/a_0^2$ with N being the number of particles trapped in the condensed state, a is the ground state scattering length, a_0 and a_1 denote the transverse (y, z) and the longitudinal (x) condensate sizes, respectively, and λ is positivecoefficient ofnonlinear interactionfor condensation with self-attraction (negative scattering length).The coherent regime was observed in Boson-Einstein condensation in lithium.

The specific form of the interactional potential V (x')in Eq.(6.7) depends on the details of the experimental setup. Work on Boson-Einstein condensationbased on the above Hamiltonianmodel, were carried out and are reported by C. F. Barenghi et al. [29]. Therefore we can say that Eqs.(6.6)-(6.7) embody and characterize the basic features of the BEC state. We will use them here to reveal and elucidate the essences and natures of the BEC state.

It is not surprising to see that Eq. (6.6) is exactly the same with dynamic equation (2.66),or (3.1), and the corresponding Hamiltonian density function in Eq. (6.7) is also the same, with the Hamiltonian density function in Eq. (2.74) in Pang's nonlinear quantum mechanics. This indicated clearly that Pang's nonlinear quantum mechanical theory could be used to describe and characterize the properties of the Boson-Einstein condensation state. In this case, it is quite necessary to demonstrate that the Boson particles denoted by Eqs.(6.6)-(6.7) are really condensed using the nonlinear quantum mechanical theory.

We here employ the Bogoliubov method, which was used to study the superfluid feature of liquid helium ^{4}He [11-12], to prove the appearance of the condensation starting from Eqs.(6.6)- (6.7). To do that, we here represent the wave function $\phi(x,t)$ in Eq. (6.6) by

$$\phi(x)=\frac{1}{\sqrt{\bar{V}}}\sum_{\mathrm{p}} q_{\mathrm{p}}e^{\mathrm{ipx}}, q_{\mathrm{p}}=\frac{1}{\sqrt{2\varepsilon_{\mathrm{p}}}}\left(b_{\mathrm{p}}+b^{+}_{\mathrm{-p}}\right) \tag{6.8}$$

by introducing creation and annulment operator , b_q^+ and b_q , with wavevector q, in the second quantum representation, where $\overline{\mathrm{V}}$ is the volume of the system, and ε_q is the dimensionless energy of the particle, since $\phi(x,t)$ describes a Boson in Boson condensed systems, such as the Cooper electron pair in a superconductor [30-33]. Applying the following traditional method in quantum field theory we further express b_q^+ and b_q by

$$b_{\mathrm{p}}=\sqrt{N_0}\delta(p)+\gamma_{\mathrm{p}}, b_{\mathrm{p}}^{+}=\sqrt{N_0}\delta(p)+\beta_{\mathrm{p}} \tag{6.9}$$

where N_0 is the number of Bosons in the system and

$$\delta(p)=\begin{cases}0\ , if\, p\neq 0\\ 1\ ,\ if\, p=0\end{cases}$$

Substituting Eqs. (6.8) and (6.9) into Eq. (6.7) at V(x')= α , we can arrive at the Hamiltonian operator of the system represented by γ_{p} and β_{p} , which is

$$H = \left(\frac{4\lambda N_0^2}{\varepsilon_0^2 \overline{V}} - \frac{\alpha}{\varepsilon_0}\right)\sqrt{N_0}\left(\gamma_0 + \gamma_0^+ + \beta_0 + \beta_0^+\right) + \sum_{\mathrm{P}}\left(\frac{4\lambda}{\varepsilon_0 \varepsilon_{\mathrm{P}}}\frac{N_0}{\overline{V}} - \varepsilon_{\mathrm{p}}\right)\left(\gamma_{\mathrm{p}}^+ \beta_{\text{-p}}^+ + \gamma_{\mathrm{p}} \beta_{\text{-p}}\right) + \frac{4\lambda N_0^2}{\varepsilon_0^2 \overline{V}} -$$
$$\frac{2N_0\alpha}{\varepsilon_0} + \frac{\lambda N_0}{\varepsilon_0 \overline{V}}\sum_{\mathrm{p}}\frac{1}{\varepsilon_{\mathrm{p}}}\left(\beta_{\mathrm{p}}^+ \beta_{\text{-p}}^+ + \beta_{\mathrm{p}}\beta_{\text{-p}} + \gamma_{\mathrm{p}}^+ \gamma_{\text{-p}}^+ + \gamma_{\mathrm{p}}\gamma_{\text{-p}} + 2\gamma_{\mathrm{p}}^+ \beta_{\mathrm{p}}^+ + 2\beta_{\mathrm{p}}^+ \gamma_{\mathrm{p}}\right) +$$
$$\sum_{\mathrm{p}}\left(\varepsilon_{\mathrm{p}} - \frac{\alpha}{2\varepsilon_0} + \frac{4\lambda N_0}{\varepsilon_0 \varepsilon_p \overline{V}}\right)\left(\gamma_{\mathrm{p}}^+ \gamma_{\mathrm{p}} + \beta_{\mathrm{p}}^+ \beta_{\mathrm{p}}\right) + \sum_{P}\frac{4\lambda}{\varepsilon_0 \varepsilon_{\mathrm{P}}}\frac{N_0}{\overline{V}} + O\left(\frac{\sqrt{N_0}}{\overline{V}}\right) + O\left(\frac{N_0}{\overline{V}^2}\right) \quad (6.10)$$

Because the condensed density $N_0/\overline{V}$ must be finite, it is possible that the higher order terms $O\left(\sqrt{N_0}/\overline{V}\right)$ and $O\left(N_0/\overline{V}^2\right)$ may be neglected. Next, we perform the following canonical transformationusing the Bogoliubov method [11-12]

$$\gamma_{\mathrm{p}} = u_{\mathrm{p}}^* c_{\mathrm{p}} + \upsilon_{\mathrm{p}} c_{\text{-p}}^+, . \beta_{\mathrm{p}} = u_{\mathrm{p}}^* d_{\mathrm{p}} + \upsilon_{\mathrm{p}} d_{\text{-p}}^+ \quad (6.11)$$

where v_p and u_p are real and satisfy $\left(u_{\mathrm{p}}^2 - \upsilon_{\mathrm{p}}^2\right) = 1$. In this case, weagain introduce the following transformation:

$$\varsigma_{\mathrm{p}} = \frac{1}{\sqrt{2}}\left(u_{\mathrm{p}}\gamma_{\mathrm{p}}^+ - \upsilon_{\mathrm{p}}\gamma_{\text{-p}} + u_{\mathrm{p}}\beta_{\mathrm{p}}^+ - \upsilon_{\mathrm{p}}\beta_{\text{-p}}\right), \eta_{\mathrm{p}}^+ = \frac{1}{\sqrt{2}}\left(u_{\mathrm{p}}\gamma_{\mathrm{p}}^+ - \upsilon_{\mathrm{p}}\gamma_{\text{-p}} - u_{\mathrm{p}}\beta_{\mathrm{p}}^+ + \upsilon_{\mathrm{p}}\beta_{\text{-p}}\right) \quad (6.12)$$

where the following commutations hold

$$\left[\varsigma_{\mathrm{p}}, H'\right] = g_{\mathrm{p}}\varsigma_{\mathrm{p}} + M_{\mathrm{p}}\varsigma_{\text{-p}}^+, \ \left[\eta_{\mathrm{p}}, H'\right] = g_{\mathrm{p}}'\eta_{\mathrm{p}} + M_{\mathrm{p}}'\eta_{\text{-p}}^+ \quad (6.13)$$

where

$$\begin{cases} g_{\mathrm{p}} = G_{\mathrm{p}}\left(u_{\mathrm{p}}^2 + \upsilon_{\mathrm{p}}^2\right) + F_{\mathrm{p}} 2u_{\mathrm{p}}\upsilon_{\mathrm{p}}, \ \mathrm{M}_{\mathrm{p}} = F_{\mathrm{p}}\left(u_{\mathrm{p}}^2 + \upsilon_{\mathrm{p}}^2\right) + G_{\mathrm{p}} 2u_{\mathrm{p}}\upsilon_{\mathrm{p};} \\ g_{\mathrm{p}}' = G_p'\left(u_{\mathrm{p}}^2 + \upsilon_{\mathrm{p}}^2\right) + F_{\mathrm{p}}' 2u_{\mathrm{p}}\upsilon_{\mathrm{p}}, \ \mathrm{M}_{\mathrm{p}}' = F_{\mathrm{p}}'\left(u_{\mathrm{p}}^2 + \upsilon_{\mathrm{p}}^2\right) + G_{\mathrm{p}}' 2u_{\mathrm{p}}\upsilon_{\mathrm{p}} \end{cases} \quad (6.14)$$

with

$$G_{\mathrm{p}} = \varepsilon_{\mathrm{p}} - \frac{\alpha}{2\varepsilon_{\mathrm{p}}} + 6\xi_{\mathrm{p}}', \ \mathrm{F}_{\mathrm{p}} = -\frac{\alpha}{2\varepsilon_{\mathrm{p}}} + 6\xi_{\mathrm{p}}', \ G_{\mathrm{p}}' = \varepsilon_{\mathrm{p}} - \frac{\alpha}{2\varepsilon_{\mathrm{p}}} + 2\xi_{\mathrm{p}}', \ \mathrm{F}_p' = \frac{\alpha}{2\varepsilon_{\mathrm{p}}} - 2\xi_{\mathrm{p}}' \quad (6.15)$$

here $\xi'_p = \dfrac{\lambda N_0}{\varepsilon_0 \varepsilon_p \overline{V}}$ and H'= $\mathcal{H}$.

The following two cases are studied and illustrated in this case.

(**A**) $M'_\text{p} = 0$. In this case, we know from Eq. (6.13) that η_p^+ is the creation operator of an elementary excitation, its energy is given by

$$g'_\text{p} = \sqrt{\varepsilon_\text{p}^2 + 4\varepsilon_\text{p}\xi'_\text{p} - 2\alpha} \tag{6.16}$$

Based on this result we can obtain the following representations from Eqs. (6.14) and (6.15)

$$\left(u'_\text{p}\right)^2 = \frac{1}{2}\left(1 + \frac{G'_\text{p}}{g'_\text{p}}\right)$$

and

$$\left(\upsilon'_\text{p}\right)^2 = \frac{1}{2}\left(-1 + \frac{G'_\text{p}}{g'_\text{p}}\right) \tag{6.17}$$

This means that ς_p^+ in Eq. (6.13), is *not* a creation operator of the elementary excitation. Thus, the following transformation must be made

$$B_\text{p} = \chi_\text{p}\varsigma'_\text{p} + \mu_\text{p}\varsigma'^+_\text{p},\ \left|\chi_\text{p}\right|^2 - \left|\mu_\text{p}\right|^2 = 1 \tag{6.18}$$

In this case, we can prove that

$$[B_\text{p}, \mathcal{H}] = E_\text{p} B_\text{p} \tag{6.19}$$

where $E_\text{p} = \sqrt{12\varepsilon_\text{p}\xi'_\text{p} + \varepsilon_p^2 - 2\alpha}$

Inserting Eqs. (6.11), (6.18)-(6.19) and $M'_p = 0$ into Eq. (6.10), and after some reorganization, we have

$$H = U + E_0 + \sum_{\text{p}>0}\left[E_\text{p}\left(B_\text{p}^+ B_\text{p} + B_{\text{-p}}^+ B_{\text{-p}}\right) + g'_\text{p}\left(\eta_\text{p}^+\eta_\text{p} + \eta_{\text{-p}}^+\eta_{\text{-p}}\right)\right] \tag{6.20}$$

where

$$U=\frac{\lambda N_0^2}{\varepsilon_0^2\bar{V}}-\frac{2\alpha N_0}{\varepsilon_0}+\sum_{\mathrm{p}}4\xi_{\mathrm{p}}'+\sum_{\mathrm{p}}\left(\varepsilon_{\mathrm{p}}+\frac{\alpha}{2\varepsilon_{\mathrm{p}}}+4\xi'\right)4\upsilon_{\mathrm{p}}'^2+\sum_{\mathrm{p}>0}4\eta\cdot 2u_{\mathrm{p}}'\upsilon_{\mathrm{p}}'$$

$$E_0=-2\sum_{\mathrm{p}>0}E_{\mathrm{p}}\left|\mu_{\mathrm{p}}\right|^2=-\sum_{\mathrm{p}>0}\left(g_{\mathrm{p}}'-E_{\mathrm{p}}\right) \tag{6.21}$$

Both U and E_0 are now independent of the creation and annihilation operators of the Bosons, $U+E_0$ gives the energy of the ground state, and N_0 can be determined from the condition

$$\frac{\delta\left(U+E_0\right)}{\delta N_0}=0$$

which gives finally

$$\frac{N_0}{\bar{V}}=\frac{\alpha\varepsilon_0}{4\lambda}=\frac{1}{2}\varepsilon_0\phi_0^2 \tag{6.22}$$

This representation in Eq.(6.22) manifests clearly the condensed density of the ground state ϕ_0 . This means that the system is now in a condensation state, thus the Boson condensation occurs in this case. Then we demonstrated the existence of Boson condensation by using Eqs.(6.6)-(6.7).Equations(6.6)-(6.7)truly characterize the feature and state of BEC.

From Eqs. (6.17) and (6.18) and (6.21), we can arrive at:

$$g_{\mathrm{p}}'=\sqrt{\varepsilon_{\mathrm{p}}^2-\alpha},\quad E_{\mathrm{p}}=\sqrt{\varepsilon_{\mathrm{p}}^2-\alpha} \tag{6.23}$$

Thus, theyrepresent the energy spectra of η_p^+ and B_p^+ ,respectively, which then are similar to the energy spectra of the Cooper pair and phonon in the BCS theory in the superconductor [30-33].

Substituting Eq. (6.23) into Eq. (6.17) we can obtain:

$$u_{\mathrm{p}}'^2=\frac{1}{2}\left(1+\frac{2\varepsilon_{\mathrm{p}}^2-\alpha}{2\sqrt{\varepsilon_{\mathrm{p}}^2-\alpha e_{\mathrm{p}}}}\right),\quad \upsilon_p'^2=\frac{1}{2}\left(-1+\frac{2\varepsilon_{\mathrm{p}}^2-\alpha}{2\sqrt{\varepsilon_{\mathrm{p}}^2-\alpha\varepsilon_{\mathrm{p}}}}\right) \tag{6.24}$$

(**B**) In the case of $M_{\mathrm{p}}=0$, if a similar approach is used, then we can arrive at the energy spectrum corresponding to ξ_{p}^{+} , which is $E_{\mathrm{p}}=\sqrt{\varepsilon_{\mathrm{p}}^2+\alpha}$, but for $A_{\mathrm{p}}^+=\chi_{\mathrm{p}}\eta_{\mathrm{p}}^++\mu_{\mathrm{p}}\eta_{\mathrm{-p}}$, we obtained $g_{\mathrm{p}}'=\sqrt{\varepsilon_{\mathrm{p}}^2+\alpha}$, where

$$u_{\mathrm{p}}^2 = \frac{1}{2}\left(1+\frac{2\varepsilon_{\mathrm{p}}^2+\alpha}{2\varepsilon_{\mathrm{p}}\sqrt{\varepsilon_{\mathrm{p}}^2+\alpha}}\right), \quad \upsilon_{\mathrm{p}}^2 = \frac{1}{2}\left(-1+\frac{2\varepsilon_{\mathrm{p}}^2+\alpha}{2\varepsilon_{\mathrm{p}}\sqrt{\varepsilon_{\mathrm{p}}^2+\alpha}}\right) \tag{6.25}$$

Based on experiments in quantum statistical physics, we can determine the occupation number at the level with energy ε_{p} in the thermal equilibrium system with finite temperature $T(\neq 0)$, which is shown by

$$N_{\mathrm{p}} = \left\langle b_{\mathrm{p}}^{+} b_{\mathrm{p}} \right\rangle = \frac{1}{e^{\varepsilon_{\mathrm{p}}/K_{\mathrm{B}}T} - 1} \tag{6.26}$$

where $\langle \cdots \rangle$ denotes the Gibbs average, defined as $\langle \cdots \rangle = \dfrac{SP\left[e^{-H/K_B T} \cdots\right]}{SP\left[e^{-H/K_B T}\right]}$, here SP denotes the trace in a Gibbs statistical description. Equation (6.26) indicated that at low temperatures, or $T \to 0$ K, the majority of the Bosons or Cooper pairs in a superconductorare condensed to the ground state with $p=0$. Then $\left\langle b_0^{+} b_0 \right\rangle \approx N_0$, where N_0 is the total number of Bosons in the system or Cooper pairs in superconductor [30-33] and $N_0 >> 1$, then $\left\langle b^{+} b \right\rangle = 1 << \left\langle b_0^{+} b_0 \right\rangle$.This shows clearly that the Boson condensations of the microscopic particlestruly occur at low temperatures in the systems.

On the other hand, we can decide that the number of particles is extremely largein the condensed state. Thus, from Eqs. (6.8) and (6.9) we obtain in this case

$$\phi_0 = \phi_{p=0} = \frac{1}{\sqrt{2\varepsilon_0 \overline{V}}}\left(b_0 + b_0^{+}\right) \tag{6.27}$$

Because $\gamma_0 \left|\phi_0\right\rangle = 0$ and $\beta_0 \left|\phi_0\right\rangle = 0$, b_0 and b_0^{+} can be taken to be $\sqrt{N_0}$. Then the average value of $\phi^{*}\phi$ in the ground state becomes

$$\left\langle \phi_0 \left| \phi^{*}\phi \right| \phi_0 \right\rangle = \left\langle \phi^{*}\phi \right\rangle_0 = \frac{1}{2\varepsilon_0 \overline{V}} \cdot 4N_0 = \frac{2N_0}{\varepsilon_0 \overline{V}} \tag{6.28}$$

Substituting Eq. (6.22) into Eq. (6.28), we can see that:

$$\left\langle \phi^{*}\phi \right\rangle_0 = \alpha / 2\lambda \text{ or } \left\langle \phi^{*} \right\rangle_0 = \pm\sqrt{\alpha / 2\lambda} \tag{6.29}$$

which is the ground state of the condensed phase. Equations(6.27)-(6.29) indicated the BEC appears in the ground state of the system, which has a minimal energy, and its condensed density is $N_0/\bar{V}$, which shows that all microscopic particles are condensed into the ground state. Therefore, both particle number and condensed density of particles are all maximal in this condensation state. Thus, we here used the nonlinear quantum mechanics to reveal clearly the essences of the BEC and to elucidate its feature.

Meanwhile, we can demonstrate that the Boson-Einstein condensation is formed by means of nonlinear interaction $-\lambda|\phi|^4/2=-(\lambda/2)|\phi|^2.|\phi|^2$ in Eq.(6.7), which is an attractive interaction. This means that the properties of the Boson particle cannot be described by linear quantum mechanics in Chapter 1. However, where does the attractive interaction come from? As it is known, there is an attractive interaction among these Boson particles, according to Boson-Einstein statistical theory; thus, many Boson particles can exist simultaneously on one energy level due to the attractive interaction. Just so, the mechanism of generation of the nonlinear interaction is called self-condensation, asintroduced in Chapter 4.In fact, it should be the attractive self-condensation mechanism of nonlinear interaction. Obviously, this nonlinear interaction plays a key role in the condensation. There is no BEC state without the nonlinear interaction because this nonlinear interaction is related to the state wave-function of the Boson particles, then it can balance out or suppresses the dispersed effect of kinetic energy in Eq.(6.6) and the dissipated effect of heat energy of motion of the particles, thus the state and features of the particles are changed and become as some solitons with bell-type occurred in Fig.6.3, which are the solutions of Eq.(6.6), in this case, these Boson particles are spontaneously condensed to the lower energy state through the depression of energy of the system at low temperature. This implies that lower temperature and the Boson feature of the particles are the fundamental conditions of generation of the BEC phenomenon. On the contrary, the condensation state will be destructed if the temperature is lifted because the attractive nonlinear interaction cannot again suppress the dispersed effect of kinetic energy in Eq.(6.6) and the dissipated effect of heat energy of motion of the particles in this case. Hence, Pang's nonlinear quantum mechanics can completely explain the Boson-Einstein condensation phenomenon.

6.4. The Essences of Boson-EinsteinCondensed State of Microscopic Particles

6.4.1. The Spontaneous Breakdown of Symmetry in the BEC Systems

We can also demonstrate that the Boson-Einstein condensation is a result of the symmetry of the system, which is spontaneously broken down due to the nonlinear interaction $-\lambda|\phi|^4/2$.

In fact, the Hamiltonians in Eq. (6.7) at V(x')= α possess the U (1) symmetry. That is, they remain unchanged while undergoing the following transformation:

$$\phi(\vec{r},t)\rightarrow\phi'(\vec{r},t)=e^{-iQ\theta}\phi(\vec{r},t)$$

where Q_j is the charge of the particle, θ is a phase and, in the case of one dimension. Obviously, each term in the Hamiltonian in Eq. (6.7) contains the product of the $\phi_j(\vec{r},t)s$, but they are invariant in the above transformation because the neutrality is required for the Hamiltonian; there must be $(Q_1 + Q_2 + \cdots + Q_n) = 0$ in such a case. At the same time, since θ is independent of x, it is necessary that $\nabla\phi_j \to e^{-i\theta Q_j}\nabla\phi_j$. Thus, each term in the Hamiltonian in Eq. (6.2) is invariant under the above transformation. Its detailed demonstration is described in Chapter 2. Thus we say the Hamiltonian possesses the $U(1)$ symmetry [34-35].

If we now rewrite Eq. (6.2) the following

$$H' = -\frac{1}{2}(\nabla\phi)^2 + U_{eff}(\phi), U_{eff}(\phi) = -\alpha\phi^2 + \lambda\phi^4 \tag{6.30}$$

where H'= $\mathcal{H}$.

We can see that the effective potential energy, $U_{eff}(\phi)$, in Eq. (6.30) has two sets of the extreme, $\phi_0 = \pm\sqrt{\alpha/2\lambda}$ and ϕ_0=0, but the minimum is now located at

$$\phi_0 = \pm\sqrt{\alpha/2\lambda} = \langle 0|\phi|0\rangle, \tag{6.31}$$

rather than at ϕ_0=0. This means that the energy at $\phi_0 = \pm\sqrt{\alpha/2\lambda}$ is lower than that at ϕ_0=0. Therefore, ϕ_0=0 corresponds to the normal ground state, while $\phi_0 = \pm\sqrt{\alpha/2\lambda}$ *is* the ground state of the nonlinear quantum mechanical state. This means that the stable state of the system is in the macroscopic quantum mechanical state.. Thus, we obtain that the Hamiltonian of a normal state differs from that of the nonlinear quantum mechanical state due to the existence of nonlinear interactions, in which then the two ground states have the relation of $\langle 0|\phi|0\rangle \neq -\langle 0|\phi|0\rangle$ under the transformation, $\phi \to -\phi$. That is, the ground states no longer have the U(1) symmetry. In other wordsthe symmetry of the ground states has been destroyed in this case. The reason for this is evidently due to the nonlinear term $\lambda\phi^4$ in the Hamiltonian of the system. Therefore, this phenomenon is referred to as a spontaneous breakdown of symmetry. According to Landau's theory of phase transition, the system undergoes a second-order phase transition, in such a case the normal ground state ϕ_0=0 is changed to the nonlinear quantum mechanical ground state $\phi_0 = \pm\sqrt{\alpha/2\lambda}$. Proof will be presented in the following example.

In order to make the expectation value in a new ground state be *zero* in the nonlinear quantum mechanics, the following transformation [34-35] is done:

$$\phi' = \phi + \phi_0 \tag{6.32}$$

so that

$$\langle 0|\phi'|0\rangle = 0 \tag{6.33}$$

After this transformation, the Hamiltonian density of the system becomes

$$H'(\phi+\phi_0) = \frac{1}{2}|\nabla\phi|^2 + (6\lambda\phi^2 - \alpha)\phi^2 + 4\lambda\phi_0\phi^3 + (4\lambda\phi_0^3 - 2\alpha\phi_0)\phi + \lambda\phi^4 - \alpha\phi_0^2 + \lambda\phi_0^4 \tag{6.34}$$

Inserting Eq. (6.31) into Eq. (6.34), we have

$$\langle \phi_0|4\lambda\phi_0^2 - 2\alpha|\phi_0\rangle = 0 \ .$$

Consider now the expectation value of the variation $\delta H'/\delta\phi$ in the ground state, i.e. $\langle 0|\frac{\delta H'}{\delta\phi}|0\rangle = 0$, then from Eq. (6.34), we get

$$\langle 0|\frac{\delta H'}{\delta\phi}|0\rangle = \langle 0|-\nabla^2\phi + 2\alpha\phi - 4\lambda\phi^3|0\rangle = 0 \tag{6.35}$$

After the transformation Eq. (6.32), it becomes

$$\nabla^2\phi_0 + (4\lambda\phi_0^2 - 2\alpha)\phi_0 + 12\lambda\phi_0\langle 0|\phi^2|0\rangle + 4\lambda\langle 0|\phi^3|0\rangle - (2\alpha - 12\lambda\phi_0^2)\langle 0|\phi|0\rangle = 0 \tag{6.36}$$

where the terms $\langle 0|\phi^3|0\rangle$ and $\langle 0|\phi|0\rangle$ are both zero, but the fluctuation $12\lambda\phi_0\langle 0|\phi^2|0\rangle$ of the ground state*is not zero*. However, for a homogeneous system, at T=0K, the term $\langle 0|\phi^2|0\rangle$ is very small and can be neglected.

Then Eq. (6.35) can be written as

$$-\nabla^2\phi_0 - (4\lambda\phi_0^2 - 2\alpha)\phi_0 = 0 \tag{6.37}$$

Obviously, two sets of solutions, $\phi_0 = 0$, and $\phi_0 = \pm\sqrt{\alpha/2\lambda}$, can be obtained from the above equation, but we can demonstrate that the former is unstable, and that the latter is stable.

If the displacement is very small, i.e. $\phi_0 \to \phi_0 + \delta\phi_0 = \phi_0'$, then the equation satisfied by the fluctuation $\delta\phi_0$ is relative to the normal ground state $\phi_0 = 0$ and is

$$\nabla^2\delta\phi_0 - 2\alpha\delta\phi_0 = 0 \tag{6.38}$$

Its' solution attenuates exponentially indicating that the ground state, $\phi_0 = 0$ is unstable [18].

On the other hand, the equation satisfied by the fluctuation $\delta\phi_0$, relative to the ground state $\phi_0 = \pm\sqrt{\alpha/2\lambda}$, is

$$\nabla^2\delta\phi_0 + 2\alpha\delta\phi_0 = 0. \tag{6.39}$$

Its solution, $\delta\phi_0$, is an oscillatory function and thus the ground-state $\phi_0 = \pm\sqrt{\alpha/2\lambda}$ in nonlinear quantum mechanics is stable [36]. Further calculations show that the energy of the ground state in nonlinear quantum mechanics is lower than that of the normal state by $\varepsilon_0 = -\alpha^2/4\lambda < 0$. Therefore, the ground state of the normal phase and that of nonlinear quantum mechanics are separated by an energy gap of $\alpha^2/(4\lambda)$, so then, we verified thatat T=0K, all particles will spontaneously condense to the ground state of the nonlinear quantum mechanics rather than filling the ground state of the normal phase in this case . Based on this energy gap, we can conclude that the specific heat of the system described by the nonlinear quantum mechanics has an exponential dependence on the temperature, and the critical temperature is given [16-17]by:

$$T_c = 1.14\omega_p \exp[-1/(3\lambda/\alpha)N(0)].$$

This is a result of feature of the second-order phase transition. The above results are in agreement with those of superconductivity [30-33].

Therefore, the transition from the state $\phi_0 = 0$ to the state $\phi_0 = \pm\sqrt{\alpha/2\lambda}$ and the corresponding condensation of particles are due to the second-order phase transitions. This is necessarily the result of a spontaneous breakdown of symmetryarising from the nonlinear interaction, $\lambda\phi^4$.

6.4.2. The Coherent Features of Boson-EinsteinCondensation

Proof that the above Boson condensation state is a kind of coherent state is described as follows [34-35].

As discussed abovewhen $\delta H'/\delta\phi$ =0 from Eq. (6.7) at V(x')= α, we have

$$\nabla^2\phi - 2\alpha\phi + 4\lambda|\phi|^2\phi = 0 \tag{6.40}$$

It is a time-independent nonlinear Schrödinger equation in Eq.(3.1) or Eq.(6.6),where ϕ can be represented in terms of the creation and annihilation operators, b_p^+ and b_p, which is denoted in Eq.(6.8), but we now represent it by

$$\phi = \frac{1}{\sqrt{V}} \sum_p \frac{1}{\sqrt{2\varepsilon_p}} (b_p e^{-ipx} + b_p^+ e^{ip.x}) \tag{6.41}$$

After a spontaneous breakdown of symmetry, ϕ_0 , the ground-state of ϕ , for the system is no longer zero, but $\phi_0 = \pm\sqrt{\alpha / 2\lambda}$, as mentioned above. The operation of the annihilation operator on $|\phi_0\rangle$ no longer gives zero, but has the feature of

$$b_p |\phi_0\rangle \neq 0 \tag{6.42}$$

This implies that a new field ϕ' can be defined according to the transformation Eq. (6.32), where ϕ_0 is a scalar field and satisfies Eq. (6.33) in such a case. Evidently, ϕ_0 can be expanded into

$$\phi_0 = -\frac{1}{\sqrt{V}} \sum_p \frac{1}{\sqrt{2\varepsilon_p}} (\zeta_p e^{-ip.x} + \zeta_p^+ e^{ip.x}) \tag{6.43}$$

The transformation between the fields ϕ and ϕ' is obviously a unitary transformation, that is

$$\phi' = U\phi U^{-1} = e^{-s}\phi e^{s} = \phi + \phi_0 \tag{6.44}$$

where

$$S = i\int dx' [\phi'(x', t)\phi_0(x', t) - \phi_0(x', t)\phi(x', t)] \tag{6.45}$$

ϕ and ϕ' satisfy the following commutation relation

$$[\phi'(x', t), \phi(x, t)] = i\delta(x' - x) \tag{6.46}$$

From Eq. (6.33) we now have

$$\langle 0|\phi'|0\rangle = \phi_0' = 0 \, .$$

The ground state $|\phi_0'\rangle$ of the field ϕ' thus satisfies

$$b_p\left|\phi_0'\right\rangle = 0 \tag{6.47}$$

From Eq. (6.32), we can obtain the following relationship between the annihilation operator a_p of the new field ϕ ' and the annihilation operator b_p of the ϕ field

$$a_p = e^{-S} b_p e^{S} = b_p + \zeta_p \tag{6.48}$$

where

$$\zeta_p = \frac{1}{(2\pi)^{3/2}}\int \frac{dx}{\sqrt{\varepsilon_p}}[\phi_0(x,t)e^{ip.x} + i\phi_0^*(x,t)e^{-ip.x}] \tag{6.49}$$

Therefore, the new ground state $\left|\phi_0'\right\rangle$ and the old ground state $\left|\phi_0\right\rangle$ are related through $\left|\phi_0'\right\rangle = e^{S}\left|\phi_0\right\rangle$.

Thus we have

$$a_p\left|\phi_0'\right\rangle = (b_p + \zeta_p)\left|\phi_0'\right\rangle = \zeta_p\left|\phi_0'\right\rangle \tag{6.50}$$

According to the definition of the coherent state, we can judge from Eq. (6.50) that the new ground state $\left|\phi_0'\right\rangle$ is a coherent state. Because such a coherent state is formed after the spontaneous breakdown of symmetry of the systems, thus, it is referred to as a spontaneous coherent state. But when $\phi_0 = 0$, the new ground state is the same as the old state, which is not a coherent state. Thus, we can affirm that the Boson condensation state formed after the symmetry of the system is spontaneously broken down, due to the nonlinear interaction, is a kind of spontaneous coherent state.

The same conclusion can be directly derived from the BCS theory [30-33] in the superconductor. In fact, the wave function of the ground state of a superconductor the BCS theory is written as

$$\left|\phi_0'\right\rangle = \prod_k(\mu_k + \upsilon_k \hat{a}_k^+ a_{-k}^+)\left|\phi_0\right\rangle = \prod_k(\mu_k + \upsilon_k \hat{b}_{k-k}^+)\left|\phi_0\right\rangle \sim \eta' \exp\left(\sum_k \frac{\upsilon_k}{\mu_k}\hat{b}_{k-k}^+\right)\left|\phi_0\right\rangle \tag{6.51}$$

where $\hat{b}_{k-k}^+ = \hat{a}_k^+\hat{a}_{-k}^+$. This equation shows that the superconducting ground state is a coherent state. Hence, we can conclude that the spontaneous coherent state is formed in superconductors after the spontaneous breakdown of symmetry.

By reconstructing a quasiparticle-operator-free new formulation of the Bogoliubov-Valatin transformation parameter dependence [37], W. S. Lin et al., [38] demonstrated that the BCS statein superconductoris not only a coherent state of single-Cooper-pairs, but also the squeezed state of the double-Cooper- pairs, which is the coherent state of two quanta, and thus reconfirmed thus the coherent feature of BCS superconductive state [30-33].

The above results showed clearly that the Boson-Einstein condensation statecould be described by our nonlinear quantum mechanical theory.

6.5. Experimental Observation of Property of the Boson-EinsteinCondensation

6.5.1. The Properties of Boson-EinsteinCondensation in Helium Atom

It is well known that Landau [5-6] thought thatthe phenomena of superfluidity of a Boson gas and superconductivity of a strongly-correlated Fermi gas (a gas of Cooper pairs) are tightly connected to Boson-Einstein condensation. Under corresponding conditions, below the temperature of phase transition, these phenomena were observed in helium-4 and different classes of superconductors. In this sense, we can think that the superconductivity of materials is often called simply as the superfluidity of Fermi gas.In the simplest form, the origin of supefluidity can be seen from the weakly interacting bosons model.

As it is known, in 1938, Pyotr Kapitsa, John Allen and Don Misener discovered that helium-4 became a new kind of fluid, which is just the superfluid we now said, at temperatures less than 2.17 K (the lambda point).

Helium is a common inert gas. It is also the most difficult gas to liquefy. There are two isotopes of helium, ^{4}He and ^{3}He, with the former being the majority in a normal helium gas. The boiling temperatures of ^{4}He and ^{3}He are 4.2 K and 3.19 K, respectively. The critical pressure for ^{4}He is 1.15 atm. Because of their light mass, both ^{4}He and ^{3}He have extremely high zero-point energies, and remain in gaseous form from room temperature down to a temperature near the absolute zero. Helium becomes solid due to cohesive force only when the interatomic distance becomes sufficiently small under high pressure. For example, a pressure of $25 - 34$ atm is required in order to solidify ^{3}He. For ^{4}He, when it is crystallized at a temperature below 4 K, it neither absorbs nor releases heat, i.e. the entropies of the crystalline and liquid phases are the same and only its volume is changed in the crystallization process. However, ^{3}He absorbs heat when it is crystallized at a temperature T <0.319 K under pressure. In other words, the temperature of ^{3}He rises during crystallization under pressure. Such an endothermic crystallization process is called the Pomeranchuk effect. This indicates that the entropy of liquid ^{3}He is smaller than that of its crystalline phase. That is, the liquid phase represents a more ordered state [17-18,39-41]. These peculiar characteristics are due to the unique internal structures of ^{4}He and ^{3}He. Both ^{4}He and ^{3}He can crystallize in a body-centered cubic or hexagonal close-stacked structure. A phase transition for ^{4}He will occurs at the pressure of 1 atm and the temperature of 2.17 K. Above this temperature, ^{4}He has no difference from a normal liquid and this liquid phase is referred to as He I. However, when the temperature is below 2.17 K, the liquid phase, referred to as He II, is completely different from HeI and becomes a superfluid state. This superfluid state can pass through capillaries with a diameter of less than 10^{-6} cm, without experiencing any resistance,. The superfluid state has a low viscosity ($<10^{-11}$ P) and its fluid velocity is independent of the pressure difference over the capillary and its length [11-12]. If a test tube is inserted into liquid He II contained in a container, the level of liquid He II inside the test tube is the same as that in the container. If the test tube is pulled up, the He II inside the test tube would rise along the inner wall of the tube, climb over the mouth of the tube and then flow back to the container along the outer wall of the tube, until the liquid level inside the test tube reaches the same level as that in the container. On the other hand, if the test tube were lifted up above the container, the liquid in test tube would drip directly into the container until the tube becomes empty. Such a property is called the superfluidity of ^{4}He [41-44].

The superfluid helium has many unusual properties, including zero viscosity (the ability to flow without dissipating energy) mentioned above and the existence of quantized vortices. It was quickly believed that the superfluidity was due to partial Boson–Einstein condensation of the liquid. In fact, many of the properties of superfluid helium also appear in the gaseous Boson–Einstein condensates created by Cornell, Wieman, and Ketterle (see below). In the superfluid helium-4, the interactions between the atoms are relatively strong; then the original theory of Boson–Einstein condensation must be heavily modified in order to describe it. Boson–Einstein condensation remains, however, fundamental to the superfluid properties of helium-4. Note that helium-3, consisting of fermionsenters also a superfluid phase at low temperature, which can be explained by the formation of bosonic Cooper pairs of two atoms.

The first "pure" Boson–Einstein condensate was created by Eric Cornell, Carl Wieman, and co-workers at JILA on June 5, 1995. They did this by cooling a dilute vapor consisting of approximately two thousand rubidium-87 atoms to below 170 nK using a combination of laser cooling (a technique that won its inventors Steven Chu, Claude Cohen-Tannoudji, and William D. Phillips the 1997 Nobel Prize in Physics) and magnetic evaporative cooling. About four months later, an independent effort led by Wolfgang Ketterle at MIT created a condensate made of sodium-23. Ketterle's condensate had about a hundred times more atoms, allowing him to obtain several important results such as the observation of quantum mechanical interference between two different condensates. Cornell, Wieman and Ketterle won the 2001 Nobel Prize in Physics for their achievements [45]. A group led by Randall Hulet at Rice University announced the creation of a condensate of lithium atoms only one month following the JILA work [46]. Lithium has attractive interactions, which causes the condensate to be unstable and to collapse for all but a few atoms. Hulet and co-workers showed, in a subsequent experiment, that the condensate could be stabilized by the quantum pressure from trap confinement for up to about 1000 atoms.

The Boson–Einstein condensation(BEC) applies also to quasiparticles in solids. A magnon in an antiferromagnet carries spin 1 and thus obeys Boson–Einstein statistics. The density of magnons is controlled by an external magnetic field, which plays the role of the magnon chemical potential. This technique provides access to a wide range of boson densities from the limit of a dilute Boson gas to that of a strongly interacting Boson liquid. A magnetic ordering observed at the point of condensation is the analog of superfluidity. In 1999, Boson condensation of magnons was demonstrated in the antiferromagnet TlCuCl3. [39]. The condensation was observed at temperatures as large as 14 K. Such a high transition temperature (relative to that of atomic gases) is due to the greater density achievable with magnons and the smaller mass (roughly equal to the mass of an electron). In 2006, condensation of magnons in ferromagnets was even shown at room temperature, [40-41] where the authors used pumping techniques.

The velocity-distribution data graph in the BEC was also researched. This result indicates the formation of a Boson–Einstein condensate out of a gas of rubidium atoms [42].

6.5.2. The Quantum Vortices in the Boson-EinsteinCondensation States

Plenty of experiments showed that the vortices could exist in BECs. These can be created, for example, by 'stirring' the condensate with lasers, or by rotating the confining trap. The vortex created is a quantum vortex. The phenomenon can be explained by the nonlinear term$\left|\phi(\vec{r},t)\right|^2$ in Eq.(6.4).As the vortices must have quantized angular momentum then the wavefunction may have the form $\phi(\vec{r})=\phi'(\rho,z)\exp[il\theta]$, where ρ, z and θ are as in the cylindrical coordinate system, and ℓ is the angular number. This is particularly likely for an axially symmetric (for instance, harmonic) confining potential, which is commonly used. To determine $\phi(\rho,z)$,the energy of $\phi(\vec{r})$ must be minimized, according to the constraint $\phi(\vec{r})=\phi'(\rho,z)\exp[il\theta]$. This is done computationally, but in a uniform medium, the analytic form $\phi'(X)=nX/\sqrt{2+X^2}$, where n^2 is the density far from the vortex, $X=\rho/l\xi$, ξ is healing length of the condensate, demonstrates the correct behavior, and is a good approximation. A singly charged vortex ($\ell=1$) is in the ground state, with its energy ε_v given by $\varepsilon_v=(n\pi\hbar^2/m)Ln(1.464b'/\xi)$, where b' is the farthest distance from the vortex considered.To obtain an energy, which is well defined, it is necessary to include this boundary b' .For multiply charged vortices ($\ell>1$) the energy is approximated by $\varepsilon_v=(l^2n\pi\hbar^2/m)Ln(b/\xi)$,which is greater than that of ℓ singly charged vortices, indicating that these multiply charged vortices are unstable to decay. The results indicated that they are metastable states, so they have relatively long lifetimes.

The experiments show also that the dark solitonscan occur in one-dimensional BECs, which accompany the creation of vortices in BECs. These topological objects mark the existence of a phase gradient across their nodal plane, which stabilizes their shape even in propagation and interaction. Although solitons carry no charge and are thus prone to decay, relatively long-lived dark solitonshave been produced and studied extensively [43].

6.5.3. The Properties of BEC with the Attractive Interactions

Randall Hulet's experiments at Rice University in 1995 through 2000 showed that lithium atoms condensates with attractive interactions could stably exist, but only up to a certain critical atom number. Beyond this critical number, the attraction interaction overwhelmed the zero-point energy of the harmonic confining potential, causing the condensate to collapse in a burst reminiscent of a supernova explosion where an explosion is preceded by an implosion. By quench cooling the gas of lithium atoms, they observed the condensate to first grow, and subsequently collapse when the critical number was exceeded.

The JILA team consisting of Cornell, Wieman and coworkers performed in 2000 further experimentations on attractive condensates. They originally used rubidium-87, an isotope whose atoms naturally repel each other, making a more stable condensate. Their instrumentation had better control over the condensate so the experiment was made on naturally attracting atoms of another rubidium isotope, rubidium-85 [47], which has a negative atom–atom scattering length. Through a process called Feshbach resonance involving a sweep of the magnetic field causing spin flip collisions, they lowered the characteristic, discrete energies at which the rubidium atoms bond into molecules, making their Rb-85 atoms repulsive and creating a stable condensate. The reversible flip from attraction to repulsion stems from quantum interference among condensate atoms.

When the magnetic field strength was raised still further in this experiment, the condensate suddenly reverted to attraction, imploded and shrank beyond detection, and then exploded, expelling off about two-thirds of its 10,000 or so atoms. About half of the atoms in the condensate seemed to have disappeared from the experiment altogether, not being seen either in the cold remnant or the expanding gas cloud. [45]. Carl Wieman explained that under current atomic theory this characteristic of Boson–Einstein condensate could not be explained because the energy state of an atom near absolute zero should not be enough to cause an implosion. But,mean field theories have been subsequentlyused to explain it. In this case, the atoms that seem to have disappeared almost certainly still exist in some form, just not in a form that could be accounted for in that experiment. Most likely, they formed molecules consisting of two bonded rubidium atoms [44]. The energy gained by making this transition imparts a velocity sufficient for them to leave the trap without being detected.

6.5.4. The Features of Isotopic Effect in the BEC

Theisotopic effect was observed on alkaline atoms, which have nuclear properties particularly suitable for working with traps. In 2012, using ultra-low temperatures of 10^{-7} K or more below, thus the Boson–Einstein condensates were obtained fromplenty of isotopes, such as alkaline, alkaline earth, and lanthanoid atoms (^{7}Li, ^{23}Na, ^{39}K, ^{41}K, ^{85}Rb, ^{87}Rb, ^{133}Cs, ^{52}Cr, ^{40}Ca, ^{84}Sr, ^{86}Sr, ^{88}Sr, ^{174}Yb, ^{164}Dy, and ^{168}Er). The properties of the Boson-condensation research have been successfully performed even with hydrogen with the aid of special methods. However, the superfluid state of ^{4}He at temperatures below 2.17 K is not a good example of Boson–Einstein condensation, because the interaction between the ^{4}He bosons is too strong. Only 8% of the atoms are in the single-particle ground state near zero temperature, rather than the 100% expected of a true Boson–Einstein condensate[44-47].

As it is known, the above bosonic, rather than fermionic, behavior of some of these alkaline gases appears odd at first sight, because their nuclei have only half-integer total spin. The bosonic behavior arises from a subtle interplay of electronic and nuclear spins: at ultra-low temperatures and corresponding excitation energies, the half-integer total spin of the electronic shell and the half-integer total spin of the nucleus of the atom are coupled by a very weak hyperfine interaction. The total spin of the atom, arising from this coupling, is an integer value leading to the bosonic ultra-low temperature behavior of the atom. The chemistry of the systems at room temperature is determined by the electronic properties, which is essentially fermionic, since at room temperature, thermal excitations have typical energies much higher than the hyperfine values[44-47].

Meanwhile,the related experiments in cooling fermions rather than bosons to extremely low temperatures have created degenerate gases, where the atoms do not congregate in a single state due to the Pauli exclusion principle. But, to exhibit Boson–Einstein condensation, the fermions must "pair up" to form compound particles with Boson feature (e.g. molecules or

Cooper pairs). Thus, a Boson-Einstein condensation occurs possibly in these matters in a specific case. The first molecular Boson–Einstein condensates were observed in November 2003 by the groups of Rudolf Grimm at the University of Innsbruck, Deborah S. Jin at the University of Colorado at Boulder and Wolfgang Ketterle at MIT. Jin quickly went on to create the first fermionic condensate composed of Cooper pairs [30-33].

6.6. The Results of Gross–PitaevskiiTheory of Boson-EinsteinCondensation

Gross and Pitaevskii [13-15] discussed the ground state of a quantum system of identical bosons using the Hartree–Fock approximation and the pseudopotential interaction model.In the Hartree–Fock approximation they represented the total wave-function Φ of the system composed of N bosons using a product of single-particle functions ϕ ,which[13-15,48-50] is

$$\Phi(r_1, r_2, r_3 \ldots r_n) = \phi(r_1)\phi(r_2)\phi(r_3)\ldots\phi(r_n) \tag{6.52}$$

where r_i is the position of the i-th boson.

In this case, they represented the Hamiltonian operator of the system in the pseudopotential model, which is

$$H = \sum_{i=1}^{N}\left(-\frac{\hbar^2}{2m}\frac{\partial^2}{\partial r_i^2} + V(r_i)\right) + \sum_{i<j}\frac{4\pi\hbar^2 a_s}{m}\delta(r_i - r_j) \tag{6.53}$$

where m is the mass of the boson, V is the external potential, a_s is the boson-boson scattering length, and ϕ is the Dirac delta-function.

If the single-particle wave-function satisfies the Gross–Pitaevski equation,

$$\mu\frac{\partial\phi(r)}{\partial t} = -\frac{\hbar^2}{2m}\frac{\partial^2}{\partial r^2}\phi(r) + \frac{4\pi\hbar^2 a_s}{m}|\phi(r)|^2\phi(r) + V(r)\phi(r) \tag{6.54}$$

the total wave-function minimizes the expectation value of the model Hamiltonian under normalization condition $\int dV|\phi(r)|^2 = N$.

It is a model equation for the wavefunction of one particle in a Boson–Einstein condensate. It is similar in form to the Ginzburg–Landau equation orthe nonlinear Schrödinger equation.

Gross and Pitaevskii [13-15] thought that the Boson–Einstein condensate is a gas of bosons that are in the same quantum state, and thus can be described by the same wavefunction. A free quantum particle is described by a single-particle Schrödinger equation. Interaction between particles in a real gas is taken into account by a pertinent many-body Schrödinger equation. If the average spacing between the particles in a gas is greater than the scattering length in the so-called dilute limit, then one can approximate the true interaction potential that is featured in this equation by a pseudopotential. The nonlinearity of the Gross–Pitaevskii equation has its origin in the interaction between the particles. This is made evident by setting the coupling constant of interaction in the Gross–Pitaevskii equation to zero, thereby, the single-particle Schrödinger equation, describing a particle inside a trapping potential, is recovered.

The equation has the form of the Schrödinger equation with the addition of a nonlinearinteraction term. The nonlinear coupling constant, $g = 4\pi\hbar^2 a_s / m$, is proportional to the scattering length a_s of two interacting bosons, where $\hbar$ is the reduced Planck's constant and m is the mass of the boson[13-15,48-50]. The energy density is

$$\varepsilon = \frac{\hbar^2}{2m}|\nabla\phi|^2 + V(r)|\phi|^2 - \frac{1}{2}g|\phi|^4 \tag{6.55}$$

where ϕ is the wavefunction, or order parameter, and V is an external potential. The time-independent Gross–Pitaevskii equation, for a conserved number of particles, is

$$\mu\phi(r) = -\frac{\hbar^2}{2m}\nabla^2\phi(r) + g|\phi(r)|^2\phi(r) + V(r)\phi(r) \tag{6.56}$$

where μ is the chemical potential. The chemical potential is found from the condition that the number of particles is related to the wavefunction by

$$\int d^3r|\phi(r)|^2 = N$$

From the time-independent Gross–Pitaevskii equation, we can find the structure of a BEC in various external potentials (e.g. a harmonic trap).

The time-dependent Gross–Pitaevskii equation can also be represented by

$$i\hbar\frac{\partial}{\partial t}\phi(r) = -\frac{\hbar^2}{2m}\nabla^2\phi(r) + g|\phi(r)|^2\phi(r) + V(r)\phi(r) \tag{6.57}$$

From the time-dependent Gross–Pitaevskii equation we can look at the dynamics of the BEC or Boson particles. It is used to find the collective modes of a trapped gas.

Since the Gross–Pitaevskii equation is a nonlinear, partial differential equation; exact solutions are hard to come by. As a result, solutions must be approximated via myriad techniques.

The simplestsolution is the free particle solution[13-15,48-50], with, $V(r)=0$ and $\phi(r) = (N/V)^{1/2}\exp[i\vec{k}.\vec{r}]$. This solution is often called the Hartree solution. Although it satisfies the Gross–Pitaevskii equation in Eq.(6.57), it leaves a gap in the energy spectrum due to the interaction:

$$E(k) = N[\hbar^2 k^2 / 2m + gN / 2V] .$$

According to the Hugenholtz–Pines theorem, an interacting Boson gas does not exhibit an energy gap (in the case of repulsive interactions).

A one-dimensional soliton can form in a Boson–Einstein condensate, and depending upon whether the interaction is attractive or repulsive, there is either a bright or dark soliton. Both solitons are local disturbances in a condensate with a uniform background density

If the BEC is repulsive, so that $g > 0$, then a possible solution of the Gross–Pitaevskii equation (6.57) is,

$$\phi(x) = \phi_0 \tanh(x / \sqrt{2}\xi) ,$$

where ϕ_0 is the value of the condensate wavefunction at ∞, and $\xi = \hbar / \sqrt{2mn_0 g}$ is the coherence length. This solution represents the dark soliton, since there is a deficit of condensate in a space of nonzero density. The dark soliton is also a type of topological defect, since ϕ flips between positive and negative values across the origin, corresponding to a π phase shift[13-15,48-50].

For $g < 0$, the solution of Eq.(6.57) is

$$\phi(x,t) = \phi(0) \exp[-i\mu t / \hbar] / \cosh(\sqrt{2m|\mu| / \hbar^2}\, x) \tag{6.58}$$

where the chemical potential is $\mu = g|\phi(0)|^2 / 2$ .This solution represents the bright soliton, since there is a concentration of condensate in a space of zero density.

It is possible to make some variational approximation for finding the solutions of Eq.(6.57). The basic idea is to make a variational ansatz for the wavefunction with free parameters, plug it into the free energy, and minimize the energy with respect to the free parameters. In the Thomas–Fermi approximation, if the number of particles in a gas is very large, the interatomic interaction becomes large so that the kinetic energy term can be neglected from the Gross–Pitaevskii equation (6.57). Thus, its solution is the form of

$$\phi(x,t) = \sqrt{(\mu - V(x)) / nU_0} .$$

In the Bogoliubov approximation, which treated the solution of Gross–Pitaevskii equation(6.57) using a method that finds the elementary excitations of a Boson–Einstein condensate, the condensate wavefunction is approximated by a sum of the equilibrium wavefunction $\phi_0 = \sqrt{n} \exp[-i\mu t]$ and a small perturbation $\delta\phi$ .The solution of Eq.(6.57) may be represented by

$$\phi = \phi_0 + \delta\phi$$

Then this form is inserted in the time dependent Gross–Pitaevskii equation and its complex conjugate, and linearized to the first order in $\delta\phi$

$$i\hbar\frac{\partial\delta\phi}{\partial t}=-\frac{\hbar^2}{2m}\nabla^2\delta\phi+V\delta\phi+g(2|\phi_0|^2\delta\phi+\phi^2\delta\phi^*)$$

$$i\hbar\frac{\partial\delta\phi^*}{\partial t}=-\frac{\hbar^2}{2m}\nabla^2\delta\phi^*+V\delta\phi^*+g(2|\phi_0|^2\delta\phi^*+\phi^2\delta\phi)$$

If assuming the following solution of $\delta\phi$

$$\delta\phi=\exp[-i\mu t][(u(r)\exp(-i\omega t)+v^*(r)\exp(i\omega t)]$$

one finds the following coupled differential equations for u and v

$$(-\frac{\hbar^2}{2m}\nabla^2+2gn+V-\hbar\omega-\mu)u-gnv=0$$

$$(-\frac{\hbar^2}{2m}\nabla^2+2gn+V-\hbar\omega-\mu)v-gnu=0$$

For a homogeneous system, i.e. for $V(r)=0$, one can assume u and $v$to be plane waves of momentumq, which leads to the energy spectrum

$$\hbar\omega=[\frac{\hbar^2q^2}{2m}(\frac{\hbar^2q^2}{2m}+2gn)]^{1/2}=\in_q$$

For large q, the dispersion relation is quadratic in q,as one would expect for usual non-interacting single particle excitations. For small q, the dispersion relation is linear $\in_q=s\hbar q$ with $s=\sqrt{ng/m}$ being the speed of sound in the condensate. The fact that $\in_q/(\hbar q)>s$ shows, according to Landau's criterion, that the condensate is a superfluid, meaning that if an object is moved in the condensate at a velocity inferior to s, it will not be energetically favorable to produce excitations and the object will move without dissipation, which is characteristic of a superfluid[13-15,48-50]. Experiments have been done to prove this superfluidity of the condensate, using a tightly focused blue-detuned laser. The same dispersion relation is found when the condensate is described from a microscopical approach using the formalism of second quantization.

6.7. The Nonlinear Quantum-Mechanical Features of BEC States

We study further the dynamic features of Boson particles in BEC statesusing nonlinear quantum mechanics. In fact, after the first experimental observation of this condensation phenomenon was carried out, it was suggested that the coherent dynamics of the condensed macroscopic wave function in Eqs.(6.6)-(6.7) or (6.54) and (6.56)could lead to the formation of nonlinear solitary waves. For example, self-localized bright, dark and vortex solitons, formed by increased (bright) or decreased (dark or vortex) probability density respectively, were experimentally observed, particularly for the vortex solution. These experimental results were in concordance with the results of the above theory. In the section, we will study the soliton motions of quasiparticles in the Boson condensation systems, such as the superfluid system. We will see that the dynamic equation (6.6) in the systems have soliton solutions with nonlinear quantum mechanical properties, where the superfluid helium [11-12,39-40]can serve as a real sample of a Boson condensation system to study the soliton motions of quasiparticles in it.

6.7.1. The Properties of BosonParticles in BEC System in a Non-Relativistic Case and Experimental Evidences

Studies show that the superfluid state of the*He* atom below the critical temperature is a result of BEC, which is very similar to the superconducting state. However, $\phi(\vec{r},t)$ in Eq.(6.4) is also referred to as a macroscopic wave functionor order parameter in the superfluid heliumstate or the effective wave function of helium atoms.

It is known that the effective wave function of Boson particle, $\phi(\vec{r},t)$, satisfies the Gross-Pitaerskii (GP) equation (6.54) [13-16] as mentioned above, which is now represented by

$$i\hbar\frac{\partial\phi}{\partial t}=-\frac{\hbar^2}{2m}\nabla^2\phi+\lambda|\phi|^2\phi-\mu'\phi \tag{6.59}$$

where $\lambda=g$ and $\mu'=V(r)$ are constants. Equation (6.59) is similar to Eq.(6.4) and was extensively used by Ventura etal. [51], in their studies of superfluidity. A similar equation was derived by Dewitt, in 1966 [52]. According to Dewitt [52] λ in Eq. (6.59) should be a negative value. Equation (6.59) is also a typical nonlinear Schrödinger equation, which is similar to Eq.(2.66) in nonlinear quantum mechanics. Therefore, we can conclude that the motion of helium atoms in superfluid liquid helium is a nonlinear quantum mechanical problem, and its rules of motion are same as those in Chapters 3-4.

The Lagrangian function density of the system corresponding to Eq.(6.59) can be obtained and is given by

$$L' = \frac{i\hbar}{2}(\phi\frac{\partial\phi^*}{\partial t} - \phi^*\frac{\partial\phi}{\partial t}) - \frac{\hbar^2}{2m}|\nabla\phi|^2 - \frac{\lambda}{2}|\phi|^4 + \mu'|\phi|^2 \quad (6.60)$$

where L'=$\mathcal{L}$. In a one-dimensional case if we let $t' = t/\hbar$ and $x' = x\sqrt{(2m/\hbar^2)}$, then Eq. (6.59) at $V(x) = \mu'$ becomes similar to Eq. (2.66), thus its solution can be obtained following the same procedure as the one used in Chapter 3. Its solution can be directly obtained from Eq.(3.25) at $\alpha = 0$ and $\mu' = c$ [34-38,56-57], its result is now represented by

$$\phi = \sqrt{\frac{2\beta}{|\lambda|}}\,\mathrm{sech}\{\sqrt{\beta}[x' - \upsilon_e(t' - t_0')]\}\exp\{i[\upsilon_e x'/2 - (\beta + \upsilon_e^2/4 - \mu')t']\} \quad (6.61)$$

where β is an arbitrary constant. Obviously, it differs from the solutions mentioned above.

Therefore, the atomic helium system undergoes a second-order phase transition and changes from the normal He-I state to the superfluid He-II state at 2.17 K.In this case, the helium atoms behave as solitons due to the spontaneous Boson condensation and as a result of nonlinear interaction, $\lambda|\phi|^2\phi$,in the system. The nonlinear interaction suppresses the dispersion effect of helium atoms. Because the solitons can preserve their energy momentum, wave form, and other properties as quasiparticles throughout their motion, thus the superfluidity of the liquid helium occurs naturally when the helium atoms move as solitons in such a case.

As a matter of fact, we can prove that the superfluid helium atom (soliton) moves with a uniform speed and can determine its speed furtherin this movement from the solutionin Eq.(6.61), the Hamilton equations (3.67) or (3.70)and classical motion-equation (3.59) or (3.60) in nonlinear quantum mechanics.

First, from (6.61) we can determine that the wave number of the helium-soliton is $k = \partial\theta/\partial x' = \upsilon_e/2$ and its frequency is $\omega = -\partial\theta/\partial t' = \beta + \upsilon_e^2/4 - \mu' = \beta - \mu' + k^2$. Thus the acceleration of the helium-soliton is [34-36]

$$\frac{dk}{dt'} = -\frac{\partial\omega}{\partial x'}\Big|_k = 0,\ \upsilon_g = \frac{dx'}{dt'} = \frac{\partial\omega}{\partial k} = 2k = v_e.$$

The group speed of the helium-soliton is a constant, υ_e, and the helium atoms move in the form of soliton with constant speed in the superfluid state. This is just a basic feature of superfluidity. Thus, we can use these results to explain the properties of superfluid helium and its BEC, mentioned above.

The effective mass of the helium-soliton can be determined from Eq. (6.61)

$$M = \int_{-\infty}^{\infty}|\phi|^2 dx = \frac{4}{\lambda h}\sqrt{m\beta} = \mathrm{const}$$

The effective energy of the helium-soliton is

$$E=\int_{-\infty}^{\infty}[\left|\frac{\partial\phi}{\partial t'}\right|^2+\frac{1}{2}\lambda|\phi|^4-\mu'|\phi|^2]dx'=(\frac{1}{2}\upsilon_e^2-\mu')M+\frac{4\sqrt{2m\beta}}{3\lambda h}(4+\beta) \quad (6.62)$$

Here the first term is the kinetic energy of the soliton, the second term is that of its binding energies, and the last term is the energy of interaction.

We now discuss properties of circulation (vortex line) produced by motion of the helium atoms in the superfluid state using the above result. This property is defined by the velocity of the superfluid helium atom

$$Q=\oint_r \upsilon_s dr . \quad (6.63)$$

In terms of the phase, θ (x,t), of the macroscopic wave function, the velocity of the superfluid can be written as

$$\upsilon_s=\frac{\hbar}{m}\nabla\theta . \quad (6.64)$$

Earlier we concluded from Eq. (6.61) that the velocity of the superfluid is equal to 2 times the group velocity of the soliton i.e., $\upsilon_s=2\upsilon_e$. This indicates that the motion of the helium-soliton is the motion of the superfluid and the vortex lines in the superfluid is a result of movement of helium-soliton in the superfluid state. The phase difference along a closed path is given by the line integral corresponding to the circulation of the velocity υ_s:

$$\Delta\theta=\oint_r \nabla\theta dr=\frac{m}{\hbar}\oint_r \upsilon_s dr \quad (6.65)$$

Thus, the circulation is related to the phase difference $\Delta\theta$ of the helium-soliton. If the path of integration lies in a multiple connected domain or it encloses a vortex line, then $\nabla\theta(r)\neq 0$. Furthermore, if $\phi(r)\neq 0$ and is single valued, we have $\Delta\theta=2\pi n$ and

$$\oint_r \upsilon_s dr=n\frac{h}{m}, n=1,2,3,\ldots\ldots \quad (6.66)$$

This indicates that the circulation of the whirl (or energy of the vortex lines) is quantized, where n is an integer. Further, equation (6.66) implies that whenever the velocity of the rotating superfluid helium exceeds a critical velocity, the vortex can be produced in the superfluid [57]. The circulation (the vortex) is quantized and is given a value by an integer multiple of h/m. Therefore, the nonlinear Gross-Pitaevskii equation indeed gives an adequate description of "quantum vortex" in the superfluid helium. Thus, the superfluid can be viewed as a Boson condensate with local interactions (Pitaevskii 1961 [14-16], Gross 1963) [13]. This is consistent with the experimental result. An experiment was done in 1963 to measure the energy of the vortex lines, and the results obtained were consistent with the theoretical prediction. The quantization of circulation was thus proved. Firbake and Maston in the U.S.A. observed the macroscopic quantum effect of ^{4}He II experimentally once again. When the superfluid ^{4}He II was set into rotational motion in a cup, a whirl would be formed when the temperature of the liquid was reduced to below the critical temperature. In this case, an effective viscosity develops between the fluid and the cup, which is very similar to normal fluid in a cup being stirred. The surface of the superfluid becomes inclined at a certain angle and the cross-section of the liquid surface is in a shape of a parabola, due to the combined effects of gravitation and centrifugal forces. Fluid away from the center has a tendency to converge towards its center, which is balanced by the centrifugal force, and a dynamic equilibrium is reached. The angular momentum of such a whirl is very small and consists of only a small number of discrete quantum packets. The angular momentum of the quantum packets can be obtained via quantum theory. In other words, the whirl can exist only in discrete form over a certain range in certain materials such as superfluid helium. Firbake and Maston managed to obtain sufficiently large angular momentum in their experiment, which are able to observe the whirl's surface shape using visible light. In this experiment, they used a thin layer of rotating superfluid helium. While the rotating superfluid was illuminated, from both top and bottom, by laser beams with wavelength of 6328 $\overset{0}{A}$ from a He-Ne laser, thus the whirls were observed. Alternating bright and dark interference fringes were formed when the reflected beams were focused on an observing screen. The analysis of the interference pattern showed that the surface was indeed inclined at an angle. Based on this result, the shape of the surface can be correctly constructed. The observed interference pattern was found in excellent agreement with those predicted by the theory. This experiment further confirmed the existence of quantized vortex rings in the superfluid ^{4}He[57].

The concept of "quantum vortex" was proposed first by Pitaevskii et al., [15, 16], but quantization of vortices in a superfluid was first suggested by Onsager (1949) [55] on the basis of classical vortex flows and turbulence. The superfluidity of ^{4}He and its quantum vortex lines and loops, weak turbulence and dissipative vortex dynamics in the superfluids were studied by Barenghi, et al., [56], Roberts et al., Pismeu and Rica, and many others, all using the nonlinear Schrödinger equation or the Gross-Pitaerskii equation given above (see Barenghi et al., [56], Donnely [57], Avenel et al., [58], Lindensmitth et al., [59]. Experimental observations of these vortices were reported early by Yamchuk [60], Packard [61], and Zieve et al. [62-63]. Numerical simulations were presented by Frish [64], Pomeau and Rica [65], and Schwarz [66-67]. The energy of the vortex lines was also measured.

If the superfluid liquid does not rotate, then $\nabla\times\upsilon_s=(h/m)\nabla\times\nabla\theta=0$, then the superfluid velocity field becomes a conservative field without rotation. This suggests that, when the superfluid liquid helium flows through a tube with a gradually decreasing diameter, the pressure inside the tube is the same everywhere and it does not depend on the diameter of the tube. This is completely different from that of a normal fluid, but it has been demonstrated experimentally. Moreover, the macroscopic wave function, $\phi(x,t)$ given in Eq.(6.61) approaches 0 when x approaches ∞. That is, $\phi(x,t)$ vanishes at the boundary. This implies that the superfluid density $\rho_s(\propto|\phi|^2)$ should also approach 0 at the boundary. The value of ρ_s was measured in 1970 and it was found that its value dropped from the value in the bulk to zero over a few atomic layers. This result provided a direct verification for the above theoretical data.

6.7.2. Properties of Motion of Particles in the Relativistic Case

The GP equation (6.61) is neither relativistic nor does it consider the gravitational field.Anandan[68] and others extended the theory to include the relativistic effect for the superfuid liquid. The generalized relativistic equation of motion for the quantum superfluid liquid helium is given by

$$\frac{\partial^2\phi}{\partial t^2}-\nabla^2\phi+\alpha^2\phi=-\lambda'|\phi|^2\phi \tag{6.67}$$

where $\alpha^2=\frac{m^2c^2}{\hbar^2}$, $\lambda'=\frac{2m\lambda}{\hbar^2}$.

Equation (6.67) is called the Gross-Pitaevskii-Anandan (GPA) equation. It is, in essence, a type of ϕ^4 -equation in Eq.(2.68), which describes the properties of microscopic particles in relative theory in nonlinear quantum mechanics. Anandan did not find its solutions. Instead, he gave an order of magnitude estimate of $\phi=\varphi e^{i\theta}$ for all types of solutions using the Einstein-Planck Law. As will be shown below, an exact solution of Eq. (6.67) is actually possible.

Let's assume the following trial solution [66-67]

$$\phi(x,y,z,t)=\varphi(Z)e^{i\theta} \tag{6.68}$$

where $Z=\vec{p}.\vec{r}-\Omega t$, $\theta=\vec{k}.\vec{r}-\omega t=k_1x+k_2y+k_3z-\omega t$.

Substituting Eq. (6.68) into Eq. (6.67),, the latter can be written as

$$(\Omega^2-p^2)\frac{d^2\varphi}{dZ^2}+(\alpha^2+k^2-\omega^2)\varphi+\lambda'\varphi^3=0 \tag{6.69}$$

in terms of $\vec{k}=(k_1,k_2,k_3)$ and $\vec{p}=(p_1,p_2,p_3)$, and $\omega\Omega=\vec{k}.\vec{p}$. In the integrate Eq. (6.69), we finally obtain the solution of Eq. (6.69) as follows [69-70]

$$\phi(x,y,z,t)=f(Z)e^{i\theta}=\sqrt{\frac{w}{R}}\operatorname{sech}[\sqrt{w}(\vec{p}.\vec{r}-\Omega t)]e^{i(\vec{k}.\vec{r}-\omega t)} \tag{6.70}$$

where

$$w = \frac{\omega^2 - \alpha^2 - k^2}{\Omega^2 - p^2}, R = \frac{\lambda^{'}}{2(\Omega^2 - p^2)} \tag{6.71}$$

This is a soliton solution of wave packet type, and its group velocity is υ.

From the above study, we see that the time-dependent motion of superfluid helium atoms still behave like a soliton. Thus, we can conclude that the superfluid helium atoms are, in essence, a soliton. Then we concluded that the soliton motion of the superfluid helium atoms causes the superfluidity. Because the solitons can move over macroscopic distances retaining its amplitude, velocity, and energy, thus the motion of the liquid helium necessarily results in superfluidity. In such a case, the motions of the superfluid helium atoms are described in a nonlinear Schrödinger equation (6.59). According to the soliton theory the superfluid helium atoms are localized and have a wave-corpuscle duality due to the nonlinear interaction, which suppresses the dispersive effect of the kinetic energy in Eq. (6.59). Obviously, the nonlinear interaction is due to the self-condensed mechanism, the helium atoms, as mentioned in Chapter 4. Therefore, this mechanism is carried out truly in the superfluid, thus the correctness of nonlinear quantum mechanics is again confirmed.

6.8. DIFFERENT KINDS OF BOSON-EINSTEIN CONDENSATION AND THEIR PROPERTIES

6.8.1. Dynamics of Boson-Einstein Condensates with Long-Range Attractive Interactions

The laboratory observation of Boson-Einstein condensates created from cold atoms [68] openednew opportunities for theoretical investigation of those fascinating systems. The old Gross-Pitaevskii equation [13-15,72], is probably the most beautiful example of the density functional theory [73, 74].It was rediscovered to provide information both about the condensate equilibriumdensity and dynamical properties[75, 76].

As it is known, new results have recently emerged fromnovel experimental techniques for condensed matter experiments. For example, in Figure 6.4, the classical picture of the Boson-Einstein condensate in the trap was shown, in which all atoms try to occupy the same quantum state, which was obtained in the mean field. The elementary excitations[77], the wave matter lasingeffect [78, 79], as well as solid-state modeling and control in optical lattices [80, 81] were observed. Meanwhile, the combination of a few laser beamswas used to generate attractive long-range interactions within the Boson-Einstein condensate due to the interplay between dipole-dipole forces[82], which are equivalent to thegravity forces with an enormously large modeled gravitational constant. The quantum phasetransition of the density modulation [83]by the dipole-dipole optical interaction was predicted from the variational calculations[84]. The question arose whether it is the artificial effect ofthe variational methods. Matt Kalinski, et al., solvedalso the time-dependent Gross-Pitaevskii equation [13-15] for the Boson-Einstein condensate withnon-local dipole-dipole interaction potential, as well as with the attractive gravity-like potentialnumerically. They observed formation of supersolid structure above the critical intensity in harmonic traps, in which sample linear combination of Gaussian orbitals(LCGO) theory and self-bound structures for the condensate with gravity-like potentials were proposed.In the following, we exhibited their results of the existence of supersolid ground states by means of the direct integration of the Gross-Pitaevskii (G-P) equation, using the imaginary time integration method[73]inspatial dimensions, and without any prior assumption on the symmetry of the condensate. Non-dispersing structures of the condensate are also observed numerically withoutthe external trap for soft-core model potentials.

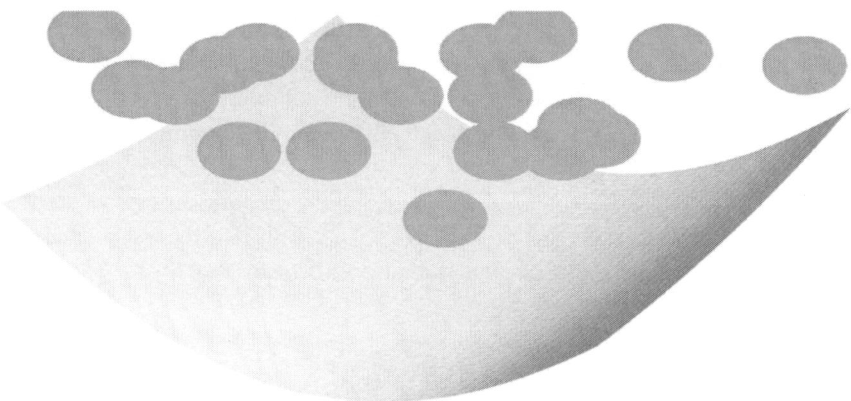

Figure 6.4. Classical picture of the Boson-Einstein condensate in the trap. All atoms try to occupy the same quantumstate described by the mean field [85].

Matt Kalinski et al.[85]startedtheir studies from the time dependent G-P equation[82,83, 86,87] for the BEC with non-local inter-particle interaction. Inthe second quantization representation,the Hamiltonian of system they used is represented by

$$\hat{H}=\int\Phi^+(x)[\nabla^2/2-V(x)]\Phi(x)+\alpha\int\Phi^+(x)\Phi^+(x)\Phi(x)\Phi(x)+g\int\Phi^+(x)\Phi^+(x)V_{dd}(x-y)\Phi(x)\Phi(x) \tag{6.72}$$

where $\Phi^+(x)$ and $\Phi(x)$ are the bosonic field operators obeying the standard commutation relations $[\Phi(x),\Phi^+(y)]=\delta(x-y)$ inspace. One can obtain the mean-field equation by neglecting the following correlations $<\Phi^+(x)\Phi^+(x)\Phi(x)\Phi(x)>=\rho(x)^2$ with the density of the condensate $\rho(x)=<\Phi^+(x)\Phi(x)>=<\psi^+(x)\psi(x)>$.

In this study, they used the dipole-dipole induced potential, generated by a circularly polarized externalfield, which is denoted by

$$V_{dd}=g[\frac{2z^2-x^2-y^2}{r^5}(\cos(r)+r\sin(r))-\frac{2z^2+x^2+y^2}{r^3}\cos(r)]\cos(z)$$

They justifiedthat the time-dependent G-P equation can be obtained within the density functional theory [74]from the variational principle

$$\frac{\delta < H - id/dt >}{\delta \rho} = 0 .$$

This G-P equation in this case is represented by

$$-[\Delta/2 + V(X) + \alpha|\phi(x,t)|^2]\phi(x,t) + g\int \rho(y,t)V_{dd}(x-y)\phi(x,t) = i\,{}^{d\phi(x,t)}\!\big/_{dt} \tag{6.73}$$

Matt Kalinski, et al., [85],used the above equations to discuss the supersolid formation, in which the potentialV_{dd} in Eq.(6.73) is quasi-periodic as shown in Figure 6.4. They thought that the solid formation for the minima, according to the modified Bloch theory. Thus, they expressed the periodic part of the Bloch function, which now becomes quasi-periodic

$$\phi_k(x) = e^{ikx}u(x), u(x) = u(x + \pi n) \tag{6.74}$$

They assumed that the Bloch function is alinear combination of Gaussian orbitals (LCGO) with originally unspecified widths, i.e.,

$$\phi_0(x) = \sum e^{ikz_n}\psi_n(x - z_n) \tag{6.75}$$

But,present Gaussian orbitals are different from the classic LCAO method due to predicted deviation from the perfect translationalsymmetry

$$\psi_n(x) = C_n \exp[-a_n^2(x^2 + y^2)/2]\exp[-b_n^2(z - z_n)^2/2] \tag{6.76}$$

They made the zero nearest-neighbors approximation for the condensate density by keeping only the self-overlap

$$\rho(x) = \sum |\psi_n(x - z_n)|^2 \tag{6.77}$$

Therefore, they obtained the equation:

$$-[\Delta/2 + V(x) + \alpha\rho(x)]\psi_i(x - z_i) + \sum \int V_{dd}(x - y)|\psi_n(x - z_n)|\psi_i(x - z_i) = E\psi_i(x - z_i) \tag{6.78}$$

They again assumed the harmonic trap potential, which is prolate and cylindrically symmetric

$$V(x)=\frac{1}{2}[\omega_r^2(x^2+y^2)+\omega_z z^2$$

and

$$\int V_{dd}(x-y)|\psi_n(x-z_n)|^2\,\psi_n(x-z_i)=[\frac{1}{2}\omega_{nri}^2(x^2+y^2)+\frac{1}{2}\omega_{nri}(z-z_i)^2]\psi_n(x-z_i) \tag{6.79}$$

where

$$\omega_{nri}^2=\left.\frac{d^2\int V_{dd}(x-y)|\psi_n(y-z_i)|^2}{dx^2}\right|_{00z_i},\ \omega_{nzi}^2=\left.\frac{d^2\int V_{dd}(x-y)|\psi_n(y-z_i)|^2}{dz^2}\right|_{00z_i} \tag{6.80}$$

and assumedthe hard self-interaction

$$|\psi_n(y-z_i)|^2=-C_n a_n^2(x^2+y^2)-C_n b_n^2 z_i^2 \tag{6.81}$$

which closes the system of equation forz_i anda_n,b_z.

When the frequencies generatedby the self-interaction are much larger, the longitudinal frequency of the harmonic trap, namely

$$\Omega_{nzi}^2=\omega_{nri}^2-\alpha C_n b_n^2>\omega_z^2 \tag{6.82}$$

and the period of the solid satisfied $\Omega_{nzi}^{-1/2}<\pi$ as well as $|\psi_n(x-z_i)|^2=\delta(x-z_i)$, hold they are obtained from summation of the potential,z_0=0.0000, $z_{\pm1}$=±3.1358, $z_{\pm2}$=±6.2701,$z_{\pm3}$ =±9.4001, $z_{\pm4}$ =±12.5173, and $z_{\pm5}$ =15.5874[85], wherethe soft corepotential used in the numerical simulationfora solid, consisting of 11 minima and the initial perfectly periodicz_i =iπ solid. Thus, they concluded that the solid is self-sustained when the density distribution generates componentsof the interaction potential shifted by the lattice constant of the solid.

Meanwhile, they also studied the self-binding of the BEC using the above formulae, in which they considered a more general form of the dipole-dipole interaction potential between atoms ofthe condensate state induced by the external laser field with the intensity I, which is denoted by

$$IU(r)=\frac{I}{4\pi c}\in_0^{-2}\alpha^2(q)e_i e_i^* V_{ij}(\vec{q}.\vec{r})\cos(\vec{q}.\vec{r}) \tag{6.83}$$

where $V_{ij}(\vec{q}.\vec{r})=\frac{1}{r^3}[(\delta_{ij}-3r_i r_j)(\cos(qr)+qr\sin(qr)]-(\delta_{ij}-r_i r_j)(qr)^2\cos(qr)$.

The existence of non-dispersing, self-bound structures of the condensate density canbe obtained from two simple approximations, i.e., one may use the sharp density approximationfor the purpose of self-interaction integral in the stationary G-P equation:

$$\{-\Delta/2 + U(x) + \alpha|\phi(x)|^2 \phi(x) + \int \rho(y) V_{dd}(x\_y)\phi(xt) = E\phi(x) \quad (6.84)$$

where $V_{dd}(x - y) = -u/|x - y|$ and assume

$$|\phi(x)|^2 = \delta(x) \quad (6.85).$$

Then equation (6.84) (were assume $\alpha = 0$ for simplicity) has a simple solutionbecause it describes the stationary atom of hydrogen, which is expressed by

$$\phi(x) = c\exp[-u(x^2 + y^2 + z^2)^{1/2}] \quad (6.86)$$

This is a self-bound non-dispersing condensate density with an exponential profile.

On the other hand, they demonstrated that there is a Gaussian condensate profile. However, the true ground state, which is expected to be between the Gaussian and exponential, can be obtained and confirmed numerically.

They have further solved the Gross-Pitaevskii equationusing the imaginary integration time method[88] to findits numerical solution, which they assumed is the existence of the instantaneous spectrumof the density dependent Hamiltonian H at the certain time $t = i\tau$, which is

$$H(\rho, t)\varphi_i = \in_i \varphi_i \quad (6.87)$$

In the series representing the infinitesimal solution is

$$\varphi(0 + \delta t) = C_0\varphi_0 \exp[\in_0 \delta\tau] + ... + C_n\varphi_n \exp[\in_n \delta\tau] \quad (6.88)$$

But only the term with the lowest instantaneous energy $\in_0$ is magnified and the procedure leadsto finding the ground state of the system. For the Gross-Pitaevskii propagator, they used the second order split operator method [89,90]. In this case, the nonlinear propagator was approximated by

$$\exp[-i\int H(t)dt] \approx \exp[-iV(\rho)dt/2]\exp[-iTdt]\exp[-iV(\rho)dt/2] \quad (6.89)$$

here

$$V(\rho) = \omega_r^2 (x^2 + y^2)/2 + \omega_z^2 z^2 / 2 + \alpha\rho(xt) + \gamma \int V_{dd}(x-y)\rho(yt) \tag{6.90}$$

In this researchthey used the soft-core dipole-dipole potential in Eq.(6.80) with its Coulomb version $1/\sqrt{x^2+1}$ in the one dimensional strong field [91], which allowsthe observation of supersolids without the hard core repulsion($\alpha = 0$), where the initial state was chosen to be a Gaussian ground state of the harmonic oscillator of the atom trap.

$$\psi(xo) = C\exp[-\omega_r(x^2+y^2)/2 + \omega_z z^2/2], \tag{6.91}$$

to cover many wavelengths of the self-interaction potential. Figure 6.5 shows the volumetric plot of the condensate density for the value of the coupling content γ, which is sufficient to create the supersolid. This figure showed a clear density modulation in the longitudinal direction of the condensate. The separation of peaks is approximately by π, in units of qz, as predicted by simple theory. The condensate density obtained from the simulations with gravity-like potentialis shown in Figure 6.6, which indicated clearly that a self-bound and stable structure is formed during the imaginary-time evolution.

This result manifested thatthe mean-field Gross-Pitaevskii theory is feasible for the numerical treatment of the BEC statein three dimensions. Using this technique, they observed the transition tosupersoliddensity modulation in the Boson-Einstein condensate using the imaginary time split-operateor method. The simulations were free from the variational constraints. Meanwhile, they have also observed the self-binding of the condensate with modeled gravitational attraction.

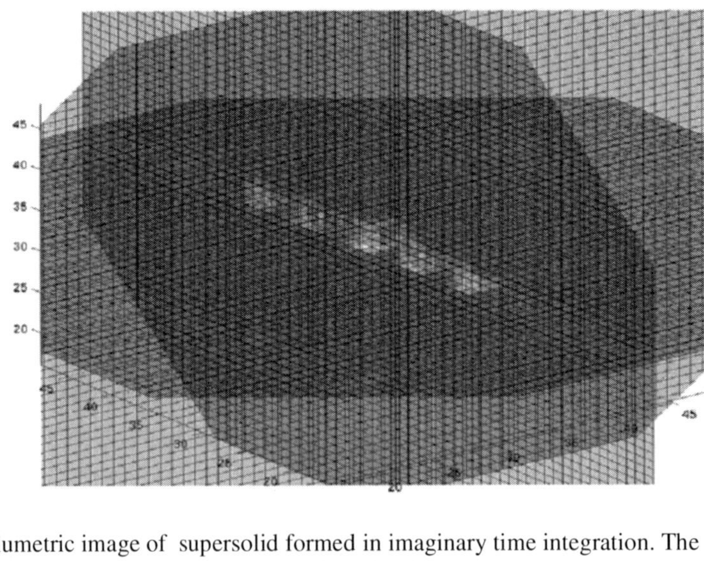

Figure 6.5. Volumetric image of supersolid formed in imaginary time integration. The peaks of the condensate density are separated approximately by π [85].

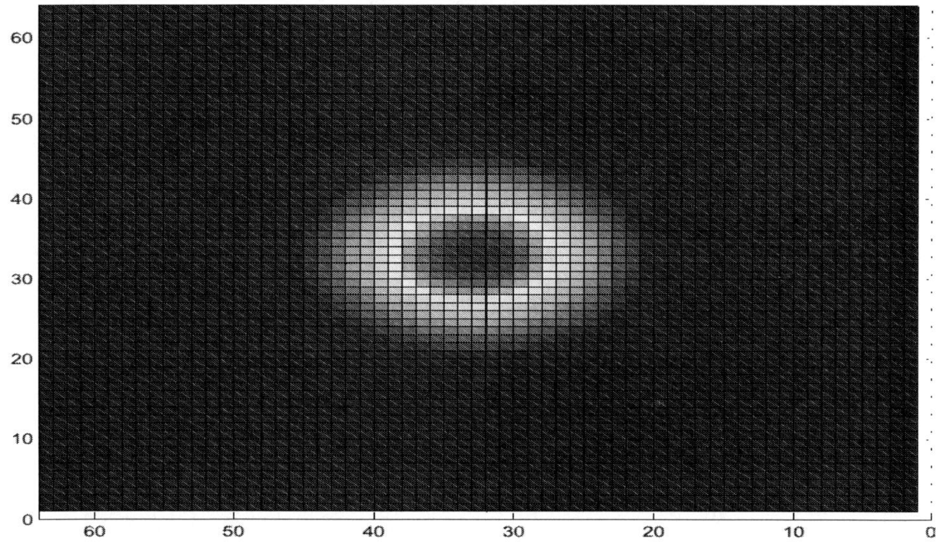

Figure 6.6. Volumetric image of self-bound condensate formed in imaginary time integration at u=1.0. The distribution will stably oscillate if perturbed from the equilibrium [85].

6.8.2. The Properties of the Boson–Einstein Condensate with a Parabolic and Other Traps

As mentioned above, the different types of Boson–Einstein condensation in atomic clouds were experimentally observed [92], whichbuilt the foundation of a new direction in the study of macroscopic quantum phenomena, and can be used to verify the validity of nonlinear quantum mechanics. The dynamics of gases of cooled atoms confined in a magnetic trap at very low temperatures was studied using the Gross–Pitaevskii(GP) equation [13-16, 93], in which the effects of the particle interaction through an effective mean field is considered. Kivsher et al.[94]investigated the properties of the Boson–Einstein condensate in a parabolic trap asa macroscopic quantum oscillator. Meanwhile, the similar models of the confined dynamics of macroscopic quantum systems also occurred in the case of an electron gas confined in a quantumwell [95]. Optical modes in a photonic microcavity were used already[96]. In these systems, confined single-particlestates are restricted to a set of discrete energies thatform a set of eigenmodes, likely the harmonic quantum oscillator with equally spaced energy levels [97]. Kivsher et al., [97]also discussed the properties of the quasi-classical system of interacting bosons in amacroscopic ground state confined in an external potential, in which a standard application of the mean-field theory allows introduction of a macroscopic wave function as a classical field $\phi(r,t)$, having the meaning of the order parameter. They thought that the function $\phi(r,t)$ includes the effect of interparticle interaction, which was considered as a mean-field nonlinear term. Thus, they analyzed nonlinear eigenmodes of a macroscopic quantum oscillator as a set of nonlinear stationary states that extend the Hermite-Gauss eigenfunctions, and made a link between different approximations, where the Thomas-Fermi approximation and the perturbation theory are developed for the case of weak nonlinearity. They further proved a close connection between the nonlinear modes and (bright and dark) multi-soliton stationary states for both attractive and repulsive interactions, in which Kivsher et al., [94]considered the macroscopic dynamics of condensed atomic clouds in a three-dimensional, strongly anisotropic, external parabolic potential created by amagnetic trap. From these investigations they obtained that the BEC collective dynamics can be also described by the GP equation in Eq.(6.4). Thus, these problems can also be described by the dynamic equation of nonlinear quantum mechanics [13-16].

They havealso derived from Eq. (6.4) an effective one-dimensional modelof the system with a highly anisotropic (cigar-shaped) trap of the axial symmetry $V(r)=(m\omega_\perp^2/2)(r_\perp^2+\lambda X^2)$, where $r_\perp=\sqrt{y^2+z^2}$ .This implies that $\lambda=\omega_{11}^2/\omega_\perp^2<<1$ and the transverse structure of the condensate, being close to a Gaussianin shape, is mostly defined by the trapping potential [98]. In units ofthe longitudinal harmonic oscillator length $a_{ho}=(\hbar/m\omega\sqrt{\lambda})^{1/2}$, and the wavefunction amplitude in the units $(\hbar\omega_\perp/2U_0\sqrt{\lambda})^{1/2}$, they obtained the following dynamic equation for the condensate particles from Eq.(6.4)

$$i\frac{\partial\phi}{\partial t}=-\nabla^2\phi+[(y^2+z^2)/\lambda+x^2]\phi-\sigma|\phi|^2\phi \tag{6.92}$$

where time is measured in the units of $(2/\omega_\perp\sqrt{\lambda})$ and(x, y, z) =(X,Y,Z)/ a_{ho} , and the sign $\sigma=sgn(a)=\pm1$ in front of the nonlinear term is defined by the sign of the s-wave scattering length of the two-body interaction. When the nonlinear interaction is weak relative to the trapping potential force in the transverse dimensions, i.e., $\lambda<<1$, then the solution of Eq.(6.92) is

$$\phi(r,x,t)=\Phi(r)\varphi(x,t)\exp(-i2\kappa t)$$

where $r=\sqrt{y^2+z^2}$ and $\Phi(r)$ is a solution of the auxiliary problem for the 2D radially symmetric quantum harmonic oscillator and satisfied the equation

$$\nabla_\perp^2\Phi+2\kappa\Phi-(r^2/\lambda)\Phi=0$$

In this case they took the form of the no-node groundstate, $\Phi_0(r)=C\exp(-\kappa r^2/2)$, where $\kappa=1/\sqrt{\lambda}$. Topreserve all the information about the structure of the 3D condensate in an asymmetric trap describing its properties by the effective GP equation for the longitudinal profile, they imposed the normalization for $\Phi_0(r)$ that yields $C^2=\kappa/\pi$.

After substituting such a factorized solution into Eq.(6.92),dividing by $\Phi(r)$ and integrating over the transverse cross-section of the cigar-shaped condensate, they derived at the following 1D non-stationary GP equation:

$$i\frac{\partial\varphi}{\partial t}+\frac{\partial^2\varphi}{\partial x^2}-x^2\varphi+\sigma\left|\varphi\right|^2\varphi=0 \tag{6.93}$$

We [99-101], and Chen and Liu [102-103], solved the soliton solutions of Eq.(6.93) using different analytic methods, respectively. However, Kivsher et al., [97]used another way to study the features of solution of Eq.(6.93). In their study the number of the condensate particlesNis represented by $N=(\hbar\omega/2U_0\sqrt{\lambda})Q$, where $Q=\int_{-\infty}^{\infty}\left|\varphi(x,t)\right|^2 dx$ is the integral of motion for the normalized non-stationary GP equation (6.93). In fact, equation (6.93) includes all the terms of the same order, andit presents a longitudinal profile of the condensate state in a highly anisotropic trap. In the linear limit,i.e.,when formally $\sigma\to 0$, the equation (6.93) becomes the equation for a harmonic quantum oscillator. Its stationary localized solutions may be expressed by

$$\varphi(x,t)=\psi(x)\exp(-i\Omega t), \tag{6.94}$$

which exists only for discrete values of Ω, such as $\Omega_n=1+2n, n=0,1,2,...$, they gave the solution by the Hermite–Gausspolynomial

$$\psi_n(x)=C_n\exp(-x^2/'2)H_n(x),\ C_n=(2^n n!\sqrt{\pi})^{-1/2},$$

where $H_n(x)=(-1)^n\exp(x^2/2)(d^n\exp(-x^2/2)/dx^n)$, $H_0=1$, $H_1=2x$, etc..

At the same time, they also solved Eq. (6.93) numerically. The results are shown in Figure 6.7, in which Figs.(a)–(b) presented examples of the numericallyfound ground-state solutions of Eq. (6.93) in the form Eq.(6.94) as continuous functions of the dimensionless parameter Ω for both negative (a) and positive (b) scatteringlength. For $\Omega\to 0$, i.e., in the limit of the harmonic oscillator ground-state mode, the solution is close to Gaussian typesin both cases. When Ω deviates from 1, the solution profile is defined by the type of nonlinearity. For attraction ($\sigma=+1$), the profile approaches the Sech-type soliton [99], whereas for repulsion ($\sigma=-1$) the solution flattens. Therefore, it is better described by the Thomas–Fermi approximation, which is good except at the edge points.

Figures 6.7(c)–(f) show examples of the first- andsecond-order modes for both negative and positive scattering length, respectively. In the limit $\Omega\to\Omega_n$,all those modes transform into the corresponding eigenfunctions of a linear harmonic oscillator.

As it is well known, nonlinearity has different effects. For the negative scattering length (attraction) ($\sigma=+1$), the higher order modes are transformed into multi-soliton states consisting of a sequence of solitary waves with alternating phases (as Figs. 6.7(c)–(e)), which can be confirmed from the analysis of the invariant Q vs Ω, where all the branches of the higher-order modes asymptotically approach the soliton dependencies $Q_n\approx(n+1)Q_s$, where n is the order of the mode (n =0,1,2,...), because the higher-order stationary modes exist in the case of attractive interaction due to a balance between repulsion of out-of-phase bright NLS solitons and attraction imposed by the trapping potential. This depends on the global stability of such higher-order multihump multi-soliton modes [104]. For the positive scattering length ($\sigma=-1$), the higher-order modes are transformed into a sequence of dark solitons (or kinks) [105], so that the first-order mode corresponds to a single dark soliton, the second-order mode, to a pair of dark solitons, etc. [See Figs. 6.7(d)– (f)]. These stationary solutions satisfy a forced balance condition, i.e., repulsion between dark solitons is exactly compensated by an attractive force of the trapping potential [94].

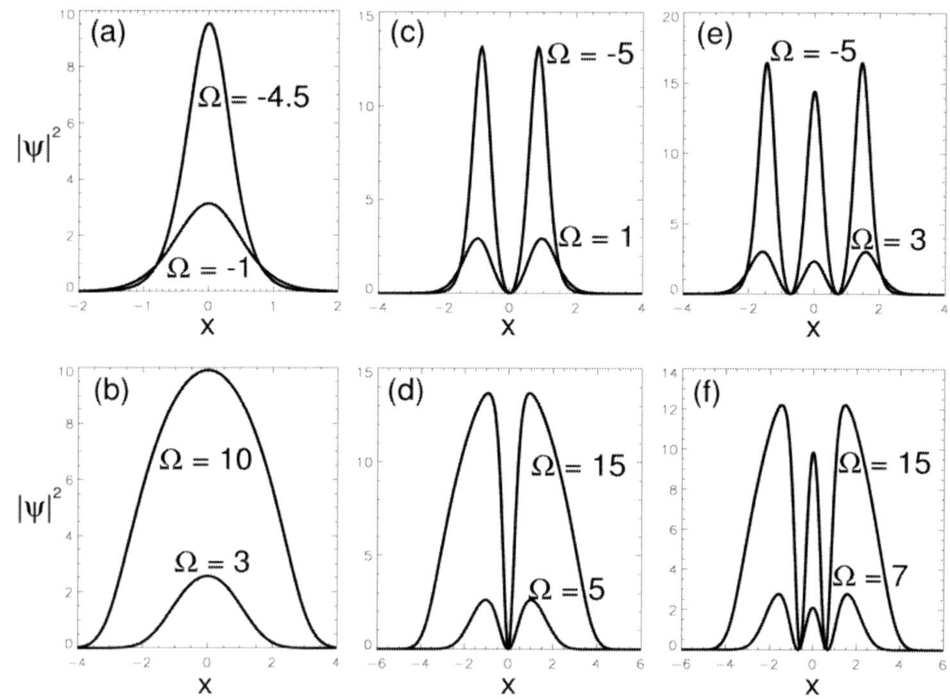

Figure 6.7. The examples of the condensate density $|\varphi(x,t)|^2$ for the first three nonlinear modes described by the stationary solutions of Eq. (6.93) for the negative ($\sigma=+1$, upper row), and positive($\sigma=-1$, lower row) scattering length, respectively [94].

Kivsher et al., [94] thought that the model structure of the condensate macroscopic states, described above, for both positive and negative values of the scattering length, allows us to draw an analogy between BEC in a trap, and the guided-wave optics, where the condensate dynamics in time corresponds to the stationary mode propagation along an optical waveguide, with the parameter Ω as the propagation constant. As it is known for the guided-wave optics, in the presence of interaction, the guided modes become coupled and the coupling can lead to both power exchange and nonlinear phase shifting between the modes. In application to the BEC theory, the mode coupling resembles a kind of internal Josephson effect [106].

The above theory of nonlinear stationary modes of a macroscopic quantum oscillator, developed above, for the 1D analog of BEC can extend to 2D and 3D cases. In the meanwhile, the coupled-mode theory for a single condensate is closely connected to the dynamics of strongly coupled two-component BECs[107], in this case there are excitations of an antisymmetric collective mode in the form of collapses and revivals [108].

Otherwise, the above theoretical model can be applied to explain some physical phenomena, for example the effect of nonlinearity can lead to a mode coupling in the "photonic atom", a micrometer-sized piece of semiconductor that traps photons inside [96], or two such"photonic atoms" coupled together, a "photonic molecule" [109]. In such photonic microcavity structures, the macroscopic nature of the states may lead to different nonlinear effects including the mode mixing and power exchange.

6.8.3. Dynamic Properties of Photo-Induced Boson Condensate in Optical Superlattices

As it is known, the influence of the series plane waves of monochromatic radiation brings a diffraction pattern in which the electric field intensity periodically varies in space. The ensued periodic system, designated as an optical lattice (OL), is an array of microscopic potentials induced by the high frequency Stark effect in the field of interfering laser beams. Ultra-cold bosons trapped in such periodic potentials have been widely used as a model system for the study of some fundamental concepts of quantum physics like the Josephson effect, squeezed state, Landau-Zener tunneling, Bloch oscillation, and superfluid-Mott insulator transition. The system of optical traps with the drops of BEC dilute in ultracold gas represents, in fact, a sort of artificial media, where the matter resides in the coherent state, and the parameters of the BEC array can be effectively managed by virtue of the Feshbach resonances or by an external field. Because of the quantum interference between BECs, their lattice can provide collective excitations, which, in the approximation of strong coupling between Bosons at the OL site, are nonlinear solitons. Under specific conditions the ensemble of BEC drops placed in the nodes of the OL are able to produce the echo-effect.

In the mean-field approximation, there are two ways to delineate the dynamic features of atoms in the OL. The first method is based on the nonlinear Schrödinger equation with the periodic potential cited as the Gross–Pitaevskii (GP) equation in Eq.(6.4) [13-16, 93] for the BEC. The second approach uses the tight-binding approximation where operators of creation and annihilation of Fermions in the Habbard model are replaced by operators satisfying the Boson commutation relations. The resulting model is referred as to the Boson Habbard model. The OL with two sorts of microscopic potentials, with wells of different depths per unit cell, have attracted concerned attention. By analogy with the theory of solid-state crystals, such a periodic system of microscopic potentials is called an optical superlattice(OSL). The superlattice structures are featured by the difference in energy levels of neighbor microscopic potentials. So, the quantum tunneling between the nearest-neighbor sites is suppressed at ultra-low temperatures. The thermo-activated transfer of atoms from one micro-potential site to another is improbable as well. However, the photo-induced transport of Bosons along the OSL is possible under the conditions of Raman resonances. A similar process is the coherent population transfer in nonlinear optics and the coherent transfer of electrons in solid-state physics of low dimensional system or excitons in the systems of coupled quantum dots.

An ensemble of condensate drops dispersed over the array of asymmetric optical micropotentials is able to support the various sorts of nonlinear excitations induced by dichromatic resonance CW electromagnetic radiation, different from the rays constituting the lattice sites. The parameters of excitions can be controlled by this external radiation, which particularly determines the hopping rate of Bosons between the adjacent sites. The model of nonlinear photon-induced lattice excitations permits the solutions in the form of a steady state wave. But there could be definite conditions. Therefore Sorokina et al., [110]seriously researched the properties of photo-induced dynamics of the Boson condensate of cold atoms in a one-dimensional optical superlattice, in which they observed and found that the steady state distribution of the probability amplitudes, and the site population in a one-dimensional optical lattice are formed. At the same time, it also shows that this solution of dynamic equations, which describes the temporal behavior of BEC in the lattice, is unstable at sufficiently high levels of boson density. The expression for the increment of modulational instability is obtained based on the linear stability analysis. Their numerical examples of non-stationary solutions for boson density in a superlattice for the general model are also discussed, as applied to both the attraction and repulsion potentials of boson interaction.

On the other hand, Elyutin [111]also discussed the photo-induced waves of the Boson condensate in optical superlattices, in which the dynamics of ultracold atoms loaded in an optical lattice (superlattice) is described based on the Boson–Hubbard model. Both linear and circular lattices are considered by virtue of attractive and repulsive types of interaction between the bosons. The numerically observed quantum-mechanical current state of the Boson condensate atoms is due to a hopping transfer of the ultracold atoms, are caused by the excitation of the system, by the biharmonic electromagnetic field under the Raman resonance condition. When the energy of interaction between the bosons is close to the energy of photo-activation, the boson current decreases, which may be the sign of the superfluidity-insulator phase transition. In the disordered linear chain of potentials, where the depth of optical potentials is changed from one site to another, the current spreading over the sites of optical lattice is suppressed in the same way as in the Anderson localization effect.

6.8.4. Other Investigations on the Boson –Einstein Condensate in Different Systems

Bullough et al.,[112]studied quantum integrable and nonintegrable nonlinear Schrödinger models for realizable Boson–Einstein condensate in d+1 dimensions (d = 1, 2, 3) and evaluated finite-temperature equilibriumcorrelators by new methods of functional integration in **d**=1, 2, 3 dimensions, and with the trap potentials $V(\mathbf{r})\neq 0$, in the Gross–Pitaevskii equation in Eq.(6.4). As in the translationally invariant cases, asymptotic behaviors fall to longer-range condensate values only for $\mathbf{d} = 3$ in agreement with experimental observations; but there are generally significant corrections due to the presence of the traps. For $\mathbf{d} = 1$, they regain the exact translationally invariant results as the trap frequencies $\Omega \to 0$. In analyzing the attractive cases, they investigated the time-dependent **c**-number Gross–Pitaevskii equation with the trap potential for a generalized nonlinearity $-2c\psi|\psi|^{2n}$and$\mathbf{c} < 0$. For $\mathbf{n} = 1$, the stationary form of the GP equation appears in the steepest-descent approximation of the functional integrals. They showed also that the collapsein the sense of Zakharov can occur for $\mathbf{c} = 0$ and $\mathbf{n} \geq 2$, and a functional $\mathbf{E}_{NLS}[\psi] \leq 0$, even when $V(r) \neq 0$. The singularities typically arise as δ-functions centered on the trap origin $\mathbf{r}$=0.

Bogoliubov et al. [113] studied the features of finite-temperature correlations in the one-dimensional trapped and untrapped Boson gases and calculated further the dynamic single-particle and many-particle correlation functions at non-zero temperature in one-dimensional trapped repulsive Boson gases. The decay for increasing the distance between the points of these correlation functions is governed by a scaling exponent that has a universal expression in terms of observed quantities. This expression is valid in the weak-interaction Gross-Pitaevskii equation, as well as in the strong-interaction Girardeau-Tonks limit, but the observed quantities involved depend on the interaction strength. The confining trap introduces a weak center-of-mass dependence in the scaling exponent. They also conjectured the results for the phenomenon.

REFERENCES

[1] S. N. Boson, "Plancks Gesetz und Lichtquantenhypothese". *Zeitschrift für Physik* 26: (1924) 178.
[2] A.Einstein, "*Quantentheorie des einatomigen idealen Gases*", Sitzungsberichte der Preussischen Akademie der Wissenschaften 1 (1925) 3.
[3] C.W. Ronald and A.Einstein, The Life and Times. *Avon Books*. 1971, pp. 408.
[4] S. T. Beliaev, Zh. Eksp. Teor. Fiz. 34(1958)418 and 433 [Soviet Phys. JETP 3 (1957) 299.
[5] L. D. Landau, *J. Phys.* USSR 5: (1941) 71.
[6] L. D. Landau,. *Phys. Rev.* 60 (1941) 356.
[7] C. P. Arora, *Thermodynamics*, Tata McGraw-Hill. 2001, p.43.
[8] F. London, *Nature* 141 (1938)64.
[9] F. London, *Superfluids, Vol.I and II,* New York: Dover 1964.
[10] F. London, *Superfluids Vol.1,* Weley, New York 1950.
[11] N. N. Bogoliubov, *Quantum statistics, Nauka*, Moscow, 1949 and *J. Exp. Theor. Phys.* USSR. 34(1958) 58.
[12] N. N. Bogoliubov, V. V. Toimachev and D. V. Shirkov, *A New Method in the Theoryof Superconductivity,* AN SSSR, Moscow, 1958.
[13] E. F. Gross, Nuovo Cimento, 20 (1961) 452.
[14] L. P. Pitaerskii, *Sov. Phys. JETP* 13 (1961) 451 (*Zh Eksp, Theor. Fiz.* 40 (1961) 646.
[15] L. P. Pitaevskii, *Sov. Phys. JETP* 10 (1960) 553
[16] L. P. Pitaevskii and S. Stringari, *Boson–Einstein Condensation*, Clarendon Press, Oxford, 2003
[17] M. Schick, *Phys. Rev.* A 3 (3) (1971) 1067.
[18] E. Kolomeisky, J. Straley, *Phys. Rev.* B 46 (1992) 11749.
[19] E. B. Kolomeisky, T. J.Newman, J. P.Straley, X. Qi, *Phys. Rev. Lett.* 85 (2000) 1146.
[20] S. Chui, V. Ryzhov, *Phys. Rev.* A 69 (2004) 043607.
[21] L. Salasnich, A. Parola, L.Reatto, *Phys. Rev.* A 65 (2002) 043614.
[22] A. V. Avdeenkov, K. G. Zloshchastiev, *J. Phys.* B: At. Mol. *Opt. Phys.* 44 (2011) 195303.

[23] E. A. Cornell, W. Ketterle, C. E. Wieman, “Nobel Lecture": Boson-Einstein condensation in a dilute gas; The First 70 Years and Some Recent Experiments, www.nobelprize.org/nob, 2001-12-08.
[24] H. K. Bullough, N. M. Bogolyubov, V. S. Kapitonov, C. Malyshev, J. Timonen, A. V. Bullough, R. K. and P. T. Caudeey, *Solitons,* Plenum Press, New York, 1980.
[25] C. Huepe, and M. E. Brachet, Physica D 140 (2000) 126.
[26] P. V. Elyutin and A. N. Rogoverko, *Phys. Rev*. E63 (2001) 026610.
[27] P. V. Elyutin, P.V., A.V. Buryak, V.V. Gubernov, R.A. Sammut, and I.N. Towers, *Phys. Rev*. E64 (2001) 016607.
[28] A. J. Leggett, *Prog.Theor. Phys*. (Supp.) 69(1980) 80.
[29] C. F. Barenghi, R. J. Donnerlly and W. F.Vinen, *Quantized Vortex Dynamics andSuperfluid Turbulence,* Springer, Berlin, 2001.
[30] L. N. Bardeen, L. N. Cooper and J. R. Schrieffer, *Phys.Rev*. 108 (1957) 1175.
[31] L. N. Cooper, *Phys. Rev*. 104 (1956) 11.
[32] J. R. Schrieffer, *Superconductivity,* Benjamin, New York, 1966.
[33] J. R. Schrieffer, *Theory of Superconductivity*, Benjamin, New York, 1964.
[34] Pang, Xiao-feng, *Theory of nonlinear quantum mechanics*, Chongqing Press, Chongqing, 1994.p35-97
[35] Pang Xiao-feng, *Nonlinear quantum mechanics*, Beijing, Chinese Electronic Industry Press, Beijing, 2009,p 20-63.
[36] Pang Xiao-feng, Soliton physics, *Sichuan. Sci.Tech.Press*.Chengdu, 2002, p.714-771.
[37] J. G. Valatin, Naopa Ciminio 7(1958)843.
[38] W. S. Liu, and X. P. Li, *European Phys. J*. D2 (1998) 1.
[39] S. O. Demokritov, V. E. Demidov, O. Dzyapko, G. A.Melkov, A.A.Serga, B. Hillebrands, A.N. Slavin, *Nature* 443 (2006) 430.
[40] Demokritov, Magnon Boson Einstein Condensation made simple, Website of the"Westfähliche Wilhelms Universität Münster", Retrieved 25 June 2012.
[41] R. Baierlein, Thermal Physics. Cambridge University Press, Cambridge, 1999.
[42] E. M. Richter, J. Kronjäger, K. Bongs, K. Sengstock, *Nature Physics* 4 (2008)496.
[43] M.H.P.M. van Putten, *Phys. Lett.* A 374 (2010)3346.
[44] A. Gorlitz, "Interference of Condensates", *Cua.mit.edu*. 13 October 2009.
[45] C. C. Bradley, C. A. Sackett, J. J.Tollett, R. G.. Hulet, *Phys. Rev. Lett*. 75(1995)1687.
[46] T. Nikuni, M. Oshikawa, A. Oosawa, H.Tanaka,. *Phys.Rev. Lett*. 84 (1999)5868.
[47] M. Mackie, K.A.Suominen, J.Javanainen, *Phys Rev Lett*. 89(2002)180403.
[48] N. M.Hugenholtz and D. Pines, Physical Review**116**(1959) 489.
[49] C. J.Pethick and H. Smith,. Bose–Einstein Condensation in Dilute Gases. Cambridge: Cambridge University Press.,2002,ISBN 0-521-66580-9. .
[50] L. P.Pitaevskii and S.Stringari,. Bose–Einstein Condensation. Oxford: Clarendon Press. 2003, ISBN 0-19-850719-4. .
[51] I. Ventura, Solitons in Liquid 4He, in *Solitons and Applications* eds. V. G. Makhankov, V. K. Fedyanin and O. K. Pashaev, World Scientific, Singapore, 1990.
[52] B. S. Dewitt, *Phys. Rev. Lett.* 16 (1966) 1092.
[53] Pang, Xiao-feng, *J. Xinjian Univ. Sin*., 3 (1988) 33.
[54] Pang, Xiao-feng, *J. Xinjian Univ. Sin*., 4 (1989) 72.

[55] L.Onsager, Nuovo Cimento, 6 (2) Suppl. (1949) 246.
[56] C. F. Barenghi, R. J. Donnerlly and W. F.Vinen, *Quantized Vortex Dynamics andSuperfluid Turbulence*, Springer, Berlin, 2001.
[57] R. J., Donnely, *Quantum Vortices in Helium II,* Cambridge Univ. Press, Cambridge, 1991.
[58] O. Avenel, et al., Proc. 20th Inter. Conf. on Low Temp. Phys., *Physica B* 194196(1994) 491.
[59] C. A. Lindensmith, et al., Proc. 21st Inter.Conf. on Low Temp. Phys., *Czech J. Phys.* 46 (1996) 131.
[60] E. J. Yarmchuk, M. J. V. Gordon and R. E. Packard, *Phys. Rev. Lett.* 43 (1979) 214.
[61] R. E. Packard, *Rev. Mod. Phys.* 70 (1998) 641.
[62] R. J. Zieve, J. D. Close, J. C. Davis and R. E. Packard, *J.Low Temp. Phys.* 90 (1993) 243.
[63] R., J. Zieve, Yu. Mukhjarsky, J. D. Close, J. C. Davis and R. E. Packard, *Phys. Rev.* Lett. 68 (1992) 1327.
[64] T. Frisch, Y. Pomeau and S. Rica, *Phys. Rev.Lett.* 69 (1992) 1644.
[65] Y. Pomeau, and S. Rica, *Comptes Rendus Acad.Sci.* (Paris) 316 (Series II) (1993) 1523.
[66] K. W. Schwarz, *Phys.Rev.* B 47 (1993) 12030.
[67] K. W. Schwarz, *Phys. Rev. Lett.* 64 (1990) 1130.
[68] J. Anandan, *Phys. Rev. Lett.* 47 (1981) 463.
[69] Pang Xiao feng, *Nature Science*, 3(7)(2011) 978.
[70] Pang Xiao feng, *International Journal of Reseach and Reviews in Applied Science*, 10(2012) 349.
[71] K. B. Davis, M.O. Mewes, M. R. Andrews, N. J. van Druten, D. S. Durfee, D. M. Kurn, and W. Ketterle, *Phys. Rev. Lett.* 75 (1995) 969.
[72] A. L. Fetter and J. D. Walecka, *Quantum Theory of Many-Particle Systems*, McGraw Hill, New York, 1971.
[73] P. Hohenberg and W. Kohn, *Phys. Rev. Lett.* B 136 (1964) 864.
[74] E. Runge and E. K. U. Gross, *Phys. Rev. Lett.* 52 (1984) 997.
[75] Q. Han Park and J. H. Eberly, *Phys. Rev.* Lett. 85 (2000) 4195.
[76] S. Burger, K. Bongs, S. Dettmer,W. Ertmer, K. Sengstock, A. Sanpera, G. V. Shlyapnikov, and M. Lewenstein, *Phys. Rev. Lett.* 83 (1999) 5198.
[77] J. M. Vogels, K. Xu, C. Raman, J. R. Abo-Shaeer and W. Ketterle, *Phys. Rev. Lett.* 88 (2002) 060402-1.
[78] M O. Mewes, M. R. Andrews, D. M. Kurn, D. S. Durfee, C. G. Townsend, and W. Ketterle, *Phys. Rev. Lett.* 78 (1997).582.
[79] I. Bloch, T. W. Hansch, and T. Esslinger, *Phys. Rev. Lett.* 82 (1999) 3008.
[80] O. Morsch, J. H. M¨uller, M. Cristiani, D. Campini, and E. Arimondo, *Phys. Rev. Lett.* (2001)140402.
[81] M. Blaauboer, G. Kurizki, and V. M. Akulin, *Phys. Rev. Lett.* 86 (2001)3518.
[82] D. O'Dell, S. Giovanazzi, G. Kurzicki, and V. M. Akulin, *Phys. Rev. Lett.* 84(2000)5687.
[83] Y. Pomeau and S. Rica, *Phys. Rev. Lett.* 72 (1994) 2426.
[84] S. Giovanazzi, D. O'Dell, and G. Kurizki, *Phys. Rev. Lett.* 88 (2002).130402-1.

[85] M. Kalinski, K. Vogel, G. Kurizki, W. P. Schleich, Dynamics of Boson-Einstein Condensates with Long-Range Attractive Interactions, NIC Symposium 2004, *Proceedings*, NIC Series, 20(2003) 251.
[86] S. Giovanni, G. Kurizki, I. E. Mazetz and S. Stringari, *Europhys. Lett.* 56 (2001)1.
[87] S. Giovanazzi, D. O. Dell, and G. Kurizki, *Phys. Rev.* A 63 (2001).031603-1.
[88] M. L. Chiofalo, S. Succi, and M. P. Tosi, *Phys. Rev.* E 62 (2000)7438.
[89] R. Grobe, S. L. Haan, and J. H. Eberly, *Comp. Phys. Comm.* 117 (1999) 200.
[90] A. D. Bandrauk and H. Shen, *J. Phys.* A 27 (1994) 7147.
[91] Q. Su, and J.H. Eberly, *Phys. Rev.* A 43 (1991) 2474.
[92] M. N. Anderson et al.,*Science* 269 (1995) 198.
[93] F. Dalfovo, S. Giorgini, L.P. Pitaevskii, S. Stringari, *Rev. Mod. Phys.* 71 (1999) 463.
[94] Y. S. Kivshar, T. J. Alexander, S. K. Turitsyn, *Phys. Lett.* A 278 (2001) 225.
[95] S. V. Gaponenko, *Optical Properties of Semiconductor Nanocrystals*, Cambridge University Press, Cambridge, 1998.
[96] Reithmaier J P, Rohner M, Zull H et al., *Phys. Rev. Lett.* 78 (1997) 378.
[97] L. D. Landau, E.M. Lifshitz, *Quantum Mechanics*, Pergamon Press, Oxford, 1987.
[98] V. M. Perez-Garcia, H. Michinel, H. Herrero, *Phys. Rev.* A 57 (1998) 3837.
[99] Pang Xiao-feng, *Nonlinear quantum mechanics (Chinese),* Chinese Electronic Industry Press, Beijing, 2008.
[100] Pang Xiao feng, *Physica B* 404(2009)3125.
[101] Pang Xiao-feng, *Physica B.* 403(2008)4292.
[102] H. H. Chen and C.S.Liu, *Phys. Rev. Lett.* 37(1976)693.
[103] H. H. Chen, *Phys. Fluids* 21(1978)377.
[104] E. A. Ostrovskaya, Yu. S. Kivshar, D. V. Skryabin, W. J. Firth, *Phys. Rev. Lett.* 83 (1999) 296.
[105] Yu. Kivshar, B. Luther-Davies, *Phys. Rep.* 298 (1998) 81, and references therein.
[106] E. A. Ostrovskaya, Yu.S. Kivshar, M. Lisak, B. Hall, F. Cattani, D. Anderson, *Phys.Rev.* A 61 (2000) 031601.
[107] J. Williams, R. Walser, J. Cooper, E. Cornell and M. Holland, *Phys. Rev.* A 59 (1999) R31.
[108] J. G. Williams, S. G. Turyshev, D. H. Boggs, *Phys. Rev. Lett.* 93 (2004)261101.
[109] M. Bayer et al., *Phys. Rev. Lett.* 81 (1998) 2582.
[110] E. A.Sorokina, A. L.Maimistov and S. O Elyutin, *Physica* D 238(2009)1394.
[111] S. O. Elyutin, Kazan. Gos. Univ. *Uchen. Zap. Ser. Fiz.-Mat. Nauki,* 15(2010)69.
[112] R. K. Bullough, N. M. Bogolyubov, V. S. Kapitonov, C. Malyshev, J. Timonen, A. V. Rybin, G. G. Varzugin, M. Lindberg, *Theoretical andMathematical Physics*, 134 (2003) 47.
[113] N. M. Bogoliubov, R. K. Bullough, C. Malyshev, J. Timonen, *Phys. Rev.* A 69(2004) 023619.

Chapter 7

NONLINEAR QUANTUM DYNAMIC PROPERTIES OF ELECTRONS AND POLARONS IN ATOMS AND CONDENSED-STATE SYSTEMS

In this chapter the nonlinear quantum-mechanical properties of electrons and polarons are studied in atoms and condensed-state systems including the coupled electron-phonon systems, ionic crystals, superconductor and semiconductors, where these nonlinear interactions arise from the interactions between moved electron and moved nucleon in atoms, or moved phonons in coupled electron-phonon systems (or superconductor), or moved polarization lattice-fields in ionic crystal and semiconductors, respectively. This means that the self-trapping mechanism of nonlinear interaction holds in these physical systems. We here built and gave their dynamic equations, which are all the nonlinear Schrödinger equation in nonlinear quantum mechanics, their soliton solutions and corresponding wave-corpuscle duality are obtained, some physical phenomena are explained by these results. These results show clearly that nonlinear quantum mechanics can completely describe the properties of the microscopic particles in these systems.

7.1. THE NONLINEAR QUANTUM PROPERTIES OF THE ELECTRON IN ATOMIC SYSTEMS

7.1.1. The Problems of Linear Quantum Mechanics in the Study of the Electron and Its Direction of Improvement and the Mechanism of Generation of Nonlinear Interaction in the Atoms

As it is known, the properties of motion of the electron in the atom, such as hydrogen atom, is described linear Schrödinger equation in linear quantum mechanics, which were discussed in Chapter 1. We here re-write it by

$$i\hbar\frac{\partial\phi(\vec{r},t)}{\partial t}=-\frac{\hbar^2}{2m}\nabla^2\phi(\vec{r},t)-\frac{e^2}{\vec{r}}\phi(r,t) \tag{7.1}$$

where $e^2/\vec{r}$ is the Coulomb potentialbetween the electron and the nucleon, $\vec{r}$ is the position of the electron relative to the nucleon. Since the hydrogen atom is in the stable state of independent time, thus the state of the electron is, in general, depicted by Eq.(1.7), where the eigenwave-function of the electron is denoted in Eq.(1.18), its eigenenergy, E_1 =E, is represented in Eq.(1.19).

These results indicated clearly that the state of the electron in the atom is represented by a wave as denoted in Eq.(1.18).This result leads to the two difficulties. (1). It is contradicted by the wave-corpuscle duality of the electron, which was confirmed by many experiments described in Chapter 1 and the De Broglie relationship of $E = h\nu = \hbar\omega$ and $\vec{P} = (h/\lambda)\vec{n} = \hbar\vec{k}$ as well as that the electron is a real particle having certain sizes and masses of 9.10953x10^{-31}kg, its classical radius is 2.817938Fm.

(2).It lends to the contradiction of the first assumption with final results, i.e., the interaction between the electron and nucleon in Eq.(7.1) is represented by the Coulomb force, $F(\vec{r}) = -e^2/\vec{r}^2$ between two classical point chargesin macroscopic world.This implies that both the electron and nucleon all have a classical or corpuscle feature. But the electron obtained from Eq.(7.1) only has a wave feature. Therefore, they are contradictory. Long-time investigations indicated that the two contradictions cannot be overcome and eliminated in the linear quantum mechanical framework no matter what methods are used, i.e., they are intrinsic problems of linear quantum mechanics. Thus, the theory that described the motion of the electron in atoms must be improved. However, how does it improve? In order to solve this problem we must first know the reasons that generated these difficulties. We know from the detailed analyses for the above results that the motion of the nucleon is completely frozen or neglected because its position is always fixed at the origin of the coordinate, even though the electron always moves around the nucleon in atoms. This is completely unreasonable and cannot be accepted, it may be just the basic reason arising from the difficulties and problems. It should be abandoned for the following reasons:

(1). As it is well known, all matters are all in motion in nature, especially in the microscopic world where particles are frequently in motion because their masses are very small. No motions of microscopic particles are only an artificial assumption, which is not existent at all. This means that the motion of the nucleon must be considered seriously in the investigations of the dynamics of the electron in atoms, such as the hydrogen atom.

(2). In accordance with mechanical theory, the Coulomb interaction between the electron and nucleon is changed with varying the distance between them, $\vec{r}$, this means that it will change because the electron moves frequently, in which the size and direction of the radial vector $\vec{r}$ or Coulomb force $F(\vec{r}) = -e^2/\vec{r}^2$ varies continuously. Then it is only an artificial assumption and it is wrong that thought be invariable of the Coulomb interaction. Thus we should think that the Coulomb potential in Eq.(7.1) is changed.

(3).According to the law of action-counteraction in physics the Coulomb interaction $F(\vec{r}) = -e^2/\vec{r}^2$, can change the states of both the electron and the nucleon. Now that the Coulomb interaction can vary the states of the electron in Eq.(7.1), it is necessary to change the state of motion of the nucleon under the counteraction of the Coulomb interaction. This means that we must study the motion of the nucleon arising from the Coulomb interaction.

(4). The intrinsic zero-point vibration of the nucleon in the atoms always exists in accordance with the quantum theory. Then the nucleon is also always in movement.

(5). As it is known, the influences of other interactions, such as the interactions of orbital motions and spins of the electron with the nucleon, on the motion of the electron in the atoms were already considered and calculated in previous investigations, then these interactions also change the states of the nucleon. This implies that the nucleon acts by many forces, so it is not reasonable and wrong to freeze its movement.

(6). It is well known that the motions of nucleons, including the proton and neutron, were studied in nuclear physics [1-18], so it is not correct that the motion of the nucleon is not considered in the study of dynamics of the atom in atomic physics.

These reasons, mentioned above, result in what we must consider and study seriously: the movement of the nucleon and changes of Coulomb interaction $V(\vec{r}) = -e^2/\vec{r}$, when the electron moves in atoms. Once the motion of the nucleon is considered, it is a straightforward result to vary the representation oftheCoulomb interaction $V(\vec{r}) = -e^2/\vec{r}$. If the vector $\vec{u}(t)$ is used to represent the motion or displacement of the nucleon leaving its equilibrium position, then the distance and Coulomb potential between the electron and nucleon can be denoted by $\vec{r} \pm \vec{u}(t)$ and $V(\vec{r},\vec{u}) = -e^2/(\vec{r} \pm \vec{u})$, respectively, where $\vec{r}$ is the position of the electron relative to the equilibrium position of the nucleon. Since

$$\frac{1}{r+u} = \frac{1}{r(1 \pm u/r)} = \frac{1}{r}(1 \mp \frac{u}{r} + ...) = \frac{1}{r} \mp \frac{u}{r^2} + ... = \frac{1}{r} \pm \frac{d}{dr}(\frac{1}{r})u + ...,$$

then the dynamic equation of the electron in the hydrogen atom should, at least, be of the form

$$i\hbar\frac{\partial\phi}{\partial t} == -\frac{\hbar^2}{2m}\nabla^2\phi - \frac{e^2}{r}\phi \mp \chi\phi u \tag{7.2}$$

Evidently, equation (7.2) differs from Eq.(7.1), a new interaction term $\chi\phi u$, which is related to the states of both electron and nucleon, is added in Eq.(7.2). Thus, the natures of the electron will be changed. In this case, we still represented the states of the electron by $\phi(\vec{r},t)$. The new interaction, $\chi\phi u$,in Eq.(7.2) denoted just the variation of state of the electron arising from the change of Coulomb force because of the motion of the nucleon, where $\chi = -e^2 dr^{-1}/dr = e^2/r^2$ is a coupling coefficient between the electron and nucleon, which is just the Coulomb force between them. Therefore, equation (7.2) represents clearly that the motion of the electron is related to both the Coulomb interaction and the state of the nucleon, i.e., the motion of the nucleon results in changes of nature and state of the electron. This idea can also applied in hydrogen molecule and other atoms. We expect that the interaction term, $\chi\phi u$, which relates to the wave function of the electron, $\phi(\vec{r},t)$, can suppress the wave feature of the electron to make its state change or make it become localized.

7.1.2. The Nonlinear Theory of Motion of the Electron and Its New Dynamic Equation

However, how do we now describe the state and motion of the nucleon in a hydrogen atomusing a complete model? From the features of structure of the hydrogen atom, we know that the nucleon follows and always accompanies the electron, where the electron moves circularly around the nucleon due to the Coulomb interaction and the magnetic interactions of orbital motion and spin between them. This means that the nucleon always moves around its equilibrium position due to periodic changes in both direction and size of the interaction between it and the electron, which may approximately amount to the nucleon moves, from right to left and from up to down, relative to its equilibrium position due to the circular motion of the electron. If the intrinsic zero-point vibration of the nucleon is again considered, then it is very reasonable to assume that the nucleon vibrates harmonically around its equilibrium position, thus we could use a harmonically vibrational model to describe the motion of the nucleon. In this case, the movements of the nucleon, and the electron in the hydrogen atom, may be approximately described by these models of harmonic vibration and circular motion, respectively. Obviously, the motion models of the electron and nucleon can also guarantee the stability of structure of the hydrogen atom, or else, the hydrogen atom is not stable, which cannot be accepted. Therefore, the above models of motion of the electron and nucleon in the hydrogen atom are reasonable.

In this mode, let the origin point of the coordinate place still be at the equilibrium position of the nucleon, then the displacement of the nucleon should be represented, in general, by $\vec{u}_{n+1}$ - $\vec{u}_n$, instead of $\vec{u}_n$ in Eq.(7.2), which denotes the displacement of the nucleon at (n+1)th site relative to the nth site. Based on the above models of the nucleon and the electron, the Hamiltonian of the hydrogen atom can be represented by

$$H = H_e + H_n + H_{\text{int}} = -\frac{\hbar^2}{2m}|\nabla\phi_n|^2 - \frac{e^2}{\vec{r}}|\phi_n|^2 + \frac{1}{2}M(\frac{\partial u_n}{\partial t})^2 + \frac{1}{2}W(u_{n+1} - u_n)^2 + \chi(u_{n+1} - u_n)|\phi_n|^2 \tag{7.3}$$

where $H_e = -\frac{\hbar^2}{2m}|\nabla\phi_n|^2 - \frac{e^2}{\vec{r}}|\phi_n|^2$ is the Hamiltonians of the electron, $\frac{\hbar^2}{2m}|\nabla\phi_n|^2$ is its kinetic energy, $-\frac{e^2}{\vec{r}}|\phi_n|^2$ is its Coulomb potential given by the nucleon at the equilibrium position, where $H_n = \frac{1}{2}M(\frac{\partial u_n}{\partial t})^2 + \frac{1}{2}W(u_{n+1} - u_n)^2$ is the Hamiltonian of the nucleon in the harmonic vibration model, $\frac{1}{2}M(\frac{\partial u_n}{\partial t})^2$ is its kinetic energy, $\frac{1}{2}W(u_{n+1} - u_n)^2$ is the harmonical potential of the nucleon arising from the changes of both direction and size of the

Coulomb interaction between it and electron, M is its mass, W is the force constant of vibration of the nucleon, $H_{\text{int}} = \chi(u_{n+1} - u_n)|\phi_n|^2$ corresponds to the later term $\chi\phi u$ in the r.h.s. in Eq.(7.2), and is an interactional Hamiltonian between the electron and the nucleon, which represents the changes of Coulomb interaction between them due to the motion of the nucleon. Therefore, the physical significances of Eq.(7.3) are very clear and correct.

From the Schrödinger equation, $i\hbar\partial|\phi\rangle / \partial t = \hat{H}|\phi\rangle$, for the Hamiltonian operator in Eq.(7.3) or from Hamilton equations for the variables, $\phi(x_i, t)$ and u(x$_i$, t), and Eq.(7.3) [19-23] we can obtain the equations of motion of $\phi(x_i, t)$ and u(x$_i$, t), which are

$$i\hbar\frac{\partial\phi}{\partial t} = -\frac{\hbar^2}{2m}\nabla^2\phi - \frac{e^2}{r}\phi \pm \alpha a_0 \chi\phi\frac{\partial u}{\partial x_i}, (i = 1, 2, 3) \tag{7.4}$$

and

$$M(\frac{\partial^2 u}{\partial t^2} - v_0^2\frac{\partial^2 u}{\partial x_i^2}) = -\alpha a_0 \chi \frac{\partial}{\partial x_i}|\phi|^2 \tag{7.5}$$

respectively, where the following continuous approximations are used,

$$\begin{gathered} u_{n+1} = u_n + \alpha a_0 \nabla u_n + \frac{1}{2!}(\alpha a_0)^2 \nabla^2 u_n + ..., \\ u_n(x_i, t) \to u(x_i, t), \phi_n(x_i, t) \to \phi(x_i, t), i = 1, 2, 3, \end{gathered} \tag{7.6}$$

here α is a fraction, a_0 is the Bohr radius.

Equation (7.5) shows that the nucleon is in a forced vibration, in which the force of the vibration comes from the Coulomb interaction between the nucleon and the electron, it is related closely to the changes of density of the electron resulting from the vibration of the nucleon. From Eq.(7.5) we can obtain

$$\frac{\partial}{\partial x_i}u = -\frac{\chi\alpha^2 a_0^2}{M(v^2 - v_0^2)}|\phi|^2 \tag{7.7}$$

where $v_0 = \alpha a_0 (W/M)^{1/2}$, v is the speed of motion of the nucleon. Inserting Eq.(7.7) into Eq.(7.4) the following nonlinear Schrödinger equation for the motion of the electron is obtained

$$i\hbar\frac{\partial}{\partial t}\phi = -\frac{\hbar^2}{2m}\nabla^2\phi - \frac{e^2}{r}\phi - b|\phi|^2\phi \tag{7.8}$$

where $b=\alpha^2 a_0^2\chi^2/[M(v^2-v_0^2)]$ is the nonlinear interaction energy accepted by the electron in this case, which is proportional to the square of Coulomb force and is inversely proportional to the kinetic energy of the nucleon. Equation (7.8) is a standard nonlinear Schrödinger equation, which is the same as the dynamic equation (2.66) in Pang's nonlinear quantum mechanics, where the externally applied potential $V(\vec{r})$ should be the Coulomb $V(\vec{r})=-e^2/\vec{r}$. This indicates clearly that the motions of the electron in the atom can be described by the nonlinear quantum mechanics, where the nonlinear interaction, $b|\phi|^2\phi$, in Eq.(7.8) is due to the interaction between the moved electron and moved nucleon in the hydrogen atom, which can be confirmed from Eqs.(7.4)-(7.8)[19-27]. Therefore, the generation of nonlinear interaction in the atoms is subject to the self-trapping mechanism described in Chapter 4.

Equation (7.5), therefore, represents the vibration of the nucleon under action of the Coulomb interaction due to the change of motion of the electron. Then Eq. (7.5) concretely embodied our first assumption. Obviously, once the motion of the nucleon stops, the nonlinear interactiondisappears immediately, andthe motion of the electron should then be described by Eq.(7.1). This manifests clearly that the new theory in the nonlinear hydrogen atom is self-consistent with that in the linear case. Thus, it is correct and comprehensive. However, because the electron and nucleon in the hydrogen atom are always in motion; the Coulomb coupling interaction between them always exists also, then the nonlinear interaction $b|\phi|^2\phi$ in Eq.(7.6)always exists. Hence, the motion of the electron in the hydrogen atom should always be depicted by the nonlinear Schrödinger equationgiven in Eq. (7.8) or nonlinear quantum mechanics, instead of the Schrödinger equation given in Eq.(7.1) or linear quantum mechanics. Hence, they arean exhibition and embodiment of the essence of nonlinearity of motion of the electron in a hydrogen atom. This means that the traditional quantum mechanicsmodel does not adequately describe the true motions of the electron and the nucleon in the atom, but rather this new set of equations provides a better modelof calculation [19-27].

7.1.3. The Wave and Corpuscle Natures of the Electrons Arising from the Nonlinear Interaction in Hydrogen Atom

In order to clarify the nature of the electron described by Eq.(7.8), we first study its changes of feature arising from the nonlinear interaction. This means that the Coulomb interaction term $-e^2/r$ in Eq.(7.8) now is not temporarily considered because the externally applied potential cannot change the essence of the electron, only influence its outline and velocity of movement due to the independence of the potential on its wave function no matter what form of $-e^2/r$, in Eq.(7.8) is takenas mentioned in Chapter 3.

Then the nonlinear Schrödinger equation (7.8) is now represented by

$$i\phi_{t'}+\phi_{x'x'}+b|\phi|^2\phi=0\ , \tag{7.9}$$

in the one dimensional case, where $x' = x\sqrt{2m}/\hbar, t' = t/\hbar$. Equation (7.9) is similar to Eq.(3.1), but we here use another way to find its solution. We here assume that the solution of Eq.(7.9) is the form

$$\phi(x',t') = \varphi(\xi)e^{i\Omega t' + iQx'} \tag{7.10}$$

where $\xi = x' - v_e t'$ .If inserting Eq. (7.10)into Eq.(7.9) we obtain

$$\varphi_{\xi\xi}(\xi) + i(2Q - v_e)\varphi_\xi - (\Omega + Q^2)\varphi(\xi) - b\varphi^2\varphi = 0, (b > 0) \tag{7.11}$$

If we let the imaginary coefficient of φ_ξ equal zero, then $Q = v_e/2$, again from $A = Q^2 + \Omega$ we get $\Omega = A - v_e^2/4$. Equation (7.11) becomes $\varphi_{\xi\xi} + b\varphi^3 - A\varphi = 0$. This equation can be integrated, which results in

$$(\varphi_\xi)^2 = D + A\varphi^2 - b\varphi^4/2$$

where D is an integral constant. Its solution $\varphi(\xi)$ can be obtained by inverting an elliptic integral:

$$\int_{\varphi_0}^{\varphi} \frac{d\varphi}{\sqrt{D + A\varphi^2 - b\varphi^4/2}} = \pm\xi \tag{7.12}$$

Let $P(\varphi') = (\alpha_1 - \varphi'^2)(\varphi'^2 - \alpha_2) = -\varphi'^4 + A(2/b)^{1/2}\varphi'^2 + D$ where $\varphi' = (b/2)^{1/4}\varphi$ then Equation (7.12) becomes $[K(k) - F(\varphi',k)] = \pm\xi$, where K(k) and $F(\varphi',k)$ are the first associated elliptic integral and the incomplete elliptic integral, respectively, and $k = [(\alpha_1 - \alpha_2)/\alpha_1]^{-1/2}$, $\alpha_{1,2} = A/(2b)^{1/2} \pm [D + (A^2/2b)]^{1/2}$. Using these and $\alpha = \varphi_{1,2}'^2$, we have

$$\varphi'[(b/2)^{1/4}\xi] = \varphi_1'\{1 - [(1 - \varphi_1'^2/\varphi_2'^2)sn^2(\xi(b/2)^{1/4}, k)]\}^{1/2}$$

When $D \to 0, \varphi_1' \to \varphi_0', k \to 1, \varphi' \to \varphi_0' \sec h[\varphi_0'\xi(b/2)^{1/4}]$,

where $\varphi_0' = (2A^2/b)^{1/4}$.

Then the soliton solution of Eq.(7.9) can finally be represented [25-28] by

$$\phi_s(x',t') = \sqrt{\frac{2A}{b}}\sec h[\sqrt{A}(x'-v_e t')]\exp[i(v_e x'/2 - v_e^2 t'/4 + A(1/b)^{1/2}t')] \quad (7.13)$$

If the solution of the nonlinear Schrödinger equation in Eq. (7.9) is represented in the coordinate of (x, t), then the above solution can be eventually expressed by [25-27]

$$\phi(x,t) = A_0 \sec h\left\{\frac{A_0\sqrt{bm}}{\hbar}\left[(x-x_0)-v_e^{'}t\right]\right\}e^{i[mv_e^{'}(x-x_0)-Et]/\hbar} \quad (7.14)$$

Clearly, Eq.(7.14) is analogous to Eq.(3.7.), then its outline is the same with Figure 3.1, but here $A_0 = \sqrt{\frac{mv_e^{'2}-E}{b}} = \sqrt{\frac{m(v_e^{'2}-2v_e^{'}v_c^{'})}{b}}$, $v_e^{'}$ and $v_c^{'}$ are the group and phase velocity of the electron, respectively, and $E = 2mv_e^{'}v_c^{'} = \hbar\omega$ in the coordinate of (x, t). Thus, the soliton solution (7.14) represented the wave-corpuscle duality of the electron, where the bell-typeenvelope $\varphi(x,t) = A_0 \sec h\left\{A_0\sqrt{bm}\left[(x-x_0)-v_e^{'}t\right]/\hbar\right\}$ is a soliton and denotes thestate of the mass center of the electrons, with an amplitude A_0 and a width $W' = 2\pi\hbar/\sqrt{mb}A_0$, then its size is $A_0 W' = 2\pi\hbar/\sqrt{mb}$. Upon that, we can say that this electron is localized at x_0 in this case. The carrier waves $\exp\{i[mv_e^{'}(x-x_0)-Et]/\hbar\}$ are similar to Eq.(1.9) and a plane wave, which embodied the wave feature of the electron. In this case, the electron described by Eq. (7.4) can move freely in a uniform velocity, $v_e^{'}$ in space-time, retaining its form and amplitude according to the soliton theory [28-30].

In this case, the electron described by Eq. (7.14) can move freely in a uniform velocity $v_e^{'}$ in space-time, retaining its form and amplitude according to the soliton theory [28-30]. Its effective mass, momentum, and energyare also determinant, which can be obtained from the following formulae

$$N_s = \int_{-\infty}^{\infty}|\phi|^2 dx = 2A_0\hbar/\sqrt{bm},\ P = -i\hbar/2m\int_{-\infty}^{\infty}\left(\phi^*\phi_x - \phi\phi_x^*\right)dx = N_s v_e^{'} = const$$

$$E = \int_{-\infty}^{\infty}\left[\frac{\hbar^2}{2m}|\phi_x|^2 - \frac{1}{2}b|\phi|^4\right]dx = E_0 + \frac{1}{2}M_{sol}v_e^{'2} \quad (7.15)$$

where $E_0 = -\sqrt{bm}\hbar A_0^2/3m$ is its effective static-energy in the static state, which is also a constant. Therefore, the momentum and energy of the electron in Eq.(7.15) resembles those of classical particles.

Therefore, the solution in Eq.(7.14) represented the wave-corpuscle duality of the electron in this case. This nature of the electron has never been reported in quantum mechanics literature. It is also consistent with the de Broglie relation of wave-corpuscle duality and Davisson and Germer's experimental result of electron diffractionon double seam, in 1927 [31-35], as described in Chapter 1.

In this case, we can also demonstrate that the electron in the atom no longer disperses and is localized. This demonstration is described as follows.

We now represented the solution of Eq.(7.8) by [19-20]

$$\phi(\vec{r},t)=\varphi'(\vec{r})\exp(-iEt/\hbar) \tag{7.16}$$

Inserting Eq.(7.16) intoEq. (7.10) we can get

$$-\frac{\hbar^2}{2m}\nabla^2\varphi'(r)+V(\vec{r})\varphi'(r)-b|\varphi'|^2\varphi'=E\varphi' \tag{7.17}$$

where $V(\vec{r})=-e^2/\vec{r}$. In this case, the Hamiltonian equation with both the nonlinear interaction $b|\varphi'|^2\varphi'$ and the Coulomb potential $V(\vec{r})=-e^2/\vec{r}$ can be represented by

$$\hat{H}\varphi'(\vec{r})=E\varphi'(\vec{r}) \tag{7.18}$$

where E is the eigenenergy of the electron, the Hamiltonian operator of the system is

$$\hat{H}=-\frac{\hbar^2}{2m}\nabla^2+V(\vec{r})-b|\varphi'|^2=\hat{H}_0-b\rho\,, \tag{7.19}$$

where $\hat{H}_0=-\frac{\hbar^2}{2m}\nabla^2+V(\vec{r})$, which is the same with Eq.(1.2), $\rho(\vec{r})=|\varphi'(\vec{r})|^2$. Equation (7.19) manifests that the eigenenergy of the electron in hydrogen atom is depressed about $-b|\varphi'|^2=-b\rho$ relative to that in quantum mechanics due to the nonlinear interaction. Thus, the stability of the electron is enhanced in this case.

In fact, equation (7.19) can be written as

$$\nabla^2\varphi'=-\frac{d}{d\varphi'}V_{eff}(\varphi') \tag{7.20}$$

$$V_{eff}(\varphi')=\frac{\hbar^2}{2m}[\frac{1}{4}b|\varphi'|^4-\frac{1}{2}(V-E)|\varphi'|^2] \tag{7.21}$$

Equation (7.20) is analogous with quasi-Newton law in classical physics, if x and φ' serve as the time and position of the electron, respectively. This displays the electron's classical feature.

On the other hand, when $V > E$ and $V < E$, the relationship between $V_{eff}(\varphi')$ and φ' is similar to Figure 3.2 in a one-dimensional case. This figure shows that there are two minimum values in the effective potential, which correspond to two ground states of the electron in the system, *i.e.* $\varphi_0' = \pm\sqrt{(V-E)/b}$ .Thus, it is a double-well potential, the energies of the two ground states are negative and can be denoted as $-(V-E)^2/4b \leq 0$. This shows that the electron can be really localized and is stable due to the negative binding energy. This localization of the electron is achievedand carried out through repeated reflection of the electron in the double-well potential field. The two ground states limit the energy diffusion, thus the energy of the electron is gathered, a soliton stateis formed, and then the electron is eventually localized. Obviously, only if $b \neq 0$, the system always has two ground states, and the electron also is always localized. This indicates again that the nonlinear interaction $b|\phi|^2\phi$ plays a key role for the wave-corpuscle duality and localization of the electron in the hydrogen atom.

The above results manifest clearly the electron has a wave-corpuscle duality and is localized, then the wave function $\phi(\vec{r},t)$ or φ(x,t) and the Coulomb potential $V(\vec{r}) = -e^2/\vec{r}$ in Eq.(7.9) really represented the states and interaction potential of the electron at position $\vec{r}$ or x and time t, respectively. Therefore, the contradiction between first assumption and final results in linear quantum mechanics was eliminated in this nonlinear model. This is a very interesting result; it indicates the necessity to develop nonlinear quantum mechanics.

However, what is the size of the nonlinear interaction of the electron in the hydrogen atom on earth? In fact, we can estimate its size from Eqs.(7.2) and (7.7). For the electron in the hydrogen atom, we know its charge is e =1.6×10^{-19}C, r =0.5×10^{-10} m, m_e= 9.1×10^{-31}kg, its greatest wavelength is $\lambda_e \approx 9\times10^{-11} m$ because the electron is limited to the hydrogen atom, thus we can find the electronic kinetic energy $E = \hbar^2 k^2/2m =$ $\hbar^2/(2m_e\lambda_e^2)$ $\approx 4.83\times10^{-19} J$. But the nonlinear interaction energy of the electron is $b = Ke^2u/r^2 =$ $4.81\times10^{-19} J$,where $K = 9\times10^9 Nm^2C^{-2}$, the greatest displacement of the nucleon of $u = 0.1r$ is used. This indicates that the nonlinear interaction energy of the electron is greater and can balance its kinetic energy in the hydrogen atom, i.e., the dispersive effect of kinetic energy of the electron can be suppressed by the nonlinear interaction, thus the electron becomes a soliton and is localized in accordance with the soliton theory [28-30]. Then, it is not surprising that the electron possesses a wave-corpuscle duality in the hydrogen atom no matter what form of the externally applied potential, $-Ze^2/r$, in Eq.(7.8) is taken because the latter cannot change the nature of the electron, but only its states, such as velocity and amplitude.

Also, the soliton solution of Eq.(7.8) without the Coulomb potential, in the three dimensional case, can be also obtained using Jajiri and colleagues [36] and Gaididei et al.'s [37] methods, and so on. We can infer that the soliton nature and feature of the solution of Eq.(7.8) cannot be changed, although in dimension of the systems are changed, only its outline and representation may be varied in this case.

7.1.4. The Real Wave-Corpuscle Duality of Electron and Its Energy Quantization in the Hydrogen Atom under Influence of the Coulomb Potential

However, once the Coulomb potential is involved in Eq.(7.8) in the three dimensional case, its analytic solutions are very difficult to find, even though the methods of the inverse scattering and variable separation are used. Therefore, we have not seen it yet. The analytic solutions of Eq.(7.8) are directly obtained and listed by others. We gave its analytic soliton-solutions using a separated function method even though the Coulomb potential is now considered.

We now find the solution of the time-independent nonlinear Schrödinger equation (7.17) and demonstrate further the wave-corpuscle duality of the electron in the stable hydrogen atom.

In this case, we now express the solution of Eq.(7.18) by a product of two wave functions, $\varphi_1'(\vec{r})$ and $\varphi_2'(\vec{r})$, i.e.,

$$\varphi'(\vec{r}) = B\varphi_1'(\vec{r})\varphi_2'(\vec{r}) \tag{7.22}$$

where B is a constant, and its concrete representation can be found by normalized condition of $\varphi'(\vec{r})$. We now compel $\varphi_1'(\vec{r})$ and $\varphi_2'(\vec{r})$ to satisfy the different soluble equations, respectively. In the hydrogen atoms, let them satisfy the following linear and nonlinear equations:

$$-\frac{\hbar^2}{2m}\nabla^2\varphi_1'(\vec{r}) - (Ze^2/\vec{r})\varphi_1'(\vec{r}) = E_1\varphi_1'(\vec{r}) \tag{7.23}$$

and

$$-\frac{\hbar^2}{2m}\nabla^2\varphi_2'(\vec{r}) - b\left|\varphi_2'(\vec{r})\right|^2\varphi_2'(\vec{r}) = E_2\varphi_2'(\vec{r}), \tag{7.24}$$

respectively, where E_1, $\varphi_1'(\vec{r})$ and E_2, $\varphi_2'(\vec{r})$ are their eigenenergy and eigenfunction, respectively. If E_1, $\varphi_1'(\vec{r})$ and E_2, $\varphi_2'(\vec{r})$ can be found out from Eqs.(7.23)-(7.24), then the representation of the solution of Eq.(7.17), E and $\varphi(\vec{r})$, can be obtained from Eq.(7.22).

The form of the solution in Eq.(7.22) of Eq.(7.17) resembles every De Broglie's solution described in his nonlinear wave theory, in which he suggested and thought that the solution to the dynamic equation of microscopic particles should be composed of two functions, where one solution denoted the wave feature of the microscopic particle, but another solution represents its corpuscle feature. Sorrily, he finally has not found these dynamic equations satisfied by the particle. However, we both give their dynamic equations, Eqs. (7.23)-(7.24) and find out their concrete representations of the solutions. This is quite an interesting event.

In practice, equation (7.23) is just the linear Schrödinger equation in Eq.(7.1), then its solution $\varphi_1^{'}(\vec{r})$ is Eq.(1.18), the eigenenergy, E_1 , is just Eq.(1.19). Equation (7.24) is a time-dependent nonlinear Schrödinger equation, its solution and eigenenergy E_2 can be easily obtained by the methods mentioned above, which are [19-23]

$$\varphi_2^{'}(x) = A_0^{'} \sec h[\frac{A_0^{'}\sqrt{bm}}{\hbar}(x - x_0)] \tag{7.25}$$

and

$$E_2 = -A_0^{'2} b / 2 \tag{7.26}$$

in a one dimensional case, respectively. The solution is a non-propagating bell-type soliton with amplitude $A_0^{'}$, which is a constant. This shows that the electronis still in a binding state and localized; its binding energyis completely provided by the nonlinear interaction energy b.

In the three dimensional case we can also find the soliton solution of Eq.(7.24), its one solution can be denoted by

$$\varphi_2^{'} = A_0^{'} \sec h\{\frac{A_0^{'}\sqrt{bm}[a_1(x - x_0) + a_2(x - x_0) + a_3(x - x_0)]}{\hbar(a_1^2 + a_2^2 + a_3^2)^{1/2}}\}$$

where a_1, a_2, and a_3 are some constants, its energy is still $E_2 = -A_0^{'2} b / 2$.

Inserting now Eq.(7.22) into Eq.(7.17) and utilizing Eqs.(7.23)-(7.24) we obtain

$$\begin{aligned}
&-\frac{\hbar^2}{2m}\nabla^2\varphi'(r) + V(\vec{r})\varphi'(r) - b|\varphi'|^2\varphi' = \\
&B\varphi_2^{'}[-\frac{\hbar^2}{2m}\nabla^2\varphi_1^{'}(\vec{r}) - (e^2/\vec{r})\varphi_1^{'}(\vec{r})] + B\varphi_1^{'}[-\frac{\hbar^2}{2m}\nabla^2\varphi_2^{'}(\vec{r}) - b|\varphi_2^{'}(\vec{r})|^2\varphi_2^{'}(\vec{r})] \\
&-bB\varphi_1^{'}(\vec{r})[B^2|\varphi_1^{'}(\vec{r})|^2 - 1]|\varphi_2^{'}(\vec{r})|^2\varphi_2^{'}(\vec{r}) - \frac{\hbar^2 B}{m}\nabla\varphi_1^{'}(\vec{r})\nabla\varphi_2^{'}(\vec{r}) \\
&= E_1 B\varphi_1^{'}(\vec{r})\varphi_2^{'}(\vec{r}) + E_2 B\varphi_1^{'}(\vec{r})\varphi_2^{'}(\vec{r}) - bB\varphi_1^{'}(\vec{r})[B^2|\varphi_1^{'}(\vec{r})|^2 - 1]|\varphi_2^{'}(\vec{r})|^2\varphi_2^{'}(\vec{r}) - \\
&\frac{\hbar^2 B}{m}\nabla\varphi_1^{'}(\vec{r})\nabla\varphi_2^{'}(\vec{r}) = EB\varphi_1^{'}(\vec{r})\varphi_2^{'}(\vec{r})
\end{aligned} \tag{7.27}$$

Therefore, the eigenenergy of the electron in the hydrogen atom in this case is

$$E = E_1 + E_2 - b[B^2|\varphi_1^{'}(\vec{r})|^2 - 1]|\varphi_2^{'}(\vec{r})|^2 - \frac{\hbar^2}{m}\nabla\varphi_1^{'}(\vec{r})\nabla\varphi_2^{'}(\vec{r}) / \varphi_1^{'}(\vec{r})\varphi_2^{'}(\vec{r}) \tag{7.28}$$

Since $\varphi_1^{'}(\vec{r})$ and $\varphi_2^{'}(\vec{r})$ are determinant, the value of B in Eq.(7.22) can also be obtained, then the eigenenergy of the electron can be found and is also determinant. This implies that equation (7.22) is really a solution of Eq.(7.21), the energy of the electron is denoted in Eq.(7.28) in this case. This is the first time we can obtain the solution of the nonlinear Schrödinger equation with the center or Coulomb potential in nonlinear science and physics. Thus, we represented and exhibited the natures and state of the electron in nonlinear hydrogen atom by Eqs.(7.22) and (7.27) from the representations of $\varphi_1^{'}(\vec{r})$ and $\varphi_2^{'}(\vec{r})$ as well as the values of E_1 and E_2 mentioned above. In fact, the solution shown in Eq.(7.22) is still a bell-type soliton, which is similar to Figure 3.1. Therefore, equation (7.17) still has a nonpropagating soliton solution. Thus, the electron in the hydrogen atom described by the new nonlinear theory is, in essence, a soliton having wave-corpuscle duality and a localized feature, but the amplitude of the wave is decaying. Thus, we were the first to give the wave-corpuscle duality and localized feature of the electrons in atoms by the new nonlinear theory. This result is completely consistent with the de Broglie relation of wave-corpuscle duality and Davisson and Germer's experimental result of electron diffractionon double seam in 1927 [31-35]. This shows that considering the movement of the nucleon and the new nonlinear theory, describing the nature of the electron in the hydrogen atom is successful and correct.

On the other hand, the energy of the electron in Eq. (7.28) contains four terms, where the energy, E_1, is discrete and quantized as shown in Eq.(1.19), but the energies, E_2 , $-b[B^2|\varphi_1^{'}(\vec{r})|^2-1]|\varphi_2^{'}(\vec{r})|^2$ and $-(\hbar^2/m)\nabla\varphi_1^{'}(\vec{r})\nabla\varphi_2^{'}(\vec{r})/\varphi_1^{'}(\vec{r})\varphi_2^{'}(\vec{r})$ are continuous and negative. Therefore, the electron in the nonlinear hydrogen atom is still quantized and distributed in accordance with the energy levels, which resemble those in linear quantum mechanics, but the negatively continuous energy in Eq.(7.28) depresses the values of the original energy-levels relative to those of E_1 in Eq.(1.16) in linear quantum mechanics. Thus, the electron in the nonlinear hydrogen atom is now in a binding state and more stable relative to that in linear quantum mechanics.

It is well known that $\varphi_1^{'}(\vec{r})$ is normalized and B is an undetermined constant in Eq.(7.26), then we can make $[B^2|\varphi_1^{'}(\vec{r})|^2-1]$ in Eq.(7.28) approach zero through adjusting the value of B. Meanwhile, $(\hbar^2/m)$ in $-(\hbar^2/m)\nabla\varphi_1^{'}(\vec{r})\nabla\varphi_2^{'}(\vec{r})/\varphi_1^{'}(\vec{r})\varphi_2^{'}(\vec{r})$ is very small because $\hbar=1.054\times10^{-34}J.\text{sec.}, m=9.1\times10^{-31}kg$. Thus, we can completely neglectthe two terms of $-b[B^2|\varphi_1^{'}(\vec{r})|^2-1]|\varphi_2^{'}(\vec{r})|^2$ and $-(\hbar^2/m)\nabla\varphi_1^{'}(\vec{r})\nabla\varphi_2^{'}(\vec{r})/\varphi_1^{'}(\vec{r})\varphi_2^{'}(\vec{r})$ in Eq.(7.28). Then, the eigenenergy of the electron in Eq.(7.28) can be approximately expressed by

$$E\approx E_1+E_2=-A_0^{'2}b/2 - me^4/2\hbar^2 n^2, (n=1,2,3......) \tag{7.29}$$

This shows clearly that the eigenenergy of the electron in the hydrogen atom is quantized. Thus, we can conclude that Eq. (7.22) with Eqs.(1.15) and (7.25), which has a center potential, is really a solution of Eq.(7.18), and the eigenenergy of the electron is denoted in Eq.(7.28) or Eq.(7.29). This is the first time the soliton solution of the nonlinear Schrödinger equation with center, or the Coulomb potential in nonlinear science and physics, was found. The solution in Eq.(7.22) denotes the motion of a soliton, thus the electron in the hydrogen atom, which is described by the solution in Eq.(7.22), has a real wave-corpuscle duality, which is completely consistent with the de Broglie relation of wave-corpuscle duality, and Davisson and Germer's experimental result of electron diffraction on double seam in 1927 [31-35].Meanwhile, the soliton in Eq.(7.22) is localized at a determinant position, then the Coulomb potential, $V= -e^2/\vec{r}$, represents only the potential of the electron localized at this position. Then, this result eliminates thoroughly the contradiction between thefirst assumption and final results, as mentioned in Chapter 1. So, its meaning is quite important. Therefore, the nonlinear Schrödinger equation with a center potential, Eq.(7.8) or Eq.(7.18),thoroughly change the description and representation of the electron in atoms, which will promote the developments of atomic and molecular physics.

The eigenenergy of the electron in the hydrogen atom in Eqs.(7.28) or (7.29) is still determinant and quantized, which is the same as that in linear quantum mechanics, but its size is depressed, about $A_0'^2 b/2$ relative to $- Ze^4/2\hbar^2 n^2$ in linear quantum mechanics, due to the nonlinear interaction $b|\phi|^2\phi$, which is due to the change of interaction between the moved electron and the nucleon in the hydrogen atom. This is a new and very interesting result, which has not yet been obtained. This means that the nonlinear interaction, or the interaction between the moved electron and the nucleon enhanced the localized and stable natures of the electron in hydrogen atom, thus the electron is more stable in the nonlinear hydrogen atom than that in linear quantum mechanics. This affirms the correctness of the nonlinear theory of the hydrogen atom.

On the other hand, we can explain perfectly the spectrum features of the hydrogen atom, which were confirmed by many experiments, using the luminescence theory of materials, which is due to the transition of the energy level of the electron with a corpuscle feature, based on the distribution of energy-levels of the hydrogen atom shown in Eq.(7.29). This affirmed experimentally the validity of the wave-corpuscle natures of the electron and the nonlinear quantum theory of the hydrogen atom proposed by us. Certainly, we can also explain other phenomena of the hydrogen atom using this nonlinear theory.

7.1.5. Dynamic Properties of the Electron in Atoms with Many Electrons

The nonlinear nature of motion of the electron arising from the interaction between the moved electron and the nucleon in the hydrogen atom can be generalized to all atoms and molecules containing many electrons. In this case, we must consider again the interactions of other electrons on the studied electron, or in other words, the interaction among the electrons, except for the above nonlinear interaction between the electrons and nucleons, $b|\phi|^2\phi$. The former is an effect of many particles, in which there is self-interaction among these electrons. In accordance with the quantum-field theory [38], and the theory of many bodies-many particles [39-40], we may express the nonlinear interaction by $D|\phi|^4\phi$ in their dynamic equation [39-40]. Thus, the dynamic equation of the electron in the atoms with many electrons will be changed, and the dynamic equation of the electron in these atoms and molecules with many electrons should be denoted by the following nonlinear Schrödinger equation

$$i\hbar\frac{\partial\phi}{\partial t}=-\frac{\hbar^2}{2m}\nabla^2\phi-\frac{Ze^2}{r}\phi-b|\phi|^2\phi+D|\phi|^4\phi+A(\phi) \tag{7.30}$$

where the Coulomb potential between the electrons and nuclei, $V(\vec{r})$, should now be denoted by $-Ze^2/r$, where Z is the atom number of the atom. Thus, equation (7.30) should be a correct and universal form of motion of the electrons in the atoms and molecules with many electrons. The distinctions among the different atoms or molecules are only the sizes and forms of $V(\vec{r})$, b, D and $A(\varphi(\vec{r},t))$.

Using Eq.(7.16), and inserting it into Eq.(7.30) at $A(\varphi(\vec{r},t))=0$ we can obtain

$$-\frac{\hbar^2}{2m}\nabla^2\varphi'(r)-\frac{Ze^2}{r}\varphi'(r)-b|\varphi'|^2\varphi'+D|\varphi'|^4\varphi'=E\varphi' \tag{7.31}$$

It is quite difficult to solve Eq.(7.31). Wewill still introduce Eq. (7.23) to make Eq.(7.31) become two equations, in which the equation of $\varphi_1'(\vec{r})$ is still Eq.(7.23), and its solution and corresponding eigenenergy are still denoted by Eq.(1.18), but the equation of $\varphi_2'(\vec{r})$ becomes

$$-\frac{\hbar^2}{2m}\nabla^2\varphi_2'(\vec{r})-b|\varphi_2'(\vec{r})|^2\varphi_2'(\vec{r})+D|\varphi'|^4\varphi'=E_2\varphi_2'(\vec{r}), \tag{7.32}$$

The solution of Eq.(7.32) is still difficult to acquire. However, Barashenkov and Makhankov [40] found out the solutions of some equations, which resemble Eq.(7.32). We here give the solution of Eq.(7.32) in a one-dimensional case, using Barashenkov and Makhankov's method. For the solution of Eq.(7.32) we now represent it approximately by

$$\varphi_{dr}'(x)=e^{i\theta_0}\left\{\frac{4E_2}{1+\sqrt{1-16E_2/3}\cosh[2\sqrt{E_2}(x'-x_0')]}\right\}^{1/2} \tag{7.33}$$

where $\theta_0\to vx'/2+\theta_0$, $x'=x\sqrt{2m}/\hbar$. $\xi'=x'-vt'$

We also showed the localized solution of Eq.(7.32) in accordance with Barashenkov and Makhankov's method [40]. In the case of $\rho_0>A>0,$ the localized solution can be denoted by

$$\varphi_b'(\xi)=e^{i\theta_0}\frac{\sqrt{A}\cosh[v_{s0}(x'-x_0')/2]}{(1+A/\rho_0)si\,\mathrm{sh}^2[2v_{so}(x'-x_0')/2]}$$

where $\rho_0 = (1+\sqrt{1-4E_2})/2$, A is a constant parameter, and v_{so} is the speed of the localized soliton solution.

If in the region of $-\rho_0/2 > A > 0$ and $A < \rho_0/2$, its solution may use the following form:

$$\varphi_k^{'} = e^{i\theta_0} \frac{\sqrt{|A|} si\,\mathrm{sh}[v_{s0}(x'-x_0^{'})/2]}{(1+|A|/\rho_0)\cosh^2[v_{so}(x'-x_0^{'})/2]}$$

where $\varphi_k^{'}$ is an odd function of $(x'-x_0^{'})$ and has a kink form.

We confirmed that the above, localized solutions are stable using a numerical simulation method and linear stability theorem of the quasiparticle described in Chapter 3.

The above results indicated that the electrons in the atoms with many electrons is still localized, but its features and outlines are different from that in Eq.(7.17), i.e., its amplitude and outlines also change with varying space due to the variation of interaction among the electrons. However, in this case the properties of the electron and its changed rules still need further study.

7.2. Nonlinear Quantum Dynamic Properties of Electrons in Coupled Electron-Phonon Systems

As is known, the investigation of property of coupled election-phonon systems has always been a very important subject in the physics of condensed matter, because there are a great number of interesting phenomena and effects in the systems, which arise from the interaction between the electron density fluctuation and lattice deformation, for example, the charge density wave (CDW), polaron effect, superconductivity and dimerization of lattice and Peierls transition, and so on [41-47], which occur in different coupling strengths and materials, respectively, and are some nonlinear quantum mechanical phenomena. Therefore, it is quite necessary to study, in-depth, these problems in these systems.

In general, a coupled electron-phonon system, also including anelectron-electron interaction, is often described by the modified Hubbard modelin the systems [44]. The one-dimensional modified Hubbard model [48]was studied in detail by Lindner and Fedyanin [49], and Makhankov et al. [50-52]. In this case, the model Hamiltonianof the systems is often written in the form of

$$H = H_{ph} + H_e + H_{\text{int}} \tag{7.34}$$

where H_{ph} and H_e are the Hamiltonians of the phonon and electronsubsystems, respectively, while H_{int} describes the interaction between them. H_e corresponds to the Hamiltonian in the Hubbard model[48] in the nearest-neighbor approximation. We here assume that its ground state is antiferromagnetic and that the lattice is divided into two equivalent-ferromagnetic sub-lattices, A and B, so that the resulting total magnetic moment is zeroin the absence of an external magnetic field. Based on this, Makhankov et al., [50-52] established the following form for H_e,

$$H_e = \sum_{n\in A, j\in B;\sigma} J_{nj}(b^{A+}_{n,\sigma} b^{B}_{n,\sigma} + b^{B+}_{n,\sigma} b^{A}_{n,\sigma}) + \frac{V}{2} \sum_{n\in A,B;\sigma} N_{n,\sigma} N_{n,-\sigma} - \mu \sum_{n,\sigma} N_{n,\sigma} \tag{7.35}$$

where $J_{n,j} = \begin{cases} \bar{J}, (j = n \pm 1), \\ 0, (j \neq n \pm 1), \end{cases}$

$b^{+}_{n,\sigma}(b_{n,\sigma})$ is the creation (annihilation) operator for an electron with spin σ in a Wannier state at site n, $N_{n,\sigma} = b^{+}_{n,\sigma} b_{n,\sigma}$ is the number operator, J is the transition amplitude between nearest neighbors, V is the repulsive interaction between electrons in the same atom, and μ denotes the chemical potential of the system. The lattice phonon part of the Hamiltonian, H_{ph}, is given in the harmonic approximation, which is denoted by

$$H_{ph} = \frac{M}{2} \sum_{n} \left[(R_n)_t \right]^2 + \frac{W}{2} \sum_{n} (R_{n+1} - R_n)^2 \tag{7.36}$$

where M is the atomic mass, W denotes the force constant of the system, $R_n = R_n(t) = \overline{R}_n + u_n(t)$ is the position of the nth atom, and $\overline{R}_n$ represents its equilibrium position. In the linear approximation with respect to small deviations $u_n(t)$, the interaction between the electron and the lattice subsystems in Eq.(7.34) can be written as

$$H_{\text{int}} = I \sum_{n,\sigma} (u_{n,\sigma} - u_{n+1,\sigma}) \left[b^{B+}_{n,\sigma} b^{A}_{n+1,\sigma} + b^{A+}_{n+1,\sigma} b^{B}_{n,\sigma} + b^{B+}_{n+1,\sigma} b^{A}_{n,\sigma} + b^{A+}_{n,\sigma} b^{B}_{n+1,\sigma} \right] \tag{7.37}$$

where

$$I = \left\langle (n+1) \left| \frac{\partial V(x - \tilde{R}_n)}{\partial x} \right| n \right\rangle$$

denotes the matrix element of the local "force" between the Wannier states of neighboring lattice atoms, corresponding to the local potential V = V (x − $\overline{R}_n$). Effects of anharmonicity in the lattice vibrations, and nonlinearity in the expansion of electron integration force in this model have been studied, and many interesting features of the microscopic particles (solitons) can be revealed.

Since the ground state of the system is antiferromagnetic, thus, we are here interested in the low-lying excitations, and in the equations describing the amplitudes of spin up or down states $(\phi_{n\uparrow}, \phi_{n\downarrow})$ at the nth lattice site. In this case, we may eliminate these terms of the type $(u_{n+1} - u_n) b^{B+}_{n,\sigma} b^{A}_{n-1,\sigma}$, etc.

To simplify the calculation, Makhankov et al. [50-52] employed a method similar to the adiabatic approach related to the ground state, and examined the change of the state $\|0\rangle$ under the action of the evolution operator, which is defined on single-particle states. Thus, Lindner et al. [49], and Makhankov et al. [50-52], in this case defined

$$\phi^{p}_{n,\sigma}(t) = \langle \varphi(t) | b^{p}_{n,\sigma}(t) | \varphi(t) \rangle = \langle 0 | b^{p}_{n,\sigma}(t) | 0 \rangle = \langle 0 | U b_{n,\sigma} U | 0 \rangle \tag{7.38}$$

Making use of (7.36) – (7.38) at $\hbar = 1$ we obtain the following equation for $b^{A}_{n,\sigma}(t)$

$$\begin{aligned} i\left(b^{A}_{n,\sigma}(t)\right)_t &= \left[b^{A}_{n,\sigma}, H\right]_- = V b^{A}_{n,\sigma}(t) n_{n-\sigma} - \mu b^{A}_{n,\sigma}(t) + \\ &\left[\bar{J} + I(u_n - u_{n-1})\right] b^{B}_{n+1,\sigma}(t) + \left[\bar{J} + (u_{n-1} - u_n)\right] b^{B}_{n-1,\sigma}(t) \end{aligned} \tag{7.39}$$

The corresponding equation for $b^{B}_{n,\sigma}$ is the same as (7.39) except that A and B areinterchanged. Applying (7.38) and (7.39), we can get

$$\begin{aligned} i\left(\phi^{A}_{n,\sigma}(t)\right)_t &= V \langle 0 | b^{A}_{n,\sigma}(t) n_{n,-\sigma} | 0 \rangle - \mu \phi^{A}_{n,\sigma}(t) + \\ &\left[\bar{J} + I(u_n - u_{n-1})\right] \phi^{B}_{n+1,\sigma}(t) + \left[\bar{J} + I(u_{n-1} - u_n)\right] \phi^{B}_{n-1,\sigma}(t) \end{aligned} \tag{7.40}$$

and a similar equation for $\phi^{B}_{n,\sigma}(t)$, which is the same as the above except that Aand B are interchanged. Applying the following decoupling procedure to (7.40),

$$\begin{aligned} &\langle 0 | b^{p}_{n,\sigma}(t) b^{p+}_{n,-\sigma}(t) b^{p}_{\eta,-\sigma}(t) | 0 \rangle \to \langle 0 | b^{p}_{n,\sigma}(t) | 0 \rangle \langle 0 | b^{p+}_{n,-\sigma}(t) | 0 \rangle \langle 0 | b^{p}_{\eta,-\sigma}(t) | 0 \rangle \\ &= \phi^{p}_{n,\sigma}(t) \phi^{p*}_{n,-\sigma}(t) \phi^{p}_{n,-\sigma}(t) \end{aligned} \tag{7.41}$$

where p = A,B. Makhankov et al., [50-52] finally got a system of nonlinear equations for the four complexfunction $\phi^{p}_{n,\sigma}(t)$ and the real function u_n. For the "pure" Hubbard model(I = 0),similar nonlinearequations can be obtained for $\phi_{n,\sigma}(t)$ as well.

Makhankov et al., [50-52] also discussed the influence of spin ordering on the lattice vibrations, in which they constructed the functional:

$$F\left(\phi_{n,\sigma}^0, u_n\right) = \langle\phi(t)|H|\phi(t)\rangle = -\mu \sum_{n\in A\cup B;\sigma} \left|\phi_{n,\sigma}\right|^2 - \sum_{n,\sigma}\left[\bar{J} - I\left(u_n - u_{n+1}\right)\right]\times(\phi_{n,\sigma}^{A*}\phi_{n+1,\sigma}^{B} + \phi_{n+1,\sigma}^{B*}\phi_{n,\sigma}^{A} +$$

$$\phi_{n,\sigma}^{B*}\phi_{n+1,\sigma}^{A} + \phi_{n+1,\sigma}^{A*}\phi_{n,\sigma}^{B}) - \frac{1}{2}\sum_{n\in A\cup B}\left|\phi_{n,\sigma}\right|^2\left|\phi_{n,-\sigma}\right|^2 + \frac{M}{2}\sum_n\left[\left(R_n\right)_t\right]^2 + \frac{W}{2}\sum_n\left(R_{n+1} - R_n\right)^2$$

From the Hamilton equations,

$$M\left(R_n\right)_{tt} = P_n, P_n = -\frac{\delta F}{\delta R_n}$$

they follow that

$$M\left(R_n(t)\right)_{tt} = W\left[R_{n+1}(t) - 2R_n(t) + R_{n-1}(t)\right] - I\sum_\sigma \begin{Bmatrix} \phi_{n,\sigma}^{A*}(t)\left[\phi_{n+1,\sigma}^{B}(t) - \phi_{n-1,\sigma}^{B}(t)\right] - \left[\phi_{n+1,\sigma}^{B*}(t) - \phi_{n-1,\sigma}^{B*}(t)\right]\phi_{n,\sigma}^{A}(t) - \phi_{n,\sigma}^{B*}(t)\left[\phi_{n+1,\sigma}^{A}(t) - \phi_{n-1,\sigma}^{A}(t)\right] - \\ \left[\phi_{n+1,\sigma}^{A*}(t) - \phi_{n-1,\sigma}^{A*}(t)\right]\phi_{n,\sigma}^{B}(t) \end{Bmatrix} \tag{7.42}$$

This equation is a discrete analogue of the harmonic oscillator equation under the action of an external force, which does not vanish if variation in the initial spin ordering takes place. The usual condition for antiferromagnetism $\left\langle N_{n,\sigma}^A\right\rangle = \left\langle N_{n\pm1,\sigma}^B\right\rangle$ now becomes $\phi_{n,\sigma}^A = \phi_{n\pm1,\sigma}^B$ which causes the last term (−I)in Eq. (7.42) to vanish.

If we assumed that the variation of magnetic order is described by a sufficientlysmooth function of coordinate along the chain, then we can use the continuum approximation:

$$u_n(t) = R_n(t) - \bar{R}_n \to \left(\chi(\xi,t) - \xi\right)\gamma_0, \phi_{n,\sigma}^p(t) \to \phi_\sigma^p(\xi,t)$$

to find the solution. Then, expanding up to the second derivative and taking into account, up to the cubic nonlinear terms, we can obtain

$$R_{n+1}(t) \to \gamma_0\left(\chi + \chi_\xi + \frac{1}{2}\chi_{\xi\xi} + \cdots\right),$$

$$\phi_{n\pm1,\sigma}^B(t) \to \phi_{-\sigma}^A(\xi,t) \pm \phi_{-\sigma}^{A'}(\xi,t) + \frac{1}{2}\phi_{-\sigma}^{A''}(\xi,t) + \cdots,$$

$$\phi_{n\pm1,\sigma}^A(t) \to \phi_{-\sigma}^B(\xi,t) \pm \phi_{-\sigma}^{B'}(\xi,t) + \frac{1}{2}\phi_{-\sigma}^{B''}(\xi,t) + \cdots,$$

where $\phi_\sigma^B(\xi,t) = 0$ if $n \in A$, and similarly $\phi_\sigma^A(\xi,t) = 0$, if $n \in B$.

Makhankov et al., [50-52] finally obtained the following equations for $\phi_\sigma(\xi,t)$ and $\chi(\xi,t)$:

$$i\left(\phi_\sigma(\xi,t)\right)_t=\left[\left(\phi_{-\sigma}^n(\xi,t)\right)_{\xi\xi}+2\phi_{-\sigma}(\xi,t)\right]\left[\tilde{J}+r_0I\left(1-\chi_\xi(\xi,t)\right)\right]+I\left|\phi_{-\sigma}\right|^2\phi_\sigma-\mu\phi_\sigma \quad (7.43)$$

$$\ddot{\chi}(\xi,t)=v_0^2\chi_{\xi\xi}(\xi,t)+\frac{2I}{r_0M}\sum_\sigma\frac{\partial}{\partial\xi}\left(\phi_\sigma^*\phi_{-\sigma}\right), \quad (7.44)$$

respectively, where $v_0^2=Wr_0^2/M$. Equation(7.43) is a discrete nonlinear Schrödinger equation.

In the quasi-static approximation, or setting $\chi(\xi,t)=\chi(\xi-vt)$, we can integrate Eq.(7.44) to obtain

$$\chi_\xi(\xi,t)=\frac{\tilde{I}}{2Ir_0}\sum_\sigma\phi_\sigma^*\phi_\sigma+K \quad (7.45)$$

where $\tilde{I}=(2I)^2/M\left(v^2-v_0^2\right)$ with $v^2<<v_0^2$ and the constant K, depending on initial conditions. Using Eq. (7.45), they obtained from Eq. (7.43)

$$\begin{aligned}i\left(\phi_\sigma(\xi,t)\right)_t=T\left(\phi_{-\sigma}^n(\xi,t)\right)_{\xi\xi}+\left[2T-\tilde{I}\sum_\sigma\phi_\sigma^*(\xi,t)\phi_{-\sigma}(\xi,t)\right]\phi_{-\sigma}(\xi,t)-\\ \mu\phi_\sigma(\xi,t)+V\left|\phi_{-\sigma}(\xi,t)\right|^2\phi_\sigma(\xi,t)\end{aligned} \quad (7.46)$$

where $T=\bar{J}-r_0I(1-K)$.

In terms of even and odd combinations of

$$\phi_\sigma=\{\phi_\uparrow,\phi_\downarrow\},\ \varphi^{(1)}=\phi_\uparrow+\phi_\downarrow,\quad \varphi^{(2)}=\phi_\uparrow-\phi_\downarrow,$$

where $\phi_\uparrow=(\varphi^{(1)}+\varphi^{(2)})/2$, $\phi_\downarrow=(\varphi^{(1)}-\varphi^{(2)})/2$, they got

$$\begin{aligned}i\varphi_t^{(m)}=T\varphi_{\xi\xi}^{(m)}-\left\{\delta_{m,1}2T-\delta_{m,2}2T-\mu-\frac{V-\hat{I}}{2}\left[\left|\varphi^{(1)}\right|^2+\left|\varphi^{(2)}\right|^2-2\left|\varphi^{(m)}\right|^2\right]\right\}\varphi^{(m)},\\ -\hat{I}\left\{-\frac{1}{2}\left[\left|\varphi^{(1)}\right|^2+\left|\varphi^{(2)}\right|^2\right]\varphi^{(m)}+\frac{3}{4}\left|\varphi^{(m)}\right|^2\varphi^{(m)}+\frac{1}{4}\left|\varphi^{(m)}\right|^2\left[\delta_{m,1}\varphi^{*(2)}+\delta_{m,2}\varphi^{*(1)}\right]\right\}\end{aligned} \quad (7.47)$$

where m= 1, 2. Assuming again that the average occupation number $n=\left|\phi_\uparrow\right|^2+\left|\phi_\downarrow\right|^2$ is a constant, and neglecting further the term $\varphi_\sigma^*\varphi_{-\sigma}\varphi$, compared to $\left|\varphi_{-\sigma}\right|^2\varphi_\sigma$, they got two decoupled nonlinear Schrödinger equations for ' $\varphi^{(1)}$ and $\varphi^{(2)}$, respectively. This demonstrated that the coupled electron-phonon systems described by Eq.(7.34) could be studied by nonlinear quantum mechanics. In this case, the soliton solutions of the two decoupled nonlinear Schrödinger equations can be represented by

$$\varphi^{(m)}(\xi,t)=\sqrt{2n}\,\mathrm{sec}\,h\left[\sqrt{\frac{n(V-\hat{I})}{T}}(\xi-vt)\right]\exp\left(\frac{iv\xi}{T}-i\omega_i t\right),(m=1,2)$$

where $\omega_1=2T-\mu-\frac{v^2}{4T},\omega_2=-2T-\mu-(v-\hat{I})+\frac{v^2}{4T},$

and v is the soliton's velocity. Therefore, the electron in the Hubbard modelwith electron-electron and electron-phonon interactions satisfies the nonlinear Schrödinger equationor laws in the nonlinear quantum mechanics. The soliton consists of a "drop" associated with a "bubble", and it is also localized in the region of

$$W_s \approx \sqrt{T(V-\bar{I})/n}$$

which is accompanied by a lattice distortion wave,whichis also bounded in the same region, and $\chi(\xi,t)$ in Eq. (7.44) or Eq. (7.45) can be represented approximately by $\chi \approx \tanh[(\xi-vt)/W_s]$ under appropriate initial conditions. Very clearly, the above soliton also possesses the wave-corpuscle duality properties described by nonlinear quantum mechanics in Chapter 3.

However, for the pure Hubbard model, where $\bar{I}=0,T=\bar{J}$, and a half-filled band, i.e., n = 1, it follows that $\omega_1=\omega_2$ for $\mu=V/2$ .Then the basic features of the soliton do not change.

In the more general case, it is difficult to find the solution of Eq. (7.47). But if we neglect the last term on the right-hand side in Eq. (7.47), then it becomes

$$i\varphi_t^{(m)}+\delta_{i,1}T\varphi_{xx}^{(m)}-\delta_{i,2}T\varphi_{xx}^{(m)}-\frac{1}{2}V\left(\eta\left|\varphi^{(m)}\right|^2-\left|\varphi^{(m+1)}\right|^2\right)\varphi^{(m)}=0, \tag{7.48}$$

where m= 1, 2 and $\eta=1-\hat{I}/2V,\varphi^{(1)}=\varphi^{(2)}$

This is a vector nonlinear Schrödinger equation, which occurs in Chapter 2. In the case of η = 1 (or $\bar{I}$ = 0), which corresponds to the situation when the electron-phonon interaction is "switched off". Makhankov et al., [50-52]proved that (7.48) is integrable. They can also obtain the Lax pair, examined its symmetry property, and constructed further the soliton solutions forgiven values of η . These results are important in condensed state physics.

7.3. New State-Ansatz for Microscopic Particles and Corresponding Uncertainty Relationship in Strong Electron-Phonon Coupling Systems

7.3.1. New State-Ansatz and Its Features in the Systems

As it is known, the concept of "small polaron" was introduced to describe the electron-phonon coupling systems in the weak coupled case. Then the ground state properties of the coupled systems are obtained self-consistently from the minimization of total elastic plus electronic energies in the mean-field theory, by freezing out the lattice distortion. However, this method cannot generalize into the intermediately and strongly coupled regions, in which the motion of the electron is always accompanied by a phonon cloud, which instantaneously follows the electron and fluctuates. Thus, the polarons occur in the systems. This is a nonadiabatic effect. In this case, the adiabatic and nonadiabatic dynamics can be studied, for example, Q. Niu et al. [53], studied the adiabatic quantum dynamics of the local spin moment in itinerant magnets, and H. Li et al. [54], studied the nonadiabatic quantum dynamics of a vortex in a Josephson junction. However, in strongly coupled electron-phonon systems, such as the Holstein model[42], a basic problem is how to treat the nondiabatic effect and polaronic behavior within the whole range of parameters.

It is well known that the Holstein model[42] is a prototype model to describe the electron-phonon dynamics, and has usually been used to study the antagonistic tendencies(localization-delocalization) in the formation of the polarons and bipolarons in the systems. In this model the Hamiltonian of the molecular crystal is usually represented by [42]

$$H=\sum_n\left(P_n^2/2M+\frac{W}{2}q_n^2\right)-\bar{t}\sum_n\left(b_n^+b_{n+1}+b_{n+1}^+b_n\right)-\lambda\sum_n b_n^+b_nq_n \tag{7.49}$$

where M and W are the mass of one molecule and the force constant of the molecular crystals, respectively, $\bar{t}$ and λ are the hopping and transfer integral of the electron and the electron-phonon coupled constant, respectively, and q_n and P_n are operators of ordinary coordinate and momentum of molecule at site n, respectively. $n_n=b_n^+b_n$ is the number operator of the electron, and $b_n^+(b_n)$ is the creation (annihilation) operator of the electron at site n. If displacement q_n in Eq.(7.49) is replaced by relative displacement ($q_{n+1}-q_n$) for the phonons and the kinetic energy term of the electron $\varepsilon_0 b_n^+b_n(\varepsilon_0=\hbar\omega_0)$,is added into Eq.(7.49), then the Hamiltonian of the coupled electron-phonon systems in Eq.(7.49) are, in essence, similar to Eq.(7.35). Therefore, Eq.(7.49) can also give the soliton solution, mentioned above, thus, the Holstein systems can be described by nonlinear quantum mechanics.

In this model, the motion and quantum tunneling of "heavy"polarons, arising from the postponement effect, result in strong fluctuations and squeezing of multiphonon states, which resemble that of quantum optics. The squeezed state of the phonons is a two-phonon coherent state, which is induced by the motions of the polarons, and it is deeply related to anomalous lattice fluctuation [55-56]. When the phonons are squeezed, the state and properties of the polarons, including its binding energy, effective mass, and bandwidth, are simultaneously changed due to the counteraction of squeezed phonons, on the latter, according to the rule of action-counteraction in physics. The anti-squeezed effects, thus, can also induce significant changes of properties of the systems involving the ground state. This shows that there is interaction and dependence between squeezed phonons and polarons. Therefore, the properties of the states of intermediately and strongly coupled electron-phonon systems are very complicated. In such a case, a difficult question is how we describe the above states in the systems.

The concept of squeezed coherent states of quantumoptics is often used for a quantumtreatmentof the coherent two-phonon states in the coupled electron-phonon systems. The resulting nonadiabatic effects imply the formation of bipolarons [41],quantum fluctuations of polarons [42,45] or phonon-assistedtunneling [44], and so on. The coherent two-phonon states are usually represented by a product of unitary operators, generating displaced and squeezed states,e.g., for a coupled electron-phonon system it is denoted by

$$U=U_c(\alpha)U_s(\zeta)=\exp[\sum_n(\alpha a_n^+-\alpha^* a_n)b_n^+b_n)\times\exp(\sum_n(\zeta^* a_n^2-\zeta a_n^{+2})]=U_s(\zeta)U_c(\overline{\alpha}) \tag{7.50}$$

where $\overline{\alpha}=\alpha\cosh(2r)+\alpha^* e^{i\theta}\sinh(2r)$, $\zeta=re^{i\theta}$, $a_n^+(a_n)$ is the creation (annihilation) operator of the phonon at site n, $U_c(\overline{\alpha})$ represents the Lang-Firsov ansatz for a unitaryoperator generating a linear displacement of atoms [55-76] due to their coupling to the electron density, and $U_s(\zeta)$ represents a two-phonon (squeezing) unitary operator. However, in the coupled electron-phonon system, the effects of displacement and squeezing occur simultaneously, and are mutually dependent and closely related, especially in the strong and intermediate coupling regimes. Therefore, the state formed by these physical mechanisms should be described more rigorously by a unified wave function or state ansatz, but not being resolved into individual and factorized operators as it is represented by Eq. (7.50), which are used in many condensed matter systems.

Based on the above notions, Pang et al., [77-84]proposed the new state ansatz for the coherent two-phonon state related to the polaron effect in the form:

$$D=\exp\left(\sum_n(\zeta^* a_n^2-\zeta a_n^{+2})+(\alpha a_n^+-\alpha^* a_n)b_n^+b_n\right) \tag{7.51}$$

maintaining the number of parameters. The operators in the exponent do not commute, i.e., $D\neq U_cU_s$, because

$$[L_2,L_1]=-2r|\alpha|\left[a_n^+e^{i(\theta-\gamma)}-a_ne^{-i(\theta-\gamma)}\right]b_n^+b_n,$$

where $L_1=(\alpha a_n^+-\alpha^* a_n)b_n^+b_n$, $L_2=(\zeta^* a_n^2-\zeta a_n^{+2})$, $\alpha=|\alpha|e^{ir}$, $\zeta=re^{i\theta}$.

Clearly, the ansatz (7.51) accounts for an additional correlation of the Lang-Firsov type functionaloperator with the two-phonon (squeezed) state operator. This correlation is expected to influence both the displacement and squeezing and related effects in the dimerization and polaron characteristics, namely, i.e., (i) the Debye-Waller factor determining the extension of the polaron (parameter of tunneling)related with, e.g., the possibility of bipolaron formation, (ii) quantum fluctuations of the ground state of a Holstein polaron [42] and of the dimerization order parameters[72]. The weaker the electron-phonon coupling, the weaker become the correlation effects, and the quantum fluctuations can overcome the localizinginfluence of the electron-phonon coupling.

To clearly understand the physical relevance of the ansatz (7.51), let us factorize it as follows,

$$D=\exp\sum_n(\zeta^* a_n^2-\zeta a_n^{+2})\times\exp\sum_n\left[\left(\Gamma a_n^+-\Gamma^* a_n\right)+i\theta'\right]b_n^+b_n \tag{7.52}$$

where

$$\Gamma=\frac{|\alpha|}{2r}\left[\left(\cosh 2r-1\right)e^{i(\theta-r)}+\sinh 2re^{ir}\right] \tag{7.53}$$

Γ^* is a complex conjugate of Γ, and θ' is a c-number phase factor, which cancels at unitary transformations and tends to zero for real α and ζ. In the last case, L_1 and L_2 become Lie operators,

$$\left[L_2,L_1\right]=2r\alpha\left[a_n^\dagger-a_n\right]=2rL_1 \text{ and } \Gamma=\Gamma_0=\alpha\left(e^{2r}-1\right)/2r$$

can be obtained easily [47]. Eqs. (7.52) clearly shows that the correlated ansatz in Eq.(7.51) results in a renormalization of the displacement effect by squeezing and, consequently, of the related quantum effects on the polaron characteristics. The effective displacement Γ includes the correlation of the displacement α and of the squeezing ζ and, in view of $\left(e^{2r}-1\right)/2r>1$, it implies an effective enhancement of $|\alpha|$.

There are approaches including, e.g., correlations between different phonon modes [48] or special types of squeezed coherent states in quantum optics [56], which introduce additional variational parameters. Here we shall briefly summarize the properties of the unitary operator$D$, $D^{-1}D=I$. One can easily find

$$Da_nD^{-1}=a_n\cosh 2r+e^{i\theta}a_n^+\sinh 2r-\Gamma b_n^+b_n \tag{7.54}$$

and an analogous expression for Da_nD^{-1}. Here, Γ is given by Eq. (7.53).For comparison, in the case of the factorized displacementand squeezing generators, it reads

$$U_s U_c a_n U_c^{-1} U_s^{-1} = a_n \cosh 2r + a_n^+ \sinh 2r - \alpha b_n^+ b_n \quad (7.55)$$

and similarly for a Hermitian conjugate $U_s U_c a_n^+ U_c^{-1} U_s^{-1}$. When comparing Eqs. (7.54) and (7.55) we see that the difference between the two transformations lies in replacing the displacement α in Eq. (7.55) by a joint correlation factor of the two effects mentioned above, i.e.,

$$\alpha \to \Gamma(\alpha, \zeta) \quad (7.56)$$

In other words, when going from the correlated to the uncorrelated case one should set $\Gamma / \alpha \to 1$.

If the squeezed effect of the phononsis further taken into account in the strong coupling electron-phonon systems, then the squeezed and coherent states of the phonons and the polaron state are now represented by the following product of operators of respective displacement, and the squeezing and polaron parts: [55-56]

$$U = U_C(\zeta) U_l(\alpha) U_S(\xi) = \exp\left[\sum_n (-1)^i (\zeta a_n^+ - \zeta^* a_n)\right] \exp\left[\sum_n (\alpha a_n^+ - \alpha^* a_n) b_n^+ b_n\right] \exp\left(\sum_n (\xi^* a_n^2 - \xi a_n^{+2})\right) \quad (7.57)$$

for the coupled electron-phonon systems, where $a_n^+(a_n)$ and $b_n^+(b_n)$ are the creation (annihilation) operators of the phonon and electron, respectively; ξ, ζ, and α are some reales, which represent the squeezed and coherent effects of the phonons and polaronic effect of the electrons (α is related to the binding energy of the polaron, i.e., $\alpha^2 = E_p / \hbar\omega$, E_p is the binding energy of polaron, and ω is the phonon frequency), respectively. U_C and U_S are the ansatzs for the one-phonon coherent state and the two-phonon squeezed state, respectively. U_l is an ansatz of Lang-Firsor small polaron state, which associates the electron with the local lattice deformation. Evidently, Eq.(7.57) is an approximate representation of the coupled electron-phonon systems because it denotes only the squeezed effect of the phonons induced by the polaron motion, but does not exhibit its anti-squeezed effect on the polaron state, or in other words, it ignores the correlated effect between them. In fact, in practically physical processes, the squeezing and antisqueezing effects occur simultaneously, influence mutually, and relate closely, i.e., the motion of the polaron results in the squeezing of the phonons, but when the phonons are in the squeezed process the state of the polarons will also be changed due to the antisqueezed effect of the phonons on it according to the action-counteraction rule in physics, especially in strongly and intermediately coupled regions. Thus, the binding energy, effective mass, and band width of the polarons are also simultaneously changed, due to the counteraction of squeezed phonons, as mentioned above. Therefore, we cannot use Eq.(7.57), but should use a unified state-asatz to more rigorously describe the states of strongly coupled electron-phonon systems. However, an appropriate state-ansatz for the systems has not yet occurred. Pang [77-84] proposed a new ansatz of state by introducing three variational parameters, r, ξ, and α,which includes correlation effects between the squeezed-phononic and polaronic states to study the properties of the coupled electron-phonon systems [77-84]. The new state-ansatz proposed by Pang [77-84] is represented by

$$D = \exp\left\{\sum_n \left[\left(\xi^* a_n^2 - \xi a_n^{+2}\right) - \left(\alpha a_n^+ - \alpha^* a_n\right)\left(b_n^+ b_n + (-)^n \zeta\right)\right]\right\} \tag{7.58}$$

It shows the mutual correlations among the one-phonon coherent state, polaronic state, and the two-phonon squeezed state, or squeezing-antisqueezed effects appearing in the systems resulting from the variations of electron density and motion of the polarons and nonadiabatic phonon-fluctuation in strongly and intermediately coupling regions in the systems. Evidently, the new ansatz has the relation of $D^{-1}D=I$, but the operators in the exponents in Eq.(7.58) are not mutually commuted, i.e.,

$$[L_1\ L_2] = 2r\,|\alpha|\,[a_n^+ e^{i(\theta-\gamma)} - a_n e^{-i(\theta-\gamma)}](b_n^+ b_n + (-1)^n \zeta)$$

where $L_1 = [\alpha a_n^+ - \alpha^* a_n](b_n^+ b_n + (-1)^n \zeta)$, $L_2 = \xi^* a_n^2 - \xi a_n^{+2}, \alpha = |\alpha| e^{i\gamma}, \xi = re^{i\theta}$

Thus, $D \neq U = U_S U_l U_C$ or $U_C U_l U_S$. Therefore, Eq.(7.58) is a new wavevector of state, and completely different from U in Eq.(7.57) and has more deeply physical-meanings.

7.3.2. The Uncertainty Relationship of Position and Momentum in the New State-Ansatz

Let us now transform the coordinate and the momentum operators

$$q_i = \sqrt{\hbar/2M\omega}\left(a_n^\dagger + a_n\right) \text{ and } P_n = i\sqrt{\hbar M\omega/2}\left(a_n^\dagger - a_n\right), i = \sqrt{-1} \tag{7.59}$$

where $\omega = \sqrt{W/M}$. It yields from Eq.(7.51) that [77-84]

$$Dq_n D^{-1} = \sqrt{\frac{\hbar}{2M\omega}}\left(\left(a_m^+ + a_n\right)\cosh 2r + \left(a_n^+ e^{i\theta} + a_n e^{-i\theta}\right)\sinh 2r - \frac{|\alpha|}{r}\left[\left(\cosh 2r - 1\right)\cos\left(\theta - r\right) + \sinh 2r \cos r\right] b_n^+ b_n\right) \tag{7.60}$$

$$DP_n D^{-1} = i\sqrt{\frac{\hbar M\omega}{2}}\left(\left(a_n^+ - a_n\right)\cosh 2r - \left(a_n^+ e^{i\theta} - a_n e^{-i\theta}\right)\sinh 2r + i\frac{|\alpha|}{r}\left[\left(\cosh 2r - 1\right)\sin\left(\theta - r\right) + \sinh 2r \cos r\right] b_n^+ b_n\right) \tag{7.61}$$

If $|0\rangle_D$, is the ground state of an electron-phonon system, then for the mean square fluctuations in our stategenerator ansatz $|D\rangle = D|0\rangle_D$ one obtains

$$\langle D|\Delta^2 q|D\rangle = \frac{\hbar}{2Mw} e^{4r}\left(1 - \left(1 - e^{-8r}\right)\sin^2\left(\frac{1}{2}\theta\right) + \frac{|\alpha|^2}{r^2} e^{-4r}\left[\left(\cosh 2r - 1\right)\cos\left(\theta - r\right) + \sinh 2r \cos r\right]^2 \langle \Delta^2 n\rangle\right) \tag{7.62}$$

$$\langle D|\Delta^2 P|D\rangle = \frac{\hbar M\omega}{2} e^{-4r}\left(1+\left(e^{8r}-1\right)\sin^2\left(\frac{1}{2}\theta\right)+\frac{|\alpha|^2}{r^2}e^{4r}\left[\left(\cosh 2r-1\right)\sin\left(\theta-r\right)+\sinh 2r\sin r\right]^2\left\langle\Delta^2 n\right\rangle\right) \tag{7.63}$$

where $\left\langle\Delta^2 n\right\rangle = \left\langle n^2\right\rangle - \left\langle n\right\rangle^2$ and $\left\langle n\right\rangle = \frac{1}{N}\sum_n {}_D\left\langle 0\left|b_n^\dagger b_n\right|0\right\rangle_D$

The minimum uncertainty principle then yields

$$\left\langle\Delta^2 P\right\rangle\left\langle\Delta^2 q\right\rangle = \frac{\hbar^2}{4}\left(1+\left(e8r-1\right)\sin 2\left(\frac{1}{2}\theta\right)+\frac{|\alpha|^2}{r^2}e^{4r}\left[\left(\cosh 2r-1\right)\sin\left(\theta-r\right)+\sinh 2r\sin r\right]^2\left\langle\Delta^2 n\right\rangle\right)$$

$$\times\left(1-\left(1-e^{-8r}\right)\sin^2\left(\frac{1}{2}\theta\right)+\frac{|\alpha|^2}{r^2}e^{-4r}\left[\left(\cosh 2r-1\right)\cos\left(\theta-r\right)+\sinh 2r\cos r\right]^2\left\langle\Delta^2 n\right\rangle\right) \tag{7.64}$$

We compare Eqs. (7.62)-(7.63) with the respective expressions for the uncorrelated case in Eq.(7.57),which are

$$\langle U|\Delta^2 q|U\rangle = \frac{\hbar}{2Mw} e^{4r}\left(1-\left(1-e^{-8r}\right)\sin^2\left(\frac{1}{2}\theta\right)+4|\alpha|^2 e^{-4r}\cos^2 r\left\langle\Delta^2 n_0\right\rangle\right) \tag{7.65}$$

$$\langle U|\Delta^2 P|U\rangle = \frac{\hbar Mw}{2} e^{-4r}\left(1+\left(e^{8r}-1\right)\sin^2\left(\frac{1}{2}\theta\right)+4|\alpha|^2 e^{4r}\sin^2 r\left\langle\Delta^2 n_0\right\rangle\right) \tag{7.66}$$

where $\left\langle\Delta^2 n_0\right\rangle = \left\langle n_0^2\right\rangle - \left\langle n_0\right\rangle^2$ and $\left\langle n\right\rangle = \frac{1}{N}\sum_n {}_U\left\langle 0\left|b_n^\dagger b_n\right|0\right\rangle_U$, $|U\rangle = U|0\rangle_U$

and find that the correlations between the displacement and the squeezing of phonons enhance the quantum uncertainties of the phonon coordinate and momentum due to the coupling with the electron density fluctuations. The term related with the electron density correlations in Eq. (7.66) vanishes for real $\alpha(r=0)$, but in the correlated case in Eq.(7.63) this term is not so, due to the correlations, except if $\theta=0$. The absolute minimum of the r.h.s. of equation (8.64) occurs if the density fluctuation of electrons $\Delta^2 n$ is zero, i.e., $\left\langle\Delta^2 P\right\rangle\left\langle\Delta^2 q\right\rangle \to \frac{1}{4}\hbar^2$ (a harmonic oscillator). In fact, an enhancement of fluctuations beyond the harmonic oscillator fluctuations has occurred, and was found by means of an exact numerical diagonalization of a one-dimensional model of electrons coupled to phonons, which was obtained by Borghi et al.[85].

We again calculate the uncertainty relation of the phonons in the systems represented by Eq.(7.58). In this case, the mean square fluctuations of the coordinate and momentum of the phonon in the new state wave function | D>= D | 0>$_D$ are, respectively, denoted by

$$\langle D|\Delta q^2|D\rangle = \frac{\hbar}{2M\omega}e^{4r}[1-(1-e^{-8r})\sin^2\frac{\theta}{2}+\frac{|\alpha|^2}{r^2}(1-e^{-2r})^2\cos^2(\gamma-\theta)(\langle\Delta n^2\rangle+\zeta m_e-(-1)^i\zeta\langle n\rangle)] \quad (7.67)$$

$$\langle D|\Delta P^2|D\rangle = \frac{\hbar M\omega}{2}e^{-4r}[1+(e^{8r}-1)\sin^2\frac{\theta}{2}+\frac{|\alpha|^2}{r^2}(e^{2r}-1)^2\sin^2(\gamma-\theta)(\langle\Delta n^2\rangle+\zeta m_e-(-1)^i\zeta\langle n\rangle)] \quad (7.68)$$

From Eqs.(7.67)-(7.68) we can obtain the sizes of the uncertainty between the position and momentumof the phonons, arising from the strong electron-phonon coupling in the systems, which are, respectively, expressed by

$$Q = \langle D|\Delta q^2|D\rangle\langle D|\Delta P^2|D\rangle = \frac{\hbar^2}{4}\{1+\sinh^2(4r)\sin^2\theta+2|\alpha|^2((\langle\Delta n^2\rangle+\zeta m_e -$$
$$(-1)^i\zeta\langle n\rangle).[(e^{2r}-1)/2r]^2[2e^{-4r}+\sinh(4r)(\cos^2(\gamma-\theta)-\sin^2\frac{\theta}{2})]+$$
$$\frac{|\alpha|^4}{r^4}\sinh^4(r)e^{4r}\sin^2[2(\gamma-\theta)][(\langle\Delta n^2\rangle+\zeta m_e-(-1)^i\zeta\langle n\rangle]^2\}$$

(7.69)

$$\langle\Delta n^2\rangle = \langle n^2\rangle - \langle n\rangle^2, \langle n\rangle = \frac{1}{N}\sum_i {}_D\langle 0|b_i^\dagger b_i|0\rangle_D, m_e = \frac{1}{N}\sum_i(-1)^i\langle b_i^+ b_i\rangle = \frac{1}{N}\sum_i(-1)^i\langle n_i\rangle$$

(7.70)

From Eq.(7.69) we see that the term related to the electron density fluctuation vanishes for a real a (here $\gamma=0$), but the corresponding term in Eq.(7.68) does not vanish in such a condition, except for θ =0. The last terms in Eqs.(7.67)-(7.68) show that the correlation effect enhances the fluctuations of coordinate and momentum, when compared with that of the uncorrelated case. This conclusion is also consistent with the result of exact numerical diagonalization of a one–dimensional model for coupled electron-phonon systems by Borghi et al. [85].

However, the corresponding representations for the uncorrelated case in Eq.(7.57) are as follows

$$\langle U|\Delta q^2|U\rangle = \frac{\hbar}{2M\omega}e^{4r}[1-(1-e^{-8r})\sin^2\frac{\theta}{2}+4|\alpha|^2e^{-4r}\cos^2\gamma(\langle\Delta n^2\rangle+\zeta m_e-(-1)^i\zeta\langle n\rangle)] \quad (7.71)$$

$$\langle U|\Delta P^2|U\rangle = \frac{\hbar M\omega}{2}e^{-4r}[1+(e^{8r}-1)\sin^2\frac{\theta}{2}+4|\alpha|^2e^{4r}\sin^2\gamma(\langle\Delta n^2\rangle+\zeta m_e-(-1)^i\zeta\langle n\rangle)] \quad (7.72)$$

Thus

$$Q_o = \langle U|\Delta q^2|U\rangle\langle U|\Delta P^2|U\rangle = \frac{\hbar^2}{4}\{1+\sinh^2(4r)\sin^2\theta + 2|\alpha|^2(\langle\Delta n_0^2\rangle + \zeta m_e^0 - (-1)^i\zeta\langle n_0\rangle)[2e^{-4r} + \sinh(4r)(\cos^2(\gamma-\theta) - \sin^2\frac{\theta}{2})] + |\alpha|^4\sin^2[2(\gamma-\theta)][\langle\Delta n_0^2\rangle + \zeta m_e^0 - (-1)^i\zeta\langle n_0\rangle]^2\} \tag{7.73}$$

$$\langle\Delta n_0^2\rangle = \langle n_0^2\rangle - \langle n_0\rangle^2, \langle n_0\rangle = \frac{1}{N}\sum_i {}_U\langle 0|b_i^\dagger b_i|0\rangle_U, m_e = \frac{1}{N}\sum_i(-1)^i\langle b_i^+ b_i\rangle = \frac{1}{N}\sum_i(-1)^i\langle n_0\rangle \tag{7.74}$$

where m_e in Eqs. (7.70) and (7.74) are the order parameter of CDW, where $\langle\cdots\cdots\rangle$ represents the average over the ground state of the systems in correlated and uncorrelated cases, respectively. Obviously, Eqs.(7.62)-(7.64) and Eqs.(7.67)-(70) differ from Eqs.(7.65)-(7.66) and Eqs.(7.71)-(7.74), respectively. Thus, we can obtain from these comparisons that the uncertainty representations of coordinate and momentum of the phonons are enhanced due to the mutual correlation and squeezing-antisqueezing effects between the phonons and polarons. The difference of the uncertainty between the correlated and uncorrelated case can be obtained from Eqs.(7.69) and (7.73), which is represented by

$$\Delta Q = Q - Q_0 = \frac{\hbar^2}{4}\{|\alpha|^2[2e^{-4r} + \sinh(4r)][\langle\Delta n^2\rangle[(e^{2r}-1)/2r]^2 - \langle\Delta n_0^2\rangle] + \zeta[m_e[(e^{2r}-1)/2r]^2 - m_m^0] - (-1)^i\zeta[\langle n\rangle((e^{2r}-1)/2r)^2 - \langle n_0\rangle]\} \tag{7.75}$$

Therefore, $\Delta Q = Q - Q_0 \neq 0$, i.e., the sizes of the uncertainty are different between correlated and uncorrelated state-ansatzsdue to the squeezing-antisqueezing effect $(e^{2r}-1)/2r > 1$.This manifests that the squeezing-antisqueezing effect results in quantum fluctuations of the phonons.

From the above studies, we can gain the following properties of the uncertainty relation of the phonons for both the correlated and uncorrelated state-ansatzs in Eqs.(7.50)-(7.51), and Eqs.(7.57)-(7.58) in strong electron-phonon coupling systems.

(1) Their sizes are not a constant, but change with varying characteristic parameters of the systems, i.e., their sizes increase with decreasing nonlinear interaction λ^2 and increasing phonon energy $\hbar\omega$. They approach the minimum value $\hbar^2/4$,when λ^2 is quite large and $\hbar\omega$ is very small. This means that the strong coupling between the electron and phonon depresses the quantum uncertainty and fluctuations of the phonons, thus the localized features of the phonons are enhanced in this case.

(2) Only if the density fluctuation of the electron, its average values, and order parameters of CDW vanish simultaneously, as well as ζ and α are all real, then the absolute minimum of the r. h. side of Eqs.(7.67)-(7.68) can appear, thus, we can obtain the minimum uncertainty relation of $\langle\Delta q^2\rangle\langle\Delta P^2\rangle = \hbar^2/4$ in this case.

(3) For the case of one-phonon coherent state (i.e., r=0, $\theta = 0, \gamma = 0$ and $\alpha = 0$), we can obtain from Eq.(7.69) or Eq.(7.73) that $\langle \Delta q^2 \rangle \langle \Delta P^2 \rangle = \hbar^2/4$. For the pure squeezing state of phonon (i.e., $\theta = 0, \gamma = 0, \zeta = 0$ and $\alpha = 0$), we can get $\langle \Delta q^2 \rangle = (\hbar / 2M\omega)\exp[4r]$ and $\langle \Delta P^2 \rangle = (\hbar M\omega / 2)\exp[-4r]$, thus $\langle \Delta q^2 \rangle \langle \Delta P^2 \rangle = \hbar^2/4$can be obtained. These results agree with that in Chapter 4. Meanwhile, when the one-phonon coherent and two-phonon coherent states occur simultaneously in the systems, then the minimum uncertainty relation exists only if $\theta = 0$.

(4) From Eqs.(7.67)-(7.74) we can obtain the uncertainty of coordinate and momentum for the phonons in the systems, in which the electrons have no spin, are larger than that for spin -1/2, when λ^2 and $\hbar\omega$ are the same. This means that the electrons with spins cause the reduction of the uncertainty of the phonons, thus the localized feature of the phonons is enhanced in this case.

(5) Equations (7.67)-(7.74) indicated that the squeezing-antisqueezing effect changes the anomalous fluctuation of the coordinate and momentum of the lattice oscillator. However, for real ξ and α, the electron density fluctuations do not contribute to the momentum uncertainty. This is favorable for the physical well-founding of the present ansatz.

In order to understand the physical meanings of new state-ansatz in Eq.(7.58) we can, through calculation, factorize it as in following form:

$$D = \exp\sum_n (\xi^* a_n^2 - \xi a_n^{+2}) \exp\sum_n -\{\frac{|\alpha|}{2r}(e^{2r}-1)[\cos(\gamma-\theta) + ie^{2r}\sin(\gamma-\theta)](a_n^+ - a_n) b_n^+ b_n\}. \qquad (7.76)$$
$$\exp\sum_n -(-1)^n \{\frac{|\alpha|}{2r}(e^{2r}-1)[\cos(\gamma-\theta) + ie^{2r}\sin(\gamma-\theta)](a_n^+ - a_n)\zeta\}$$

If Eq.(7.58) is compared with Eq.(7.57) we find that the parameter denoting the feature of the one-phonon coherent state, ζ, and the parameter exhibiting the characteristic of the polaron effect, $a\zeta$, which are all related with the squeezed factor of the phonons, ξ, in Eq.(7.57), which can be now replaced by $\Gamma = \frac{|\alpha|}{2r}(e^{2r}-1)[\mathrm{con}(\gamma-\theta) + ie^{2r}\sin(\gamma-\theta)]$ and $\Gamma\zeta$ in Eq.(7.76),respectively, i.e., the polaronic and one-phonon coherent states are all affected by the antisqueezed effect of the phonons. Therefore, Eq.(7.58) represents exactly the squeezing-antisqueezed effect of the phonons and polarons in the coupled electron-phonon systems. Thus, we refer to Eq.(7.58) as the correlated ansatz of state. A straightforward result, arising from this ansatz, is that the binding energy of the polaron is changed to $\bar{E}_p = -|\alpha|^2 \Gamma^2 \hbar\omega$ from $E_p = -|\alpha|^2 \hbar\omega$ in the uncorrelated case. When *a and* ζ are real, then $\bar{E}_p = -|\alpha|^2 (\frac{e^{2r}-1}{2r})^2 \hbar\omega$, hence $|\bar{E}_p| > |E_p|$ due to r>0 and $(\frac{e^{2r}-1}{2r}) > 1$. Thus, the stability of the polaron is significantly enhanced in such a case, when compared with the uncorrelated case. Therefore, we can conclude that the new ansatz can result in considerable changes of property of the systems.

7.4. Properties of Nonadiabatic Quantum Fluctuations for Phonons and Electrons in the Strongly Coupled Electron-Phonon Systems

7.4.1. Changes of Properties of Ground State of the Systems

As same as the squeezed effect of the phonons, the antisqueezed effect can also create the occurrence of a lot of new phenomena in the systems. We now study the squeezing-antisqueezed effect on the properties of the ground state of the systems by Pang's ansat, mentioned above [77-84].

The Holstein model in Eq.(7.49) describes a one-dimensional and linearly coupled electron-phonon system, in which the total number of electrons is N/2 for the half-filling band and spinless, where N is the total number of lattice sites. It is obvious that in Eq.(7.49) the equilibrium position of every oscillator depends on whether there is an electron on it or not. If n=1, the equilibrium position of q_i is $\lambda/2N$, otherwise it is $-\lambda/2N$. Under the standard transformation in Eq.(7.59), the Hamiltonianin Eq.(7.49) is replaced by

$$H=\frac{N}{2}\hbar\omega+\sum_n \hbar\omega a_n^+ a_n-\sum_n \lambda\left(\frac{\hbar}{2M\omega}\right)^{1/2}(a_n^+ + a_n)(b_n^+ b_n-\frac{1}{2})-t\sum_n (b_n^+ b_{n+1}+b_{n+1}^+ b_n) \tag{7.77}$$

Pang introduced three new parameters, r, f, and m_0 into Eq.(7.58) [80],then Eq.(7.58) becomes

$$D=\exp\left\{\sum_n\left[(\xi^* a_n^2-\xi a_n^{+2})-(\alpha a_n^+-\alpha^* a_n)(f(b_n^+ b_n-\frac{1}{2})+(-1)^i\zeta m_0)\right]\right\} \tag{7.78}$$

The variational parameters *r, f, m₀*represent the degree of squeezing of the lattice phonons, the size of the phonon cloud brought by the electron, or the degree of polaron effect, the magnitude of phonon-staggered ordering or staggered lattice deformationin the squeezing-antisqueezed cases, respectively. We can find out the change of energy of the ground state of the systems in the squeezing-antisqueezed cases, which is

$$E_g(f,\rho)/N={}_{ph}\langle 0|DHD^{-1}|0\rangle_{ph}/N=\frac{\hbar\omega}{4}(\rho^2+\rho^{-2})-\frac{\lambda^2}{8W}f(\frac{e^{2r}-1}{2r})[2-f(\frac{e^{2r}-1}{2r})]+$$
$$\frac{\lambda^2}{2}(\frac{e^{2r}-1}{2r})^2 Wm_o^2-\frac{1}{\pi}\int_0^{\pi/2}\{4\beta^2t^2\cos^2 k+(\frac{e^{2r}-1}{2r})^2[2-f(\frac{e^{2r}-1}{2r})]^2 m_0^2\}^{1/2} \tag{7.79}$$

where

$$\beta(f,\rho)=\exp\left[-\frac{\lambda^2}{2W\hbar\omega}f^2\rho^2\left(\frac{e^{2r}-1}{2r}\right)^2\right] \tag{7.80}$$

is the Debye-Waller factor, $|0>_{ph}$ is the vacuum state of the phonon, and the variational parameters, $\rho=e^{-2r}$, f and m_0, can be determined self-consistently from

$$\frac{\partial E_g}{\partial \rho}=\frac{\partial E_g}{\partial f}=\frac{\partial E_g}{\partial m_o}=0, \tag{7.81}$$

with,

$$m_e=Wm_o[(e^{2r}-1)/2r]\{[1-f((e^{2r}-1)/2r)]\lambda\}^{-1}\ , \tag{7.82}$$

where $0\leq f\leq 1$ and $0\leq\rho\leq 1$.We here used the relation, $(n_i-\frac{1}{2})(n_i-\frac{1}{2})=\frac{1}{4}$ in calculation for the spinless case. Therefore, the properties of the ground state of the systems are determined by the phonon energy $\hbar\omega$ and the polaron's binding energy $E_p=\lambda^2/2W$ and t.

As far as the energy of stable ground state in the variational method is concerned, we should first determine the minimum value of energy of the ground state by adjusting the variational parameters, f and ρ, to their optimum values. For each f in the ranges of $0\leq f\leq 1$ we can determine local minimum values of $E_g(f,\rho)$ by adjusting ρ to its local optimum value. Thus, we can also get the curves of f-dependence of ρ. We show the relation of $E_g(f,\rho)$/N in Eq.(7.79) versus f and ρ in Figure 7.1. From this figure, we know that the energy of the ground state is negative in the regions of $0.2\leq f\leq 1$and $0.8\leq f\leq 1$. This is an interesting result. Again,through the comparison of these local optimum values for the energies of the ground state, we can determine the absolute minimum values of energy of the ground state, $E_g(f_m,\rho_m)/N$, and the corresponding optimum values of these variational parameters, f_m and ρ_m. In this way, we finally find out the absolute minimum value of energy of stable ground-state, which is $E_g(f_m,\rho_m)/N=-0.2659$ at $f_m=0.42$ and $\rho_m=0.913$in the cases of $J=\lambda/2Wt=4$, $\hbar\omega=2$, $W=0.25$ and $t=1$. In Figure 7.1 we also give the dependences of energy of the ground state, $E'_g(f',\rho')/N$, for the uncorrelated case in Eq.(7.68)onf andρ in the dashed lines. From this figure, we see that the absolute minimum value is $E'_m(f'_m,\rho'_m)/N=-0.208$ at $f'_m=0.56$, $\rho'_m=0.951$. Obviously, this value is larger than $E_g(f_m,\rho_m)/N$. This shows that the correlation and antisqueezed effects of the one-phonon coherent state, polaronic state, and two-phonon squeezed state, are represented by the new ansatz in Eq. (7.78) or Eq.(7.58), which result exactly in the dropping of energy of the ground state of the systems. In such a case, we can also find out the dependence of $E_g(f_m,\rho_m)/N$ on the $\hbar\omega$ and J, which is shown in Figure 7.2, where W=0.25, t=1. We see from this figure that $E_g(f_m,\rho_m)/N$ increases with increasing $\hbar\omega$ and decreasing J. In order to understand further the features of correlation and squeezing-antisqueezed effects, we also give the changes of energy of ground state, resulting from these effects, $\Delta E_g/N=E'_g(f^0_m,\rho^0_m)/N-E_g(f_m,\rho_m)/N$ versus the $\hbar\omega$ and J relation, which is shown in Figure 7.3, where $E'_g(f'_m,\rho'_m)/N=\langle 0|UHU|0\rangle/N$ is the energy of the ground state in the uncorrelated case. From this figure we see that there are always $\Delta E_g/N>0$ for all values of J and $\hbar\omega$. This shows that the squeezing-antisqueezed effect always results in the decreases of energy of the ground state of the systems as compared with that in the squeezed case due to the competition between the grown quantum fluctuations of the phonons and reduction of binding-energy of the localized polaron. Thus, we can conclude that the correlation and squeezing-antisqueezed effects can enhance the stability of the system. Then we can decide that the new ansatz is relevant and appropriate to the systems.

Also, we see from the above result that the polaron binding-energy, E_p, increases to

$$\frac{-\lambda^2}{8W}\left(\frac{e^{2r}-1}{2r}\right)f\left[2-\left(\frac{e^{2r}-1}{2r}\right)f\right]$$

from $-\lambda^2 f'(2-f')/8W$ due to the squeezing-antisqueezed effect.

This comparisonindicated that the stability of the polaron is also enhanced because $[(\exp(2r)-1)/2r] > 1$. However, the Debye-Waller factor, β, measuring the polaron narrowing of the electron band, is increased due to the squeezing-antisqueezed effects, i.e., $\Delta\beta = \beta(f_m, \rho_m) - \beta'(f_m', \rho_m') > 0$, in such a case, where $\beta'(f_m', \rho_m') = \exp\left[-\lambda^2 \rho_m'^2 f_m'^2 / 2\hbar\omega W\right]$ is the Debye-Waller factor in the uncorrelated case in Eq.(7.57).

In Figure 7.4 we plot the $\Delta\beta$ versus J and $\hbar\omega$ functionalrelation. This result shows that the increase of classicalpolaron narrowing of the electron band, arising from the squeezed effect of the phonons, are also suppressed due to the antisqueezed effect. Thus, the quantum-tunneled effect of the polarons is also weakened. This shows that the correlation and squeezing-antisqueezed effects are not favorable to increases of the quantum tunneling effect and dimerization of the polarons. This is an interesting result in physics.

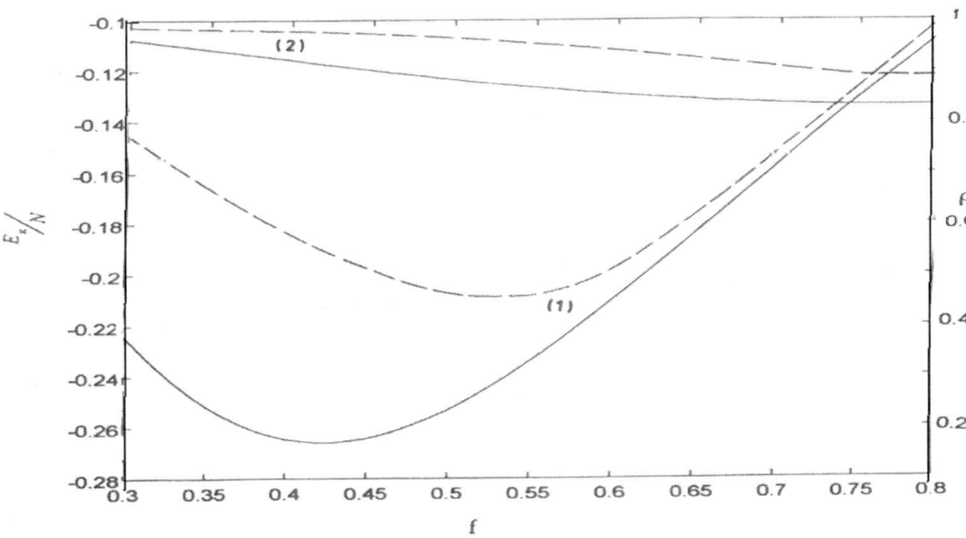

Figure7.1.The dependences of $E_g(f_m, \rho_m)/N$ (real line) and $E'_g(f', \rho')/N$ (dashed lines)on f and ρ, respectively.

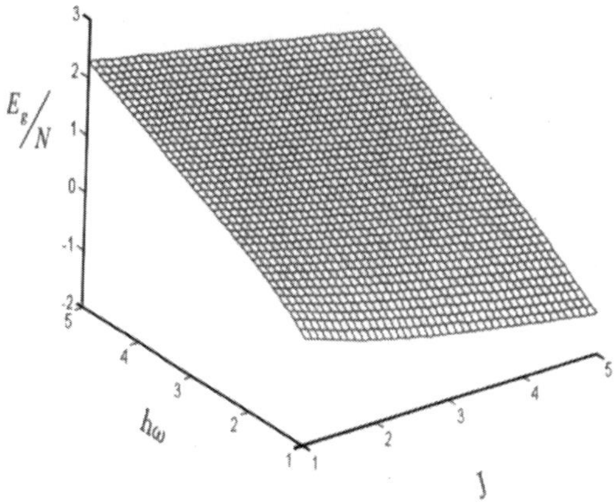

Figure7.2. Changes of $Eg(f_m,\rho_m)/N$ with $\hbar$ ω and J.

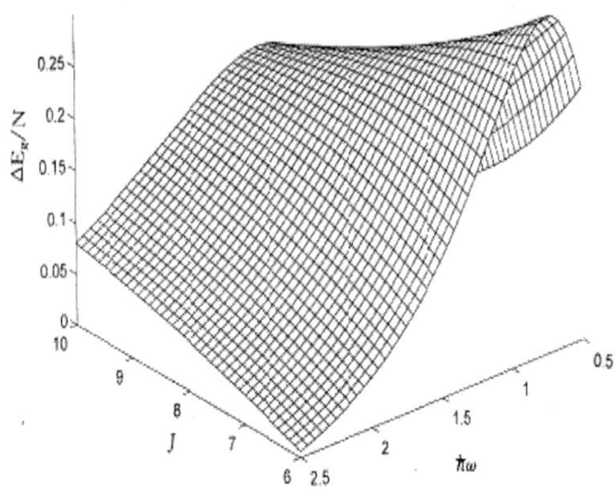

Figure7.3. Dependence of Eg / N on J and $\hbar\omega$.

7.4.2. Quantum Fluctuations of Phonons and CDW Order Effects

We now determine the phonon-staggered ordering parameter m_p, which can be determined in experiments. It is defined as $m_p = \frac{1}{N}\sum_n (-1)^n \langle q_n \rangle$ where q_n is the position operator of the phonon mode and $\langle \cdots\cdots \rangle$ represents an average over the ground state of the systems. After finishing aunitary transformation of the new ansatz and making an average over the vacuum state of the transformed phonon subsystem we can easily obtain $m_p = \lambda m_e / W$. Since $m_0 = \lambda m_e \{1 - f[(e^{2r}-1)/2r]/[(e^{2r}-1)/2r]/W$, then $m_p > m_0$. This means that m_p contains the contribution of the polaron effect, arising from the antisqueezed effect, apart from the static staggered ordering parameter m_0. We find that the relation between the phonon-staggered ordering parameter m_p and the CDW ordering parameter m_e is the same as that in the adiabatic limit and uncorrelated cases. Meanwhile, we see also from the above representations that m_p and m_e increase simultaneously with $\hbar$ ω, but the m_p increases and m_e decreases, when λ increases.

Since the lattice deformation, which follows the electronic motion is proportional to the size of the charge density, then it fluctuates as the electrons move. Thus, the anomalous fluctuations thatappeared are larger than the zero-point fluctuations if $a$or $J > 1$. The fluctuation in every phonon mode q_n can be represented by $\langle \Delta q_n^2 \rangle = \langle q_n^2 \rangle - \langle q_n \rangle^2$. If the average is done over the ground state of the harmonic oscillator with $\omega = \sqrt{W/M}$, then $\langle \Delta q_n^2 \rangle_h = \frac{\hbar}{2M\omega}$ .When its average is found over the new ansatz in Eq.(7.78) in the coupled electron-phonon system at $\theta = 0$ and $\gamma = 0$, then the quantum fluctuation of the phonon mode can be represented by

$$R = \langle \Delta q_n^2 \rangle - \langle \Delta q_n^2 \rangle_h = \frac{\hbar}{2M\omega}(\rho^{-2}-1) + (\frac{\lambda}{2W})^2 f_m^2 (\frac{1-\rho_m^{-1}}{\ln \rho_m})^2 (1-4m_e^2) \quad (7.83)$$

where the first term $(\rho^{-2}-1)$ in Eq.(7.83) exhibits the squeezed effect, the second term shows the anomalous fluctuation due to the polaron motion, the electronic density fluctuation and the antisqueezed effect. From the above representations, we know that the m_e decreases with decreasing λ^2 or J and increasing $\hbar\omega$, but the m_p increases with increasing λ^2 or J and $\hbar\omega$. However, there is a non-monotonic up-down-up tendency in the $\hbar\omega$-dependences of quantum fluctuation of the phonon mode, R, i.e., R decreases with increasing $\hbar\omega$, and reaches a minimum at mid-value of $\hbar\omega$, subsequently it rises again with increasing $\hbar\omega$. On the other hand, R increases slowly with increasing λ, subsequently it rises fast with increasing λ. This means that the weakened effect of larger ordering of the system is now suppressed by the fluctuation of phonons. This is interesting, which differs from the uncorrelated case. In order to verify it, we show the differences of the above physical quantities between the correlated ansatz, Eq.(7.58), and uncorrelated ansatz, Eq.(7.57), i.e.

$$\Delta m_p = m_p(f_m, \rho_m) - m_p'(f_m', \rho_m')$$

and

$$\Delta m_e = m_e(f_m, \rho_m) - m_e'(f_m', \rho_m'), \qquad \Delta R = R(f_m, \rho_m) - R'(f_m', \rho_m'),$$

where $m_e(f_m', \vartheta_m'), m_p'(f_m', \rho_m'), R'(f_m', \rho_m')$ are corresponding values in the uncorrelated case in Eq.(7.57). These results are shown in Figures 7.5, 7.6 and 7.7, respectively.

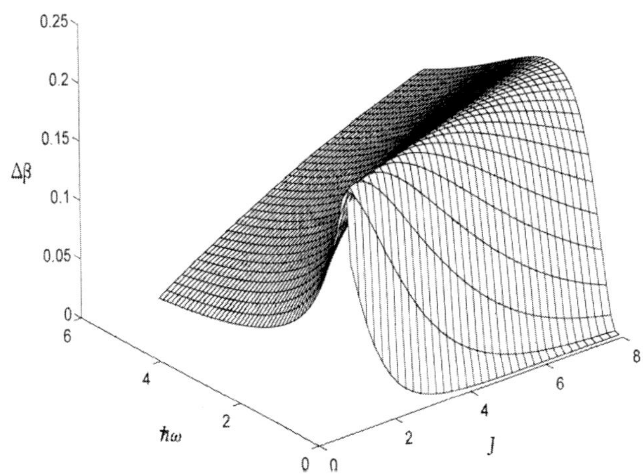

Figure 7.4. Dependences of Δβ on J and $\hbar\omega$.

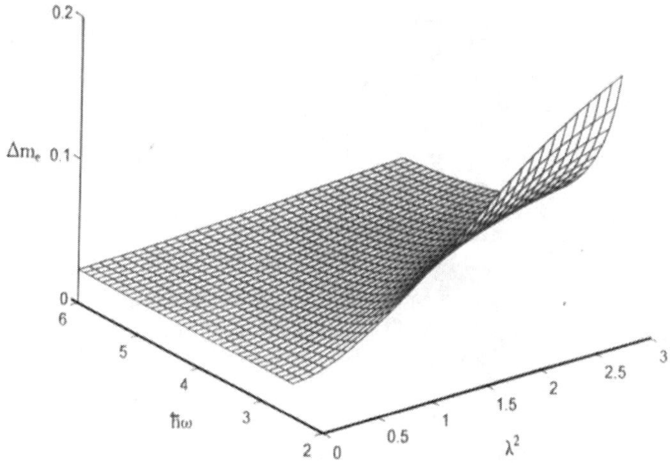

Figure 7.5.Dependences of Δm_e on λ^2 and $\hbar\ \omega$.

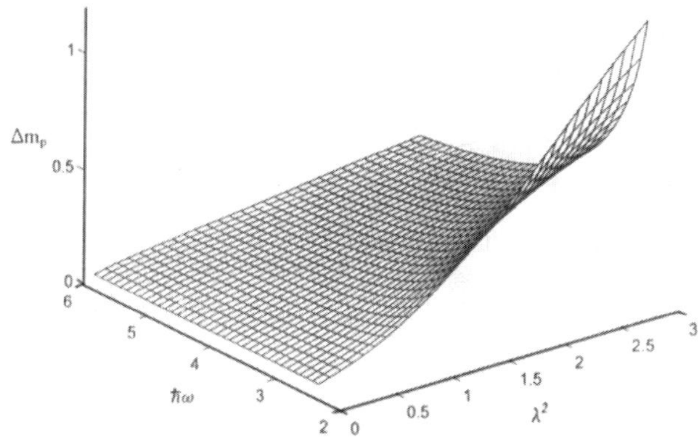

Figure 7.6. Changes of $\Delta m_p$with λ^2 and $\hbar\ \omega$.

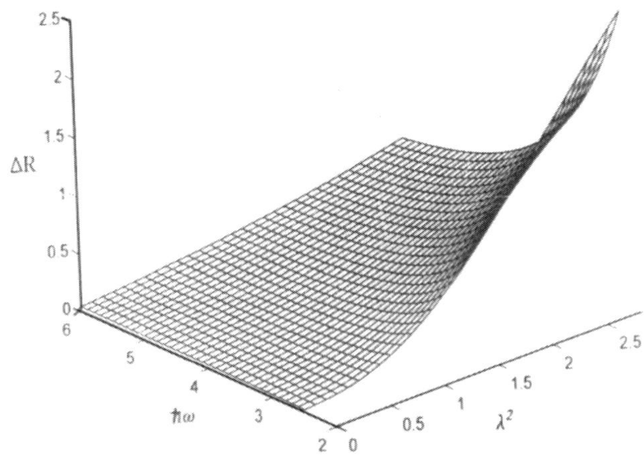

Figure 7.7. The influences of ΔR on λ^2 and $\hbar\ \omega$

From these figures we see clearly that there are always the relations of $\Delta m_e>0$, $\Delta m_p>0$ and $\Delta R>0$ in the correlation and squeezing-antisqueezed case, when compared with the squeezed and uncorrelated case due to the increases of CDW ordering and quantum fluctuations of phonons, resulting from the correlation and squeezing-antisqueezed effects. This shows that the increases of quantum fluctuation of the lattice phonons, arising from the squeezed effect of the phonons, are now suppressed due to the antisqueezed effect. This isa very interesting result in physics, which are worth further study.

We here study the influences of correlation and squeezing-antisqueezed effects on the properties of the coupled electron-phonon systems with a half-filled band by using the new stateansatz and variational method [77-84]. In this method we successively introduce three variational parameters,m_0, f and r (or ρ), which describe the staggered lattice deformation without any polaron effect, the size of lattice deformation instantaneously following the electron (polaron effect) and the squeezing effect of multiphonons due to the electron band motion and its density fluctuation, respectively. f indeed measures the on-site correlation between the electronic density and lattice deformation. The squeezed and antisqueezed effects are purelydynamical effect associated with the nonadiabatic and anomalous quantum fluctuations, resulting from the changes of electronic density and motions of the polarons. The new state-ansatz we here introduce shows accurately the influences of electron motion on the multiphonons (which cause the squeezed effect of multiphonons) and of the antisqueezed effect of the phonons on the polaron band and one-phonon coherent state. We get some new phenomena and effects, which are as follows.

(1) The energy of the ground state of the systems drops greatly. The increase of phonon displacement due to the correlation is manifested to be responsible for the stabilization of strongly coupled polarons, when compared with the uncorrelated case. The increase of quantum zero fluctuation in Figure 7.3 is overcome by the growth of the negative contribution of polaron binding energy. These contributions become more relevant for strong electron-phonon coupled systems with largeλ.

(2) The binding energy of the polaron is increased, but the increases of the band narrowing and tunnelingeffect of the polarons, determined by the Debye-Waller factor, $\beta'(f_m', \rho_m') = \exp\left[-\lambda^2 \rho_m'^2 f_m'^2 / 2\hbar\omega W\right]$, are suppressed due to the antisqueezed effect, when compared with the uncorrelated case. This effect implies that the following phenomena occur. (a)The correlation effect induces the increase of quantum enhancement of the polaron pairing due to the squeezing-antisqueezed effect. Thus, the probability forming the bipolarons increases or the two-polaron resonance effect is enhanced in such a case. (b) The additional quantum tunnelingeffect of the systems increases [71-72]. (c) The dimerization order parameters [58-60,45-46], for example, the phonon staggered parameter m_p,is enhanced.

(3) The correlation effect enhances the fluctuations of coordinate and momentum, in other words, the quantum uncertainty of the phonons is enhanced when compared with that of the uncorrelated case. This is due to its enhancement of quantum fluctuations of the phonons. Thusa new uncertainty relation occurs.

(4) The CDW ordering parameter $m_e$is reduced, andthe staggered ordering parameter $m_p$is suppressed in this case.

7.5. Nonlinear Quantum-Mechanical Properties of Electrons in Superconductors

7.5.1. Guinzburg and Landanu Theory of Superconductivity and its Physical Meanings

As it is known, the superconductivity of material is caused by the electron-phonon interaction in the condition of low temperature [44,86-89]. Therefore, a superconductor is a electron-phonon coupled system, in which the electrons become Cooper pairs or superconductive electrons due to the electron-phonon interaction, which is a nonlinear interaction. Based on this mechanism Bardeen, Cooper and Schrieffer built a BCS theory of low-temperature superconductivity [44,86-89]. In the phenomenological theory of superconductivity, the London equations, $m\partial \vec{j} / n_s e^2 \partial t = -\partial \vec{A} / c\partial t, \nabla^2 \vec{B} = \vec{B} / \lambda_L^2$, cannot explain the physical phenomena of nonuniform superconductivity and the superconductive problems in a strong magnetic field, where e and n_s are the charge and density of the superconductive electron, respectively, c is light velocity, $\vec{A}$ is electric potential, $\vec{j}$ is superconductive electric density, $\vec{B}$ is the induced strength of the magnetic field, and $\lambda_L^2 = mc^2 / 4\pi n_s e^2$, λ_L is the penetration depth of superconductivity. Thus, Guinzburg and Landau [90-98] introduced a macroscopic wave function $\phi(\vec{r},t)$ to describe the states of the superconductive electron, which are called its effective wave function, or complex order parameter, or macroscopic wave function. It is often denoted by

$$\phi(\vec{r},t) = \sqrt{\rho(\vec{r},t)} \mathrm{e}^{\mathrm{i}\theta(\vec{r},t)}, \text{ or } \phi(\vec{r},t) = f(\vec{r},t)\phi_0 e^{i\theta(\vec{r},t)} = \left|\phi(\vec{r},t)\right| e^{i\theta(\vec{r},t)},$$

which is the same as Eq.(3.2) in nonlinear quantum mechanics, where $\phi_0^2 = \alpha / 2\lambda$. Its square of absolute value is assumed as the density of the superconductive electron, i.e., $\left|\phi(\vec{r},t)\right|^2 = n_s(\vec{r}) = \rho(\vec{r})$ .These results are all the same with those in nonlinear quantum mechanics in Chapter 3. If $\phi(\vec{r},t) = 0$, then it represents a normal state.

Based these assumptions they established the Guinzburg-Landau (G-L) theory of superconductivity [90-98]. In this theory the dynamic equation of the wave function, $\phi(\vec{r},t)$,can be obtained from Landau et al.'s free energy for the superconductor.

In practice, Landau et al. [90,91], used the wave function to represent the free energy density function, f, of a superconducting system, i.e., he thought that the free energyis closely related to the density of the superconductive electron and can be represented by

$$f_s = f_n - \frac{\hbar^2}{2m}|\nabla\phi|^2 - \alpha(T)n_s + \frac{1}{2}\lambda'(T)n_s^2 = f_n - \frac{\hbar^2}{2m}|\nabla\phi|^2 - \alpha(T)|\phi|^2 + \lambda(T)|\phi|^4 \tag{7.84}$$

in the absence of any external field. If the system is subjected to an electromagnetic field, specified by a vector potential $\vec{A}$, where 2m is the mass of the superconductive electron, the free energy density of the system is of the form:

$$f_s = f_n - \frac{\hbar^2}{2m}\left|(\nabla - \frac{ie^*}{c\hbar}\vec{A})\phi\right|^2 - \alpha|\phi|^2 + \lambda|\phi|^4 + \frac{1}{8\pi}\vec{\mathrm{H}}^2 \tag{7.85}$$

where $e^*=2e$, $\vec{\mathrm{H}} = \nabla \times \vec{A}$, α and λ are some interactional constants related to the features of the superconductor, m is the mass of the electron, e^*=2e is the charge of the superconductive electron, c is the velocity of light, h is the Planck constant, and $\hbar = h/2\pi$, f_n is the free energy of normal state. The free energy of the system is $F_s = \int f_s d^3x$. In terms of the conventional field, $F_{jl}=\partial_j A_l - \partial_l A^j$, (j, l=1, 2,3), the term $\vec{H}^2/8\pi$ can be written as $F_{jl}F^{jl}/4$. Equations (7.84)-(7.85) show the nonlinear features of the free energy of the systems because it is the nonlinear function of the wave function of the particles, $\phi(\vec{r},t)$. Thus, we can predict that the superconductive electrons have many new properties, which differs from normal electrons. From $\delta F_s / \delta\phi = 0$ we get

$$\frac{\hbar^2}{2m}\nabla^2\phi - \alpha\phi + 2\lambda\phi^3 = 0 \tag{7.86}$$

and

$$\frac{\hbar^2}{2m}(\nabla - \frac{ie^*}{c\hbar}\vec{A})^2\phi - \alpha\phi + 2\lambda\phi^3 = 0 \tag{7.87}$$

in the absence and presence of an external fields respectively, and the number current is of the form

$$\vec{J} = +\frac{e^*\hbar}{2mi}(\phi^*\nabla\phi - \phi\nabla\phi^*) - \frac{e^*}{mc}|\phi|^2\vec{A} \tag{7.88}$$

Equations (7.86) - (7.88) are just well-known in the Ginzburg-Landau (G-L) equation [92-98] in a steady state, it is, in essence, the same withthe time-independent nonlinear Schrödinger equation in Eq.(2.66), which is the dynamic equation of nonlinear quantum mechanics [19-22], where the nonlinear interaction, $2\lambda\phi^3$,in Eqs.(7.86)-(7.88) is obviously due to the electron-phonon interaction. Thus, the superconductor is a nonlinear quantum-mechanical system. This means that the properties of superconductive electrons can be completely described by nonlinear quantummechanics.

We here define firstly some characteristic parameters of the superconductor from the parameters, $\alpha(T)$ and $\lambda(T)$ in the above G-L equation. Its first parameter is the coherent length of the superconductor, which is denoted by $\xi'(T)$, which is represented by

$$\xi'^2 = \hbar^2 / 2m\alpha, or, \xi'^2(T) = \hbar^2 / 2m(T_c - T)|\alpha(T_c)| \tag{7.89}$$

Obviously, it depends on temperature. For a uniform superconductor, we can find out the coherent length of the superconductor, which is $\xi(T) = 0.94\xi_0[T_c / (T_c - T)]^2$, where T_c is the critical temperature, and ξ_0 is the coherent length of superconductive electrons at T=0.

Meanwhile, we can determine the penetration depth of superconductivity, which is related to the temperature and magnetic-field; it is the second characteristic parameter and denoted by

$$\lambda_L(T) = [mc^2\lambda(T_c) / 4\pi(e^*)^2(T_c - T)\alpha'(T_c)]^{1/2} \text{ with } \alpha'(T_c) = d\alpha / dT\Big|_{T_c}.$$

Therefore, the two characteristic parameters, $\xi(T)$ and $\lambda_L(T)$, change linearly with $(T_c - T)^{-1/2}$. When $(T_c - T) << T_c$, the characteristic parameter $\kappa = mc/\hbar e^*(\lambda(T)/2\pi)^{1/2}$ can be defined, it is independent with the temperature. If $\phi_0^2 = |\alpha|/\lambda$ and $f_{no} - f_{so} = \alpha^2(T)/2\lambda(T) = \vec{H}_c^2(T)/8\pi$, where f_{no} and f_{so} are the free energies of the normal and superconductive states at zero magnetic-field, respectively, From the above representations we can find out that $|\alpha|\phi_0^2 = \vec{H}_c^2(T)/4\pi$, there $\kappa = (\sqrt{2}e^*/\hbar c)\vec{H}_c(T)\lambda_L^2(T)$ can be directly measured by the experiments. We can also use κ to distinguish the superconductive and normal states as well as different types of superconductor.

In practice, there is a transition layer between the superconductive and normal domains, and its thickness is changed with varying temperature.This means that there is a penetrated depth of the magnetic-field on the interface between the superconductor and normal state, this transition layer is referred to as N-S junctions, in which the energy of the junction can be obtained and denoted by $\varepsilon_{ns} = [\xi(T) - \lambda_L(T)]\vec{H}_c^2 / 8\pi$. When $\xi(T) > \lambda_L(T)$, then $\varepsilon_{ns} > 0$, inversely, $\varepsilon_{ns} < 0$ at $\xi(T) < \lambda_L(T)$. From the results of calculation of interface energy [44], we know $\varepsilon_{ns} = 0$ at $\kappa = 1/\sqrt{2}$. Then we referred to it as the first type of superconductor at $\kappa < 1/\sqrt{2}, \varepsilon_{ns} > 0$, inversely, it is called the second type of superconductor at $\kappa > 1/\sqrt{2}, \varepsilon_{ns} < 0$, in which the magnetic-field penetrated already into the first type of superconductor to forma N-S junction at $\kappa < 1/\sqrt{2}$. Because of $\varepsilon_{ns} > 0$ is not advantageous to retain the superconductive state. Therefore, the order parameter is a constant in the first type of superconductor. In this case, the externally applied magnetic-field cannot penetrate into the superconductor. This is just the complete diamagnetism of the superconductor. Inversely, the second type of superconductor does not have the feature because $\varepsilon_{ns} < 0$ at $\kappa > 1/\sqrt{2}$. Therefore, we explained the feature of complete diamagnetism, which is a basic feature of the superconductor using this parameter.

On the other hand, we can also elucidate the quantization of magnetic flux of the superconductor,in this case, using $\phi(\vec{r},t) = \left|\phi(\vec{r},t)\right| e^{i\theta(\vec{r},t)}$. In fact, from Eq.(7.88) we can find out

$$-\vec{J}\, mc / (e^*)^2 \left|\phi\right|^2 - (\hbar c / e^*)\nabla\theta = \vec{A}$$

From the path integral of $\vec{A}$ along the close curve C we can obtain

$$\oint_C \vec{A}.dl = -(mc / (e^*)^2)\oint_C \vec{J}.dl / \left|\phi\right|^2 - (\hbar c / e^*)\oint_C dl.\nabla\theta$$

Because the order parameter of superconductor, $\phi(\vec{r})$, is the single value function of $\vec{r}$, then we can require that the path integral along close curve C, $\oint_C dl.\nabla\theta$, take only the multiples of 2π . Utilizing again the Stokes theorem, then the integral $\oint_C \vec{A}.dl$ for the close curve C can be transferred as the integral for the area S surrounded by the close curve C, i.e.,

$$\oint_C \vec{A}.dl = -\int_S d\vec{S}.(\nabla\times\vec{A}) = \int_S \vec{B}.d\vec{S}\,.$$

Then it is not difficult to gain

$$\int_S \vec{B}.d\vec{S} = -(mc / (e^*)^2)\oint_C \vec{J}.dl / \left|\phi\right|^2 + 2n\pi\hbar C / e^*$$

If the integral circuit is in the position where there is no superconductive current, then the first term on the right side in the above equation is zero. Thus, we finally obtained

$$\Phi = \int_S \vec{B}.d\vec{S} = n(2\pi\hbar C / e^*) = n\Phi_0\,,$$

where $\Phi_0 = (2\pi\hbar C / e^*)$ is a quantum of the magnetic flux, n is a integer. This equation indicates that the magnetic flux of the superconductor is quantized.

Otherwise, because $e^* = 2e$,then we can say that the superconductive current or the quantization of the magnetic flux are the result of motion of the Copper electron pair.

7.5.2. The Properties of Motion of Superconducting Electrons in Steady States without Electromagnetic-Field

The above results tell us that the G-L theory can completely describe the basic features of a superconductor, or superconductive electrons, such as the complete diamagnetism and quantization of magnetic flux of the superconductor. However, can it explain the complete electric-conductivity without the resistance, which is another basic feature of the superconductor? This is worth investigation. In order to solve this problem we have to find the solutions of G-L equations (7.86)-(7.88) using the nonlinear quantum mechanical method. If the G-L equations have soliton solutions, in other words, if the superconductive electrons with 2e charge are a soliton, then we can elucidate clearly the complete electric-conductivity of the superconductor because the soliton possesses the feature that can retain its form, velocity, and energy to propagate longer macroscopic distances [31-32], which embody just the property of complete electric-conductivity. Then we can conclude that the superconductivity is caused by the movement of the soliton of the superconductive electron. Therefore, it is very necessary to find the solutions of the above G-L equation and to elucidate further the Josephson effects that occurred in S-N and the superconductive junctions by the G-L equations.

We now study the properties of motion of superconductive electrons in the case of no external field. In this case, we should find first the solutions of the G-L equations (7.86)-(7.89) in the one-dimensional superconductor [99-100].

To assume $\phi(\vec{r},t)=\phi_0\varphi(\vec{r},t)$ with $\phi_0=\sqrt{\alpha/\lambda}$ and $x'=x/\xi'(T)$ with $\xi'^2(T)=\hbar^2/2m|\alpha|$ and using the boundary conditions of φ (x'=0)=1, and φ ($x'\to\pm\infty$) =0,we easily find the solutionof Eqs. (7.86) and (7.89) [19-20,31-32], which is

$$\varphi=\pm\sqrt{2}\sec h\left[\frac{x-x_0}{\xi'(T)}\right] \tag{7.90}$$

or

$$\phi=\pm\sqrt{\frac{\alpha}{\lambda}}\sec h[\frac{x-x_0}{\xi'(T)}]=\pm\sqrt{\frac{\alpha}{\lambda}}\sec h[\frac{\sqrt{2m\alpha}}{\hbar}(x-x_0)] \tag{7.91}$$

This is a well-known wave packet-type soliton solution. Therefore, the superconductive electrons have a wave-corpuscle duality, which are the property of microscopic particles described by nonlinear quantum mechanics in Chapter 3. It is also the same with the bright soliton in the Bose-Einstein condensate found by Perez-Garcia et al. [101].

The energy of the soliton, (7.91), is given by

$$E_{sol}=\int_{-\infty}^{\infty}H'dx=\int_{-\infty}^{\infty}\left[\frac{\hbar^2}{2m}(\frac{d\phi}{dx})^2-\alpha\phi^2-\lambda\phi^4\right]dx=\frac{4\hbar\alpha^{3/2}}{3\lambda\sqrt{2m}} \tag{7.92}$$

We assume here that the lattice constant, r_0=1. The energy of the above soliton can be compared with the ground state energy of the superconducting state, $E_{ground} = -\alpha^2/4\lambda$. Their difference is

$$E_{sol} - E_{ground} = \alpha^{3/2}\left(\sqrt{\alpha} + \frac{16\hbar}{3\sqrt{2m}}\right)/2\lambda > 0 .$$

This indicates clearly that the soliton is not in the ground state, but in an excited state of the system, therefore, the soliton is a quasiparticle.

From the above discussion, we see that, in the absence of external fields, the motion of superconductive electrons can be described by a nonlinear Schrödinger equation, or nonlinear quantum mechanics, its essence is a soliton in a uniform system. The soliton is formed by a nonlinear interactionarising from the electron-phonon interaction, which suppresses the dispersive behavior of electrons. The solitons with determinant charges can move over macroscopic distances retaining its shape, amplitude, and energy along the superconductors, thus a superconducting phenomenon without resistance occurs in the systems.

7.5.3. The Features of Motion of Superconducting Electrons in Steady States with Electromagnetic-Field and Its Comparison with Experiments

We now consider the dynamic feature of superconductive electrons in an electromagnetic field $\vec{A}$; its equation of motion is denoted by Eqs. (7.87)-(7.88). Assuming now that field $\vec{A}$ satisfies the London gauge $\nabla\cdot\vec{A}=0$, and that the substitution of $\phi(\vec{r},t)=\varphi(\vec{r},t)\phi_0\exp[i\theta(\vec{r})]$ into Eqs. (7.87) and (7.88) yields [102-113]:

$$J=\frac{e^*\phi_0^2}{m}(\hbar\nabla\theta-\frac{e^*}{c}\vec{A})\varphi^2 \tag{7.93}$$

and

$$\nabla^2\varphi-[(\nabla\theta-\frac{e^*}{\hbar c}\vec{A})^2\varphi]-\frac{2m}{\hbar^2}(\alpha-2\lambda\phi_0^2\varphi^2)\varphi=0 \tag{7.94}$$

For bulk superconductors, J is a constant (permanent current) for a certain value of $\vec{A}$, and it thus can be taken as a parameter. Let $B^2=m^2J^2/\hbar^2(e^*)^2\phi_0^4$, $b=2m\alpha/\hbar^2=\xi'^{-2}$, from Eqs. (7.93) and (7.94), we can obtain [114-115]:

$$(h\nabla\theta-\frac{e^*}{c}\vec{A})=\frac{Jm}{e^*\phi_0^2\varphi^2} \tag{7.95}$$

$$\frac{d^2\varphi}{dx^2}=-\frac{d}{d\varphi}U_{\text{eff}}(\varphi),U_{\text{eff}}(\varphi)=\frac{B^2}{2\varphi^2}-\frac{1}{2}b\varphi^2+\frac{1}{4}b\varphi^4 \tag{7.96}$$

where U_{eff} is the effective potential of the superconductive electron in this case and it is schematically shown in Figure 7.8. Comparing this case with those without the external fields, we found that the equations have the solution, which is a similar to Eq.(7.91). This means that the electromagnetic field only changes the effective potential of the superconductive electron. When $\vec{A}=0$, the effective potential well is characterized by double wells. In the presence of an electromagnetic field, there are still two minima in the effective potential, corresponding to the two ground states of the superconductor in this condition. This shows that the spontaneous breakdown of symmetrymay still occur in the superconductor, in this case, thus the superconductive electrons also move in the form of solitons. To obtain the soliton solution, we integrate Eq. (7.96) and can get:

$$x=\int_{\varphi_1}^{\varphi}\frac{d\varphi}{\sqrt{2[E-U_{\text{eff}}(\varphi)}} \tag{7.97}$$

where E is a constant of integration, which is equivalent to the energy, and the lower limit of the integral, φ_1 , is determined by the value of φ at x=0, i.e. $E=U_{\text{eff}}(\varphi_0)=U_{\text{eff}}(\varphi_1)$. Introduce the following dimensionless quantities $\varphi^2=u$, $E=\frac{b}{2}\varepsilon, 2\tilde{d}=\frac{4J^2m\lambda}{(e^*)^2\alpha^2}$, and equation (7.97) can be written as follows upon performing the transformation$u\to-u$,

$$-\sqrt{2b}x=\int_{u_1}^{u}\frac{du}{\sqrt{u^3-2u^2-3\varepsilon u-2\tilde{d}^2}} \tag{7.98}$$

It can be seen from Figure 7.8 that the denominator in the integrand in Eq. (7.98) approaches zero linearly when $u=u_1=\varphi_1^2$, but approaches zero gradually when $u=u_2=\varphi_0^2$. Thus, we give [102-109]

$$u(x)=\varphi^2(x)=u_0-g\sec h^2\left(\sqrt{\frac{1}{2}gb}x\right)=u_1+g\tan h^2\left(\sqrt{\frac{1}{2}gb}x\right) \tag{7.99}$$

where $g=u_0-u_1$ and satisfies

$$(2+g)^2(1-g)=27\tilde{d}^2\ ,\ 2u_0+u_1=2\ ,u_0^2+2u_0u_1=-2\varepsilon\ ,\ u_1u_0^2=2\tilde{d}^2 \tag{7.100}$$

It can be seen from Eq. (7.99) that for a large part of sample, u_1 is very small and may be neglected; the solution$u$ is very close to u_0. We then get from Eq. (7.99) that

$$\varphi(x) = \varphi_0 \tan h\left(\sqrt{\frac{1}{2}gbx}\right) \tag{7.101}$$

Substituting the above into Eq. (7.95), the electromagnetic field $\vec{A}$ in the superconductors can be obtained

$$\vec{A} = -\frac{\vec{J}mc}{(e^*)^2\phi_0^2}\frac{1}{\varphi^2} - \frac{\hbar c}{e^*}\nabla\theta = \frac{\vec{J}mc}{(e^*)^2\phi_0^2\varphi_0^2}\cot h\left(\sqrt{\frac{1}{2}gbx}\right) - \frac{\hbar c}{e^*}\nabla\theta$$

For a large portion of the superconductor, the phase change is very small. Using $\vec{H} = \nabla \times \vec{A}$, the magnetic field can be determined and is given by [102-109]

$$\vec{H} = \frac{\vec{J}mc\sqrt{2gb}}{(e^*)^2\phi_0^2\varphi_0^2}[\cot h^3\left(\sqrt{\frac{1}{2}gbx}\right) + \cot h\left(\sqrt{\frac{1}{2}gbx}\right)] \tag{7.102}$$

Equations (7.101) and (7.102) are analytical solutions of the GL equation(7.63) and (7.64) in the one-dimensional case, which are shown in Figure 7.9. Equation (7.101) or (7.99) shows that the superconductive electron in the presence of an electromagnetic field is still a soliton. However, its amplitude, phase, and shape are changed, when compared with those in a uniform superconductor*and* in the absence of external fields, Eq. (7.100).

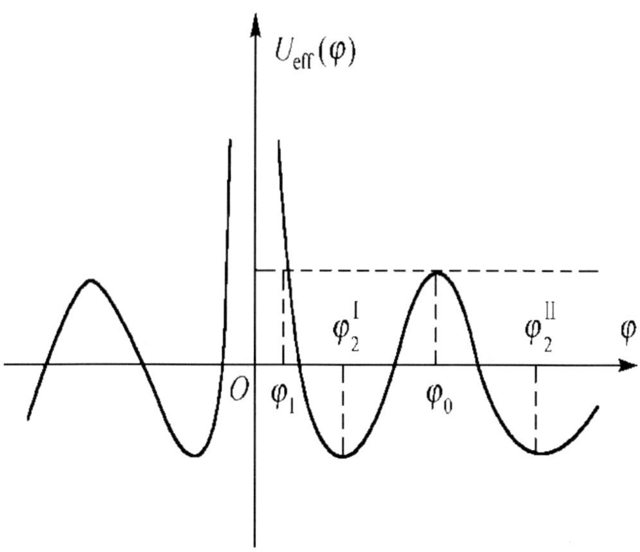

Figure 7.8. The effective potential energy.

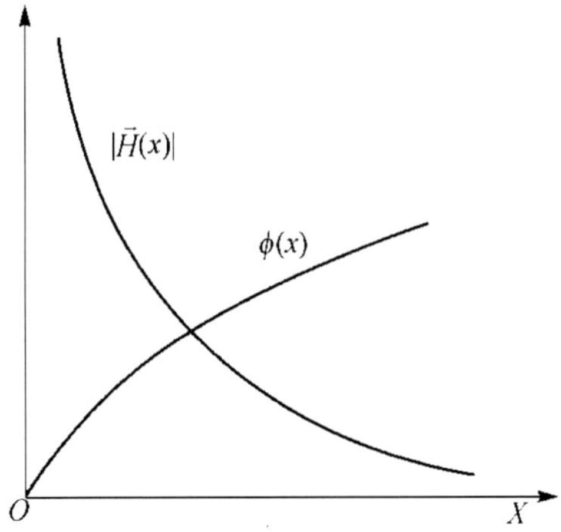

Figure 7.9. Changes of $\phi(x)$ and $\left|\vec{H}(x)\right|$ with x in Eq. (7.101). in Eqs. (7.101)-(7.102).

The soliton here is obviously influenced by the electromagnetic field, as reflected by the change in the form of a solitary wave. This is why a permanent superconducting current can be established by the motion of superconductive electrons along a certain direction in such a superconductor, because solitons have an ability to maintain their shape and velocity while in motion. Therefore, we can obtain that the superconductivity of materials is caused by soliton motion of superconductive electrons.

It is clear from Figure 7.9 that $\tilde{H}(\mathrm{x})$ is larger where $\phi(x)$ is small, and vice versa. When $x \to 0$, $\tilde{H}(x)$ reaches a maximum, while ϕ approaches zero. On the other hand, when $x \to \infty$, ϕ becomes very large, while $\tilde{H}(x)$ approaches zero. These are exactly the well-known behaviors of vortex lines-magnetic flux lines in type-II superconductors [102-109]. Experiments discover that the external magnetic-field can partially penetrate inside type-II superconductors (here the Ginzburg-Landau parameter is greater than 1/2), such as the NbTi alloy and Nb_3Sn compounds, when the external field $\vec{H}$ is greater than the lower critical magnetic field $\vec{H}_{cl}$, but less than the upper critical field $\vec{H}_{c2}$ [110-115]. The penetration of a magnetic field results in some small regions changing from superconductive to normal states. These small regions, having the feature of normal state, are of cylindrical shape and regularly arranged in the superconductor. Its physical image is the same as Figure 7.9. Each cylindrical region is called a vortex (or magnetic field line) [102, 110-120]. The vortex lines are similar to the vortex structure formed in a turbulent flow of fluid and Bose-Einstein condensation. Both theoretical analysis and experimental measurements have shown that the magnetic flux associated with one vortex is exactly equal to one magnetic flux quantum, ϕ_0 when the applied field $\vec{H} \geq \vec{H}_{cl}$, the magnetic field penetrates into the superconductor in the form of vortex lines, increased one-by-one. This is just a macroscopic quantum effect occurring in type-II superconductors, as mentioned above. The results in Figure 7.9 give the structure of vortex lines. This means that nonlinear quantum mechanics is correct.

Garadoc-Daries et al.,[121], Matthews et al. [122], and Madison et al. [123], observed vertex solitons in the Boson-Einstein condensates. Tonomure [124] experimentally observed magnetic vortexes in superconductors. These vortex lines in type-II-superconductors are quantized. Thus, we can determine that the macroscopic quantum effects are in essence a nonlinear quantum effect and can be well described by nonlinear quantum mechanics.

We now proceed to determine the energy of the soliton given by (7.101). From the earlier discussion, the energy of the soliton is given by:

$$E=\int_{-\infty}^{+\infty}\left[\frac{1}{2}\left(\frac{d\varphi}{dx}\right)^2+\frac{b}{2}\varphi^2-\frac{\mathrm{b}}{4}\varphi^4-\frac{B^2}{2\varphi^2}\right]dx\approx\varphi_0^{\ 2}\left[\frac{2b\varphi_0^2}{3}-1+\frac{b}{2}\left(1-\frac{\varphi_0^2}{2}\right)\right]-\frac{B^2}{2\varphi_0^2}$$

which depends on the interaction between superconductive electrons and the electromagnetic field.

From the above discussion, we understand that for a bulk superconductor, the superconductive electrons behave as solitons, regardless of the presence of external fields. Thus, the superconductive electrons are a special type of soliton. Obviously, the solitons are formed due to the fact that the nonlinear interaction $\lambda|\phi|^2\phi$, which is generated by the electron-phonon interaction, suppresses the dispersive effect of the kinetic energy in Eqs. (7.86) and (7.87). Thus, the superconductive electrons move in the form of a solitary wave in the superconducting state. In the presence of external electromagnetic fields, we demonstrate theoretically that a permanent superconductive current is established and the vortex lines or magnetic flux lines also occur in type-II superconductors. These results are consistent with the experimental data.

7.5.4. The Application of the Above Theory and Josephson Effect in Superconductive Junctions

7.5.4.1. Proximity Effect in Superconductive Junctions

Proximity effect refers to the phenomenon that a normal conductor (N) in contact with a superconductor (S) can be superconductive, which was observed experimentally. This is obviously the result of the long-range coherent property of superconductive electrons. It can be regarded as the penetration of electron pairs from the superconductor into the normal conductor or a result of diffraction and transmission of the superconductive electron wave.In this phenomenon, superconductive electrons can occur in the normal conductor, but their amplitudes are much too small compare to that in the superconductive region, thus the nonlinear term $\lambda|\phi|^2\phi$ in GL equations (7.87)-(7.88) can be neglected. Because of these, GL equations in the normal and superconductive regions have different forms. On the S side of the S-N junction, the GL equation is [103,125]

$$\frac{\hbar^2}{2\mathrm{m}}\left(\nabla-\frac{\mathrm{ie}^*}{\mathrm{ch}}\vec{\mathrm{A}}\right)\phi-\alpha\phi+2\lambda\phi^3=0 \qquad (7.103)$$

while that on the N side of the junction it is

$$\frac{\hbar^2}{2m}(\nabla - \frac{ie^*}{ch}\vec{A})\phi - \alpha'\phi = 0 \tag{7.104}$$

Thus, the expression for $\vec{J}$ remains the same on both sides.

$$\vec{J} = \frac{e^*\hbar}{2mi}(\phi^*\nabla\phi - \phi\nabla\phi^*) - \frac{(e^*)^2}{mc}|\phi|^2\vec{A} \tag{7.105}$$

In the S region, we have obtained a solution of (7.103) in the previous section, and it is given by (7.99) or (7.101) and (7.102). In the N region, from Eqs. (7.104)- (7.105) we can easily obtain

$$\begin{cases} \varphi^2 = \frac{1}{2}\sqrt{(\varepsilon')^2 - 4\vec{d}^{2}}\sin(2\sqrt{b'}x) + \frac{\varepsilon'}{2} \\ \phi_N^2 = \varphi^2\phi_0^2 e^{-2i\theta} = \frac{1}{2}\sqrt{(\varepsilon')^2 - 4\vec{d}^{2}}\sin(2\sqrt{b'}x)e^{-i2\theta} + \frac{\varepsilon'}{2}\phi_0^2 e^{-i2\theta} \end{cases} \tag{7.106}$$

where

$$b' = \frac{2m\alpha'}{\hbar^2} = \frac{1}{\bar{\xi}'^2},\quad 2\bar{d}^2 = \frac{4\vec{J}^2 m\lambda}{(e^*)^2\alpha'^2},\quad E' = \frac{b'}{2}\varepsilon'$$

here ε' is an integral constant. A graph of ϕ vs. x in both the S and the N regions, as shown in Figure 7.10, coincides with that obtained by Blackbunu [126]. The solution given in Eq. (7.106) is the analytic solution in this case. On the other hand, Blackbunu's result was obtained by expressing the solution in terms of elliptic integrals and then integrating it numerically. From this, we see that the proximity effect is caused by diffraction or transmission of the superconductive electrons.

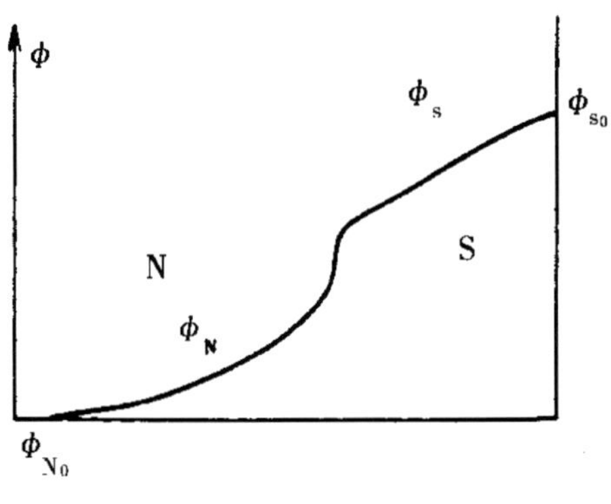

Figure 7.10. Proximity effect in S-N junction.

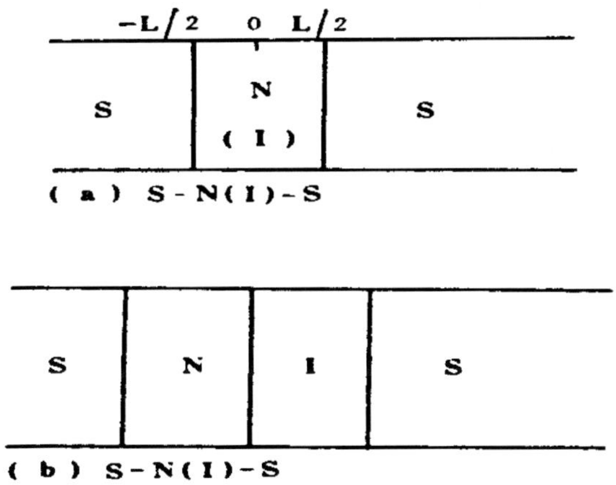

Figure 7.11. Superconductive junction for S-N-S and S-I-S.

7.5.4.2. The Josephson Effect in S-J-S and S-N-S Junctions

A superconductor-normal-superconductor junction (S-N-S), or a superconductor-insulator-superconductor junction (S-I-S), consists of a normal conductor or an insulator sandwiched between two superconductors as is schematically shown in Figure 7.11a.The thickness of the normal conductor or the insulator layer is assumed as L, and we choose the z coordinate such that the normal conductor or the insulator layer is located at $-L/2 \le x \le L/2$. The features of S-I-S junctions were studied by Jacobson et al. [127]. We will treat this problem using the above results of dynamic equation and method [102, 128-129].

The electrons in the superconducting regions ($|x| \ge L/2$) are depicted by GL equation (7.103). Its solution was given earlier in Eq.(7.101). After eliminating $u_1$from Eq.(7.100), we have [126-127] $J=(e^*\alpha u_0/2)\sqrt{\alpha(1-u_0)/m\lambda}$ .Setting $dJ/du_0=0$,we get the maximum current $J_c=(e^*\alpha/3)\sqrt{\alpha/3m\lambda}$. This is the critical current of a perfect superconductor, corresponding to the three-fold degenerate solution of (7.100), *i.e.*,$u_1=u_0$.
From (7.105),we have

$$\vec{A} = -\frac{m\vec{J}c}{(e^*)^2\phi_0^2\varphi^2} + \frac{hc}{e^*}\nabla\theta .$$

Using the London gauge, $\nabla.\vec{A}=0$, we can get [123-125]

$$\frac{d^2\theta}{dx^2} = \frac{mJ}{e^*\phi_0^2\hbar}\frac{d}{dx}\left(\frac{1}{\varphi^2}\right) .$$

Integrating the above equation twice, we get the change of the phase to be

$$\Delta\theta = \frac{mJ}{e^*\phi_0^2\hbar}\int(\frac{1}{\varphi^2}-\frac{1}{\varphi_\infty^2})dx \tag{7.107}$$

where $\varphi^2 = u$, and $\varphi_\infty^2 = u_0$. Here we have used the following de Gennes boundary conditions in obtaining Eq. (7.107)

$$\left.\frac{d\phi}{dx}\right|_{|x|\to\infty} = 0, \left.\frac{d\theta}{dx}\right|_{|x|\to\infty} = 0, \ \phi(|x|\to\infty) = \phi_\infty \tag{7.108}$$

If we substitute Eqs.(7.98) - (7.101) into Eq. (7.107), the phase shift of the wave function from an arbitrary point x to an infinite can be obtained directly from the above integral, and takes the form of:

$$\Delta\theta_L(x\to\infty) = -\tan^{-1}\sqrt{\frac{u_1}{u_0-u_1}}+\tan^{-1}\sqrt{\frac{u_1}{u-u_1}} \tag{7.109}$$

For the S-N-S or S-I-S junction, the superconducting regions are located at $|x| \geq L/2$ and the phase shift in the S region is thus

$$\Delta\theta_s = 2\Delta\theta_L(\frac{L}{2}\to\infty) \approx 2\tan^{-1}\sqrt{\frac{u_1}{u_s-u_1}} \tag{7.110}$$

According to the results in (7.104) - (7.105) and the above similar method, the change of the phase in the I or N region of the S-N-S or S-I-S junction may be expressed as [102, 123-126]

$$\Delta\theta_N = -2\tan^{-1}[\frac{2e^*h'}{\dot{j}}\sqrt{\frac{\alpha^2}{8m\lambda}}\tan(\frac{\sqrt{b'}L}{2})]+\frac{m\dot{J}L}{2e^*h'\mu_0} \tag{7.111}$$

where $h' = \sqrt{\frac{8m\lambda}{\alpha^2}}\frac{\dot{j}}{2e^*}\frac{\tan(\Delta\theta_N/2)}{\tan(\sqrt{b'}L/2)}$, $\frac{m\dot{J}L}{2e^*h'\mu_0}$ is an additional term to satisfy the boundary conditions (7.108),and may be neglected in the case being studied.

Near the critical temperature ($T<T_c$), the electric current passing through a weakly linked superconductive junction is very small ($\dot{j} << 1$).We then have

$$\mu_1' = \frac{4\dot{J}^2 m\lambda}{(e^*)^2\alpha^2} = 2\overline{A}^2, \text{ and } g'=1$$

Since $\eta\varphi^2$ and $d\varphi^2/dx$ are continuous at the boundary x=L/2, we have

$$\frac{d\mu_s}{dx}\Big|_{x=L/2} = \frac{d\mu_N}{dx}\Big|_{x=L/2}\ \eta_s\mu_s\Big|_{x=L/2} = \eta_N\mu_N\Big|_{x=L/2},$$

where η_s and η_N are the constants related to features of superconductive and normal phases in the junction, respectively. These give [130-131]

$$2\sqrt{b'}\,A\sin(2\Delta\theta_N) = \varepsilon_1[1-\cos(2\Delta\theta_s)]\sin(\sqrt{b'}L)$$
$$\cos(\sqrt{b'}L)\sin(2\Delta\theta_s) = \varepsilon\sin(2\Delta\theta_N) + \sin(2\Delta\theta_s + \Delta\theta_N)$$

where $\varepsilon_1 = \eta_N/\eta_S$. From the two equations, we can get

$$\sin(\Delta\theta_s + \Delta\theta_N) = \frac{2\sqrt{2m\lambda}\dot{J}}{e^*\alpha}\sqrt{b'}\sin(\sqrt{b'}L)$$

Thus,the Josephson current in superconductive junctions can be represented as

$$\dot{J} = \dot{J}_{max}\sin(\Delta\theta_s + \Delta\theta_N) = \dot{J}_{max}\sin(\Delta\theta) \tag{7.112}$$

where

$$\dot{J}_{max} = \frac{e^*\alpha_s}{2\sqrt{2m\lambda b'}}\cdot\frac{1}{\sin(\sqrt{b'}L)}, \Delta\theta = \Delta\theta_s + \Delta\theta_N \tag{7.113}$$

Equation (7.112) is the well-known example of the Josephson current, which was obtained from the Josephson effect of superconductivity junctions by Josephson [130-131], which is a macroscopic quantum effect. At present, this effect can be explained by nonlinear quantum mechanics. This again shows that this effect is a nonlinear quantum phenomenon and can be described by nonlinear quantum mechanics.

Also, from Eq. (7.113) we can see that the Josephson critical current [130-131]is inversely proportional to sin ($\sqrt{b'}L$), which means that the electric-current increases suddenly whenever $\sqrt{b'}L$ approaches $n\pi$, suggesting some resonant phenomena occurs in the system.

This has not been observed before. Moreover, $\dot{J}_{max}$ is proportional to $e^*\alpha_s/2\sqrt{2m\lambda b'} = (e^*\hbar\alpha_S/4m\sqrt{\lambda\alpha_N})$, which is related to $(T - T_c)^2$.

Finally, it is worth mentioning that no explicit assumption was made in the above on whether the junction is a potential well (α <0) or a potential barrier (α >0). The results are thus valid and the Josephson effect, in Eq. (7.112), occurs for both potential wells and for potential barriers.

7.5.4.3. Josephson Effect in SNIS Junctions

A superconductor-normal conductor-insulator-superconductor junction (SNIS) is shown schematically in Figure 7.11(b).

It can be regarded as a multilayer junction consisting of the S-N-S and S-I-S junctions. If appropriate thicknesses for the N and I layers are used (approximately 2nm- 3nm), the Josephson effect, similar to that discussed above, can occur in the SNIS junction. Since the derivations are similar to that in the previous sections, we will skip much of the details and give the results in the following. The Josephson current in the SNIS junction [130-131] is still given by

$$\dot{J} = \dot{J}_{max} \sin(\Delta\theta) \tag{7.114}$$

but, where $\Delta\theta = \Delta\theta_{s1} + \Delta\theta_N + \Delta\theta_I + \Delta\theta_{s2}$ and

$$\dot{J}_{max} = \frac{1}{\sqrt{b'_N}}\{\frac{\varepsilon_1 \sinh(\sqrt{b'_N}L)}{2[\cosh(\sqrt{b'_N}L) - \cos(2\Delta\theta_N)]}\}\times$$

$$\frac{1}{\sqrt{[1+\cos(2\Delta\theta_N)][1+\cos(2\Delta\theta_I)]} - \sqrt{[1-\cos(2\Delta\theta_N)][1-\cos(2\Delta\theta_I)]}} -$$

$$\frac{1}{\sqrt{b'_N}}\{\frac{\varepsilon_1\sqrt{[1-\cos^2(2\Delta\theta_N)]}\sinh(\sqrt{b'_N}L)}{2[\cosh(\sqrt{b'_N}L) - \cos(2\Delta\theta_N)]^2 - 1 + \cos^2(2\Delta\theta_N)}\}\times$$

$$\frac{1}{\sqrt{[1-\cos(2\Delta\theta_N)][1-\cos(2\Delta\theta_I)]} + \sqrt{[1+\cos(2\Delta\theta_N)][1+\cos(2\Delta\theta_I)]}}$$

The phase shifts $\Delta\theta_{s1}, \Delta\theta_N, \Delta\theta_I$ and $\Delta\theta_{s2}$ satisfy the following relations

$$4\sqrt{b'_N A}\sin(2\Delta\theta_N)] = \varepsilon_1[1 - \cos(2\Delta\theta_{s1})\sinh(\sqrt{b'_N}L)]$$

$$4\sqrt{b'_I A}\sin(2\Delta\theta_I)] = \varepsilon_1[1 - \cos(2\Delta\theta_{s2})\sinh(\sqrt{b'_I}L)]$$

$$\tan(\Delta\theta_I)] = \varepsilon_2 \frac{\tan(\sqrt{b'_N}L/2)}{\tanh(\sqrt{b'_N}L/2)}\tan(2\Delta\theta_N)(\sqrt{b'_N/b'_I}$$

$$\cosh(\sqrt{b'_N}L)\sin(2\Delta\theta_{s1}) = \sin(2\Delta\theta_N) + \sin(2\Delta\theta_{s1} + 2\Delta\theta_N)$$

$$\cos(\sqrt{b'_I}L)\sin(2\Delta\theta_{s2}) = \sin(2\Delta\theta_I) + \sin(2\Delta\theta_{s2} + 2\Delta\theta_I)]$$

It can be shown that the temperature dependence of J_{max} is $J_{max} \propto (T_c - T_0)^2$,which is quite similar to the results obtained by Blackburm et al., [126, 102, 132] for the SNIS junction and those by Romagnan et al., [133] using the Pb-PbO-Sn-Pb junction. Here, we obtained the same results using a completely different approach. This indicates again that we can obtain some results, which agree with the experimental data, by nonlinear quantum mechanics, thus, it is correct.

7.5.5. The Fundamental of Microscopic Theory for G-L equation

As it is known, G-L equation is a phenomenon theory, but it has a fundamental microscopic theory, i.e., the parameters of the G-L equation can be represented by physical quantities in microscopic theory or BCS microscopic theory, or in other words, the BCS microscopic theory can clarify the microscopic fundamentals of the G-L parameters.

As it is known, the superconductivity is closely related to the form of Cooper electron-pairs, the variation of the energy gap $\Delta(T)=V\sum <C_{-k}C_k>_T$ with the temperature in BCS theory exhibits, in essence, the changes of the order parameters for the superconductor, i.e., $\Delta \neq 0$, at $T<T_C$. But if $T>T_C$, then $\Delta=0$, where C_k and C_{-k} are the annihilation operators of the electron with contrary spins, and V is their interactional potential.

The energy gap for the uniform superconductor is defined by $\Delta=V\sum_k <C_{-k}C_k>_T=V\sum_k u_k v_k[1-2f(\xi_k)]$ in BCS theory [44,86,87], which is a real quantity and independent with the position vector $\vec{r}$, where $f(\xi_k)=[\exp(\xi_k/K_BT)+]^{-1}$ is the residue numberofquasi-particles with the excited energy ξ_k.

In the nonuniform superconductor the energy gap is defined by $\Delta(\vec{r})=V<\psi_\downarrow(\vec{r})\psi_\uparrow(\vec{r})>$, in which the field quantity $\psi_\downarrow(\vec{r})$ or $\psi_\uparrow(\vec{r})$ are now unfolded by the plane wave. In this case, the energy gaps obtained by these unfolded representations are all the same for the ground state and the excited state without the superconductive current; they are represented by $\Delta(\vec{r})=V\sum_k <C_{-k}C_k>$, which is not related to the position.

Generally, $\Delta(\vec{r})$ is a complex and depends on the $\vec{r}$. Therefore, $\Delta(\vec{r})$ should correspond to the complex order parameter in the G-L theory mentioned above. Gorkov [95,96] demonstrated that the dynamic equation of $\Delta(\vec{r})$ is the same as the G-L equation mentioned above, where the G-L order parameter $\phi(\vec{r})$ is positively proportional with the energy gap function $\Delta(\vec{r})$, which may be denoted by $\phi(\vec{r})=[7\varsigma(3)n]^{1/2}\Delta(\vec{r})/4\pi K_BT$, where $\varsigma(3)$ is a Riemann function.

In this investigation, the relationships of $m^*=2m, e^*=2e$ have also been acquired. This means that the G-L order parameters characterized and marked the form and existences of Cooper electron pairs in the superconductors.

On the other hand, the phenomenon parameters, $\alpha, \lambda, \lambda_L(t), \xi(T)$ in G-L equations can be found out from microscopic theory or BCS theory, they are, respectively,

$$\alpha=1.83\hbar^2(T-T_c)/2m\xi_0T_c,$$

$$\lambda=0.35\hbar^4/4m^2\xi_0^2g(0)(K_BT_c)^2, \lambda_L(T)=\lambda(0)T_c^{1/2}/[2(T_c-T)]^{1/2},$$

$\xi(T) = 0.74\xi_0[T_c/(T_c - T)]^{1/2}$.

Therefore, these results indicated clearly that the G-L parameters are closely related to the wave function and characteristic parameters in microscopic BCS theory, or in other words, the former can be derived from the latter. Thus, we can say that the microscopic BCS theory can serve as a microscopically theoretical fundamental of the macroscopic G-L theory; namely, the latter has a good theoretical foundation.

7.6. Nonlinear Quantum-Mechanical Properties of Polarons in Ionic Crystals

7.6.1. Form of Large Polaron and Its Properties in Ionic Crystals

The concept of the polaron was first proposed by Landau to describe the motion of an additional electron in ionic crystals. It is suggested in ionic crystals,constructed by positive and negative ions, that the electron would polarize the lattices by attracting positive ions and repulsing negative ions, thus the crystal's energy is lowered and the state of the electron is also changed in the process. The localized electron and the polarized lattice field, which surrounds the electron through electron-phonon attractive interaction, incorporate to form a stable complex, which is called the polaron. In essence, it is an element excitation or quasi-particle formed by the combination of the slower electron with longitudinal-optics (LO) phonon [81-87, 134].It is a carrier in ionic crystal. Itis used often to research and explain the luminescence and transportation phenomena in ionic crystals and polarized seminconductors.

The polaron's scale is determined by the size of the distorted or polarized domain of the lattice around the electron. If the size of the domain is larger than the lattice constant, then it is referred to as a great polaron, on the contrary, it is called a small polaron. Its size is denoted by $L' = 1/\Delta k = \sqrt{(\hbar/2m\omega_L)}$, where m and k are the mass and wavevector of the electron, respectively, and ω_L is the frequency of the LO phonon. This representation can be obtained using the uncertain relation. If the energy of the LO phonon is $\hbar\omega_L$, then the uncertainty of the energy of the electron is $\Delta E = \hbar\omega_L$, when it is absorbed or radiated in LO phonon. In this case, the uncertainty of the wavenumber of the electron is $\Delta k = (2m\omega_L\hbar)^{1/2}$, which is obtained from $E = \hbar^2 k^2/2m$ and $\Delta E = \hbar\omega_L$ .Thus, the uncertainty of the position or size of the electron is $\Delta r = 1/\Delta k = \sqrt{(\hbar/2m\omega_L)}$, where m is the mass of the electron. For the ionic crystal, such as a halide with alkaline metal, the effective mass of the electron in the energy-band is about the same asthe masse of free electron. Then we can obtain that the size of the polaron is $L' = 1/\Delta k = 10\,\overset{o}{A}$ through calculation. In the polarized semiconductors, which are composed of the elements between II-VI and III-V periods, in which the effective mass of the electron is about 100 times smaller than that of the free electron, then the size of the polaron is about 10nm, which is larger than the lattice constant of the semiconductor, thus it is called the large polaron. However, in the semiconductors with narrow energy-band, such as NiO, the excitedquasiparticles are small polarons because their sizes are smaller than the constant of the lattice.

If the interaction between the electron and LO phonon is considered, then the effective mass of the electron will be increased with increasing coupling strength between them, thus the scale of the polaron will be decreased. Therefore, the small polaron always occurs in matter with a narrow energy-band and strong coupling, in which the electron is bound by means of its own polarized field, thus the localized state of the polaron can be formed. This is Landan's self-trapped state of electron [134]. The mobility of the small polaron is lower, so it will transfer or shift from one lattice point to another under influence and excitation of thermal phonons. This is just the mechanism of electric-conductivity of small polarons.

Therefore, the properties of motion of the large and small polarons are completely different due to their effective masses, moved velocities, bonding strengths and formed conditions. These problems were investigated by Pekar [135], Frohlech [136-137], and Holstein [138], and many others [139]. In this section, we first discuss the nonlinear quantum –mechanical properties of large polarons by a new method.

7.6.2. Frohlech's Theoretical Description of Large Polarons

Frohlech [136-137] gave first the Hamiltonian of large polarons in ionic crystal, which is

$$H = H_e + H_p + H_{\text{int}} = p^2/2m + \sum_q \hbar\omega_L a_q^+ a_q + \sum_q (V_q a_q e^{iq.r} + h.c) \tag{7.115}$$

where $H_e = p^2/2m$ is the Hamiltonian of slower electrons in the conductive band, m is its effective mass, p is its momentum, $H_p = \sum_q \hbar\omega_L a_q^+ a_q$ is the Hamiltonian of LO phonons, a_q^+ and a_q are its creation and annihilation operator, which satisfied the commutation relation of the Boson particle, $H_{\text{int}} = \sum_q (V_q a_q e^{iq.r} + V_q^+ a_q^+ e^{-iq.r})$ is the interaction Hamiltonian between the slower electrons and LO phonons in the Lee-Low-Pines representation [140], where

$$V_q = i\frac{\hbar\omega_L}{q}\left(\frac{\hbar}{2m\omega_L}\right)^{1/2}(4\pi\alpha)^{1/2}, \alpha = \frac{e^2}{2\hbar\omega_L}\left(\frac{\hbar}{2m\omega_L}\right)^{-1/2}\left(\frac{1}{\epsilon_\infty} - \frac{1}{\epsilon_0}\right),$$

α is a dimensionless coupling constant.

The energy of the polaron can be found using the perturbation method, which is

$$E_k = -\alpha\hbar\omega_L + \hbar^2 k^2 / 2m, m^* = m(1-\alpha/6)^{-1}$$

at T=0K in the case of weak coupling $\alpha << 1$.

However, the coupling constant in most of ionic crystal is between 3 and 5, in which the coupling of the electron with LO phonon is strong, thus the perturbation method can not be used, which means that we must use the non-perturbation method to treat the feature of the system.

Pines et al., [141-142] used the median coupling theory and the transformation of the displacement vibrator to give the free energy of the polaron and its other feature, which are represented as

$$E(P) = -\alpha\hbar\omega_L + P^2 \big/ 2m[1+\alpha/6)] + \ldots,$$

its energy in ground state is $E(0) = -\alpha\hbar\omega_L$, its effective mass is $m^* = m[1+\alpha/6)]$.

Low et al., [143] and Mahan [144] built the shift-theory of large polarons and the Chew-Low scattering theory of polarons with phonons and found the state density of the polaron, which is

$$\rho_R d\Omega = \frac{1}{(2\pi\hbar)^2} P_f^2 (\frac{dP_f}{dE}) d\Omega_f = \frac{m^* P_f}{(2\pi\hbar)^2} d\Omega_f ,$$

the average hysteresis time of scattering of the polaron by LO phonon, which is approximately represented by

$$\tau = (\frac{m}{m^*})^2 \frac{f(\alpha)}{2\alpha\omega_L} [\exp(2\omega_L / K_B T) - 1] ,$$

and the polaron's migration rate in the medium in the case of the median coupling of $f(\alpha) \approx 5/6$, where $3 < \alpha < 6)$, which is denoted in

$$\mu = \frac{e\tau}{m^*} = \frac{f(\alpha)}{2\alpha\omega_L} (\frac{e}{m})(\frac{m}{m^*})^3 [\exp(2\omega_L / K_B T) - 1] = \frac{f(\alpha)}{2\alpha\omega_L (1+\alpha/6)^2} (\frac{e}{m}) [\exp(2\omega_L / K_B T) - 1]$$

respectively.

7.6.3. Nonlinear Dynamic Equation of Large Polarons in Ionic Crystals

We now discuss the behavior of an additional electron in a one-dimensional polar crystal in the framework of the Holstein model [136], which takes into consideration the appearance of soliton-like excitations of the additional electron in such systems. Assume in this case that the system consists of N identical diatomic molecules, with atomic masses of m_1 and m_2, respectively, and located at their equilibrium positions, $x_n^{(1)}$ and $x_n^{(2)}$ of the nth molecular cell. We here will consider only the variations of the interatomic distance due to the longitudinal optical oscillation with frequency ω_0, and ignore the longitudinal acoustic oscillations. The Hamiltonian operator for vibration of lattice molecules is given by

$$H_{ph} = -\frac{\hbar}{2M}\sum_n \frac{\partial^2}{\partial u_n^2} + \frac{1}{2}M\omega_0^2\sum_n u_n^2 + \frac{1}{2}M\omega_1^2\sum_n u_n u_{n+1} \tag{7.116}$$

with $M = \dfrac{m_1 m_2}{m_1 + m_2}$, $u_n = x_n^{(2)} - x_n^{(1)} = a$

The last term in Eq.(7.116) represents the nondiagonal dispersion effect of the optical phonons, with $\omega_1^2 << \omega_0^2$. According to Fedyanin et al., [145-146], the Hamiltonian of an electron interacting with the lattice in the systems is given by

$$H_e + H_{\text{int}} = (\varepsilon_0 + W')\sum_n N_n - J\sum_n (b_n^+ b_{n+1} + b_{n+1}^+ b_n) + V\sum_n N_n N_{n+1} \tag{7.117}$$

in the second quantization representation, where W' is the shift of the energy level ε_0 of the electron at the nth site due to the influence of the electron at other sites (we assume there is only one electron level ε associated with a possible bound state). Also, in (7.117), J is the overlap integral (J >0) and V is the strength of the electron-phonon interaction (having the dimensions [energy] × [length]$^{-1}$), and $b_n^+(b_n)$ is the creation (annihilation) operator of the additional electron $N_n = b_n^+ b_n$.

For small V, the Hamiltonian H_{ph}+ H_e+ H_{int} can be transformed to the well known Pekar-Fröhlich type, which allows us to take into account the long-range interaction between the electron and the polarization field caused by the lattice deformation, whereas the Holstein Hamiltonian contains only a short-range interaction. Applying the Bogolyubov variational principle, Fedyanin et al., [145-146] transformed the Hamiltonian H_{ph}+ H_e+ H_{int} into the following form, with "separation" of the phonon and electron subsystems:

$$H = \widetilde{H}_{ph} + \widetilde{H}_e$$

with $\widetilde{H}_{ph} = \hbar\sum_q \omega_q\left(a_q^+ a_q + \frac{1}{2}\right)$,

$$\widetilde{H}_e = \left(\varepsilon_0 + W' - \frac{V^2}{2M\omega_0^2}\right)\sum_n N_n - Je^{-F_\theta}\sum_n\left(b_n^+ b_{n+1} + b_{n+1}^+ b_n\right) + \frac{V^2}{2M\omega_0^2}\sum_n N_n N_{n+1}, \tag{7.118}$$

where

$$\left[a_q, a_{q'}^+\right] = \delta_{qq'}, \left[b_n, b_{n'}^+\right]_+ = \delta_{nn'}, \left[\widetilde{H}_e, \widetilde{H}_{ph}\right] = 0,$$

$$N_n = b_n^+ b_n, \omega_q^2 = \omega_0^2 + \omega_1^2 \cos q r_0,$$

$$F_\theta = \frac{V^2}{4MN\hbar\omega_0^4} \sum_q \omega_q \left|S_q\right|^2 \coth\left(\frac{\hbar\omega_q}{2K_B T}\right), \ (\theta = K_B T).$$

In the above,

$$S_q = \omega_0^2 \sum_n \frac{\Delta_{n+1}^m - \Delta_n^m}{\Delta} e^{iqmr_0}, \quad \Delta_n^m = \left(-\frac{\omega_1^2}{2\omega_0^2}\right)^{|n-m|} \frac{\Delta}{\omega_1^2} \left(1 - 2\frac{\omega_1^2}{\omega_0^2}\right)^{-2}$$

Here Δ is the determinant of the matrix constituted by ω_1^2 and ω_o^2, Δ_n^m is its minor obtained by eliminating the *m*th column and the *n*th row.

For $\theta \to 0$, the basic contribution to Δ_n^m comes from the diagonal terms ($n = m$) of

$$S_\theta = \frac{V^2\theta}{2M\hbar^2\omega_0^4} + O\left(\frac{\omega_1^2}{\omega_0^2}\right).$$

Thus, the parameters corresponding to the phonon subsystem ω_q and *V* renormalize the HamiltonianH_e, due to the application of the Bogolyubov variational principle. Obviously, the Hamiltonian given in Eq. (7.118) is a nonlinear operator. Therefore, the polaron undergoes a change from a linear elementary excitation to a nonlinear soliton-excitation through the nonlinear interaction arising from the electron-phonon interaction.

Using the Heisenberg equation, Makhankov et al., [51-52] showed that

$$i\hbar(\phi_n)_t = \left(\varepsilon_0 + W' - \frac{V^2}{2M\omega_0^2}\right)\phi_n(t) - Je^{-F_0}\left[\phi_{n+1}(t) + \phi_{n-1}(t)\right] + \frac{V^2}{M\omega_0^2}\left[1 - \phi_{n+1}^*(t)\phi_{n+1}(t) + \phi_{n-1}^*(t)\phi_{n-1}(t)\right]\phi_n(t) \quad (7.119)$$

In this case, the trial wave function can be written as

$$|\phi(t)\rangle = U|0\rangle = \sum_n \phi_n(t) b_n^+ |0\rangle_{po} \quad (7.120)$$

where $|0\rangle_{P0}$ is the ground state of the polaron. In the continuum approximation,Makhankov and Fedyanin [52] obtained the following from Eq. (7.119) by neglecting terms involving derivatives higher than the second.

$$i\hbar\phi_t = \left(\varepsilon_0 + W' - \frac{V^2}{2M\omega_0^2}\right)\phi(x,t) - Je^{-F_\theta}\left[2\phi + r_0^2\phi_{xx}\right] + \frac{V^2}{M\omega_0^2}\left[1 - |\phi|^2 - r_0^2|\phi_x|^2 - \frac{r_0^2}{4}|\phi_{xx}|^2 - \frac{r_0^2}{2}\left(\phi_{xx}^*\phi + \phi_{xx}\phi_x^*\right)\right]\phi(x,t). \quad (7.121)$$

If we only retain terms with orders less than the fourth, with respect to dispersion and nonlinearity in Eq. (7.121), we can get

$$i\hbar\phi_t(x,t) = \left(\varepsilon_0 + W' - \frac{V^2}{2M\omega_0^2} - 2Je^{-F_\theta} + \frac{V^2}{M\omega_0^2}\right)\phi(x,t) - Jr_0^2 e^{-F_\theta}\phi_{xx} - \frac{V^2}{M\omega_0^2}|\phi|^2\phi(x,t) \quad (7.122)$$

This is a standard nonlinear Schrödinger equation in nonlinear quantum mechanics. Therefore, the properties of the polarons can be described by nonlinear quantum mechanics. Thus the polarons in ionic crystals all have properties of microscopic particles in nonlinear quantum mechanics, described in Chapter 3, where the nonlinear interaction, $(V^2/M\omega_0^2)|\phi|^2\phi(x,t)$, is due to the interaction between the moved electron and phonon or the moved polarons because it contains the interaction potential V. In this case, the soliton solution of Eq.(7.122) is given by

$$\varphi(x,t) = \frac{r_0 \exp\left[i(kx - \omega t + \theta_0)\right]}{2W_S \cosh\left[(x - vt - x_0)/W_S\right]} \quad (7.123)$$

and its energy is

$$E_S = \varepsilon_0 + W_s + \frac{V^2}{2M\omega_0^2} - 2Je^{-F_\theta} + \frac{M_S v^2}{2} - \frac{eF_\theta}{24J}\left(\frac{V^2}{M\omega_0^2}\right)^2 \quad (7.124)$$

where $M_S = \dfrac{\hbar^2 e^{F_\theta}}{2Jr_0^2}$, $W_S = \dfrac{4r_0 JM\omega_0^2}{V^2 e^{F_\theta}}$.

The polaron-soliton in Eq.(7.123) is localized in the region $\approx W_s$. The polaron effect is defined by the last two terms in Eq.(7.124). The third term describes the renormalization of the electron level shift due to its translation jumps. Note that there arises a temperature renormalization of the conductive electron mass $m_e = \hbar^2/2r_0^2 J$ because$J$ is replaced by $J\exp[-F_\theta]$, at present. Therefore, the polaron now gets heavier, i.e., $M_s(\theta) > M_s(0)$, and the effective mass becomes temperature dependent in such a case. If $\theta \to 0$, then

$M_s(\theta) > M_s(0)(1+3\chi\theta/8\omega_0)$, where χ is a coupling constant between the electron and phonon and is given by $\chi = 4V^2/3M\hbar^2\omega_0^3$.

The same effect, i.e. increase of mass, was observed for $\theta \neq 0$ in cyclotron resonance experiments by Rodriges and Fedyanin [147]. Such a correspondence is naturally associated with the similar structures of the Pekar-Frolich and Holstein Hamiltonians [135-138].

For a polaron with a small radius, a canonical transformation method has been well developed. By means of this transformation, Makhankov and Fedyanin[52] examined the main contribution to the electron mass renormalization defined by its strong interaction with the lattice vibration, and obtained the Hamiltonian of a "residual" (weak) polaron lattice vibration interaction. In this case, the lattice Hamiltonian in Eq. (7.116) is rewritten in terms of the normal coordinates, Q_q and ξ_q via the unitary operator $U = \exp(-iL)$, where

$$L = i\sum_q \xi_q^0 \frac{\partial}{\partial \xi_q} \qquad (7.125)$$

with $\xi_q^0 = -\dfrac{U}{M\omega_q^2}\sqrt{\dfrac{2}{N}}\sum_n N_n \sin\left(qn+\dfrac{\pi}{4}\right)$

The U transformations of the normal coordinates ξ_q viz, $\bar{\xi}_q = U\xi_q U^{-1} = \xi_q + \xi_q^1$ are such that ξ_q^1 may be regarded as the operator of lattice deformation. The overall Hamiltonian, $\bar{H}$ is then divided into three parts: $\bar{H}_{ph}$, which describes harmonic oscillations of lattice sites about their equilibrium positions, the polaron Hamiltonian $\bar{H}_p$

$$\bar{H}_p = (\varepsilon_0 + W')\sum_k N_k - Je^{-F_T}\sum_n \left(b_{n+1}^+ b_n + b_{n+1}^+ b_{n+1}\right) + V\sqrt{\frac{2}{N}}\sum_n N_n \sum_q \xi_q^0 \sin\left(qn+\frac{\pi}{4}\right) + \sum_q \frac{M\omega_q^2}{2}(\xi_q^0)^2 \qquad (7.126)$$

where

$$F_T = \frac{2V^2}{4\hbar MN}\sum_q \frac{1}{\omega_q^3}\left\{\sin\left(qn+\frac{\pi}{4}\right)\sin\left[q(n+1)+\frac{\pi}{4}\right]\right\}\coth\left(\frac{\hbar\omega}{2K_BT}\right)$$

and the Hamiltonian $\bar{H}_{int}$ of the residual interaction of the polaron with the oscillation of the deformed-lattice. The polaron motion described by Eq. (7.126) is adiabatic such that the lattice reconstructs completely and is related to the oscillations with new equilibrium positions. Hence, $\bar{H}_{int}$ describes the "friction process", i.e., the polaron motion is accompanied by a change of the phonon number, and the polaron loses its energy and slows down. Such effects decrease the energy gain due to soliton formation.

Using $||\varphi(t)\rangle$ in Eq.(7.120),they [52] replaced the lattice deformation operator ξ_q^0 by

$$\bar{\xi}_q^0 = \langle\varphi(t)|\xi_q^0|\varphi(t)\rangle = -\frac{V}{M\omega_q^2}\sqrt{\frac{2}{N}}\sum_n|\varphi_n(t)|^2\sin\left(qn+\frac{\pi}{4}\right)$$

Then $\overline{H}_p$ is transformed into $\overline{H}_p'$, which satisfies the following equation

$$i\hbar\frac{\partial}{\partial t}|\phi(t)\rangle = \overline{H}_p'|\phi(t)\rangle \tag{7.127}$$

From Eq. (7.127) and the relation $\overline{H}_p' = \overline{H}_p(\xi_q^0)$ they [52] obtained the equation satisfied by $\varphi(t)$ in the continuum approximation. Up to $0(\omega_1^2/\omega_0^2)$, the equation is

$$i\hbar\varphi_t(x,t) = (\varepsilon_0 + W + \tilde{\varepsilon}_0 - 2Je^{-F_T})\varphi(x,t) - Je^{-F_T}\tau_0^2\varphi_{xx} - \frac{V^2}{M\omega_0^2}|\varphi(x,t)|^2\varphi(x,t) \tag{7.128}$$

where $\tilde{\varepsilon}_0 = \dfrac{V^2}{2Mr_0^2\omega_0^2}\int|\varphi(x,t)|^2 dx$

Equation (7.128) is similar to Eq. (7.122), and therefore, has a similar soliton solution as Eqs.(7.123) – (7.124). The canonical transformation adopted here is valid for small V so that the renormalization of the band width remains small, i.e. $J\exp(-F_T) >> V^2/M\omega_0^2$ (This is the condition for continuum approximation). If $\exp(-F_T) << 1$, we can neglect the electron translational motion, and arrive at the small radius polaron.

On the other hand, for the electron-acoustic phonon interaction, Whitfield and Shaw [148] also showed that the polaron state becomes a soliton using the deformation potential technique.

These results indicated clearly that the states and features of the polarons, by virtue of its interactions with the longitudinal optical phonons, or the interaction of “residual” polarons with optical phonons and the electron-acoustic phonon interaction, can all be described using the nonlinear Schrödinger equation in nonlinear quantum mechanics, thus we affirm that the polarons are a soliton and have a wave-corpuscle duality in the ion crystal.

7.6.4. The Nonlinear Features of Polarons in Ionic Crystals

Using the adiabatic approximation, Fedyanin et al. [146], studied the properties of polarons in the case where the kinetic energy of the lattice oscillation can be neglected. In this approximation, the electron is assumed to move rapidly in the slowly varying field of the lattice deformation, which may take place if the time between two successive translation leaps of the electron, $t_1 \approx \hbar / J$, is less than the characteristic relaxation time of the lattice deformation, $t_2 \approx \omega_0^{-1}$, i.e., $\hbar\omega_0 << J$. This is reasonable for a sufficiently wide electron band *J*. In this case, the self-consistent electron amplitude, $\varphi(x,t)$ in Eq. (7.128), satisfies the following equation

$$i\hbar\varphi_t(x,t) = (\varepsilon_0 + W - 2J)\varphi(x,t) - Jr_0^2\varphi_{xx} - \frac{V^2|\varphi|^2\varphi}{M\omega_0^2} \tag{7.129}$$

in the continuum limit.

Equation (7.129) is similar to Eq. (7.122), but the coefficients of the linear terms are constant. According to the above approach, Eqs. (7.122) – (7.124), the total energy of the electron and the lattice in the adiabatic approximation is given by

$$E_S = \frac{M_S v^2}{2} - \frac{1}{48J}\left(\frac{V^2}{M\omega_0^2}\right)^2$$

where $M_S = \hbar^2 / 2Jr_0^2$ is the soliton effective-mass, which is the same as that of the electron in the conduction band in the given approximation, i.e., there is no temperature renormalization of the energy shift due to the translation leaps (–2*J*) and the soliton localization area $W_s^0 = (2Jr_0^2 / V^2)M\omega_0^2$. This implies the electron being weakly connected with the lattice, since $V \to 0(W_s^0 \to \infty)$, so thatthere are $\hbar\omega = \varepsilon_0 + W - 2J + m_z v^2 / 2$ and $\varphi(x,t) \approx \exp[i(\omega t - kx)]$ in this case, i.e., the solitary wave degenerates into a plane wave, in other words, the soliton is delocalized as an electron.

In this study, Fedyanin et al., [104] used the normal coordinate ξ_q and its conjugate momentum P_q, and introduced an energy functional $E = \langle\varphi(t)|H|\varphi(t)\rangle$, with $|\varphi(t)\rangle$ given by Eq.(7.120), which is a function of ξ_q and P_q for a given $\varphi_n(t)$. The equation satisfied by ξ_q determines the motion of a conventional oscillator under a constant external force. The solution of the equation in such a case can be simply written as

$$\xi_q(t) = \xi_q(0)\cos(\omega_q t) - \frac{V^2}{M\omega_q^2}\sqrt{\frac{2}{N}}\sum_n |\varphi_n(t)|^2$$

Inserting the above into the equation

$$\frac{\partial}{\partial t}|\varphi(t)\rangle = \hat{H}|\varphi(t)\rangle$$

yields, in the continuum approximating, the following equation for ' $\varphi_n(x,t)$

$$i\hbar\varphi_t(x,t) = (\varepsilon_0 + B - 2J)\varphi(x,t) - Jr_0^2\varphi_{xx}(x,t) - \frac{V^2}{M\omega_0^2|\varphi|^2}\varphi(x,t) \tag{7.130}$$

where $B = \sum_q \frac{V^2}{MN\omega_q^2}\left[\sum_n \frac{\partial}{\partial t}|\varphi_n(t)^2|\sin\left(qn + \frac{\pi}{4}\right)\right]^2$

is the kinetic energy of lattice deformation consistent with the electron motion, and

$$\varepsilon_0 = \sum_q \frac{V^2}{MN\omega_q^2}\left[\sum_n |\varphi_n(t)^2|\sin\left(qn + \frac{\pi}{4}\right)\right]^2$$

is the elastic energy of the lattice. Note that in the above adiabatic approximation, the following energy of harmonic oscillations about the shifted equilibrium positions was neglected

$$E' = T' + V' = \sum_q \frac{1}{2}M\omega_q^2\xi_q^2,$$

which would lead to a constant correction to the excitation dispersion. Equation (7.130) is similar to Eqs. (7.122) and (7.129), and thus has soliton solution similar to Eqs.(7.123)-(7.124) for the Eq. (7.130). It indicates that the polaron in the nonlinear quantum system is, in essence, a soliton and satisfies the nonlinear Schrödinger equation. It describes a quasiparticle of mass $M_s = m\exp(F_\theta)$, localized in the region of $W_s = W_0\exp(-F_\theta)$ .The energy of the soliton can be represented by $\varepsilon + W - 2J\exp(-F_\theta) + V^2/(2M\omega_0^2)$, which is less than that of the electron level, $\varepsilon + W + V^2/(2M\omega_q^2)$ [104], where

$$m = \frac{\hbar^2}{2Jr_0^2},\ W_S^0 = \frac{4JM\omega_0^2}{v^2},\ F_\theta = \frac{V^2\theta}{2M\hbar^2\omega_0^4} + O\left(\frac{\omega_1^2}{\omega_0^2}\right)$$

These properties of the polaron are independent of the method of calculation. For the soliton satisfying the nonlinear Schrödinger equation (7.130), its "kinetic energy" is

$$E = \frac{1}{2}(m + \Delta m)v^2 \text{ and } m = \frac{\hbar^2}{2Jr_0^2},\ \Delta m = \frac{4}{15}\frac{V^2}{M\omega_q^2}\left(\frac{r_0}{W_S^0}\right).$$

with the latter being the temperature independent renormalization of the polaron mass. Using $W_s^0 = 4JM\omega_0^2 / V^3$, the mass change, Δm of the polaron is in this case given [104] by

$$\Delta m = \frac{(V^2 / M\omega_0^2)^4}{240J^3 r_0^2 \omega_0^2}.$$

7.7. The Nonlinear Localization of Small Polarons and Its Properties in Semiconductors

7.7.1. Form of Small Polaron and Its Properties in Semiconductors

It is well known that the band electrons are moved in the small polaron in narrow-semiconductors, such as NiO, its effective mass is great, which means that the distance Δr between the electrons is small relative to that of lattice constant, and the polarons are called the small polarons. If the coupling of the electron with LO photons is stronger, then the electrons will be bound as some small polarons because of the effect of the polarized field generated by its self motion. In this case, we must study the influences of periodic fields, and the interaction between the electron and LO phonon on the energy states of the carriers (electron or hole). If fact, the small polarons can move, by virtue of the band conductivity and hopping processes. In the case of low temperature, the small polarons move in the mechanism of band conductivity, in which the phonon configurations and the translation symmetry of the system have not been changed, when the small polarons transfer from a lattice point to a neighbor lattice point. This resembles the motions of the electrons in the case of a tight bound approximation (TBA), their difference is only that the former considered the influence of lattice vibration, which involves the affects of temperature on the TBA bands of the lattice.

If the temperature of the systems is lifted, the numbers of the phonon around the small polarons will be increased, thus, the effective mass of conductive electrons in the energy band is correspondingly increased, and the width of the energy-band of the small polaron will be decreased. When the temperature $T>T_t$, where T_t is the transformation temperature, then the width of energy-band of the small polaron becomes so much smaller so that it is smaller than the uncertainty of the energy, $\hbar / \tau$, where τ is the lifetime of the small polaron. In this case, the energy band structure of the small polaron breaks and failures. Thus, the movement of the small polaron will be changed. In this case the small polarons can only move in the hopping manner from one lattice point to another by means of the excitation of thermal phonons, thus its movement is analogous to the diffusion motion of lattice defects in the system, which is referred to as the hopping mechanism of the small polarons. Mahan [144] studied the properties of energy-band conductivity and hopping processes of the small polarons using the lattice model of coupling the electron with LO phonon and gave the Hamiltonian of three dimensional lattice, including the interaction of the electron with LO phonon, which is represented [142,144] by

$$H=H_e+H_p+H_{int}=[\varepsilon_0\sum_l C_l^+C_l-J\sum_{l\kappa}C_{l+\kappa}^+C_l]+[\sum_q\hbar\omega_q(a_q^+a_q+1/2)]+[\sum_{ql}M_q e^{iq.l}(a_q+_{-q}^+)C_l^+C_l] \quad (7.131)$$

where $H_e=[\varepsilon_0\sum_l C_l^+C_l-J\sum_{l\kappa}C_{l+\kappa}^+C_l]$ is the TBA Hamiltonian of the electron in the movement of the lattice field, where C_l^+ and C_l are its creation and annihilation operator in Wannier representation, κ is its positive vector of most neighbor lattice point, J denotes the interaction between the two neighbor lattices, and ε_0 is the energy of the molecular orbit state of the electron around the any lattice point.

$$H_p=\sum_q\hbar\omega_q(a_q^+a_q+1/2)$$

is the Hamiltonian of LO phonons with $\omega_q=\omega_{-q}$, a_q^+ and a_q are the creation and annihilation operator of the LO phonons. $\sum_{ql}M_q e^{iq.l}(a_q+_{-q}^+)C_l^+C_l$ with $M_q^+=M_{-q}$ is the electron-LO phonon interaction Hamiltonian. Based on this model he gave the energy-band representation of the small polaron in the case of finite temperature, which is represented by

$$E_k(T)=E_0(T)-Je^{-S_T}\sum_\kappa e^{-i\vec{k}.\vec{\kappa}},$$

and its half-width of the energy-band, which is $ZJ\exp(-S_T)$, where

$$S_T=\sum_q\{2|M_q|^2(\hbar\omega_q)^{-2}\sin^2[(\vec{q}.\vec{\kappa})/2]\coth[((\hbar\omega_q)/2k_BT],$$

as well as the hopping probability of the small polarons in the case of high temperature, which is $\rho_T=(J/\hbar)^2(\pi/4E_ak_BT)^{1/2}\exp(-E_a/2k_BT)$, and the diffraction coefficient of the small polaron in sample cubic lattice, which is denoted by $D=a^2\omega_T=(Ja/\hbar)^2(\pi/4E_ak_BT)^{1/2}\exp(-E_a/2k_BT)$ and its hopping mobility, which is $\mu=eD/k_BT=(ea^2/k_BT)^2\omega_T\approx\exp(-E_a/2k_BT)$, where E_a is the excited energy of the hopping motion of the small polaron, a is the lattice constant. The results indicated clearly that the hopping mobility of the small polaron increases with increasing temperature, which resembles that of the lattice defects in the system.

As it is known, the mechanical and dynamic properties of small polarons have been studied by many researchers. For example, Brown et al. [149-150], proposed a theory of polaron formation based on the Fröhlich Hamiltonian in the transportless limit [136-137]. It was found that two related processes are involved in this formation. In the first process, energy interpretable as the polaron binding energy is released by the exciton system into the normal modes of the lattice. Another process results in the dispersion of this energy throughout the lattice and the development of a persistent deformation around the region occupied by the exciton. This indicates that a polaron cannot exist until the polaron binding energy is dispersed and the lattice deformation is completed. The amount of energy released into the lattice during polaron formation and the time dependence of this energy transfer are independent of the detail with which the initial bare exciton is distributed. Ivic and Brown [151 -152] further studied the nonlinear localization of the small polarons in this case. These ideas are interesting and worth further study.

7.7.2. The Nonlinear Dynamic Properties of the Small Polaron and Its Properties in Semiconductors

It is well known that a foreign particle or excitation in a deformable solid often causes local distortions of the host, which can significantly affect the character and dynamics of the quasiparticle. For example, interstitial hydrogen isotopes in metal hydrides can cause large volume dilations; mobilities of photon injected charge carriers in organic molecular crystals, such as anthracene and naphthalene, exhibit novel temperature dependences; excitonic spectra in organic molecular crystals such as pyrene and α -perylene are significantly affected by local distortions; and certain features in the Raman spectra of biological materials, such as *l*-alanine have been identified with local vibrational modes. The quasi-particle transport in these deformable media was investigated by Ivic and Brown [151-152], using the Fröhlich Hamiltonian:

$$H=\sum_n E_n b_n^+ b_n - \sum_{nm} J_{mn} b_m^+ b_n + \sum_q \hbar\omega_q a_q^+ a_q + \sum_{qn} \hbar\omega_q \left(x_n^q a_q^+ + x_n^{q*} a_q\right) b_n^+ b_n \quad (7.132)$$

where $J_{nm}$is a transfer integral between different sites in the medium. For the translationally invariant acoustic chain model, they choose the phonon dispersion relation ω_q and the dimensionless coupling function χ_n^q to be

$$\omega_q = \omega_B \sin\left(\frac{1}{2}|qr_0|\right) \text{ and}$$

$$x_n^q = x^q e^{-iqR_n} = \frac{2ix\sin(qr_0)}{\sqrt{2NM\hbar\omega_q^3}} e^{-iqR_n} ,$$

respectively, where r_0 is the lattice constant, b_n and a_q are the annihilation operators of the small-polaron and phonon with wave vector$q$ and frequency ω_q, respectively.

Exact solutions corresponding to the Fröhlich Hamiltonian are known only in the limit of infinite effective mass, wherein the transfer integrals between different sites in the medium are all zero. In general the solutions are found by the most approximate approaches, which are either small-polaron theories, usually employing perturbation methods based on a set of polaron basis states, or large-polarontheories, usually employing variational methods to obtain nonlinear evolution equations. Among the latter are theories, which describe quasiparticle transport in terms of envelope solitons. Ivic et al. [151-152], used the latter to display the unification of polaron and soliton theories of quasiparticle transport in the small-polaron regime based on Eq.(7.132).

Following the notation adopted by Ivic et al. [151-152],they denoted the wave function related to the small-polaron by a tilde,

$$|\overline{D}(t)\rangle = \sum_{a} \varphi_n(t)\overline{b}_n^+ |0\rangle \otimes |\tilde{\beta}(t)\rangle,$$

$$|\tilde{\beta}(t)\rangle \equiv \exp\left\{\sum_{n} \tilde{\beta}_q(t)\overline{a}_q^+ - \tilde{\beta}_q^*(t)\overline{a}_q^+\right\}|0\rangle, \tag{7.133}$$

where $\overline{b}_n = Ub_nU^+ = b_n \exp[\sum_q (\beta_{nq}a_q^+ - \beta_{nq}^* a_q)]$

and $\overline{a}_q = Ua_qU^+ = a_q + \sum_q \beta_{nq} b_n^+ b_n$

are annihilation operators of the small-polaron and the dressed phonon, respectively, with

$$U = \exp\left[-\sum_{n} (\beta_{nq}a_q^+ - \beta_{nq}^* a_q) b_n^+ b_n\right]. \tag{7.134}$$

The $|\overline{D}(t)\rangle$ states contain the exact small-polaron states in the infinite-mass limit $((\beta_q(t) = 0)$.

The Hamiltonian (7.132) is not diagonal in the small-polaron states, when $J \neq 0$ .However, the interaction between the small polarons ($(\overline{b}, \overline{b}^+)$ and the dressed phonons $(\overline{a}, \overline{a}^+)$ is weaker than the interaction between the bare quasiparticles (*b*, *b*+) and the bare phonons (*a*, *a*+). The microscopic dynamics can be well described by a factored polaron state $|\overline{D}(t)\rangle$

Since the phase mixing part of $\beta_{qn}(t)$ is chosen to be time independent, Ivic and Brown [109-110] did not consider soliton, or polaron, formation from bare states; they considered instead the dressed polarons or solitons. To determine the equations of motion of polarons, Ivic et al. [151-152], applied the variational principle of time-dependent quantum mechanics,

$$\delta \int_{t_1}^{t_2} dt \left\langle \Psi(t) \right| (i\hbar \frac{d}{dt} - H) \left| \Psi(t) \right\rangle = 0,$$

which was also used by others (Skrinjar, Kapor, and Stojanovic [153], Zhang, Romero-Rochin [154]) and obtained the following evolution equations,

$$i\hbar \frac{d\varphi_n(t)}{dt} + i\hbar \sum_a \left[x_n^q \tilde{\beta}_q^*(t) - x_n^{q*} \tilde{\beta}_q(t) \right] \varphi_n(t) = \frac{\partial \langle H \rangle}{\partial \varphi_n^*(t)}, \quad (7.135)$$

$$i\hbar \frac{d\tilde{\beta}_q(t)}{dt} - i\hbar \sum_n x_n^q \frac{d}{dt} \left| \varphi_n(t) \right|^2 = \frac{\partial \langle H \rangle}{\partial \tilde{\beta}_q^*(t)} \quad (7.136)$$

where $\langle H \rangle$ is the expectation value of the Hamiltonian, (7.132), in the state $\left| \overline{D} \right\rangle$.From Eqs. (7.135) and (7.136), we can eliminate the explicit phonon variables in the latter, to obtain an integro-differential equation for the small-polaron probability amplitudes,

$$i\hbar \frac{d\varphi_n(t)}{dt} = (E - \sum_q |\chi^q|^2 \hbar\omega_q)\varphi(t) - \sum_m \tilde{J}[\varphi_{n+1}(t) + \varphi_{n-1}(t)] + \int_0^t d\tau \sum_j G_{nj}(t-\tau) \frac{d}{dt} |\varphi_j(\tau)|^2 \varphi_n(t) + g_n(t)\varphi_n(t) \quad (7.137)$$

where

$$\tilde{J} = J \exp\left[-4 \sum_{-q} |x^q|^2 \sin^2\left(\frac{1}{2} q r_0 \right) \right]$$

is the renormalized tunneling matrix element, obviously it is reduced from the bare value, *J*, by interactions with phonons, and

$$G_{nj}(t) = 2 \sum_q x_n^q x_j^{q*} \hbar\omega_q \cos(\omega_q t),$$

$$g_n(t) = \sum_a \hbar\omega_q \left[x_n^q \tilde{\beta}_n^*(0) e^{i\omega_q t} + x_n^{q*} \tilde{\beta}_q(0) e^{-i\omega_q t} \right]$$

The kernels $G_{nj}(t)$ are related to the thermal part of the fluctuations $g_n(t)$, as discussed by Wang et al., [155-156]. Ivic et al. [151-152] expressed a small-polaron Bloch state by

$$|\psi(t)\rangle = \sum_n \tilde{u}(k) \exp\left\{ i \left[kR_n - \frac{\tilde{E}(k)}{\hbar} \right] t \right\} \tilde{b}_n^+ |0\rangle \quad (7.138)$$

with

$$\tilde{E}(k) = \left(E - \sum_q |x^q|^2 \hbar\omega_q \right) - 2\tilde{J} \cos(k r_0). \quad (7.139)$$

They chose

$$\varphi_n(t) = \tilde{u}(k)\exp\left\{ikR_n - \frac{i\tilde{E}(k)t}{\hbar}\right\}$$

and $\overline{\beta}_q(0) = \beta_q(0) = 0$. Then Eq. (7.137) is linearized, indicating that the small-polaron, Bloch states are exact stationary solutions of the nonlinear evolution equation.

In the continuum limit, Ivic et al. [151-152], obtained the nonlinear Schrödinger equation given below by searching for solutions with the D'Alembert property $|\varphi(y,\tau)|^2 = |\varphi(y - v\tau)|^2$, and choosing $\beta_q(0)$, corresponding to a dressed soliton at zero temperature,

$$i\hbar\frac{d\varphi(x,t)}{dt} = -\frac{\hbar^2}{2\tilde{m}}\frac{\partial^2}{\partial x^2}\varphi(x,t) + \tilde{E}(0)\varphi(x,t) - \frac{\upsilon^2}{\upsilon_0^2}G(\upsilon)|\varphi(x,t)|^2\varphi(x,t), \tag{7.140}$$

where $\overline{m} = \hbar^2/2Jr_0^2$ with $J = J\exp(-\delta^2 F)$. $F = 4\sum_q |x^q|^2 \sin(r_0 q/2)$ is the effective mass of the small-polaron, $\tilde{E}(0) = E - \delta(2-\delta)E_0 - 2J\exp(-\delta^2 F)$ is the bottom of the energy band of the small-polaron, as given by Eq. (7.138), and

$$G(v) = \frac{4\chi^2 r_0}{\omega}\frac{v_0^2}{v_0^2 - v^2}$$

is a velocity-dependent nonlinearity parameter. In the above, δ is defined by $\beta_{nq}(t) = -\delta\chi_n^q + \beta_q(t)$ and $\beta_{nq}(t) = \delta\chi_n^q$.

The above result shows that the small polaron is a soliton, which satisfies the nonlinear Schrödinger equation (7.140). This shows clearly that the dynamic properties of the small polarons can also be described by nonlinear quantum mechanicsin this case, where the nonlinear interaction is $(\upsilon^2/\upsilon_0^2)G(\upsilon)|\varphi(x,t)|^2\varphi(x,t)$. Obviously, it is due to the interaction between moved polarons and moved phonons because the interaction coefficient χ is involved in the nonlinear interaction energy $G(\upsilon)$. Therefore, it belongs to the self-trapping mechanism of the nonlinear mechanism. In this case, the soliton solution of Eq.(7.140) is denoted by

$$\varphi(x,t) = \sqrt{\frac{\mu}{2}}\frac{e^{-i[\overline{E}(0)/\hbar]t}e^{i(kx-\omega t)}}{\cosh[\mu(x-ut)]} \tag{7.141}$$

where $\hbar k \equiv \tilde{m}v$, $\hbar\mu = \frac{\tilde{m}G(v)}{2\hbar}\frac{v^2}{v_0^2}$ and $\hbar\omega = \frac{1}{2}\tilde{m}v^2 - \frac{v^4}{v_0^2}\frac{\tilde{m}G(v)^2}{8\hbar^2}$.

The solution in Eq. (7.141) of this equation are not bare Bloch states, but are small-polaron solitons with a larger effective mass $(\overline{m} > m)$ and a lower energy $[\overline{E}(0) < E(0)]$ compared to a bare Block state, due to the dressing of the quasiparticle[151-152]. Therefore, the properties of the small polaron-soliton can be described completely by nonlinear quantum mechanics and are similar to those of microscopic particles in nonlinear quantum mechanics listed in Chapter 3.

7.8. The Theory of Motion of the Polarons and Its Properties in One-Dimensional Molecular Crystals

We now assumed the one-dimensional molecular crystal of the polaron is a linear chain composed of N molecules with two atoms, in which the interaction potential of electrons in each pair is expressed by $V(x\text{-}R_n, u_n)$, where $R_n = nr_0$ is the position of nth molecule, u_n is the displacement of the atom relative to other atom in nth molecule, then the Hamiltonian of the system can be represented [142] by

$$H = H_e + H_p + H_{\text{int}} = \frac{\hbar^2}{2m}\frac{d^2}{dx^2} + \sum_n -\frac{\hbar^2}{2M}\frac{\partial^2}{\partial x^2} + \frac{1}{2}M\omega_0^2 u_n + \frac{1}{2}M\omega_i^2 u_n u_{n+1} + \sum_n V(x - R_n, u_n) \quad (7.142)$$

where M is the effective mass of the molecule, ω_0 and ω_1 are the diagonal and nondiagonal elements of dynamic matrix of the polaron, ω_0 is also the Einstein frequency , and $M\omega_i^2 u_n u_{n+1}$ denotes the dispersed effect of vibration of L0 lattice.

In this case, the wave function of the system can be denoted[142] by

$$\Phi = \sum_n \beta_n(u_1, u_2 \ldots u_N)\varphi_n(x, u_n) \quad (7.143)$$

where $\varphi_n(x, u_n)$ expresses the stateof the localized molecular-orbit in this system, it satisfies

$$[\frac{\hbar^2}{2m}\frac{d^2}{dx^2} + V(x - R_n.u_n)]\varphi_n(x - R_n.u_n) = E(u_n)\varphi_n(x - R_n.u_n) \quad (7.144)$$

In the linearity approximation we can denote $E(u_n)$ [142]by

$$E(u_n) = -Bu_n + \text{constant} \quad (7.145)$$

But the $\beta_n(u_1,u_2....u_N)$ in Eq.(7.143) is the probability amplitude of the electron in the molecular orbits around the nth lattice point, where the subscript n denotes the quantum number of state of the electron, or the coordination of the lattice point.

From the Schrödinger equation, $i\hbar\partial\Phi/\partial t = H\Phi$, we can obtain the equation of motion for function $\beta_n(t)$ [104,105,157-158], which is expressed as

$$i\hbar\frac{\partial}{\partial t}\beta_n(u_1,u_2....u_N)=(H_p - Bu_n)\beta_n(u_1,u_2....u_N)+A(u_n,u_{n\pm1})\beta_n(u_1,u_2....u_N) \quad (7.146)$$

where

$$A(u_n,u_m)=\int\varphi^*(x-R_n,u_n)V(x-R_n,u_n)\varphi(x-R_m,u_m)dx \quad (7.147)$$

denotes the overlapping integrate with two centers, but in which the overlapping integrate, with three center integrates, and the derivative of $\varphi(x-R_m,u_m)$ with respect to u_n as well as the dependence of J on u are all neglected perylene [159]. In this case $A(u_n,u_{n\pm1})$ can be replaced by -A and $\omega_1 = 0$ is also assumed. Thus, we can approximate Eq.(7.146) by

$$i\hbar\frac{\partial}{\partial t}\beta_n(u_1,u_2....u_N)=\sum\nolimits_m(-\frac{\hbar^2}{2M}\frac{\partial^2}{\partial u_m^2}+\frac{1}{2}M\omega_0^2u_m^2)\beta_n(u_1,u_2....u_N)-A(\beta_{n+1}+\beta_{n-1})-Gu_n\beta_n(u_1,u_2....u_N) \quad (7.148)$$

As it is known, in the case of strong coupling the self-trapping state of the electronor polaron occurs under action of the nonlinear attractive potential well, which is produced by the interaction between the deformation or displacement of the lattice and the motion of the electron, this implies that we should use the method of adiabatic approximation to calculate the change of $\beta_n(u_1,u_2....u_N)$. In this case, $\beta_n(u_1,u_2....u_N)$ is a time-independent solution, its eigenenergy is $E(u_1,u_2....u_N)$ [142,105]. Thus, the following eigenequation holds

$$[\frac{1}{2}\sum\nolimits_m M\omega_0^2u_m^2 - Gu_n]\beta_n - A(\beta_{n+1}+\beta_{n-1}) = E(u_1,u_2....u_N)\beta_n \quad (7.149)$$

If to find the inner-integrate of β_n in Eq.(7.149), and to use normalization condition

$\int|\Phi|^2dx=\sum\nolimits_n|\beta_n|^2=1$ we can obtain the value of $E(u_1,u_2....u_N)$.

If to find the variation of $E(u_1,u_2....u_N)$, with respect to u_m we can gain the vibrational states $(u_1^{(0)},u_2^{(0)}....u_N^{(0)})$,corresponding to the extreme energies, in which $u_1^{(0)}$ satisfies

$$u_m^{(0)} = Q\left|\beta_m^0\right|^2 / M\omega_0^2, (m=1,2.3,....N) \tag{7.150}$$

where β_m^0 is the solution of Eq.(7.149) at $u_m = u_m^0$. Inserting Eq.(7.150) into Eq.(7.149), we finally obtain the dynamic equation of the polaron in this system[142,105], which is

$$\frac{d^2\beta_n^{(0)}}{dn^2} + 4\nu\left|\beta_n^{(0)}\right|^2 \beta_n^{(0)} = (\sigma / J)\beta_n^{(0)} \tag{7.151}$$

where the relationships of $\sigma = 1/2\sum_m M\omega_0^2 u_m^{(0)2} - 2J - E(u_1^{(0)}, u_2^{(0)}....u_N^{(0)})$, $\nu = Q^2 / 4M\omega_0^2 J$ and $\beta_{n+1}^{(0)} + \beta_{N-1}^{(0)} - 2\beta_n^{(0)} \approx \partial^2\beta_n^{(0)} \big/ \partial n^2$ are used.

Equation (7.151) is a standard nonlinear Schrödinger equation in nonlinear quantum mechanics. This indicates that the properties of the small polaron in the one-dimensional molecular crystal can be described by the nonlinear quantum mechanics. Quite clearly, the nonlinear interaction, $4\nu\left|\beta_n^{(0)}\right|^2 \beta_n^{(0)}$, is produced by the attractive interaction between the moved electron and the polarized phonon, which provided a potential-well to self-trap the electrons, thus the polaron or the electron self-traps a polaron-soliton. Its properties are described as follows.

The features of the small polaron-soliton can be obtained[142,157-158], in special cases, from Eq.(7.151). If $n \to \pm\infty$, then the solution β_n^0 and its derivative approach zero, which denote the localized state of the small polaron and can be written as

$$\beta_n^{(0)} = (\sigma / 2\nu J)^{1/2} \sec h[(\sigma / J)(n - n_0)] \tag{7.152}$$

where n_0 is an integral constant, $\sigma = J\nu^2$. In this case of $E(u_1, u_2....u_N) = -(1/3)(J\nu^2) - 2J$, where -2J is the energy of the polaron in the lowest energy-band, 2J is the half width of the TBA energy-band in the rigid crystal, then the binding energy of the polaron is $E_p = J\nu^2/3 = (Q^2 / M\omega_0^2)^2 / 48J$. Then the concrete representation of probability amplitude of the localized polaron [142,157-158] can be written as

$$u_n^{(0)} = \frac{Q}{2M\omega_0^2} \sec h^2[\nu(n - n_0)]$$

The corresponding potential-well of the self-trapped polaron in this case is

$$U(u_n^{(0)}) = E(u_n^{(0)}) = -Qu_n^{(0)} = -\frac{Q^2}{2M\omega_0^2} \nu \sec h^2[\nu(n - n_0)]$$

From Eqs.(7.152) and $\sigma = J\nu^2$ we acquired the probability amplitude of the localized polaron, which is

$$\beta_n^{(0)} = \sqrt{\nu/2}\,\mathrm{sec}\,h[\nu(n-n_0) \text{ or } \beta_n^{(0)} = \sqrt{r_0/2L}\,\mathrm{sec}\,h[\beta(n-n_0)/L] \qquad (7.153)$$

where $L = r_0/\nu = 4M\omega_0^2 Jr_0/Q^2$, r_0 is the lattice constant. Equation (7.153) describes the bell-type polaron-soliton, i.e., the polaron now becomes a bell-type soliton in this case, L is its size or the width of the bell-type solitary wave. If $L >> r_0$, then the soliton is formed from a great polaron, which contrarily is only formed by a small polaron at $L \le r_0$. We can use the above results to explain some phenomena, such as the hysteresis effect of the lattices, and so on.

REFERENCES

[1] E. Pauling and E. Wilson, *Introduction to quantum mechanics*, 1935.
[2] N. F.Mottand I.N.Sneddon, *wave mechanics and its applications*, 1948.
[3] D. Bohm, Quantum theory, University of London, London, 1954.
[4] P. A. M. Dirac, *The principles of quantum mechanics*, Clarendon Press, Oxford, 1947 and 1958.
[5] L. Schiff, *Quantum mechanics*, McGraw-Hill, New York, 1965.
[6] K. Gottfried, *Quantum mechanics*, Vol.I, New York, Benjamin (Fourth printing), 1974.
[7] A. Messianh, *Quantum mechanics*, Vol.II, NorthHolland Publishing Co. Amsterdam, 1970.
[8] E. Merzbachar, *quantum mechanics*, 2 edi. 1970 John Waley & sons, Inc., 1970.
[9] L. D.Landau, and E.M.Lfshitz, Quantum mechanics, Pregamon Press, Oxford, 1958.
[10] D. Bohm. *Quantum theory*, Prentice-Hall, New York,1951.
[11] A. S.Davydov, *Quantum mechanics*, Pergamon Press, Oxford, 1965.
[12] G. L.Trigg, *Quantum mechanics*, D.Van Nostrand Co., Inc., Princeton, 1969.
[13] P. Roman, *Advanced quantum theory*, Addisonn-Wesley, Reading, Massachusetts, 1965.
[14] A. Messiah, *Quantum mechanics*, Interscience, New York, 1961.
[15] K. Gottfried, *Quantum mechanics*, W.A.Benjamin, New York, 1966.
[16] W. Heisenberg, *The physical principles of the quantum theory*, Chicago University Press, 1930.
[17] J. Potter, *quantum mechanics*, North-Holland publishing Co., Amsterdam,1970.
[18] S. X. Zhou, *Quantum mechanics*(Chinese), People Education Press, Beijing,1979.
[19] Pang Xiao-feng and Feng Yuan ping, *Quantum mechanics in nonlinear systems*, New Jersey, World Scientific. Publishing Co., 2005.
[20] Pang Xiao-feng, *Theory of nonlinear quantum mechanics*, Chongqing, Chinese Chongqing Press, 1994.
[21] Pang Xiao-feng, *Physica B*,403(2008) 2341.
[22] Pang Xiao-feng, *Fronts of physics in China*, 3 (2008) 243.

[23] Pang Xiao-Feng, *Nature Sciences*, 3(1)(2008)29.
[24] Pang Xiao-feng, *Nature Sciences*, 2(1)(2007) 42.
[25] Pang Xiao-feng, *Mod. Phys.Lett.*B23(2009)939.
[26] Pang Xiao-feng, *Physica B* 404(2009)2353.
[27] Pang Xiao-feng,*Physica B* 404(2009)4327.
[28] Pang Xiao-feng, *Phys. Stat.* Sol.(b), 236(2003)34.
[29] Pang Xiao-feng, *J. Low. Temp. Phys.*, 58（1985） 334.
[30] Pang Xiao-feng, *Soliton Physics,* Chengdu: Sichuan Science and Technology Press, 2003.
[31] E. Schrödinger, *Collected paper on wave mechanics*, London: Blackie and Son, 1928.
[32] E. Schrödinger, *Proc. Cambridge Puil. Soc.*, 1935, 31: 555.
[33] W. Heisenberg, *Z. Phys.,* 33(1925)879.
[34] W. Heisenberg and Euler H., *Z. Phys.* 98(1936)714.
[35] M. Born and L. Infeld, *Proc. Roy. Soc.* A, 144（1934） 425.
[36] M. Tajiri, *J.Phys.Soc.*Japan, 53(1984)1634 and ibid 52(1983)1908
[37] Yu. B. Gaididei, *Phys. Rev.* E52(1995)2951
[38] P. B. Burt, *quantum mechanics and nonlinear waves*, Harwood Academic Publishers, New York, 1981
[39] I. V. Barashenkov, and V. G. Makhankov, *Phys. Lett.* A 128 (1988) 52.
[40] V. G. Makhankov, *Soliton phenomenology*, Kluwer, Amsterdam, 1990.
[41] R. E. Peierls, *Quantum theory of solids*, London: Oxford University Press, 1955. P108.
[42] L. Friedman and T.Holstein, *Ann. Phys.* 21(1963)494.
[43] H. B. Shore and L. M. Sander, *Phys. Rev. B* 7(1973)4537.
[44] J. Bardeen, L. N. Cooper and R. Schreiffer, *Phys. Rev.* 108(1957)1175.
[45] A. Alexandrov and J.Ranninger, *Phys. Rev.* B 45(1992)13109.
[46] K. Nasu, Phys, *Rev. B,* 37(1988)5075.
[47] C. G. Kuper and G. D.whitield, *Polarons and excitons*, New York: Plenum Press, 1963.
[48] L. Habbard, *Proc.Roy.Soc.* A176(1963)238.
[49] U. Lindner, V. Fedyanin, Phys.Stat.So1.(b)89(1978)123 and 95(1978)k83.
[50] V. Makhankov, N. Makhaidiani and O. Pashaev, *Phys.Lett.* A, 81 (198)167.
[51] V. G. Makhankov, *Phys.Lett.*A 81(1981)156.
[52] V. G. Makhandov, and V.K.Fedyanin, *Phys.Rep.*104(1984)1-80.
[53] Q. Niu, X.D.Wang, L. Kleimman, and W. M.Liu, et al., *Phys. Rev. Lett.* 83(1999)207.
[54] H.Li, S.Q. Shen, J.Q.Liang, W. M. Liu, *Phys. Rev*.B, 72(2005)014546.
[55] H. Zheng, *Phys. Rev*. B, 37(1988)7419.
[56] H. Zheng, *Phys. Rev*. B, 38(1988)11865.
[57] H. Zheng, *Phys. Rev*. B, 41(1990)4723.
[58] D. Feinberg, S.Ciuchi, F. de Pasquale, *Inter. J. Mod. Phys.* B, 4(1990)1317 .
[59] H. Zheng. D.Feinberg and M. Avegmon, *Phys. Rev*. B, 39(1989) 9405.
[60] J. E.Hirsch, *Phys. Rev*. B, 31(1985)6022.
[61] E. Fradkin, and J. E. Hirsch, *Phys, Rev*. B, 27(1983):1680.
[62] I. C. Lang and Y. A. Firsov, *Zh. Exsp.Ter.* Fiz.1 43(962)1843. (Sov. Phys. JETP 16 (1963)1301).
[63] H. Chen, Y. M. Zheng, X. Wu, *Phys, Rev*. B 54(1996)3273.
[64] R. Manka, *Phys. Stat. Sol.(*b) , 93(1979) 53.

[65] C. F. Lo, E. Manousakis, R. Sollie and Y. L.Wang, *Phys. Rev.* B, 50(1994)418.
[66] A. A. Eremk, Z.*Phys.*.B 104(1997)765.
[67] M. S. Kim and B.C. Sanders, *Phys. Rev.* A 53 (1996)3694.
[68] S. Sahoo, *J.Phys. Condensed* matter, 10(1998)1999.
[69] E. Jeckelman, C. L. Zhang, S. R. Mite, *Phys. Rev*.B, 60(1999)7950.
[70] Y. Zolotaryuk, P.L.Christiansen, J. J. Rasmussen, *Phys. Rev*.B 58(1998)14305.
[71] J. M.Robin, *Phys. Rev*.B 58(1998)14335.
[72] W. Z. Wang, A. R. Bishop, J. Gaammel, T and R. V. Silver, *Phys.Rev.Lett.* 80 (1998)3284.
[73] D. M. Leitner, *Phys.Rev*.B, 64 (2001)094201.
[74] P. A. Tanner, *Phys. Rev*.B, 65(2002)214305.
[75] J. L. Xiao, W.Xiao, *Phys.Rev.* B 58(1998)1678.
[76] Fan Hong-Yi and Xiao Min, *Phys. Lett* A 220(1996) 81.
[77] Pang Xiao-feng, *J. Phys. Chem.* Sol.61(2000)701.
[78] Pang Xiao-feng, *J. Mat. Sci. Technol.* 151(999)369.
[79] Pang Xiao-feng, *Chinese Phys. Lett.* 16(1998)129.
[80] Pang Xiao-feng, *Phys. Stat. Sol.* (b) 217(2000)887.
[81] Pang Xiao-feng, *Acta. Physica Sinica*, 8(1999)598
[82] E. Majernikovaand Pang Xiao -feng, *Phys. Lett*.30(1997) 89.
[83] Pang Xiao-feng, *J. Phys. Chem.solids*, 62(2001)491.
[84] Pang Xiao-feng, *Science in China*, G, 51(2008) 225.
[85] G. P. Borghi, A. Girlando, A. Painelli, and J.Voit, *Europhys. Lett.*, 24(1996)127.
[86] L. N.Cooper, *Phys. Rev.* 104 (1956) 11.
[87] J. R. Schrieffer, *Superconductivity*, Benjamin, New York, 1969.
[88] J. R. Schrieffer, *Theory of Superconductivity*, Benjamin, New York, 1964.
[89] R. D. Parks, *Superconductivity,* Marcel. Dekker,1969.
[90] L. D. Landau, and E. M. Lifshitz, *Quantum Mechanics*, Pergamon Press, Oxford, 1987.
[91] A. S. Davydov and V. N.Ermakov, *Phys. Stat.Sol*.B148(1988)305.
[92] V. I. Ginzberg, and L. D. Landau, *Zh. Eksp. Theor. Fiz.* 20 (1950) 1064.
[93] V. L.Ginzburg, and L. P. Pitayevsky, *Sov. Phys. JETP* 7 (1968) 858.
[94] V. L. Ginzberg, and D. A. Kirahnits, *Problems in High-Temperature Superconductivity*, Nauka, Moscow, 1977.
[95] L. P. Gorkov, *Sov.Phys. JETP* 9 (1958) 189.
[96] L. P, Gorkov, *Sov.Phys. JETP* 9 (1959) 1364.
[97] A. A. Abrikosov, *Zh. Eksp. Theor. Fiz.* 32 (1957) 1442.
[98] A. A. Abrikosov, and L. P. Gorkov, *Zh. Eksp. Theor. Phys.* 39 (1960) 781.
[99] Pang, Xiao-feng, *J. Kunming Tech. Sci. Univ.*, 3 (1991) 57.
[100] Pang, Xiao-feng, *J. Kunming Tech. Sci. Univ.*, Sin. 4 (1991) 13.
[101] V. M. Perez-Garcia, M. Michinel and H. Herrero, *Phys. Rev.* A 57 (1998) 3837.
[102] Pang Xiao-feng, Properties of Macroscopic Quantum Effects and Dynamic Natures of Electrons in Superconductors in 9 Chapter in <<Superconductivity-Theory and Applications>>, eds: In Tech - Open Access Publisher, Croatia, 2011, 173-212.
[103] Pang, Xiao-feng, *J. Low Temp.* Phys. 58(1985) 333.
[104] Pang Xiao feng, Establishment of nonlinear quantum mechanics and changes of property of microscopic particles as well as their experimental evidences in

Contemporary Research in quantum systems, ed. Zoheir Wzziane, Nove Publishers, New York, pp207-338, 2014.
[105] Cuo Bai-lin and Pang Xiao feng. *Solitons*, Chinese Science Press, 1987.
[106] Pang Xiao-feng, *nonlinear quantum mechanics*, LAP Lambert Publishing, Germany, 2012.
[107] Pang Xiao-feng, *Chinese. J. Low Temp. Supercond.* 10(4) (1989) 33.
[108] Pang Xiao-feng, *Nature Sciences*, 2 (2007) 42.
[109] Pang Xiao-feng, *Chinese J. Nature*, 5 (1982) 254.
[110] D. Rogovin, and M. Scully, *Ann. Phys.* 88 (1969) 371.
[111] D. Rogovin, and M. Scully, *Phys. Rep.* 25 (1976) 176.
[112] D. Rogovin, *Ann. Phys.* 90 (1974) 18.
[113] D. Rogovin, *Phys. Rep.* 251 (1976) 1786.
[114] D. Rogovin, *Phys. Rev.* B11 (1975) 1906; ibid. *Phys. Rev.* B12 (1975) 130.
[115] J. P. Ronognan, *Solid State Commun.* 14 (1974)83.
[116] A. J. Leggett, *Prog. Theor. Phys.* (Supp.) 69 (1980) 80.
[117] A. J. Leggett, in *Percolation,Localization and Superconductivity*, eds. by A. M. Goldlinan,S. Bvilf, Plenum Press, New York, 1984. pp. 1- 41.
[118] A. J. Leggett, *Low temperature physics*, Springer, Berlin, 1991, pp. 1-93;
[119] I. R. Niek, and J. C. Fraser, J. *Low Temp. Phys.* 3 (1970) 225.
[120] Pang Xiao-feng, *J. Electronic Science and Technology of China*, 6 (2008)205.
[121] B. M. Caradoc-Davies, R. J. Ballagh and K. Bumett, *Phys. Rev.Lett.* 83 (1999) 895.
[122] M. R. Matthews, B. P. Anderson, P. C. Haljan, D. S. Hall, C. E. Wieman, and E. A. Cornell. *Phys. Rev. Lett.* 83 (1999) 2498.
[123] K. W. Madison, et al., K.W. Madison, F. Chevy, W. Wholleben, and J. Dalibard, *Phys.Rev. Lett.* 84(2000) 806.
[124] A. Tonomara, *Foundation Phys.* 28 (1998) 59.
[125] Pang, Xiao-feng, *J. Southwest Inst. For Nationalities Sin.* 15 (1989) 97.
[126] J. A. Blackbunu, et al., *Phys. Rev.* B11 (1975)1053.
[127] D. A. Jacobson, *Phys. Rev.* B8 (1965) 1066 and Adv. Phys. 14 (1965) 419.
[128] Pang, Xiao-feng, *J. Science Exploration Sin.* 4 (1986) 70.
[129] Pang, Xiao-feng, J. *Southwest Inst. For Nationalities Sin.* 17 (1991) 1 and 18.
[130] B. Josephson, *Phys. Lett.* 1 (1962) 251; *Adv. Phys.* 14 (1965) 419.
[131] B. Josephson, Thesis, unpublised, Cambridge University (1964).
[132] Pang Xiao-feng, Chinese. *J. Low Temp. Supercond.* 3(3)(1982) 62.
[133] P. J. Romagnan, et al., *Solid State Commun.* 14(1974)83.
[134] L. Landau, *Phys.Z.Sowjetunion*, 3(1933)664.
[135] S. Pekar, *J.Phys. (*Moscow)10(1946)341 and 10(1946)347.
[136] H. Fröhlich, Pelzer H. and Zienau S., *Plilos.Mag.*41(1950)221.
[137] H. Fröhlich, *Adv. Phys.* 3(1954)325.
[138] T. Holstein, *Ann. Phys.* (N.Y.)and 8(1959)325 and 8(1959)343.
[139] J. T. Devreese, *Polarons in ionic crystals and polarsecaicounductors*, Amsterdam. North Holland 1972.
[140] T. D. Lee. F. E. Low and D. Pines, *Phys.Rev* .90(1953)297.
[141] D. Pines, in *palarons and excitation*, eds. C. G. Kuper and G. D. Whitfield, Plenum Press, New York, 1963.
[142] Z. Z. Lee, Solid state theory (Chinese), Advanced Educated Press, Beijing, 1985.

[143] F. E. Low, *Phys. Rev.* 97(1955)1392.
[144] G. D. Mahan, *Many particle physics*, Plenum Press, New York, 1981.
[145] V. S. Fedyanin, and V. Yushankhai, *Teor.Mat.Fiz*, 35(1978)240.
[146] V. Fedyanin, L. Yakushevich, *Phys.Lett.* A 85 (1981)100.
[147] K. Rodriges and V. Fedyanin, JINR P17-81-169, Dubna,1981; JINR p17-81-8, Dubna 1981.
[148] G. Whitfield and P.Shaw, *Phys.Rev.* B14(1976)3346.
[149] D. W. Brown, K. Lindemberg, and B. J.West, *Phys. Rev.* B37(1988)2946.
[150] D. W. Brown, K. Lindemberg, and B.J.West, *J.Chem. Phys.* 87(1987)6700 and 84(1986)1574.
[151] Z. Ivic, D.W. Brown, *Phys.Rev.Lett.* 63(1989)423.
[152] D. W. Brown, and Z. Ivic, *Phys.Rev.* B40(1989)9876.
[153] M. J. Skrinjar, D. V. Kapor, S.D. Stojanovic, *Phys. Rev.* A38(1988)6402.
[154] Q. Zhang, V.Romero-Rochin, R. Silbey, *Phys. Rev.* A38(1988)6409.
[155] X. D. Wang, D. W. Brown, K.Lindenberg, B.J. West, *Phys.Rev.* A33(1986)4101.
[156] B. W. Brown, B. J. West, and K. Lindenerg, *Phys.Rev.* A33(1986)4110.
[157] Pang Xiao feng, *Nonlinear quantum mechanics* (Chinese), Elecronic Industry Press, Beijing, 2009.
[158] Pang Xiao-feng, The behaviours and properties of microscopic particles in nonlinear systems, pp. 185-249, and The soliton theory of bio-energy transport in protein molecules, pp-251-298 in *Handbook of solitons*, Research, technology and applications, eds S.P.Lang and S.H.Bedore, Nova Science Publishers, Inc. New York, 2009.
[159] S. T. Nettel, *Phys. Rev.* 128(1982)2573.

Chapter 8

NONLINEAR QUANTUM-MECHANICAL PROPERTIES OF THE MAGNONS IN THE SPIN-SPIN INTERACTION SYSTEMS

The nonlinear quantum properties of the magnons in the ferromagnetic and antiferromagnetic system are studied in detail, using the new nonlinear quantum mechanical theory and methods, in which the Hamiltonians of the systems and dynamic equation of the magnons are first built based on their structure and distribution of the spins. The results showed the magnons satisfied either the nonlinear Schrödinger equation or the ϕ^4 – equation, thus we could affirm that the magneon is a soliton with wave-corpuscle duality, where the nonlinear interactions are due to the interactions between the moved magnon and phonon or among the moved magnons. If these nonlinear interactions are neglected, then the magnons are the quanta of the spin waves in these systems. We elucidated and explained further some phenomena in the ferromagnets and antiferromagnets by using these results, which also experimentally verified the validity of the new nonlinear theory.

8.1. THE SPIN WAVE AND ITS FEATURES IN THE SPIN-SPIN INTERACTION SYSTEMS

8.1.1. The Distribution of Spins in Different Spin-Spin Interaction Systems

As it is known, the spins at the lattice points are always orderly arranged in both ferromagnetic and antiferromagnetic systems containing the ferrite magnets, although their directions of arrangement are somewhat different in these systems. In fact, the directions of all spins are the same in the ferromagnetic system, on the contrary, the directions of neighbor spins are inverse in the antiferromagnetic and ferrite magnets, but the strengths of neighbor spins are different in ferrite magnets, as shown in Figure 8.1. Therefore, we can affirm that there are spin-spin interactions in these systems, but different systems have different properties for the interactions.

If these spin-spin interaction systems are acted on or perturbed by an externally applied magnetic field, then the direction and strength of the spins will all be changed, concretely speaking, their directions will deviate from the original direction, which will result in a collective excitation through a propagation of deviation of direction of the spins, due to the spin-spin interaction among the neighbor spins, such as exchanged interaction.

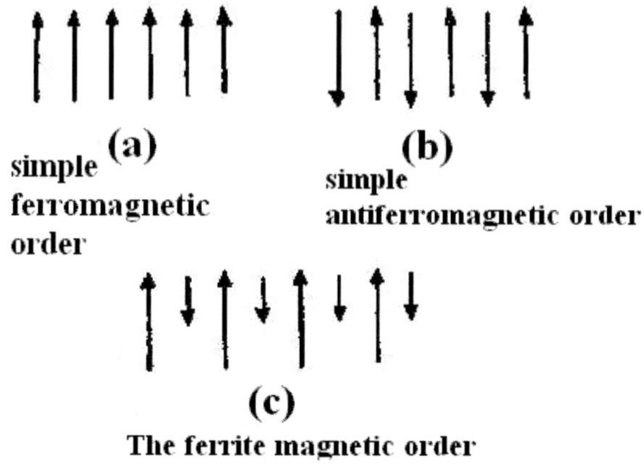

Figure 8.1. The ground states of spin-spin interactional systems [8].

In this process the action of the neighbor spin on the first perturbed spin, which makes the latter try to restore the original state, and leads to the oscillation of the spin in this case, thus the propagation of a wave arising from this oscillation occurs as shown in Figure 8.2; such wave of the spins is referred as to the spin wave. It is, in essence, a low excited state, or elementary excitation, its quantum is called the magnon. This means that the magnon is an elemental excitation or basic particles and is designated as the fundamental features in these spin-spin interaction systems, its spin is denoted by $\vec{S}_n$.

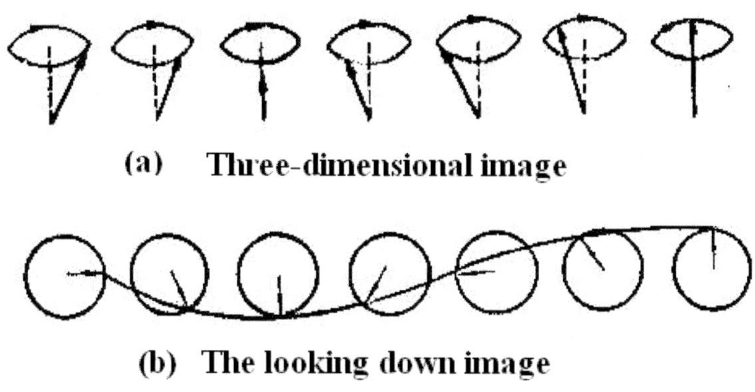

Figure 8.2. The spin wave images in spin-spin interactional systems [8].

8.1.2. Hamiltonians of Different Models in Spin-Spin Interaction Systems

(1)The Ising Model with the Symmetry

As it is known, the excitations of the margins in the spin-spin interaction systems are complicated. Many models, which are used to describe the properties of the excited states, were proposed. The Hamiltonian of the magnetic chain with Ising (easy-axis) symmetry [1], a simplest system, is firstly proposed and represented by

$$H = -J\sum_n S_n \bullet S_{n+1} - D\sum_n (S_n^z)^2 \tag{8.1}$$

where $\hat{S}_n$ is the spin operator of the electron at nth lattice point, J is the exchange integral between the electrons at the nth and n+1th lattice points, D is the anisotropy energyof the electron at nth lattice point. In a classical description this Hamiltonian has two equivalent ground states, $\hat{S}_n^z = \pm S$, where S is the spin value of the electron. The configuration describing the transition between these equivalent ground states is the static domain wall. Its structure is determined from the competition between the exchange energy J (which favors a broad wall) and the anisotropy energy D (which favors a narrow wall). In the continuum approximation, and introducing spherical coordinates, then the classical energy of the system can be denoted by

$$E\{\Theta(z), \Phi(z)\} = JS^2[\frac{1}{2}(\frac{d\Theta}{dz})^2 + \frac{1}{2}\sin\Theta(\frac{d\Phi}{dz})^2 - \frac{D}{J}\cos^2\Theta] \tag{8.2}$$

where Φ and Θ are two angles, the minimization with respect to Φ and Θ yields

$$\Phi(z) = cons\tan t, \, d^2\Theta/_{dz^2} = (D/_J)\sin(2\Theta)$$

The static sine-Gordon equation obtained is the basic equation; it described the states of the static domain wall in this simplest soliton-bearing magneticchain. Its fundamental solution describing the transition between the two equivalent ground stages is the basic kink solution, which is

$$\Theta(z) = 2\arctan\{\exp[\pm(\frac{2D}{J})(z - z_0))^{1/2}]\}$$

Its physical image is described in Figure 8.3, in which the spin configuration of the basic soliton in the easy-plan ferromagnetic chain, is shown.

This is the formulation of the ferromagnetic domain by Bloch [2], the above form was first given by Enz [3]. The occurrence of the sine-Gordon equation, which is similar to Eq.(2.68) in nonlinear quantum mechanics, should justify dealing with this simple model system.

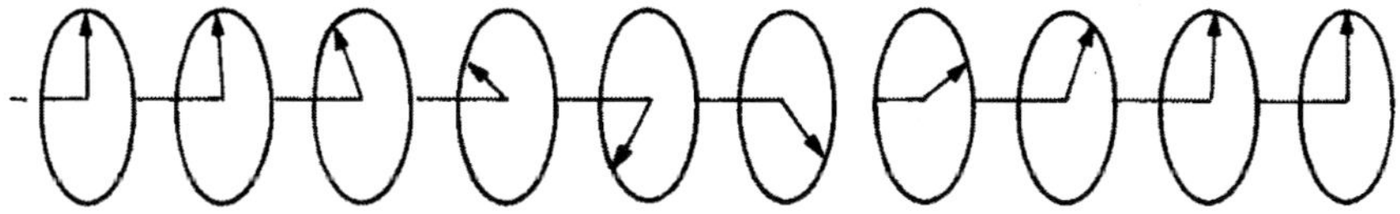

Figure 8.3. The spin configuration of the basic soliton in the easy-plan ferromagnetic chain with magnetic field pointing up in the sine-Gordon image [Ref.1].

(2) Ferromagnetic Heisenberg Chains with Various Anisotropies

The sine-Gordon chain is a model with one classical field $\Phi(z,t)$, the model, which in special limiting cases is sufficient to describe spin chains in real solids. However, a spin-spin interaction chain has to be described by two fields, which correspond to the two angles specifying the orientation of a classical vector of fixed length. In particular, the dynamics of spins always involve all three components in a non-trivial way, and one therefore expects important differences between real spin chains and the simplifying sine-Gordon system. Fortunately, however, several one-dimensional models with a vector field **S**(z, t) also allow for complete solution of their mechanical behavior. The simplest of these models is the isotropic Heisenberg chain with ferromagnetic coupling defined by the Hamiltonian (J>0) [4-6]:

$$H = -J\sum_n \hat{S}_n \bullet \hat{S}_{n+1} \tag{8.3}$$

and the Poisson brackets are

$$\{S_n^{\alpha}, S_m^{\beta}\} = i\varepsilon_{\gamma}^{\alpha\beta}\delta_{nm} \tag{8.4}$$

Representing the spin vector in polar coordinates, that is

$$S = S\{\sin\Theta\cos\Phi, \sin\Theta\sin\Phi, \cos\Theta\}$$

and using the continuum approximation, equations (8.3) and (8.4) read

$$H = JSr_0\int dz[\frac{1}{2}(\frac{\partial\Theta}{\partial z})^2 + \frac{1}{2}\sin^2\Theta(\frac{\partial\Phi}{\partial z})^2] \tag{8.5}$$

with

$$\{\cos[\Theta(z,t),\Phi(z',t)]\} = \frac{r_0}{S}\delta(z-z') \tag{8.6}$$

When J>0, the ground state of Eq. (8.3) is a ferromagnetic state, if J<0, then Eq.(8.3) describes the antiferromagnetic order state. When J<0 and neighbor lattice points are different magnetic ions(in other words, S is different), then Eq.(8.3) denotes the sub-magnetic state.

The Heisenberg Hamiltonian in (8.3) can be theoretically derived. We now assume that the crystal is composed of N lattice points, in which each lattice point has only one localized d-electron without a pair. In this case we here used the following Wallonia function: $a_n(r,l) = N^{-1/2}\sum_{k\in BZ}\exp[-ik.l]\psi_{nk}(r)$ as the basic function to denote the state wave function of this system, i.e., $\Psi(r) = \sum_{l\sigma} B_{l\sigma}a(r-l)$. In the second quantized representation, by finding an average value of Coulomb potential between the d-electrons we can obtain the exchange integral, which is expressed [5-7] by

$$H_{ex} = -\frac{1}{2}\sum_{l,l'}{}' \sum_{\sigma,\sigma'} J_{ll'} \{B^{+}_{l\sigma} B_{l'\sigma'} B^{+}_{l'\sigma'} B_{l'\sigma}\} \quad (8.7)$$

where

$$J_{ll'} = e^2 \int \frac{a^*(r-l)a(r-l')a^*(r'-l')a(r'-l)}{|r-r'|} d^3r d^3r' \quad (8.8)$$

where the Pauli spin operators have the forms as follows

$$\sigma^z = \begin{pmatrix} 1 & 0 \\ 0 & 1 \end{pmatrix}, \sigma^+ = \begin{pmatrix} 0 & 2 \\ 0 & 0 \end{pmatrix}, \sigma^- = \begin{pmatrix} 0 & 0 \\ 2 & 0 \end{pmatrix}$$

and the creative and annihilative operators of the d-electrons satisfied the following relationship:

$$\sigma_l^Z = B^{+}_{l\uparrow} B_{l\uparrow} - B^{+}_{l\downarrow} B_{l\downarrow}, \sigma_l^+ = 2B^{+}_{l\uparrow} B_{l\downarrow}, \sigma_l^- = 2B^{+}_{l\downarrow} B_{l\uparrow}$$

Utilizing the above relationship, Eq.(8.7) can be written as

$$H_{ex} = -\frac{1}{2}\sum_{l,l'}{}' J_{ll'} \{B^{+}_{l\uparrow} B_{l'\uparrow} B^{+}_{l'\uparrow} B_{l'\uparrow} + B^{+}_{l\downarrow} B_{l'\downarrow} B^{+}_{l'\downarrow} B_{l'\downarrow} + B^{+}_{l\uparrow} B_{l'\downarrow} B^{+}_{l'\downarrow} B_{l'\uparrow} + B^{+}_{l\downarrow} B_{l'\uparrow} B^{+}_{l'\uparrow} B_{l'\downarrow}\}$$

Using again the relationships:

$$B^{+}_{l\uparrow} B_{l\uparrow} = \frac{1}{2}(1+\sigma_l^z), B^{+}_{l\downarrow} B_{l\downarrow} = \frac{1}{2}(1-\sigma_l^z), \frac{1}{2}\sigma_l^+ = B^{+}_{l\uparrow} B_{l\downarrow}, \frac{1}{2}\sigma_l^- = B^{+}_{l\downarrow} B_{l\uparrow}$$

we can obtain

$$\begin{aligned} H_{ex} &= -\frac{1}{2}\sum_{l,l'}{}' \frac{1}{4} J_{ll'} [(1+\sigma_l^z)(1+\sigma_{l'}^Z) + (1-\sigma_l^z)(1-\sigma_{l'}^z) + \sigma_l^+ \sigma_{l'}^- + \sigma_l^- \sigma_{l'}^+] \\ &= -\frac{1}{4}\sum_{l,l'}{}' J_{jj'} (\sigma_l \bullet \sigma_{l'} + 1) \end{aligned} \quad (8.9)$$

If we notice that $\hat{S}_l = \sigma_l / 2$ at S=1/2, then Eq.(8.9) is just Eq.(8.3), in the Heisenberg model.

Some magnets and their features and types are shown in Table 1.

(3) The Landau-Lifshitz Model for the Ferromagnetic-Order

On the other hand, the other equations of motion in the continuum approximation are the Landau-Lifshitz equations [1-2], which are denoted by

$$\frac{\partial S}{\partial t} = S \times \frac{\partial^2 S}{\partial z^2} \tag{8.10}$$

Using polar coordinates, they are represented by

$$\frac{\partial \Theta}{\partial t} = -\sin\Theta \frac{\partial^2 \Phi}{\partial z^2} - 2\cos\Theta \frac{\partial \Theta}{\partial z}\frac{\partial \Phi}{\partial z} = \frac{1}{\sin\Theta}\frac{\partial}{\partial z}(Sin^2\Theta \times \frac{\partial \Phi}{\partial z}) \tag{8.11}$$

$$\frac{\partial \Phi}{\partial t} = \frac{1}{\sin\Theta}\frac{\partial^2 \Theta}{\partial z^2} - \cos\Theta(\frac{\partial \Phi}{\partial z})^2 \tag{8.12}$$

Table 1. The properties of one-dimensional magnets with the Hamiltorian with $H=-J\sum \hat{S}\cdot\hat{S}$ and $H=-\frac{1}{2}J\sum \bar{\sigma}\sigma=-2J\sum \tilde{S}\cdot S$ [1]

Substance	S	J/k (K)	T_c, T_N (K)	T_c/O_p	Model
$CuSO_4.5H_2O$	$\frac{1}{2}$	−2·9	0·03	0·02	Heisenberg
$CuSeO_4.5H_2O$	$\frac{1}{2}$	−1·6	0·045	0·056	Heisenberg
$Cu(NH_3)_4SO_4.H_2O$	$\frac{1}{2}$	−6·3	0·37	0·12	Heisenberg
$Cu(NH_3)_4SeO_4.H_2O$	$\frac{1}{2}$	−4·72	1·2	?	Heisenberg
$Cu(NH_3)_4(NO_3)_2$	$\frac{1}{2}$	−7·4	1·4	?	Heisenberg
$CuCl_2.2NC_5H_5$	$\frac{1}{2}$	−26	1·7	0·13	Heisenberg
CHAB	$\frac{1}{2}$	131·6	1·5		Heisenberg (weak x–y)
$KCuF_3$	$\frac{1}{2}$	−380	38	0·2	Heisenberg
$CsNiCl_3$	1	−16·6	4·65	0·13	Heisenberg
$RbNiCl_3$	1	−34	11·0	0·24	Heisenberg
$CsMnCl_32H_2O$	$\frac{5}{2}$	−7·2	4·89	0·12	Heisenberg
TMMC	$\frac{5}{2}$	−13	0·84	0·011	Heisenberg
$CsCuCl_3$	$\frac{1}{2}$	+?	10·4	?	Heisenberg
$(CH_3)_4NCuCl_3$	$\frac{1}{2}$	60	<2	?	Heisenberg
$((CH_3)_3NH)Cu_2Cl_7$	$\frac{1}{2}$	100	<2	?	Heisenberg
$(CH_3)_4NNiCl_3$	1	+4	1·2?	0·5?	Heisenberg
$K_3Fe(CN_6)$	$\frac{1}{2}$	−0·23	0·129	0·56	Ising
$CsCoCl_3$	$\frac{1}{2}$	−75	21·5	0·08	Ising
$(NH_4)_2MnF_5$	$\frac{1}{2}$	−12	7·5	0·08	Ising
$CsCoCl_3.2H_2O$	$\frac{1}{2}$		3·8	?	Ising
$CoCl_2.2NC_5H_5$	$\frac{1}{2}$	+11·5	3·2	0·37	Ising
$CsCoBr_3$	$\frac{1}{2}$	−78	28·3		Ising
$RbCoBr_3$	$\frac{1}{2}$		36	?	Ising
$Fe(N_2H_5)_2(SO_4)_2$	2	−2·0	6·0		Ising
$FeC_2O_4.2H_2O$	$\frac{1}{2}$	−82·6	11·7		Ising
$Fe(biPy)Cl_3$	$\frac{5}{2}$	−3·0	?		Ising
$RbFeCl_3.H_2O$	$\frac{1}{2}$	−39	12·0		Ising
$FeCl_2Py_2$	$\frac{1}{2}$	+25	6·6		Ising
$NiCl_2Py_2$	$\frac{1}{2}$	+21·4	6·8		Ising
$RbFeCl_3$	2		2·55	?	Planar Heisenberg
$CsNiF_3$	1	23·6	2·65	0·08	Planar Heisenberg

In Eqs. (8.10)-(8.12) we have used $(JS)^{-1}$ as our unit of time and the lattice constant is as our unit of length and we shall continue to use these units throughout this section.

The rotational symmetry of the isotropic Heisenberg chain can be reduced by adding a potential energy $\sum_n V(\Theta_n, \Phi_n)$, for example, the Zeeman energy in an external magnetic field or single ion anisotropies, to the energy in Egs. (8.11) and (8.12). The additional terms in the equations of motion are given by

$$\frac{\partial\Theta}{\partial t} = \ldots - \frac{1}{\sin\Theta}\frac{\partial V}{\partial\Phi}, \quad \frac{\partial\Phi}{\partial t} = \ldots + \frac{1}{\sin\Theta}\frac{\partial V}{\partial\Theta} \tag{8.13}$$

8.2. The Wave Features of the Magnons in the Ferromagnetic Systems

8.2.1. The Ground State of the Magnon

We first assumed that the ferromagnetic crystal with J> 0 is composed of M magnetic ions with the spin S. This system can be described by the Heisenberg Hamiltonian in Eq.(8.3), where vector operator $\hat{S}_l = (\hat{S}_l^x, \hat{S}_l^y, \hat{S}_l^z)$ meets [7-8] in the following commutation relation:

$$[\hat{S}_l^x, \hat{S}_{l'}^y] = i\hat{S}_l^z\delta_{ll'}, [\hat{S}_l^y, \hat{S}_{l'}^z] = i\hat{S}_l^x\delta_{ll'}, [\hat{S}_l^z, \hat{S}_{l'}^x] = i\hat{S}_l^y\delta_{ll'} \tag{8.14}$$

where $\hbar = 1$ is used. Because $\hat{S}_{l'}^z = \hat{S}_l \bullet \hat{S}_l = S(S+1)$, then the three components of $\hat{S}_l$ are not independent of each other. Thus, we use often $\hat{S}_l^+ = (\hat{S}_l^x + i\hat{S}_l^y)$ and $\hat{S}_l^- = (\hat{S}_l^x - i\hat{S}_l^y)$ as independent variables, they are called the rising and falling operators, respectively, and satisfy the commutation relations:

$$[\hat{S}_l^{\pm}, \hat{S}_{l'}^z] = \mp\hat{S}_l^{\pm}\delta_{ll'} \quad \text{and} \quad [\hat{S}_l^+, \hat{S}_{l'}^-] = 2\hat{S}_l^z\delta_{ll'} \tag{8.15}$$

We now assume that z is the quantized axial direction, then the spin state at one lattice point can be represented by $|S, m\rangle$, where S is the spin of the magnetic ion, m is the eigenvalue of the operator, $\hat{S}_l^{\pm}$'s effect is to increase or decrease one quantum and $m = 0, \pm 1, \pm 2, \ldots . \pm S$.

According to quantum mechanics we can obtain

$$\hat{S}_l^+ |s,m\rangle = [(s-m)(S+m+1)]^{1/2} |S,m+1\rangle$$
$$\hat{S}_l^- |s,m\rangle = [(S+m)(S-m+1)]^{1/2} |S,m-1\rangle \quad (8.16)$$

In this case, the Heisenberg Hamiltonian in Eq.(8.3) can be represented by

$$H = -J\sum_{l,\delta}{}'\{\hat{S}_l^z\hat{S}_{l+\delta}^z + \frac{1}{2}(\hat{S}_l^+\hat{S}_{l+\delta}^- + \hat{S}_l^-\hat{S}_{l+\delta}^+)\} \quad (8.17)$$

It is natural to obtain the ground state of the ferromagnetic systems, which is denoted by

$$|0\rangle = |S\rangle_1 |S\rangle_2 ....|S\rangle_l ....|S\rangle_N,$$

where $|S\rangle_l = |s,m=S\rangle_l$. Because $\hat{S}_l^+|0\rangle = 0$, then we can obtain

$$H|0\rangle = -J\sum_{l.\delta}(\hat{S}_l^z\hat{S}_{l+\delta}^z)|0\rangle = -JNZS^z|0\rangle \quad (8.18)$$

where $\hat{S}_l^z|s_l\rangle = s|s_l\rangle$ is used in the above derivation, and z is the coordination number of the lattice. From Eq.(8.18) we know that the eigenenergy of the ground state $|0\rangle$ is

$$E_0 = -JNZS^2 \quad (8.19)$$

where NZS^2 comes from the sum of the maximal eigenvalue m=S of $\hat{S}_l^z$ with respect to the state $|0\rangle$ for all lattice points. $E_0 = -JNZS^z$ is the lowest eingeenergy value of the system, here $|0\rangle$ is the ground state of the above Hamiltonian, H..

8.2.2. The Excited State of the Magnons and Its Hamiltonian

If the ferromagnetic system is perturbed by an externally applied field, then the spin at lattice point l will be somewhat rotated from S to S-1, then this state in the ferromagnetic state can be denoted by $|(S-1)_l\rangle = |S\rangle_1 ...|S\rangle_{l-1}|S-1\rangle_l|S\rangle_{l+1}....|S\rangle_N$.

When it is acted on by $\hat{S}_l^+\hat{S}_l^-$ in the Hamiltonian, then the above rotation or the excited state at l lattice point will disappear, but a new rotation state at l+1 lattice point occurs, which is expressed by $|(S-1)_{l+1}\rangle = |S\rangle_1 ...|S\rangle_l|S-1\rangle_{l+1}|S\rangle_{l+2}....|S\rangle_N$. However, the rotation of the spin cannot be stopped, but can propagate along the ferromagnetic chains through the exchange interaction between neighbor spins, thus a spin wave is formed in this case as mentioned above [7-8].

We now assumed: n=S-m, $m = 0, \pm 1, ..., \pm S$, where $n = 0,1,2,...,2S, n \le 2S$, represents the deviated value of the project of the spin on the quantized axis, or it is called the deviated number of the spin. If $|n\rangle$ denotes the spin state, then Eq. (8.16) may express by

$$\hat{S}_l^+ |n\rangle = [(n(2S-(n-1)]^{1/2} |n-1\rangle$$
$$\hat{S}_l^- |n\rangle = [(n+1)(2S-n)]^{1/2} |n+1\rangle \tag{8.20}$$

We now introduce the creation and annihilation operators with respect to $|n\rangle$, which are defined as follows [7-8]

$$a|n\rangle = \sqrt{n}|n-1\rangle, a^+|n\rangle = \sqrt{n+1}|n+1\rangle, a^+a|n\rangle = n|n\rangle \tag{8.21}$$

Their commutation relations are as follows

$$[a, a^+] = 1, [a, a] = [a^+, a^+] = 0$$

From Eqs.(8.20)-(8.21) we can obtain the Holstien-Primakoff transformations [9], which are

$$\hat{S}^+ = (\sqrt{2S - a^+a})a, \hat{S}^- = a^+(\sqrt{2S - a^+a}), \hat{S}^z = (S - a^+a) \tag{8.22}$$

Utilizing Eq.(8.22) the Heisenberg Hamiltonian in Eq.(8.3) in the second quantized representation can be denoted by

$$H = -J\sum_{l,\delta}'\{(S - a_l^+a_l)(S - a_{l+\delta}^+a_{l+\delta}) + \frac{1}{2}\sqrt{(2S - a_l^+a_l)}a_l^+a_{l+\delta}\sqrt{2S - a_{l+\delta}^+a_{l+\delta}}$$
$$+\frac{1}{2}a_l^+\sqrt{(2S - a_l^+a_l)}\sqrt{2S - a_{l+\delta}^+a_{l+\delta}}a_{l+\delta}\} \tag{8.23}$$

where the commutation relations of these operators are as follows

$$[a_l, a_{l'}^+] = \delta_{l,l'}, [a_l, a_{l;}] = [a_l^+, a_{l'}^+] = 0$$

8.2.3. The Wave Features of the Magnons

We now studied the features of the lower excited state, in this case of low temperature, which manifests that the rotations of the spins are the smaller. This means that $<a_l^+ a_l>$ is smaller than 2S. Thus we may neglect a to the fourth power term or nonlinear interaction term in Eq.(8.23). In this case, equation (8.23) becomes [8]

$$H_0 = -NZ|J|S^2 + 2Z|J|S\sum_l a_l^+ a_l - |J|S\sum_{l,\delta}{}'(a_l a_{l+\delta}^+ + a_l^+ a_{l+\delta}) \tag{8.24}$$

where the first term is the energy of the ground state, the second term denotes the deviate from the energy of the spins at the lattice point, and the third term is the interaction energy between the spins at neighbor two lattice points. We now define new operators of the magnon, and its creation and annihilation operators by

$$b_k = N^{-1/2}\sum_l \exp[-ik.l]a_l, b_k^+ = N^{-1/2}\sum_l \exp[ik.l]a_l^+ , \tag{8.25}$$

respectively, which also express the collective coordinate of the spin wave, where

$$a_l = N^{-1/2}\sum_l \exp[ik.l]b_k, a_l^+ = N^{-1/2}\sum_l \exp[-ik.l]b_k^+ \tag{8.26}$$

and

$$[b_k, b_{k'}^+] = \delta_{k,k'}, [b_k, b_{k'}] = [b_k^+, b_{k'}^+] = 0, N^{-1}\sum_l \exp[i(k-K').l] = \delta_{k,k'}$$

If using the new operators then the diagonal Hamiltonian in Eq.(8.24) becomes as

$$H_0 = -NZ|J|S^2 + 2Z|J|S\sum_k b_k^+ b_k - Z|J|S\sum_k \gamma_k (b_k b_k^+ + b_k^+ b_k) = E_0 + \sum_k \hbar\omega_k b_k^+ b_k \tag{8.27}$$

where $E_0 = -NZ|J|S^2$ and $\hbar\omega_k = 2Z|J|S(1-\gamma_k)$,but $\gamma_k = Z^{-1}\sum_\delta \exp[ik.\delta] = -\gamma_k$ with $\sum_k \gamma_k = 0$ characterizes the inversion symmetry of this system.

In the cases of low temperature, and the spin wave with long wavelength, $\hbar\omega_k = Z|J|\sum_\delta (k.\delta)^2$. For the cubic crystal of sc, bec and fcc, there are the same dispersion relations of $\hbar\omega_k = 2|J|Sa^2k^2$,because of $\sum_\delta (k.\delta)^2 = [Z^{-1}\sum_\delta Z\delta^2 \cos^2\theta(k.\delta)\}k^2$, where the average of $\cos^2\theta(k.\delta)$ is 1/3, the average of $Z\delta^2$ is $6a^2$ for the three cubic crystal with side length a [8].

Using the approximations of

$$\sqrt{(1-a^{+}a/2S)}=1-(a^{+}a/2S)/2+(a^{+}a/2S)^{2}/8+....,$$

Eq.(8.23) can be denoted by

$$H=H_0+H_1+.... \tag{8.28}$$

where

$$H_1=Z|J|/2N\sum_{kk'q}(\gamma_{k-q}+\gamma_k-2\gamma_{k-q-k'})b^{+}_{k-q}b^{+}_{k'+q}b_k b_k \tag{8.29}$$

denotes the interaction between the magnons or spin waves.

From Eq.(8.27) we can obtain the dynamic equation of the magnon of spin wave, which is [8]

$$i\hbar\frac{\partial}{\partial t}a_l=[a_l,H_0]=2Z|J|Sa_l-|J|S\sum_{\delta}{}'(a_{l+\delta}+a_{l-\delta}) \tag{8.30}$$

From Fourier transformation of Eq.(8.30) we obtain the equation of the operator of the magnon, which is of

$$i\hbar\frac{\partial}{\partial t}b_k=2Z|J|S(1-\gamma_k)b_k=\hbar\omega_k b_k \tag{8.31}$$

This is just the dynamic equation of the magnon with the frequency ω_k and energy $\hbar\omega_k$. Very clearly, it is a linear equation; its solution is a spin wave. This means that the magnon described by Eq.(8.31) is only a quasiparticle or elementary excitation having a wave feature, namely, the magnon is only a wave and does not have the wave-corpuscle duality due to the absence of the nonlinear interaction. This conclusion can be confirmed in following derivation [10].

At present, we do linear approximation for Eqs.(8.29)-(8.30) to build the dynamic equation of b_k, which is expressed as follows

$$i\hbar\,\partial b_k/\partial t=[b_k,H_0]+[b_k,H_1] \tag{8.32}$$

Through the calculation we can obtain

$$[b_k,H_0]=2Z|J|S(1-\gamma_k)b_k \tag{8.33}$$

and

$$[b_k, H_1] = Z|J|(2N)^{-1}\sum_{k',\delta}\{(\gamma_k + \gamma_{k'} - 2\gamma_{k-k'})b^+_{k'+q}b_{k+q}b_{k'} + (\gamma_{k-q} + \gamma_{k'-q} - 2\gamma_{k-k'})b^+_{k'-q}b_{k-q}b_{k'}\}$$

(8.34)

Doing the linear approximations to these nonlinear interaction terms in Eq.(8.34)

$$b^+_k b_{k'} b_{k''} \approx (\delta_{kk'} b_{k''} + \delta_{kk''} b_{k'}) < b^+_k b_k >_T, \quad < b^+_k b_k >_T = [\exp(\hbar\omega_k / KT) - 1] = < n_k >_T$$

and utilizing the relations $\gamma_{k-k'} = \gamma_{k'-k} \sum_{k'} \gamma_{k-k'} < n_k >_T = \gamma_k \sum_{k'} \gamma_{k'} < n_{k'} >_T$ we can obtain

$$i\hbar \, {\partial b_k}/{\partial t} = \hbar\omega_k(T) b_k \qquad (8.35)$$

where $\hbar\omega_k(T) = 2Z|J|S(1-\gamma_k)\{1 - N^{-1}\sum_{k'}(1-\gamma_{k'}) < n_{k'} > / S\}$.

Therefore, Eq.(8.35) with Eq.(8.32) is a linear dynamic equation in linear quantum mechanics. This verified again that the linear or wave feature of the magnons is due to the neglect of the nonlinear interaction between the spins in the ferromagnetic systems. If the nonlinear interactions in the above derivation are considered, then the dynamic equation of the magnon is a nonlinear Schrödinger equation, instead of the linear equation(8.35) [10],then the magnons will be localized and have a wave-corpuscle duality, which will be studied after this [10].

8.3. The Wave Features of the Magnons in the Antiferromagnetic Systems

8.3.1. The Dual Lattice Model of the Antiferromagnetic System

As it is known, when J<0 in Haisenberg Hamiltonian in Eq.(8.3) the spins between the neighboring lattice points form an antiparallel array. Therefore, its whole spontaneous magnetization is balances each other. Its physical images and excitation states are shown in Figure 8.4, where the degenerate ground states of antiferromagnetic chains in the 1/2 Insing model were shown in (a) and (b), its excited states with a domain wall(c), and two domain wall(d), were shown also.

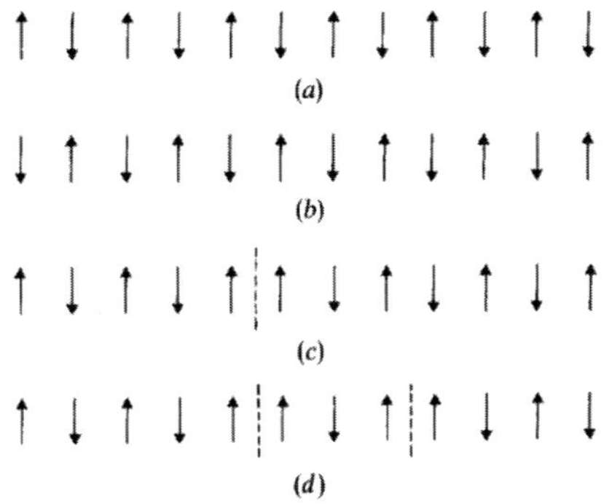

Figure 8.4. The degenerate ground states of antiferromagnetic chains in 1/2 Insing model (a),(b), (c) and (d) show its excited states with a domain wall(c) and two domain walls(d)(Ref. [1]).

Just so, we often regard the antiferromagnetas composed of two sublattices, in which the arrangement of the spins, in each sublattice, are the parallel, which is the same with that in the ferromagnetic system. For example, in the antiferromagnetic materials of MuF_2, FeF_2 and CoF_2, the magnetic ions of Mn^{2+}, Fe^{2+} and Co^{2+} array as the body centered square lattice, thus we can use dual lattice model to research the properties of the magnons and spin wave in the antiferromagnetic systems with a and b sublattices.

At present, we assume that each sublattice in the antiferromagnetic systems is composed of N spin numbers; its total magnetic ions are 2N. Then its total Hamiltonian can be represented by [1-2,8]

$$H = |J|\sum_{l,\delta}{}'\{\hat{S}^z_{al}\hat{S}^z_{b,l+\delta} + \frac{1}{2}(\hat{S}^+_{al}\hat{S}^-_{b,l+\delta} + \hat{S}^-_{al}\hat{S}^+_{b,l+\delta})\} + |J|\sum_{i,\delta}{}'\{\hat{S}^z_{bi}\hat{S}^z_{a,i+\delta} + \frac{1}{2}(\hat{S}^+_{bi}\hat{S}^-_{b,i+\delta} + \hat{S}^-_{bi}\hat{S}^+_{a,i+\delta})\} \tag{8.36}$$

where $\hat{S}_{al}$ is the spin operator at l lattice point in a sublattice, $\hat{S}_{bi}$ is the spin operator at i lattice point in b sublattice.

8.3.2. The Dynamic Features of the Magnons in the Antiferromagnetic System

We here assume further that the quantized axis in a lattice is along the +z direction, i.e., all its spins are arrayed along the +z axis, then he quantized axis in b lattice is along the -z direction, all its spins are arrayed along the -z axis. We call these states as original state, then the relation, which is similar with Eq.(8.3) is [8]

$$\hat{S}^+_{al}|origin\rangle = \hat{S}^-_{bl}|origin\rangle = 0 \tag{8.37}$$

This indicates that $\hat{S}_{al}^{+}$ corresponds to the annihilation operator in a lattice, but $\hat{S}_{bi}^{-}$ corresponds to the creation operator in b lattice. Because the director of the spin in a lattice is rising, in a lattice we can use the Holstien-Primakoff transformations in Eq.(8.22) to define its creation and annihilation operators, which are

$$\hat{S}_{al}^{+}=(\sqrt{(2S-a_l^{+}a_l})a_l,\hat{S}_{al}^{-}=a_l^{+}(\sqrt{(2S-a_l^{+}a_l}),\hat{S}_{al}^{z}=(S-a_l^{+}a_l) \qquad (8.38)$$

But, the representations of $\hat{S}^{+}$ and $\hat{S}^{-}$ in b lattice are inverse with a lattice, which are

$$\hat{S}_{bi}^{+}=b_{bi}^{+}\sqrt{(2S-b_i^{+}b_i}),\hat{S}_{bi}^{-}=(\sqrt{(2S-b_i^{+}b_i})b_i\ ,\hat{S}_{bi}^{z}=(b_i^{+}b_i-S) \qquad (8.39)$$

Since the rising of the spins in the sublattice are increased for the rotation relative to the quantized axis then $\hat{S}_b^{+}$ corresponds to the creation operator, b^{+} then defines $\hat{S}_b^{z}$ in b sublattice differs, only a negative sign in a sublattice.

Very easily, we can obtain that the following commutation relationship holds

$$[\hat{S}_{bi}^{+},\hat{S}_{bi}^{-}]=b_i^{+}b_i(2S-b_i^{+}b_i)+b_i^{+}b_i-b_i^{+}b_i(2S-b_i^{+}b_i) \qquad (8.40)$$

where the Bosen operators a_l and b_i satisfied the following commutations

$$[a_l,a_{l'}^{+}]=\delta_{ll'},[a_l,a_{l'}]=[a_l^{+},a_{l'}^{+}]=0,[b_i,b_{i'}^{+}]=\delta_{ii'},[b_i,b_{i'}]=[b_i^{+},b_{i'}^{+}]=0,$$
$$[a_l,b_i]=[a_l^{+},b_i]=[a_l,b_i^{+}]=[a_l^{+},b_i^{+}]=0, \qquad (8.41)$$

Thus we can give the following approximate representations for the lower excited states

$$\hat{S}_{al}^{+}=(\sqrt{2S}a_l,\hat{S}_{al}^{-}=\sqrt{2S}a_l^{+},\hat{S}_{al}^{z}=S-a_l^{+}a_l \qquad (8.42)$$

$$\hat{S}_{bi}^{+}=\sqrt{2S}b_{bi}^{+},\hat{S}_{bi}^{-}=(\sqrt{2S}b_i\ ,\hat{S}_{bi}^{z}=b_i^{+}b_i-S \qquad (8.43)$$

Inserting Eq.(8.41) into Eq.(8.36) and neglecting the fourth terms of a and b we obtain

$$H=-2NZS^2|J|+2ZS|J|(\sum_l a_l^{+}a_l+\sum_i b_i^{+}b_i)+2|J|S\sum_{l,\delta}[a_lb_{l+\delta}+a_l^{+}b_{i+\delta}^{+}] \qquad (8.44)$$

Using the Fourier transformation relation in Eqs.(8.25)-(8.26) we can gain

$$a_l = N^{-1/2}\sum\nolimits_k \exp[ik.R_l]a_k, a_l^+ = N^{-1/2}\sum\nolimits_k \exp[-ik.R_l]a_k^+ \tag{8.45}$$

$$b_i = N^{-1/2}\sum\nolimits_k \exp[-ik.R_j]b_i, b_i^+ = N^{-1/2}\sum\nolimits_k \exp[ik.R_j]b_k^+ \tag{8.46}$$

where R_l and R_j are the displacement vectors of a and b sublattices. After the transformations in Eqs.(8.45)-(8.46) the operators a_l *and* b_i satisfy all the Bosen commucation relations.

In this case, the Hamiltonian of the spin operator in dual lattice in the antiferromagnet becomes

$$H = -2NZS^2|J| + 2ZS|J|\sum_k(a_k^+a_k + b_k^+b_k) + 2Z|J|S\sum_k \gamma_k[a_k^+b_k^+ + b_ka_k] \tag{8.47}$$

where the first term is the energy of original state, the second term denotes the energy of the spin wave for all sublattices, the third term is the interaction energy between the two sublattices.

We now introduce the following Bogoliubov transformation [11] to eliminate the cross terms for a diagonal Hamiltonian in Eq.(8.44)

$$\begin{aligned} \alpha_k &= u_k a_k - v_k b_k^+, \alpha_k^+ = u_k a_k^+ - v_k b_k \\ \beta_k &= u_k b_k - v_k a_k^+, \beta_k^+ = u_k b_k^+ - v_k a_k \end{aligned} \tag{8.48}$$

but following commucation relations must be satisfied

$$[\alpha_k, \alpha_{k'}^+] = \delta_{kk'}, [\beta_k, \beta_{k'}^+] = \delta_{kk'}, [\alpha_k, \beta_{k'}] = [\alpha_k, \beta_{k'}^+] = [\beta_k, \alpha_{k'}^+] = [\alpha_k^+, \beta_{k'}^+] = 0 \tag{8.49}$$

In order to carry out the diagonal Hamiltonian in Eq.(8.43)the relationship

$$u_k^2 - v_k^2 = 1 \tag{8.50}$$

Utilizing Eq.(8.50) we give the inverse transformation of Eq.(8.48), which are

$$\begin{aligned} a_k &= u_k\alpha_k + v_k\beta_k^+, a_k^+ = v_k\beta_k + u_k\alpha_k^+ \\ b_k &= u_k\beta_k + v_k\alpha_k^+, b_k^+ = u_k\beta_k^+ + v_k\alpha_k \end{aligned} \tag{8.51}$$

Substituting Eq.(8.51) into Eq.(8.47) we obtain

$$H = -2NZ|J|S(S+1) + \sum_k \hbar\omega_k[(\alpha_k^+\alpha_k + 1/2) + (\beta_k^+\beta_k + 1/2)] \qquad (8.52)$$

where $\hbar\omega_k = 2Z|J|S\sqrt{1-\gamma_k^2}$ is the energy of the magnon in the antiferromagnetic system. For the three cubic crystal in a long-wavelength case (ka<<1), the dispersion relation is [8]

$$\hbar\omega_k = 2(2Z)^{1/2}|J|Ska$$

Very clearly, Eq.(8.52) is similar Eq.(8.27) in the ferromagnetic systems. Thus, we can also obtain the linear dynamic equation of the operator, α_k and β_k, of the magnons, which is similar Eq.(8.31). This indicated clearly the wave feature of the magnons. Therefore, we can conclude that the magnons only possess a wave feature in the linear case in the antiferromagnetic systems.

8.4. Nonlinear Quantum-Mechanical Properties of Magnons in Ferromagnetic Systems

8.4.1. The Nonlinear Dynamic Equation of the Magnons

We now study the properties of motion of the magnons in the nonlinearly ferromagnetic systems, by nonlinear quantum mechanics.

It is well known that the presence of magnons in ferromagnetic crystals can change the potential energy of interaction between atoms. Then the deformation created by excitation of the magnons will also result in a variation of state of the magnons due to the interaction between the spin deviation and the deformation of the lattice. In this process the nonlinear interaction appeared might make the magnons become a soliton, then the properties of the magnons are changed. The nonlinear excitations and changes of property of the magnon, in ferromagnetic chains, were studied by Pushkarovet al. [12-16], using the following Hamiltonian:

$$H = T + V - \frac{1}{4}\sum_{n\delta} J(u_n - u_{n+\delta})(S_n^+S_{n+\delta}^- + S_n^-S_{n+\delta}^+) - \frac{1}{2}\sum_{n\delta} \tilde{J}(u_n - u_{n+\delta})S_n^zS_{n+\delta}^z \qquad (8.53)$$

where u_n is the position of nth atom in the chain, δ is a parameter and turns over the nearest neighbors of a given atom, $S_n^\pm$ are the cyclic spin components given by the relations $S_n^\pm = S_n^z \pm iS_n$, $J(u_n - u_{n+\delta})$ and $\bar{J}(u_n - u_{n+\delta})$ are exchange integrals which are functions of $|u_n - u_{n+\delta}|$, $T = (m/2)\sum_m [(u_n)_t]^2$ is the kinetic energy of the atoms of mass$m$. The harmonic approximation is assumed for the interatomic interaction and the potential energy is given by

$$V = \frac{mv_o^2}{2r_o}\sum_n (u_{n+1} - u_n - r_o)^2 , \tag{8.54}$$

where v_0 is the sound velocity and r_0 the lattice constant.

At low temperature when only a few spin waves are excited, the operators $S_n^{\pm}$ and S_n^z can be expressed in terms of the Bose operators a_n and a_n^+ as

$$S_n^+ = \sqrt{2S}a_n , \; S_n^- = \sqrt{2S}a_n^+ , \; S_n^z = S - a_n^+ a_n \tag{8.55}$$

In the bilinear approximation with respect to the Bose operators, we can obtain the following by inserting Eq. (8.55) into Eq. (8.53).

$$H = T + V - \frac{S^2}{2}\sum_{n\delta}\tilde{J}(u_n - u_{n+\delta})$$

$$-\frac{S}{2}\sum_{n\delta}\left\{\tilde{J}(u_n - u_{n+\delta})[a_n^+ a_{n+\delta} + a_n a_{n+\delta}^+] - J(u_n - u_{n+\delta})[a_n^+ a_n + a_{n+\delta}^+ a_{n+\delta}]\right\} \tag{8.56}$$

From

$$i\hbar\frac{\partial}{\partial t}|\varphi\rangle = H|\varphi\rangle \tag{8.57}$$

and Eq. (8.51), we obtain

$$i\hbar\frac{\partial\varphi_n}{\partial t} = \left[T + V - \frac{S^2}{2}\sum_{n\delta}\tilde{J}(u_n - u_{n+\delta})\right]\varphi_n - S\sum_{\delta}J(u_n - u_{n+\delta})\varphi_{n+\delta} + S\sum_{\delta}\tilde{J}(u_n - u_{n+\delta})\varphi_n \tag{8.58}$$

We have assumed in Eq. (8.58) that

$$J(|u_n - u_{n+1}|) + J(|u_n - u_{n-1}|) \approx 2J , \; J(|u_n - u_{n+1}|) + \tilde{J}(|u_n - u_{n-1}|) \approx 2\tilde{J}_0 - \tilde{J}_1(u_{n+1} - u_{n-1})$$

and $\sum_{n\delta}\tilde{J}(u_n - u_{n+\delta}) \approx 2\tilde{J}_0 N - \tilde{J}_1\sum_n (u_{n+1} - u_{n-1})$, (8.59) where $\tilde{J}_1 = -\dfrac{\partial \tilde{J}}{\partial u_n} > 0$

Equation (8.58) can then be written in the following form

$$i\hbar\frac{\partial \varphi_n}{\partial t} = \left[T + V - \tilde{J}_0 S^2 N + \tilde{J}_1 S \sum_n (u_{n+1} - u_{n-1}) + 2(\tilde{J}_0 - J)S\right]\varphi_n$$

$$-J(\varphi_{n+1} + \varphi_{n-1} - 2\varphi_n) - \tilde{J}_1 S(u_{n+1} - u_{n-1}) \ . \tag{8.60}$$

Taking into account the explicit expressions of T and V as well as the Hamilton equations for u_n and $P_n = m(u_n)_t$, we get

$$m(u_n)_{tt} = mv_0^2(u_{n+1} - u_{n-1} - 2u_n) - \tilde{J}_1 S\left(|\varphi_{n+1}|^2 - |\varphi_{n-1}|^2\right). \tag{8.61}$$

In the long-wave and continuum approximations, equation (8.61) takes the form:

$$\frac{\partial^2 u(x,t)}{\partial t^2} = v_0^2\frac{\partial^2 u(x,t)}{\partial x^2} - \frac{2\tilde{J}_1 S}{m}\frac{\partial}{\partial x}|\varphi(x,t)|^2 \ . \tag{8.62}$$

In the continuum approximation Eq.(8.60) also becomes

$$i\hbar\frac{\partial \varphi(x,t)}{\partial t} = \left[T + V - \tilde{J}_0 S^2 N + \frac{(2\tilde{J}_1)^2 S}{m(v_0^2 - v^2)} + 2(\tilde{J}_0 - J)S\right]\varphi(x,t)$$

$$-JS\frac{\partial^2 \varphi(x,t)}{\partial x^2} - \frac{(2\tilde{J}_1 S)^2}{m(v_0^2 - v^2)}|\varphi\ (x,t)|^2\ \varphi\ (x,t) \tag{8.63}$$

with

$$T + V = \frac{m}{2}\left[\int_{-\infty}^{\infty}\left(\frac{\partial u}{\partial t}\right)^2 dx + v_0^2\int_{-\infty}^{\infty}\left(\frac{\partial u}{\partial x}\right)^2 dx\right] = \frac{(2\tilde{J}_1 S)^2}{2mv_0^2}\frac{1+s^2}{(1-s^2)^2}\int_{-\infty}^{\infty}|\varphi\ (x,t)|^4\ dx \tag{8.64}$$

where $s = v/v_0$. Therefore, the magnon is a soliton in nonlinear systems, which satisfies the nonlinear Schrödinger equation, where the nonlinear interaction is due to the interaction between the moved magnon and moved atom, their moved states are described in Eqs. (8.62) and (8.63), respectively. Therefore, it is called the self-trapping mechanism of nonlinear interaction, depicted in Chapter 4. Therefore, the properties of the magnon can be described by the nonlinear quantum mechanics in this case.

8.4.2. The Nonlinear Features of the Magnons in the Case of Spin-Lattice Coupling

Equation (8.63) is the nonlinear Schrödinger equation, which is the dynamic equation in nonlinear quantum mechanics; its soliton solution is represented in this case by

$$\varphi(x,t) = \frac{1}{\sqrt{2\bar{\mu}}} e^{i[\alpha x - \theta - \omega t]} \sec h\left(\frac{x - x_0 - vt}{\bar{\mu}}\right) \tag{8.65}$$

where

$$\alpha = \frac{\hbar v}{2JS}, \quad \bar{\mu} = \frac{Jmv_0^2}{\tilde{J}_1^{\,2} S}(1-s^2), \quad T + V = \frac{2(\tilde{J}_1 S)^4}{3JS(mv_0^2)^2}\frac{1+s^2}{(1-s^2)^3}$$

$$\hbar\omega = T + V - \tilde{J}_0 S^2 N + \frac{(2\tilde{J}_1)^2 S^3}{mv_0^2(1-s^2)} + 2(\tilde{J}_0 - J)S + \frac{\hbar^2 v^2}{4JS}$$

Here x_0 is the mass center of the magnon-soliton and a constants. x_0 appears as a result of the translation invariance of the system and moves together with deformation of the lattice in the velocity v, which can be determined by the initial conditions, θ depends on the initial phase of the wave function. The energy of the magnon-soliton is given by

$$\hbar\omega = \hbar\omega_0 + \frac{\hbar^2 v^2}{4JS} + \frac{(\tilde{J}_1 S)^4}{JS(mv_0^2)^2}\frac{(2 + 3s^2 - s^4)s^2}{(1-s^2)^3} \tag{8.66}$$

where $\hbar\omega_0 = -\tilde{J}_0 S\ N^2 + 2(\tilde{J}_0 - J)S - \dfrac{(\tilde{J}_1 S)^4}{JS(mv_0^2)^2}$

the total energy $\hbar\omega$ is given as a sum of the rest energy $\hbar\omega_0$ and kinetic energy. The first two terms in $\hbar\omega_0$ represent the ground state energy of the anisotropy spin system being considered and the last one describes the rest energy due to the solitary excitation.

In the case of small velocities ($v^2 << v_0^2$) the soliton moves as a particle with an effective mass $m^* = m_0^*(1 + \Delta m^*/m_0^*)$, where $m_0^* = \hbar^2/2JS$ is the effective mass of a free magnon and

$$\Delta m^* = \frac{4(\tilde{J}_1 S)^4}{m\hbar^2 v_0^4} m^*$$

is the change in the mass of the magnon due to its localization, i.e., the transformation of the magnon into a "magnon-soliton" in this case. If the velocity$v$ equals to zero, the soliton energy has the absolute minimum,

$$E_{\min} = -\tilde{J}_0 S^2 N + 2(\tilde{J}_0 - J)S - \frac{(\tilde{J}_1 S)^4}{JS(mv_0^2)^2} \tag{8.67}$$

We see from these results that the energy of the system with a soliton at rest is smaller than the energy of free magnon. Hence, such a state is more favorable and stable. If the velocity$v$ increases, then the energy of the magnon-soliton also increases. At $v > v/\sqrt{5}$, the contribution of the last term in Eq. (8.66) predominates over the last one in $\hbar\omega_0$.

In this process, the chain elongation Δ occurs due to the spin deviation. From (8.58), we obtain that

$$\Delta = \int_{-\infty}^{\infty} \frac{\partial u}{\partial x} dx = \frac{2\tilde{J}_1 S}{mv_0^2(1-s^2)}$$

Thus, Δ increases with velocity$v$. In the case of a stationary soliton, we have $\Delta = 2J_1 S / mv_0^2$.

8.4.3. The Nonlinear Characteristics of the Magnons in the Anisotropic Heisenberg Ferromagnetic Systems

Pushkarovet al., [12-16] studied the nonlinear excitations of magnons in ferromagnetic systems with a biquadratic exchange and anharmonic lattice deformation using the above method, and obtained many useful results. Davydov and Kislukha [17-18] also examined low-lying excitations of anisotropic Heisenberg ferromagnet in one-dimensional systems described by the following Hamiltonian:

$$H = -\mu\tilde{H}_Z \sum_n S_n^z + 2J\sum_n \left(\xi S_n^x S_{n+1}^x + \eta S_n^x S_n^y + S_n^z S_{n+1}^z - \frac{1}{4} \right) \tag{8.68}$$

where $B = (0, 0, \overline{H}_z), S = 1/2.$For a system of N spins, which are rigidly fixed at lattice sites, if $-1 \le (\xi, \eta) \le 1$, i.e., for any possible combination of S_n^x and S_n^y (compared to S_n^z), we have the so-called "easy axis" $0z$ model. When $\xi = \eta$, there is isotropy in the base (x-y) plane. If $\eta = 1, \xi \ne \eta$ (or alternatively $\xi = 1, \xi \ne \eta$),we have the "easy plane" model (y-z or x-z, respectively). Finally, setting ξ = 0(or η = 0), which gives the Ising model in a transverse field. The simplest variantof Eq. (8.66) is an exact "Ising model". When $\xi = \eta = 0.\xi = \eta = 1$ yields the isotropic Heisenberg model, which also includes the case of $\xi >> 1$ (or $\eta >> 1$), if the last term $S_n^z S_{n+1}^z$ in Eq. (8.68) can be treated as a small perturbation. However, in the study of low-lying excitations arising from the nonlinearity, this small term in Eq. (8.68) must beincluded. Applying the Jordan-Wigner transformation for $S_n^{\pm}$ and S_n^z

$$a_n^+ = (-2)^{n-1} S_1^z S_2^z \ldots S_{n-1}^z S_n^+ ,\quad S_n^z = S_n^+ S_n^- - \frac{1}{2}$$
$$a_n = (-2)^{n-1} S_1^z S_2^z \ldots S_{n-1}^z S_n^- ,\quad S_n^{\pm} = S_n^x \pm i S_n^y \tag{8.69}$$

where a_n^+ and a_n are the Fermi operators and have $\{a_n, a_m^-\} = \delta_{nm}, \{a_n, a_n\} = \{a_n^+, a_n^+\} = 0$, we obtain, instead of Eq. (8.68), the Hamiltonian in terms of the Fermi operators

$$H = \frac{1}{2}\mu N\tilde{H}_Z - (\mu\tilde{H}_Z + 2J)\sum_k N_k + \frac{1}{2}J(\xi+\eta)\sum_n \left(a_n^+ a_{n+1} + a_{n+1}^+ a_n\right) -$$
$$\frac{1}{2}J(\xi-\eta)\sum_n \left(a_{n-1}^+ a_n^+ - a_{n-1}a_n\right) + 2J\sum_n N_n N_{n+1} \tag{8.70}$$

where $N_n = a_n^+ a_n$. The nonlinearity is now due to the term $S_n^z S_{n+1}^z$. The anomalous term $-(1/2)J(\xi-\eta)\sum_n (a_{n-1}^+ a_n^+ - a_{n+1}a_n)$, are defined by the difference between the x and y components. Note that the model in Eqs. (8.66), like (8.70), corresponds to a system of rigidly fixed spins.From the Heisenberg equation and Eqs.(8.70) and (8.57), one can obtain

$$i\hbar\varphi_t(x,t) = -\Delta\varphi\,(x,t) + \frac{1}{2}J(\xi+\eta)r_0^2\varphi_{xx}(x,t) + \frac{1}{2}J(\xi-\eta)r_0\varphi_x^* + 2J\left|\varphi(x,t)\right|^2\varphi(x,t)$$

where

$$\Delta = 2J\left(1 - \frac{\mu\tilde{H}_Z}{2J} - \frac{\xi-\eta}{2}\right) \equiv 2Jd$$

Note that Δ can be positive (e.g. for J>0, $|\xi|<1$, and $|\eta|<1$), or negative ($\xi >> 1, \eta \approx 1$). In the case of J>0, and $0 \le \xi \le 1, 0 \le \eta \le 1, \Delta > 0, d > 0,$ letting $\hbar = r_0 = 1$, and introducing new variables $\phi = \varphi(x,t)/\sqrt{d}, t' = t\Delta, x' = \sqrt{4d/(\xi+\eta)}x$, we have

$$i\phi_{t'} - \phi + \phi_{x'x'} + \alpha\phi_{x'}^* + |\phi|^2\phi = 0, \tag{8.71}$$

where $\alpha = \dfrac{\xi-\eta}{\sqrt{4d(\xi+\eta)}}$

Equation (8.71) is still a nonlinear Schrödinger equation, and should have a soliton solution. If $\alpha < 1$, $\alpha\phi^*_{x'}$ could be treated as a small perturbation. Based on this equation, Makhankov et al., [19 -21] discussed the structural stability of microscopic particles (solitons) described by the nonlinear Schrödinger equation (8.71).

However, Fedyanin et al., [26-27] obtained an approximate solution of Eq. (8.71) using an alternative approach, in which they assumed

$$\phi = \sqrt{2}\exp(-it')F(x',t')$$

and obtained the following equation by inserting it into Eq.(8.71)

$$iF_{t'} + \frac{1}{2}F_{x'x'} + |F|^2 F = -\frac{1}{2}\alpha F^*_{x'}\exp(it') \tag{8.72}$$

Equation (8.72) has the following solution at $\alpha = 0$,

$$F_0 = v_0 \exp\left[i\left(\frac{\mu_0}{v_0}z + \theta_0\right)\right]\operatorname{sec} hz_0$$

where $z_0 = 2v_0(x' - \bar{x}'_0)$, with $\bar{x}'_0 = 4\mu t'$. If $\alpha \neq 0$ but is very small, we can assume that the parameters, $v, \bar{x}'_0$ and θ_0, change only due to this perturbation. Following the perturbative approach of Karpman and Maslov [28-29], Fedyanin et al. [26-27] obtained the first order correction in α, in which the time-dependences of the variables, $v, \mu_o, \bar{x}'$ and θ, are given by

$$\mu_{t'} = \frac{\alpha}{2}\Im\int_{-\infty}^{\infty}\frac{\sinh z}{\cosh^2 z}\left(iF^*_{z}\right)Q(z)dz\,,\quad v_{t'} = \frac{\alpha}{2}\Re\int_{-\infty}^{\infty}\frac{iF^*_{z}}{\cosh z}Q(z)dz$$

$$\bar{x}'_{t'} = 4\mu + \frac{\alpha}{4v^2}\Re\int_{-\infty}^{\infty}\frac{z}{\cosh z}\left(iF^*_{z}\right)Q(z)dz\,,$$

$$\theta_{t'} = 2\mu\bar{x}'_{t'} - 4\left(\mu^2 - v^2\right) + \frac{\alpha}{2v}\Im\int_{-\infty}^{\infty}(\cosh^{-1}z - \frac{z\sinh^2}{\cosh^2 z})\left(iF^*_{z}\right)Q(z)dz$$

where $Q(z) = \exp\left[i\left(-\frac{\mu}{v}z - \theta + 2t'\right)\right]$

In this case

$$\upsilon_{t'} \approx \alpha\Re\left[e^{2it}\int_{-\infty}^{\infty} iF^{*}F_{z}^{*}dz\right]=0$$

Thus the following relationships can be obtained

$$\mu \simeq -\frac{2}{3}\alpha v_0^2 \sin^2 t', \; \bar{x}' = -\frac{\alpha}{2}\cos 2t', \; \theta = 4v_0^2 t'.$$

Through Eq.(8.72) the perturbed solution of Eq. (8.71) can be obtained, it has

$$\phi(x,t) = A_0 \operatorname{sec}h\left\{\frac{A_0}{\sqrt{2}}\left[x' - \cos\left(2t'\right)\right]\right\}\exp\left[-i\left(1-\frac{A_0^2}{2}\right)t + \frac{\alpha A_0^2 x'}{6}\sin(2t')\right] \tag{8.73}$$

where $A_0 = 2\sqrt{1-\omega}$. This indicates that the magnon is still a soliton in this case. Certainly, the soliton solution of Eq. (8.71) can also be obtained using other approaches.

From the above discussion, we can conclude that the magnon is localized as a soliton due to nonlinear interactions occurring in the ferromagnetic system. In this case, the motion of the magnon-soliton can be described by the nonlinear quantum mechanics. Existences of the magnon-solitons has been verified by neutron scattering and NMR experiments in $CuGeO_3$ by Ronnow et al.[30], Enderle et al. [31], and Revarat et al. [32], respectively, which were described in the paper by Mikeska and Steiner(1991) [33]. So, the above results are correct.

8.5. Nonlinear Quantum Properties of Solitary Magnons in Different Ferromagnetic Chains

8.5.1. The Wave-Corpuscle Duality of Magnons in One-Dimensional Systems

Pushkarov, et al., [34] discussed the influences of the magnon-magnon interaction on the motions of the magnons in one-dimensional ferromagnetic-chains under action of an externally applied magnetic-field with a strength B. The one-dimensional systems are composed of N magnetic-ions. In the anisotropic Heisenberg model they gave the Hamiltonian of the system, which is expressed by

$$H = J\sum_{l,\delta}\frac{1}{2}(\hat{S}_l^{+}\hat{S}_{l+\delta}^{-} + \hat{S}_l^{-}\hat{S}_{l+\delta}^{+}) - \bar{J}\sum_{l,\delta}\hat{S}_l^{z}\hat{S}_{l+\delta}^{z} - \mu B\sum_{l}\hat{S}_l^{z} \tag{8.74}$$

where $\mu = g\mu_B, \bar{J} > J > 0$, the lattice constant is here assumed to be 1.

At present, we consider the nonlinear interaction between the magnon-magnon in Eq.(8.23) in this system. In this case we represent the term $\sqrt{2S-a^{+}a)}$ by

$$\sqrt{2S-a_l^{+}a_l)}=1-\frac{1}{2}(a_l^{+}a_l/2S)+\frac{1}{8}(a_l^{+}a_l/2S)^2+..... \tag{8.75}$$

and use the Holstien-Primakoff transformations [9] in Eq.(8.22) , which is now expressed as

$$\hat{S}_l^{+}=(\sqrt{2S}(1-a_l^{+}a_l/4S)a_l,\hat{S}_l^{-}=(\sqrt{2S}a_l^{+}(1-a_l^{+}a_l/4S),\hat{S}_l^{z}=(S-a_l^{+}a_l) \tag{8.76}$$

Inserting Eq.(8.76) into Eq.(8.74) and reaching the fourth power of a_l, then Eq.(8.74) is represented as

$$\begin{aligned}H=&-2\bar{J}S^2N-\mu BSN+\mu B\sum_l a_l^{+}a_l+S\sum_{l,\delta}\{\bar{J}(a_l^{+}a_l+a_{l+\delta}^{+}a_{l+\delta})\\&-J(a_l^{+}a_{l+\delta}+a_l^{+}a_{l+\delta})\}-\bar{J}\sum_{l,\delta}(a_l^{+}a_la_{l+\delta}^{+}a_{l+\delta})\\&+\frac{1}{4}J\sum_{l,\delta}(a_l^{+}a_{l+\delta}^{+}a_{l+\delta}a_{l+\delta}+a_{l+\delta}^{+}a_l^{+}a_la_l+a_l^{+}a_l^{+}a_la_l+a_{l+\delta}^{+}a_{l+\delta}^{+}a_{l+\delta}a_l)\end{aligned} \tag{8.77}$$

Then the dynamic equation of a_l is denoted as

$$\begin{aligned}&i\hbar\frac{\partial}{\partial t}a_l=(\mu B+4S\bar{J})a_l-2JS\sum_\delta a_{l+\delta}-2\bar{J}\sum_\delta a_{l+\delta}^{+}a_{l+\delta}a_l\\&J/2\sum_\delta(a_{l+\delta}^{+}a_{l+\delta}a_{l+\delta}+a_{l+\delta}^{+}a_la_l+2a_l^{+}a_la_l)\end{aligned} \tag{8.78}$$

This is a nonlinear equation of operator a_l. When $\bar{J}$=J, Eq.(8.78) may become as Eq.(8. 30) through some approximations.

We now represented the state of the magnons arising from the nonlinear interaction, which is generated by the interaction between the spin and spin or magnon and magnon. This state is denoted by the following coherent state [35, 4, 8, 10,22]

$$|\Phi\rangle=\prod_l|a_l\rangle=|\{a_l\}\rangle \tag{8.79}$$

It has the properties of $\alpha_l=\langle\Phi|a_l|\Phi\rangle$ and $a_l|\alpha_l\rangle=\alpha_l|\alpha_l\rangle$. Utilizing these properties, the dynamic equation of $\alpha_l(t)$ can be obtained, which is

$$i\hbar\frac{\partial}{\partial t}\alpha_l = (\mu B + 4S(\bar{J}-J)\alpha_l - 2JS(2\alpha_l - \alpha_{l+\delta} - \alpha_{l-\delta}) - 2\bar{J}(|\alpha^+_{l+\delta}|^2 + |\alpha^+_{l-\delta}|^2)\alpha_l$$
$$+J/2\sum_\delta(2|\alpha_l|^2 + |\alpha_{l+\delta}|^2)\alpha_{l+\delta} + J/_2\, a_l^2\sum_\delta \alpha^*_{l+\delta} \tag{8.80}$$

Because $\alpha_l(t)$ is a slowly changing function of the time t, its scale of variation in the space is larger than the lattice constant of the ferromagnetic system. Thus, we can use the continuous approximation to find out the solution of Eq.(8.80). In this case, $\alpha_l(t)$ is denoted by $\alpha(x,t)$, which can be approximately represented [35, 4, 8, 10,22] by

$$\alpha_{l\pm 1}(t) - \alpha_l(t) = \pm\frac{\partial\alpha(x,t)}{\partial x} + \frac{1}{2}\frac{\partial^2\alpha(x,t)}{\partial x^2} + ...$$

In such a case, Eq.(8.80) can be written approximately by

$$i\hbar\frac{\partial}{\partial t}\alpha(x,t) = (\mu B + 4S(\bar{J}-J)\alpha(x,t) - 2JS\frac{\partial^2\alpha(x,t)}{\partial x^2} + 4(\bar{J}-J)|\alpha(x,t)|^2\alpha(x,t) \tag{8.81}$$

This is a nonlinear Schrödinger equation, which is the same with the dynamic equation in Eq.(2.66) in nonlinear quantum mechanics [10,22-25]. This indicates clearly that if only the interaction between the magnon and magnon is considered, the states and properties of the magnon should be described by the nonlinear Schrödinger equation, or nonlinear quantum mechanics given in Chapters 2-4 [10,22-25], instead of the linear equation in Eq.(8.34), where the nonlinear interaction is just due to the interaction between the magnon and magnon. This means the interaction between the magnon and magnon change through the states and properties of the magnons in the ferromagnetic chains, or in other words, the latter should be described by nonlinear quantum mechanics, mentioned above. In this case the concrete states of the magnon may be represented by

$$\alpha(x,t) = \frac{1}{\sqrt{2W}}\exp[i(\lambda x - \omega t)]\sec h[(x - vt)/W] \tag{8.82}$$

where $W = 2JS/(\bar{J}-J), v = v_1(2JS/\hbar), \lambda = \hbar v/4JS$. Equation (8.82) is similar to Eq.(3.8) and composed of an envelope and carrier wave, therefore, it has a wave-corpuscle feature as depicted in Chapter 3, its energy now is

$$\hbar\omega = E = [\mu B + 4S(J-J) + \hbar^2 v^2/8JS - (J-J)^2/2JS = E_0 + \frac{1}{2}M_{sol}v^2 + \mu B\,.$$
$$E_0 = 4S(\bar{J}-J)[1-(\bar{J}-J)/8JS], M_{sol} = \hbar^2/8JS\,. \tag{8.83}$$

The above results showed that the properties of the magnon are thoroughly changed relative to those given in Section 8.3, due to the nonlinear interaction, in this case the amplitude of the magnon-soliton is $\sqrt{2}\pi/\sqrt{W}$, its width is W, then its size is $\pi\sqrt{W}=\pi\sqrt{2JS/(\bar{J}-J)}$, its effective mass is $M_{sol}=\hbar^2/8JS$, its static energy or formed energy is $E_0=4S(\bar{J}-J)[1-(\bar{J}-J)/8JS]$. If $\bar{J}>3J$, and the magnon-soliton is localized and very stable it becomes $E_0<0$.

At the same time, because $\bar{J}>3J>J$, then $W>>1$, this means that the size of the magnon-soliton is larger than the lattice constant of the ferromagnetic systems. Thus, the continuous approximation used in the above calculation is correct.

From Eq.(8.83) we know that $\hbar\omega$ is related to the temperature of the ferromagnetic chains because the amplitude of the magnon-soliton or v_1 depends on the temperature of this system. Thus, the energy of the magnon-soliton, $\hbar\omega(T)$ is, in general, reduced with increasing temperature. This implies that lifting the temperature of the system will result in the decrease of frequency of the spin solitary wave of the magnons, thus a stability state of the spin solitary wave will be changed or broken further, when the temperature is much higher. This is called the instability of the spin solitary wave. This phenomenon was observed in some ferromagnetic-insulation material, such as EuO [36]. Thus, we can use this result to explain some experimental phenomena. If the correct dependence of v_1 on the temperature is known, then the concrete relation of $\hbar\omega(T)$ with the temperature can be known, then we can judge the correctness of the above theoretical results by comparing the experimental data with theoretical results.

On the other hand, it is well known that the average number of the magnons excited in this system, at the temperature T, is determined by the statistical Bose distribution rule because the operators of the magnon satisfy Bose commutation relations, thus the average number of the magnon can be denoted [35, 4, 8, 10,22] by

$$\langle n_k\rangle_T=\langle b_k^+b_k\rangle_T=[\exp(\hbar\omega_k/K_BT)-1]^{-1}$$

The total number of the magnon excited in all spin ions should be represented by

$$\sum_k\langle n_k\rangle_T=\frac{V}{8\pi^3}\int[\exp(\hbar\omega_k/K_BT)-1]^{-1}d^3k$$

If $E_0=4S(\bar{J}-J)[1-(\bar{J}-J)/8JS]\to 0$ and $B=0$,then from Eq.(8.83) we can obtain

$\hbar\omega_k=v^2M_{sol}/2=Dk^2$, where $D=\hbar^2M_{sol}/2m^2$ and $x=Dk^2/K_BT$.

If the temperature of the system is lowered to $Dk_{\max}^2 >> K_B T$, in this case, the above integration can be approximately found in whole space of k, then we can finally obtain

$$\frac{1}{N}\sum_k \langle n_k \rangle_T = \frac{V}{8N\pi^3}(\frac{K_B T}{D})^{3/2}\int_0^\infty [\exp(x)-1]^{-1} x^{-3/2} dx = (\frac{K_B T}{D})^{3/2} \upsilon$$

where υ is a constant, in this calculation we used $\Gamma(3/2) = \sqrt{\pi}/2$ and the Riemann function $\zeta(3/2;1) = 2.613$.

In the case of low temperature, the magnetized strength of the ferromagnets can be denoted [8,10,22,36]by

$$M(T) = g\mu_B \sum_l (S - < a_l^+ a_l >_T) = Ng\mu_B S - g\mu_B \sum_k < b_k^+ b_k >_T$$
$$= M(0)[1-(1/NS)\sum_k < n_k >_T = M(0)[1-(\upsilon/S)(K_B T/D)^{3/2}]$$

where $M(0) = Ng\mu_B S$ is the saturation magnetized strength of the ferromagnets in the case of zero temperature. Thus, we can find out that the reduction of magnetized strength of the ferromagnets arising from the excitation of the spin-solitary wave, which is

$$\Delta M(T) = M(0) - M(T) = M(0)(\upsilon/S)(K_B T/D)^{3/2} \propto T^{3/2}$$

This result was affirmed by Bloch's experiment [37] in the ferromagnetic systems, which were called Bloch law of $T^{3/2}$.

Cordeiro et al., [38] demonstrated that there are the same magnon-soliton solutions in the ferromagnetic chains if the unfold or extension in Eq.(8.75) reaches some higher power terms, such as S^0, S^{-1} and S^{-2} terms, even in the case of the isotropy. This manifests that the states of magnon-solitons are intrinsic to the nature of the magnon or spin wave in the ferromagnetic systems.

It is well known that if there are excitations of the spin-waves in the ferromagnetic systems when perturbed, then there are interactions among the magnons in these ferromagnets, thus we can affirm that there are nonlinear interactions between them because the nonlinear interactions are just interactions between these magnons. Upon that, we can confirm that the properties and rules of motion of the magnons must be described by the nonlinear quantum mechanics, mentioned above. The above results tell us only that if the interactions between the magnons are neglected, then its properties may be described by the linear Schrödinger equation or linear quantum mechanics. However, the nonlinear interactions always exist in real spin-spin interacted systems, so the nonlinear quantum mechanics is a correct theory for describing the magnons, but the linear quantum mechanics is only an approximate theory, at nonlinear interactions equal to zero.

8.5.2. The Dynamic Properties of the Magnons in a Flat Systems of Ferromagnetic Chains

Yatsishin studied the properties of the magnons in flat ferromagnetic chains with weak interaction [39], in which N linear ferromagnetic chains are contained. To each chain, they ascribed the number n and assumed that the interaction between the spins in the chains is stronger than the interaction between the spins in different chains. To study this system, they proceeded from the ordinary Heisenberg Hamiltonian with allowance for the dynamics of the chains:

$$H = H_N + H_{Ph} \tag{8.84}$$

where

$$H_N = -\frac{1}{2}\sum_{n,m,\alpha,\beta}{}'(J_{n\alpha m\beta}S^+_{n\alpha}S^+_{m\beta} + J_{n\alpha n\beta}S^z_{n\alpha}S^z_{n\beta}) \tag{8.85},$$

here α and β are the chain numbers, n and m are the numbers of spins in the corresponding chains, $S^{\pm}_{n\alpha} = S^x_{n\alpha} \pm iS^y_{n\alpha}$, $S^v_{n\alpha}$ is the operator of the vth projection of the spin (v=x,y,z), $J_{n\alpha m\beta} = J(|r_{n\alpha} - r_{m\beta}|), J_{n\beta m\beta} = J(|r_{n\beta} - r_{m\beta}|)$ are exchange integrals between the spins with numbers n, α and m, β, where H_{Ph} in Eq.(8.84) is the energy operator describing the dynamics of the chains.

They first went over from the spin operators to the Bose operators of spin excitations by HP transformation mentioned above and expressed the operators H_{Ph} in the harmonic approximation and the coefficients of the magnetic subsystem by virtue of which the coupling between the spin excitations and the vibrations of the chains were taken into account in the linear approximation in the displacement operators. After this, the Hamiltonian in Eq.(8.84) becomes as [39].

$$\begin{aligned} H = \sum_{n\alpha}\{&\frac{P^2_{n\alpha}}{2M} + \frac{w}{2}(x_{n+1\alpha} - x_{n\alpha})^2 + 2ST_0(b^+_{n\alpha}b_{n\alpha}) - QS(b^+_{n\alpha}b_{n\alpha+1} + b^+_{n\alpha+1}b_{n\alpha}) + \\ &ST_1(b^+_{n\alpha}b_{n+1\alpha} + b^+_{n+1\alpha}b_{n\alpha}) + I_0S(x_{n+1\alpha} - x_{n-1\alpha})b^+_{n\alpha}b_{n\alpha} - I_1S(x_{n+1\alpha} - x_{n\alpha})(b^+_{n\alpha}b_{n+1\alpha} + b^+_{n+1\alpha}b_{n\alpha})\} \end{aligned} \tag{8.86}$$

where M is the effective mass displaced, together with the spin along the chain, w is the elasticity coefficient of the chain, $P_{n\alpha}$ is the momentum operator canonically conjugate to the displacement operator $x_{n\alpha}$, and $Q = I_{n\alpha,n\alpha+1}, T_0 = T(d), I_0 = I(d),$

$$T_1 = \frac{\partial T(|r_{n+1\beta} - r_{n\beta}|)}{\partial(r_{n+1\beta} - r_{n\beta})}\Big|_{r_{n+1\beta}-r_{n\beta}=1}, I_1 = \frac{\partial I(|r_{n+1\beta} - r_{n\beta}|)}{\partial(r_{n+1\beta} - r_{n\beta})}\Big|_{r_{n+1\beta}-r_{n\beta}=1}, \quad (8.87)$$

d is the lattice constant.

Please notice that the displacements of the spins from the equilibrium positions only along the chains were considered in Eq.(8.86).

In this case, they [36]introduced the following wave function [39]:

$$\Psi(t) = \sum_{n\alpha} C_{n\alpha}(t) e^{\sigma(t)} b_{n\alpha} |0\rangle, \sum_{n\alpha} |C_{n\alpha}(t)|^2 = 1, \sigma(t) = -\sum_{n\alpha} [U_{n\alpha}(t) P_{n\alpha} - V_{n\alpha}(t) x_{n\alpha}] \quad (8.88)$$

where $|0\rangle$ is the vacuum state of the spin excitations and the phonons. The complex functions $C_{n\alpha}(t)$ and the real functions $U_{n\alpha}(t)$ and $V_{n\alpha}(t)$ are to be calculated. The coefficient $C_{n\alpha}(t)$ characterizes the distribution of the excitation over the sites of chain α, while the function $U_{n\alpha}(t)$ and $V_{n\alpha}(t)$ determine the mean values of the displacements and their canonical momenta in the state in Eq.(8.88).

We now assumed that the wave function in Eq.(8.88) satisfied the dynamic equation:

$$i\hbar \frac{d}{dt} \Psi(t) = H\Psi(t) \quad (8.89)$$

In this case they also introduced the function:

$$\overline{H} = \langle \Psi | H | \Psi \rangle \quad (8.90)$$

which may be thought of as the Hamilton function of the variables $C_{n\alpha}(t)$, $U_{n\alpha}(t)$ and $V_{n\alpha}(t)$, their representations can be found from the following canonical Hamilton equations:

$$i\frac{\partial C_{n\alpha}}{\partial t} = \frac{\partial \overline{H}}{\partial C^*_{n\alpha}}, \frac{\partial U_{n\alpha}}{\partial t} = \frac{\partial \overline{H}}{\partial V_{n\alpha}}, \frac{\partial V_{n\alpha}}{\partial t} = \frac{\partial \overline{H}}{\partial U_{n\alpha}} \quad (8.91)$$

Taking into account the explicit form of the Hamiltonian in Eq. (8.86) and the trial wave function in Eq.(8.88), the Hamilton equation in Eq.(8.91) in the long-wavelength approximation, are reduced to the solution of the following nonlinear Schrödinger equation [39]:

$$i\frac{\partial C_\alpha(\varsigma,t)}{\partial t} = (\Delta + W' + \frac{\tau^2}{w'(1-s')} |C_\alpha(\varsigma,t)|^2) C_\alpha(\varsigma,t) - I_1 S \frac{\partial^2 C_\alpha(\varsigma,t)}{\partial \varsigma^2} - QS[C_{\alpha+1}(\varsigma,t) + C_{\alpha-1}(\varsigma,t)] \quad (8.92)$$

where $\Delta = 2S(T_0 - I_0), \tau = 2S(T_1 - I_1), s' = v^2 M / w = v^2 / v_1^2$, W' is the energy of the vibrational motion of the considered system, v is the velocity of the excitation or magnon.

In this case, the bell-type soliton solution of Eq.(8.92) can be obtained [39]

$$C_\alpha(\varsigma,t) = A_\alpha \sqrt{\mu'/2} \exp\{i[\frac{v}{2SI_1}(\varsigma - \varsigma_0) - \Omega t]\} \sec h[\mu'(\varsigma - \varsigma_0) - vt] \qquad (8.93)$$

where

$$(SI_1\mu'^2 + \varepsilon)A_\alpha + QS(A_{\alpha+1} + A_{\alpha-1}) = 0, 4SI_1\mu' A_\alpha - \frac{\tau'^2}{w'(1-s')} A_\alpha^2 = 0,$$
$$\sum_\alpha |A_\alpha|^2 = 1, \varepsilon = \Omega - \Delta - W' - \frac{v^2}{4SI_1}, \alpha = 1,2,...N \qquad (8.94)$$

The above results tell us that the properties of the magnon in a flat system of ferromagnetic chains and its rules of motion are still described by the nonlinear Schrödinger equation in Eq.(8.92), which is similar with Eq.(2.66) in the nonlinear quantum mechanics. This manifested clearly that nonlinear quantum mechanics mentioned in Chapters 3-4 is still a correct theory for them, and it can be still used to describe the properties and rules of motion of the magnon in a flat systems of ferromagnetic chains.

8.6. The Nonlinear Features of Magnons in Antiferromagnetic Systems

8.6.1. Dynamic Features of the Magnons under Action of Magnon-Phonon Interaction

Collective excitation and the motion of magnons, due to magnon-phonon interactions or the magnon-magnon interactions in Heisenberg antiferromagnetic systems, have been extensively studied by Pang and co-workers [40-46],and many other scientists [20-21, 47-48]. The results show that the characteristics of the collective excitation in the systems are quite different from those in ferromagnetic systems. In this section, we present some results obtained by Pang et al., based on collective excitation in anisotropic Heisenberg antiferromagnets with magnon-phonon and magnon-magnon interactions, using the double-sublattice model [45].

When the double-sublattice model is used, the Hamiltonian of the Heisenberg antiferromagnet can be expressed as [45]

$$H = T + V + \frac{1}{2}\sum_n^A \sum_\delta^A \left[\xi_{n,n+\delta} S_n^x S_{n+\delta}^x + \eta_{n,n+\delta} S_n^y S_{n+\delta}^y + J_{n,n+\delta} S_n^z S_{n+\delta}^z \right]$$

$$+\frac{1}{2}\sum_{j}^{B}\sum_{\delta}^{B}\left[\xi_{j,j+\delta}S_j^x S_{j+\delta}^x+\eta_{j,j+\delta}S_j^y S_{j+\delta}^y+J_{j,j+\delta}S_j^z S_{j+\delta}^z\right], \tag{8.95}$$

where $T=\frac{m}{2}\sum_n \dot{u}_n^2$, $V=\frac{m\upsilon_0^2}{r_0}\sum_n (u_{n+1}-u_n-u_0)$

are the kinetic and potential energies of lattice oscillations, respectively, with m being the "mass" of a spin, r_0 the lattice constant, and v_0 the sound velocity in the crystal, which we set equal to unity in subsequent calculations. $S_{n(j)}^{\kappa}(\kappa=x,y,z)$ is the spin component at site $n(j)$ in the k-direction. Applying the transformation $S_{n(j)}^{+}=(S_{n(j)}^{x}+iS_{n(j)}^{y}$ we change (8.95) into

$$H=T+V+\frac{1}{2}\sum_{n}^{A}\sum_{\delta}^{A}\left[J_{n,n+\delta}S_n^z S_{n+\delta}^z+\frac{1}{2}(\xi_{n,n+\delta}+\eta_{n,n+\delta})(S_n^+S_{n+\delta}^-+S_n^-S_{n+\delta}^+)+\frac{1}{4}(\xi_{n,n+\delta}-\eta_{n,n+\delta})(S_n^+S_{n+\delta}^++S_n^-S_{n+\delta}^-)\right]$$
$$+\frac{1}{2}\sum_{j}^{B}\sum_{\delta}^{B}\left[J_{jj+\delta}S_j^z S_{j+\delta}^z+\frac{1}{4}(\xi_{jj+\delta}+\eta_{jj+\delta})(S_j^+S_{j+\delta}^-+S_j^-S_{j+\delta}^+)+\frac{1}{4}(\xi_{jj+\delta}-\eta_{jj+\delta})(S_j^+S_{j+\delta}^++S_j^-S_{j+\delta}^-)\right] \tag{8.96}$$

We again use the Dyson-Maleev representation of the spin operators by virtue of Bose creation and annihilation operators [12.15-16,26]

$$S_a^+=\sqrt{2S}(1-\frac{a^+a}{4S})a,\ S_a^-=\sqrt{2S}a^+(1-\frac{a^+a}{4S}),\ S_a^z=S-a^+a$$
$$S_b^+=\sqrt{2S}b^+(1-\frac{b^+b}{4S}),S_b^-=\sqrt{2S}(1-\frac{b^+b}{4S})b,S_b^z=b^+b-S$$

where a+ (a) and b^+ (b) are the creation (annihilation) operators of the Heisenberg magnon field on the two sublattices, respectively. Taking into account the symmetry of the sublattices A and B,and the fact that the A and B sublattices are neighbors of each other, the Hamiltonian in Eq. (8.96) can be approximately written as [45]

$$H\approx T+V-J_0NS^2+S\sum_{n}^{A}\sum_{\delta}^{A}J_{n,n+\delta}a_n^+a_n+S\sum_{j}^{B}\sum_{\delta}^{B}J_{J,j+\delta}b_j^+b_j$$
$$+\frac{S}{2}\sum_{n}^{A}\sum_{\delta}^{A}\left(\xi_{n,n+\delta}-\eta_{n,n+\delta}\right)\left(a_nb_{n+\delta}^++b_{n+\delta}a_n^+\right)$$
$$+\frac{S}{2}\sum_{n}^{A}\sum_{\delta}^{A}\left(\xi_{n,n+\delta}+\eta_{n,n+\delta}\right)\left(a_nb_{n+\delta}+a_n^+b_{n+\delta}^+\right)$$

$$-\sum_{j}^{B}\sum_{\delta}^{B} J_{J,j+\delta} a_j^+ a_j b_{j+\delta}^+ b_{j+\delta}$$

$$-\frac{1}{8}\sum_{n}^{A}\sum_{\delta}^{A}\left(\xi_{n,n+\delta}-\eta_{n,n+\delta}\right)\left(a_n^+ a_n a_n b_{n+\delta}^+ + a_n b_{n+\delta}^+ b_{n+\delta} b_{n+\delta}\right)$$

$$+a_n^+ a_{n+\delta} a_n^+ a_n + b_{n+\delta}^+ b_{n+\delta} a_n^+ b_{n+\delta}$$

$$-\frac{1}{8}\sum_{n}^{A}\sum_{\delta}^{A}\left(\xi_{n,n+\delta}+\eta_{n,n+\delta}\right)\left(a_n^+ a_n a_n b_{n+\delta} + a_n b_{n+\delta} b_{n+\delta}^+ b_{n+\delta}\right)$$

$$+a_n^+ b_{n+\delta} a_n^+ a_n + a_n b_{n+\delta}^+ b_{n+\delta}^+ b_{n+\delta}, \tag{8.97}$$

where the last four terms are anomalous terms resulting from magnon-magnon interactions. They can be neglected because they are weak if we are interested only in effects of magnon-phonon interaction. In this case, the above Hamilltonian can become simple, which will be considered first in the following research.

We here applied the methods of Makhankov and Fedyanin et al., [20,21] and Pang [45] to study the properties of the collective excitations of the magnons in one-dimensional antiferromagnertic system. In Heisenberg representation, the equations of the operators a_f and b_f of sublattices A and B can, after some proper transformations, be written as

$$i\hbar \dot{a}_f = [a_f, H] \approx S\sum_{\delta} J_{ff+\delta} a_f + S/2\sum_{\delta}^{A}\left(\xi_{ff+\delta}-\eta_{ff+\delta}\right) b_{j+\delta} + S/2\sum_{\delta}\left(\xi_{ff+\delta}+\eta_{ff+\delta}\right) b_{j+\delta}^+ \tag{8.98}$$

$$i\hbar \dot{b}_f = [b_f, H] \approx S\sum_{\delta} J_{ff+\delta} b_f + S/2\sum_{\delta}^{A}\left(\xi_{ff+\delta}-\eta_{ff+\delta}\right) a_{j+\delta} + S/2\sum_{\delta}\left(\xi_{ff+\delta}+\eta_{ff+\delta}\right) a_{j+\delta}^+ \tag{8.99}$$

In this case, we further assume that the wave function of the collective excitation state of the quasi-particles in the system is of the form

$$|\varphi(t)\rangle = \frac{1}{\lambda}\left[2+\sum_{n}^{A}\varphi_{an}(t)a_n^+ + \sum_{j}^{A}\varphi_{bj}(t)a_j^+\right]|0\rangle, \tag{8.100}$$

where $||0\rangle$ is the vacuum state (ground state) of the system, φ_{an} and φ_{bj} are expression coefficients related to the characteristics of the magnons in a and b sublattices, which obviously are functions of time and space, and λ is a normalization constant.

Notice that $\langle\varphi(t)|a_f|\varphi(t)\rangle = \alpha_{af}/\lambda^2 = \varphi_{af}, \langle\varphi(t)|b_f|\varphi(t)\rangle = \alpha_{bf}/\lambda^2 = \varphi_{bf}$ are SchrÖdiner probability amplitudes of the magnon. Using these representations and from Eqs.(8.98)- (9.99) we have

$$i\hbar\dot{\varphi}_f = S\sum_\delta J_{ff+\delta}\varphi_{af} + S/2\sum_\delta\left(\xi_{ff+\delta}-\eta_{ff+\delta}\right)\varphi_{bj+\delta}(t) + S/2\sum_\delta\left(\xi_{ff+\delta}+\eta_{ff+\delta}\right)\varphi^*_{bj+\delta} \tag{8.101}$$

$$i\hbar\dot{\varphi}_{bf} = S\sum_\delta J_{ff+\delta}\varphi_{bf} + S/2\sum_\delta\left(\xi_{ff+\delta}-\eta_{ff+\delta}\right)\varphi_{aj+\delta}(t) + S/2\sum_\delta\left(\xi_{ff+\delta}+\eta_{ff+\delta}\right)\varphi^*_{aj+\delta} \tag{8.102}$$

If the oscillation amplitude of the lattice is small and the magnon-phonon interaction is weak, we may proceed to the continuum limit for the coefficients and the probability amplitude. Dropping the terms with higher derivatives as a result, we get

$$J_{f,f+1} \approx J_0 - J_1 r_0\frac{\partial u_f}{\partial x},\ \xi_{f,f+1} \approx \xi_0 - \xi_1 r_0\frac{\partial u_f}{\partial x},\ \eta_{f,f+1} \approx \eta_0 - \eta_1 r_0\frac{\partial u_f}{\partial x},$$

$$\varphi_{nf\pm1} \approx \varphi_{nf} \pm r_0\frac{\partial}{\partial x}\varphi_{n'f} + \frac{1}{2}r_0^2\frac{\partial^2}{\partial x^2}\varphi_{n'f} + \ldots, (n = a, b, n' = b, a),\ etc., \tag{8.103}$$

Using these relationships, we can get the following from the Heisenberg equations for φ_f

$$i\hbar\left(\varphi_f\right)_t = 2SJ_0\varphi_f + S(\xi_0-\eta_0) + S(\xi_0+\eta_0)\varphi_f^* + \frac{S}{2}(\xi_0-\eta_0)r_0^2\varphi_{fxx}$$

$$+\frac{S}{2}(\xi_0+\eta_0)r_0^2\varphi^*_{fxx} - S(\xi_1-\eta_1)r_0u_{fx}\varphi_f - S(\xi_1+\eta_1)r_0u_{fx}\varphi_f^* - 2J_1Sr_0u_{fx}\varphi_f \tag{8.104}$$

where r_0 is the average distance between neighboring sites. Using Eq.(8.103) and bearing in mind the symmetry of the sublattices, we can easily derive approximate dynamic equations of $\varphi_{af}(t)$ and $\varphi_{bf}(t)$, which can be denoted as follows

$$i\hbar\dot{\varphi}_{af} \approx 2SJ_0\varphi_{af} + S\left(\xi_0-\eta_0\right)\varphi_{bf} + S\left(\xi_0-\eta_0\right)\varphi^*_{bf} + \frac{S}{2}\left(\xi_0-\eta_0\right)r_0^2\frac{\partial^2}{\partial x^2}\varphi_{af} +$$
$$\frac{S}{2}\left(\xi_0+\eta_0\right)r_0^2\frac{\partial^2}{\partial x^2}\varphi^*_{af} + S\left(\xi_1-\eta_1\right)r_0\frac{\partial u_f}{\partial x}\varphi_{bf} - S\left(\xi_1+\eta_1\right)r_0\frac{\partial u_f}{\partial x}\varphi^*_{bf} - 2SJ_1r_0\frac{\partial u_f}{\partial x}\varphi_{af} \tag{8.105}$$

$$i\hbar\dot{\varphi}_{bf} \approx 2SJ_0\varphi_{bf} + S(\xi_0-\eta_0)\varphi_{aj} + S(\xi_0-\eta_0)\varphi_{aj}^* + \frac{S}{2}(\xi_0-\eta_0)r_0^2\frac{\partial^2}{\partial x^2}\varphi_{bf} +$$

$$\frac{S}{2}(\xi_0+\eta_0)r_0^2\frac{\partial^2}{\partial x^2}\varphi_{bf}^* + S(\xi_1-\eta_1)r_0\frac{\partial u_f}{\partial x}\varphi_{af} - S(\xi_1+\eta_1)r_0\frac{\partial u_f}{\partial x}\varphi_{af}^* - 2SJ_1r_0\frac{\partial u_f}{\partial x}\varphi_{bf} \quad (8.106)$$

We defined $\varphi_f(t) = \varphi_{af}(t) + \varphi_{bf}(t)$, then we can obtain the following equation:

$$i\hbar\dot{\varphi}_f \approx 2SJ_0\varphi_f + S(\xi_0-\eta_0)\varphi_f + S(\xi_0+\eta_0)\varphi_{bj}^* + \frac{S}{2}(\xi_0-\eta_0)r_0^2\frac{\partial^2}{\partial x^2}\varphi_f +$$

$$\frac{S}{2}(\xi_0+\eta_0)r_0^2\frac{\partial^2}{\partial x^2}\varphi_f^* + S(\xi_1-\eta_1)r_0\frac{\partial u_f}{\partial x}\varphi_f - S(\xi_1+\eta_1)r_0\frac{\partial u_f}{\partial x}\varphi_f^* - 2SJ_1r_0\frac{\partial u_f}{\partial x}\varphi_f \quad (8.107)$$

Clearly, equation (8.107) is, in essence, a nonlinear Schrödinger equation of the magnons in the antiferromagnetic system.

We [45]discuss first the part representation of the lattice oscillations. We have from Eq.(8.97)

$$\langle\varphi(t)|H|\varphi(t)\rangle = T + V - \frac{1}{2}S^2\sum_n^A\sum_\delta^A J_{nn+\delta} - \frac{1}{2}S^2\sum_n^B\sum_\delta^B J_{nn+\delta} - \sum_n^A(S\sum_\delta^A J_{nn+\delta})|\varphi_{an}|^2 +$$

$$\sum_j^B(S\sum_\delta^B J_{jj+\delta})|\varphi_{bj}|^2 + \frac{1}{2}S\sum_j^A\sum_\delta^A(\xi_{nn+\delta}-\eta_{jj+\delta})(\varphi_{an}\varphi_{bj+\delta}^* + \varphi_{an}^*\varphi_{bj+\delta}) \quad (8.108)$$

It should be pointed out that with the wave function, in the form Eq.(8.100), the anomalous terms in the original Hamiltonian have already been removed.

We here use the classical Hamilton equation

$$-M(u_f)_{tt} = \frac{\partial}{\partial u_f}\langle\varphi(t)|H|\varphi(t)\rangle \quad (8.109)$$

where M is the mass of a lattice point (atom, for example), u_f its displacement, a classical quantity. Having again in mind the symmetry of sublattices A and B as well as the fact that the neighbors of sublattice A belong to sublattice B, and vice versa, and using again

$$\sum_{n}^{A} S\sum_{\delta}^{A} J_{nn+\delta})\left|\varphi_{an}\right|^2 = S\sum_{n}^{A}\{2J_0 - J_1(u_{bn+1} - u_{bn-1})\}\left|\varphi_{an}\right|^2$$

$$\sum_{j}^{B} S\sum_{\delta}^{B} J_{jj+\delta})\left|\varphi_{bj}\right|^2 = S\sum_{j}^{b}\{2J_0 - J_1(u_{aj+1} - u_{aj-1})\}\left|\varphi_{bj}\right|^2$$

we can get from Eqs.(8.108)-(8.109)

$$-M\ddot{u}_{af} = \frac{\partial}{\partial u_{af}}\langle\varphi(t)|H|\varphi(t)\rangle = K(2u_{af} - u_{bf+1} - u_{bf-1}) + SJ_1[\left|\varphi_{bf+1}\right|^2 - \left|\varphi_{bf+1}\right|^2]$$

$$+\frac{1}{2}S(\xi_1 - \eta_1)[\varphi_{af}(\varphi^*_{bf+1} - \varphi^*_{bf-1}) + \varphi^*_{af}(\varphi_{bf+1} - \varphi_{bf-1})$$

$$-M\ddot{u}_{bf} = \frac{\partial}{\partial u_{bf}}\langle\varphi(t)|H|\varphi(t)\rangle = K(2u_{bf} - u_{af+1} - u_{af-1}) + SJ_1[\left|\varphi_{af+1}\right|^2 - \left|\varphi_{af+1}\right|^2]$$

$$+\frac{1}{2}S(\xi_1 - \eta_1)[\varphi_{bf}(\varphi^*_{af+1} - \varphi^*_{af-1}) + \varphi^*_{bf}(\varphi_{af+1} - \varphi_{af-1})$$

Assuming here $M_a \approx M_b \approx M$ we then have

$$-M\ddot{u}_{af} = K(2u_{af} - u_{bf+1} - u_{bf-1}) + SJ_1[\left|\varphi_{bf+1}\right|^2 - \left|\varphi_{bf+1}\right|^2]$$

$$+\frac{1}{2}S(\xi_1 - \eta_1)[\varphi_{af}(\varphi^*_{bf+1} - \varphi^*_{bf-1}) + \varphi^*_{af}(\varphi_{bf+1} - \varphi_{bf-1})$$

$$-M\ddot{u}_{bf} = K(2u_{bf} - u_{af+1} - u_{af-1}) + SJ_1[\left|\varphi_{af+1}\right|^2 - \left|\varphi_{af+1}\right|^2]$$

$$+\frac{1}{2}S(\xi_1 - \eta_1)[\varphi_{bf}(\varphi^*_{af+1} - \varphi^*_{af-1}) + \varphi^*_{bf}(\varphi_{af+1} - \varphi_{af-1})$$

where K is the force coefficient $K = mC_0^2/2r_0$.

Defining again $u_f = u_{af} + u_{bf}$ and taking into account the $M_a \approx M_b \approx M$, the symmetry of the sublattices, etc., we have no difficulty in obtaining the following equation in continuum approximation:

$$-Mu_{ftt} \approx -Kr_0^2 u_{fxx} + J_1 S r_0 \partial/\partial x(\left|\varphi_f\right|^2)_x + \frac{1}{4}(\xi_1 - \eta_1)Sr_0 \partial/\partial x(\left|\varphi_f\right|^2), \tag{8.110}$$

Equations (8.107) and (8.110) form a complete set of equations for the collective excitations in a Heisenberg antiferromagnetic system with magnon-phonon interactions.

Now we proceed to find out the solutions to the equations.

In this case, of a quasi-steady state, we can assume that

$$u(x,t)=u(\varsigma),\varphi(x,t)=\varphi(\varsigma)e^{i\theta}(\varsigma=x-vt) \tag{8.111}$$

Substituting Eq.(8.111) into Eq.(8.110) we have [46-48]

$$(-Kr_0^2-Mv^2)\frac{\partial^2}{\partial\varsigma^2}u_f=[J_1+\frac{1}{4}(\xi_1-\eta_1)]Sr_0\frac{\partial}{\partial\varsigma}|\varphi_f|^2 \tag{8.112}$$

We can get, by solving the above equations,

$$\frac{\partial u_f}{\partial x}=\frac{\partial u_f}{\partial\varsigma}=[J_1+\frac{1}{4}(\xi_1-\eta_1)]Sr_0(Kr_0^2-Mv^2)^{-1}|\varphi_f|^2+C' \tag{8.113}$$

where C' is an integration constant to be determined from boundary conditions. Substituting Eq.(8.113) into Eq.(8.107) and solving it we can immediately find out the characteristics of the magnons caused by magnon-phonon coupling in isotropic antiferromagnets.

For simplicity, we consider here only the anisotropic antiferromagnet with $\xi=\eta$ (the other cases, of course, can be discussed in the same way). In this case, $J>\xi$ and $J<\xi$ correspond to the easy magnetic axis (0z) and the easy magnetic plane (x0y)in an antiferromagnet, respectively. In this case, equation (8.107) reduces to

$$i\hbar\varphi_t=2J_0S\varphi+2\xi_0S\varphi^*+S\xi_0r_0^2\varphi_{xx}^*-2J_1Sr_0u_x\varphi-2\xi_1Sr_0u_x\varphi^* \tag{8.114}$$

Its conjugate equation reads

$$-i\hbar\varphi_t^*=2J_0S\varphi^*+2\xi_0S\varphi+S\xi_0r_0^2\varphi_{xx}-2J_1Sr_0u_x\varphi^*-2\xi_1Sr_0u_x\varphi \tag{8.115}$$

where we have dropped the subscript of $u(t)$ and $\varphi(t)$.

We now perform a further transformation $\varphi_\pm=\varphi\pm\varphi^*$, then the following equations [45] can be obtained from Eqs.(8.114)-(8.115)

$$\begin{aligned} i\hbar\dot{\varphi}_+ &== 2S(J_0+\xi_0)\varphi_- - S\xi_0r_0^2\frac{\partial^2}{\partial x^2}\varphi_- - 2(J_1-\xi_1)Sr_0\frac{\partial u}{\partial x}\varphi_- \\ i\hbar\dot{\varphi}_- &== 2S(J_0+\xi_0)\varphi_+ + S\xi_0r_0^2\frac{\partial^2}{\partial x^2}\varphi_+ - 2(J_1+\xi_1)Sr_0\frac{\partial u}{\partial x}\varphi_+ \end{aligned} \tag{8.116}$$

They are some coupling nonlinear Schrödinger equations of φ_+ and φ_-, their solutions are quite difficult to find out. However, in the case of v<< C_0 or $v^2 << Kr_0^2 / M$, we can obtain the following nonlinear dynamic equation of $\varphi_+(x,t)$ and $\varphi_-(x,t)$

$$-\hbar^2 \ddot{\varphi}_+ == 4S^2(J_0^2 - \xi_0^2)\varphi_+ - 4S^2\xi_0^2 r_0^2 \frac{\partial^2}{\partial x^2}\varphi_+ - 8(J_1 J_0 - \xi_1\xi_0)S^2 r_0 \frac{\partial u}{\partial x}\varphi_+ \tag{8.117}$$

$$-\hbar^2 \ddot{\varphi}_- == 4S^2(J_0^2 - \xi_0^2)\varphi_- - 4S^2\xi_0^2 r_0^2 \frac{\partial^2}{\partial x^2}\varphi_- - 8(J_1 J_0 - \xi_1\xi_0)S^2 r_0 \frac{\partial u}{\partial x}\varphi_- \tag{8.118}$$

Adding Eq.(8.117) to Eq.(8.118) and substituting Eq. (8.113) into them we finally get

$$\varphi_{tt} - A_0\varphi_{xx} - B_0\varphi - C_0|\varphi|^2\varphi = 0 \tag{8.119}$$

where

$$A_0 = \frac{4\xi_0{}^2 S^2 r_0^2}{\hbar^2} > 0, \quad C_0 = \frac{16 J_1 S^3 r_0^2 (J_0 J_1 - \xi_0\xi_1)}{\hbar^2 (K' r_0^2 - Mv^2)}$$

$$B_0 = \frac{1}{\hbar^2}[4S^2(J_0^2 - \xi_0{}^2) - 8S^2 r_0 (J_0 J_1 - \xi_0\xi_1)C], \tag{8.120}$$

Very clearly, equation(8.119) is an ϕ^4 – equation, instead of the nonlinear Schrödinger equation, but it is still the same as the dynamic equation (2.67) in nonlinear quantum mechanics [10,22-25]. Obviously, this is due to interaction between the magnons or coupling effect between the double sublattices in the antiferromagnets, although the motion of the magnons in a single ferromagnetic chain is described by a nonlinear Schrödinger equation in Eq.(2.66), which also appears in a single ferromagnetic chain, as mentioned above. However, in the antiferromagnets the features of the magnons are still described in a soliton because ϕ^4 – equation is a representation or result of the nonlinear Schrödinger equation in the relativistic case, they are self-consistent each other [10,22-25]. Therefore, they possess some of the same features. In order to verify this point, we now find the soliton solutions of Eq.(8.119).

We can prove that for the antiferromagnet, magnetized along the z-direction, the magnon -phonon coupling in this direction plays an important part in forming the localized soliton. Thus, we proceed first from Eq.(8.119) by assuming [44-50]

$$\varphi = f(x - vt)\exp[ik'x - \omega t)] \tag{8.121}$$

Inserting this into Eq.(8.119) we have [42-48]

$$K' A_0 = v\omega, (v^2 - A_0)\frac{d^2 f}{d\varsigma^2} + (B_0 + A_0 K'^2 - \omega^2) f^3 = 0$$

It becomes after integration

$$(\frac{df}{d\varsigma})^2 = -\frac{C_0}{2(A_0 - v^2)} f^4 + \frac{B_0 A_0 + \omega^2 (A_0 - v_0)}{A_0 (A_0 - v^2)} f^2 = 0$$

where Pang has set [44-45]the integration constant equal to zero because of the boundary conditions. Integrating once more, we arrive at

$$x = \int_{f(0,0)}^{f(x,t0} [\frac{C_0 f^2}{2(A_0 - v^2)} (f_0^2 - f^2)]^{1/2} df + cons\tan t \tag{8.122}$$

where

$$f_0^2 = \frac{2}{A_0 C_0^2} [B_0 A_0 + \omega^2 (A_0 - v_0)], K' = v\omega / A_0$$

In the case of $-\omega^2 < (A_0 B_0)/(A_0 - v^2)$, there are non-topological bell-type soliton solutions in Eq. (8.119) if the requirements $C_0/(A_0 - v^2) > 0$ and $f_o^2 > 0$ are satisfied, i.e., if $J_0 > \xi_0$ and either (a) $J_0 J_1 > \xi_0 \xi_1, v < \min[\sqrt{K/mr_0}, 2\xi_0 S r_0 / \hbar]$, or (b) $J_0 J_1 < \xi_0 \xi_1$, $2\xi_0 S r_0 / \hbar < v < \sqrt{K/mr_0}$, then its normalized solitary wave solution is the form of

$$\varphi = \sqrt{\frac{r_0}{2W_s}} \sec h(\frac{x - \upsilon t}{W_s}) e^{i(k' x - \omega t)}, \tag{8.123}$$

where

$$k' = \frac{\upsilon\omega}{A_0}, \ W_s = \frac{4(A_0 - \upsilon^2)}{C_0 r_0}, \ \omega^2 = \frac{A_0 B_0}{A_0 - \upsilon^2} - \frac{d^2 A_0 C_0^2}{16(A_0 - \upsilon^2)^2},$$

here v is velocity of the soliton.

Equation (8.123) indicates clearly that the magnon in the antiferromagnets still move the bell-type soliton, and its outline and features are the same with those described by the nonlinear Schrödinger equation in Eq.(2.66) or Eq.(3.9). It is clear that the magnon is still a soliton in nonlinear antiferromagnetic systems, its essence and basic features have not be changed, even though it satisfies the ϕ^4 – equation in nonlinear quantum mechanics [10,22-25] because the nonlinear interactions in the antiferromagnets are still due to the magnon-phonon or magnon-magnon interactions. Therefore, the nonlinear quantum mechanical natures of the magnons in the antiferromagnets have not been changed.

However, if $v^2 - A_0 > 0$, i.e., $C_0/(A_0 - v^2) < 0$, equation (8.117) also has another topological soliton solution, which is denoted by

$$\varphi = \sqrt{\frac{r_0}{2W_s}}\tanh\left(\frac{x-\upsilon t}{W_s}\right)e^{i(k'x-\omega t)} \triangleright \tag{8.124}$$

At the same time, it can be seen from Eq.(8.120) that $C_0=0$ and $B_0=0$ because $J_0 = \xi_0 J_1 = \xi_1$ are the isotropic antiferromagnets. Thus, there is no soliton to Eq.(8.119), in this case.

It can be seen from the above conditions that the localized soliton can be excited by the nonlinear magnon-phonon coupling only for the easy magnetic axis antiferromagnet. This was not observed before. No parallel has been obtained in ferromagnets. In the case being discussed, the coupling of the longitudinal lattice oscillations with the magnons, which results in a nonlinear interaction, causes a remarkable change in the transverse exchange integral of the antiferromagnet. It is the nonlinear interaction, caused by the coupling, which is vital to the formation of the soliton. In this case, the velocity of the soliton satisfies $v < \min[\sqrt{K/mr_0}, 2\xi_0 S r_0/h]$.

8.6.2. Properties of Motion of Magnons under Action of the Magnon-Magnon Interaction

We have so far only considered the collective excitation caused by the magnon-phonon interaction. In fact, when the magnon-magnon interaction in the system becomes too strong to be neglected, a new nonlinear interaction source will contribute to the collective excitation in the antiferromagnetic systems. The formation process, and the properties of these collective excitations, will change accordingly if this interaction is taken into consideration. The Hamiltonian of the system in this case is still given by Eq. (8.97), but the interaction term now includes direct interactions between neighboring magnons and other two-magnon effects, i.e., influences of a magnon upon the transfer of other magnons and upon the magnon "resonance", etc.

Pang [40-46] employed the following quasi-average field approximation to treat the effects of the anomalous correlation terms in Eq. (8.97) upon the soliton formation, and the quasi-particle energy in the collective excitation, that is [45]

$$a_n^+ a_n a_n b_{n+\delta}^+ = \langle a_n^+ a_n \rangle a_n b_{n+\delta}^+ + \langle a_n b_{n+\delta}^+ \rangle a_n^+ a_n - \langle a_n^+ a_n \rangle \langle a_n b_{n+\delta}^+ \rangle,$$

$$a_n^+ a_n a_n^+ b_{n+\delta}^+ = \langle a_n^+ a_n \rangle a_n^+ a_{n+\delta}^+,$$

Then the Hamiltonian of the system Eq. (8.97) now becomes

$$H = E_0 + S\sum_n^A\sum_\delta^A J_{n,n+\delta} a_n^+ a_n + S\sum_j^B\sum_\delta^B J_{j,j+\delta} b_n^+ b_n$$
$$+\frac{1}{2}S\sum_n^A\sum_\delta^A(\xi_{n,n+\delta}-\eta_{n,n+\delta})(a_n b_{n+\delta}^+ + a_n^+ b_{n+\delta})$$
$$+\frac{1}{2}S\sum_n^A\sum_\delta^A(\xi_{n,n+\delta}-\eta_{n,n+\delta})(a_n b_{n+\delta} + a_n^+ b_{n+\delta}^+)$$
$$-\sum_n^A\sum_\delta^A J_{n,n+\delta}(\langle a_n^+ a_n\rangle b_{n+\delta}^+ b_{n+\delta} + \langle b_{n+\delta}^+ b_{n+\delta}\rangle a_n^+ a_n)$$
$$-\frac{1}{8}\sum_n^A\sum_\delta^A(\xi_{n,n+\delta}-\eta_{n,n+\delta})[(\langle a_n^+ a_n\rangle + \langle b_{n+\delta}^+ b_{n+\delta}\rangle)(a_n b_{n+\delta}^+ + a_n^+ b_{n+\delta}) +$$
$$(\langle a_n b_{n+\delta}^+\rangle + \langle a_n^+ b_{n+\delta}\rangle)\bullet$$
$$(a_n^+ a_n + b_{n+\delta}^+ b_{n+\delta})] \ -\frac{1}{8}\sum_n^A\sum_\delta^A(\xi_{n,n+\delta}+\eta_{n,n+\delta})[(\langle a_n^+ a_n\rangle + \langle b_{n+\delta}^+ b_{n+\delta}\rangle)(a_n b_{n+\delta} + a_n^+ b_{n+\delta}^+) \tag{8.125}$$

where

$$E_0 = T + V - J_0 N S^2 + \sum_n^A\sum_\delta^A J_{n,n+\delta}\langle a_n^+ a_n\rangle\langle b_{n+\delta}^+ b_{n+\delta}\rangle +$$
$$-\frac{1}{8}\sum_n^A\sum_\delta^A(\xi_{n,n+\delta}-\eta_{n,n+\delta})[(\langle a_n^+ a_n\rangle + \langle b_{n+\delta}^+ b_{n+\delta}\rangle)(a_n b_{n+\delta}^+ + a_n^+ b_{n+\delta}) \tag{8.126}$$

From the second-quantum Hamiltonian in Eq.(8.126) and in the same way as mentioned above, with the aid of Makhankov et al., [20-21] and Pang's method [38-44], we obtained the apparent equations of motion for the operator a_f and b_f in the Heisenberg representation,

$$i\hbar\frac{\partial a_f}{\partial t} = [a_f, H], i\hbar\frac{\partial b_f}{\partial t} = [b_f, H],$$

Then, using the Schrödinger probability amplitude defined by Eq.(8.100)

$$\varphi_{af} = \langle\varphi(t)|a_f|\varphi(t)\rangle = \alpha_{af}/\lambda^2, \varphi_{bf} = \langle\varphi(t)|b_f|\varphi(t)\rangle = \alpha_{bf}/\lambda^2$$

We can finally derive the nonlinear equation for $\varphi_{af}(t)$ and $\varphi_{bf}(t)$, the former is as follows

$$i\hbar\dot{\varphi}_{af} = S\sum_{\delta} J_{ff+\delta}\varphi_{af} + S/2\sum_{\delta}\left(\xi_{ff+\delta} - \eta_{ff+\delta}\right)\varphi_{bj+\delta}(t) + S/2\sum_{\delta}\left(\xi_{ff+\delta} + \eta_{ff+\delta}\right)\varphi^{*}_{bj+\delta}$$
$$-\sum_{\delta} J_{ff+\delta}\left|\varphi_{bf+\delta}\right|^{2}\varphi_{aj} - \frac{1}{8}\sum_{\delta}\left(\xi_{ff+\delta} - \eta_{ff+\delta}\right)[(\left|\varphi_{af}\right|^{2} + \left|\varphi_{bf+\delta}\right|^{2})\varphi_{bf+\delta} + \left|\varphi_{af}\right|^{2}\varphi^{*}_{bf+\delta} + \left|\varphi_{af}\right|^{2}\varphi_{bf+\delta}] \tag{8.127}$$
$$-\frac{1}{8}\sum_{\delta}\left(\xi_{ff+\delta} + \eta_{ff+\delta}\right)(\left|\varphi_{af}\right|^{2} + \left|\varphi_{bf+\delta}\right|^{2})\varphi^{*}_{bf+\delta}$$

Interchanging the symbols a and b in Eq.(8.125), we can get the corresponding equation for φ_{bf}

We then proceed to the continuum limit as before, dropping the terms with higher than third derivatives and taking into account the symmetry of sublattices A and B and the distribution characteristics, mentioned above, to approximate $\varphi_f(t) = \varphi_{af}(t) + \varphi_{bf}(t)$ as

$$i\hbar\dot{\varphi} \approx 2J_0 S\varphi + (\xi_0 - \eta_0)S\varphi^{*} + S(\xi_0 - \eta_0)\varphi + \frac{1}{2}Sr_0^2(\xi_0 + \eta_0)\frac{\partial^2}{\partial x^2}\varphi^{*} + \frac{1}{2}Sr_0^2(\xi_0 - \eta_0)\frac{\partial^2}{\partial x^2}\varphi$$
$$-(2J_1 + \xi_1 - \eta_1)Sr_0\frac{\partial u}{\partial x}\varphi - (\xi_1 - \eta_1)Sr_0\frac{\partial u}{\partial x}\varphi^{*} - [2J_1 + (\xi_0 - \eta_0)/8]\left|\varphi\right|^2\varphi - \nu(\xi_0 + \eta_0)\left|\varphi\right|^2\varphi^{*} \tag{8.128}$$

where the subscript of the function in Eq.(8.126) has been dropped and a parameter ν has been introduced which equals 1/16 in the case studied. Equation (8.128) differs from Eq.(8.107) only if additional terms and dispersion effects are neglected, there will be no correlation between magnon-magnon interactions, and magnon-phonon interactions, and they become two independent nonlinear interaction sources. Meanwhile, the equation of the lattice oscillations is still expressed by Eq. (8.109). Therefore, we can solve problems of this type by combining Eq. (8.113)with Eq. (8.128).

It can be proved that there is still no soliton solution for isotropic antiferromagnets even if the interactions between magnons are taken into account. In the case of isotropy, Eq.(8.128) becomes

$$i\hbar\dot{\varphi} \approx 2J_0 S(\varphi + \varphi^{*}) + J_0 Sr_0^2\frac{\partial^2}{\partial x^2}\varphi^{*} - 2J_1 Sr_0\frac{\partial u}{\partial x}(\varphi + \varphi^{*}) - 2J_0\left|\varphi\right|^2(\varphi - \nu\varphi^{*})$$

Its conjugation equation reads

$$-i\hbar\dot{\varphi}^{*} = 2J_0 S(\varphi + \varphi^{*}) + J_0 Sr_0^2\frac{\partial^2}{\partial x^2}\varphi - 2J_1 Sr_0\frac{\partial u}{\partial x}(\varphi + \varphi^{*}) - 2J_0\left|\varphi\right|^2(\varphi^{*} + \nu\varphi)$$

Introducing $\varphi_{\pm}(t) = \varphi(t) \pm \varphi^{*}(t)$ and applied in the same way as mentioned above we obtain

$$\hbar^2 \ddot{\varphi} = -4J_0 S^2 r_0^2 \frac{\partial^2}{\partial x^2}\varphi - 8J_{01}^2 S(1-\nu)|\varphi|^2 \varphi \qquad (8.129)$$

Comparing this with Eq.(8.119), it can be seen that now B_0=0 and only nonlinear terms exist, and therefore soliton solutions similar to Eq.(8.119) cannot be found for this equation [41-45,48-49], thereby completing the proof. It can then be concluded that in the case of isotropic Heisenberg antiferromagnetic chains, neither magnon-phonon interactions nor interactions between magnons can not result in a collective excitation of the type of non-topological solitons, which is somehow similar to the situation in isotropic ferromagnetic chains [1,33].

For anisotropic antiferromagnets, we also limit our discussion to the cases of easy magnetic axis(OZ) and easy magnetic plane(xOy) where Eq.(8.128) reduces to

$$i\hbar\varphi \approx 2J_0 S\varphi + 2S\xi_0\varphi^* + \xi_0 S r_0^2 \frac{\partial^2}{\partial x^2}\varphi^* - 2J_1 S r_0 \frac{\partial u}{\partial x}\varphi - 2S\xi_0 r_0 \frac{\partial u}{\partial x}\varphi^* - 2J_0|\varphi|^2\varphi - \nu 2\xi_0|\varphi|^2\varphi^* \qquad (8.130)$$

Again introducing $\varphi_\pm(t) = \varphi(t) \pm \varphi^*(t)$, and applying the same method used in Eqs.(8.115) – (8.117), we arrive at the equation for $\varphi(x,t)$

$$-\hbar^2 \ddot{\varphi} = 4(J_0^2 - \xi_0^2)S^2\varphi - 4\xi_0^2 S^2 r_0^2 \frac{\partial^2}{\partial x^2}\varphi - 8(J_0 J_1 - \xi_0\xi_1)S r_0 \frac{\partial u}{\partial x}\varphi - 8(J_0^3 - \nu^2\xi_0^2)S|\varphi|^2\varphi \qquad (8.131)$$

Similar to the derivations of Eqs.(8.98)-(8.100) we can obtain

$$\varphi_{tt} - A\varphi_{xx} + B\varphi - g'|\varphi|^2\varphi = 0 \qquad (8.132)$$

where

$$\begin{aligned} &A = A_0 = 4\xi_0 S^2 r_0^2/\hbar^2, B = B_0 = [4S^2(J_0^2 - \xi_0^2)/\hbar^2] + 8(J_0 J_1 - \xi_0\xi_1)S r_0 C_0 \\ &g' = C_0 + 8S(J_0^2 - \nu^2\xi_0^2)/\hbar^2 \end{aligned} \qquad (8.133)$$

Therefore, equation (8.132) has the same soliton solution as Eq.(8.119) or Eq.(8.123). Simply replacing C_0, A_0, and B_0 in Eqs. (8.119) and (8.123 by g', A, and B, respectively, we get the solutions. Therefore, taking into account magnon-magnon interaction only changes the amplitude and velocity of the soliton and does not alter the fundamental nature of the magnon-solitons. The magnon-magnon interactions enhance the effects of the nonlinear interactions, thereby prompting formation of more stable solitons. This is becauseg is always greater than C_0 if the magnon-solitons can exist. Furthermore, $J_0 >> J_1$ and $\xi_0 >> \xi_1$. We can see that in the presence of magnon-magnon interactions, they are only the soliton solutions of the type in Eqs. (8.119) or (8.123), in the velocity range of $v^2 < A$. In this case

$$\omega^2 = \frac{AB}{A-\upsilon^2} - \frac{r_0^2 Ag'^2}{16(A-\upsilon^2)^2} = \frac{AB}{A-\upsilon^2} - \frac{r_0^2}{16}\frac{A}{A-\upsilon^2}[g'+8S(J_0{}^2-\upsilon^2\xi_0^2)]^2 \quad (8.134)$$

It was found that the nonlinear excitations of magnons could still be described by Eq.(8.119) or Eq.(8.132). The only differences are the coefficients in these equations. Therefore, the magnon-solitons in nonlinear antiferromagnetic systems obey the laws of nonlinear quantum mechanics.

The above investigations indicated clearly that the properties of the magnons could be described by nonlinear quantum mechanics [10,22-25], in which the magnons become solitons under action of the nonlinear interactions. In this case, the magnons have both a wave feature, a solitary wave, or the spin wave, and the corpuscle feature or a localized particle. The nonlinear interactions are generated by virtue of theinteraction between the moved magnon and magnon or between movedmagnons and phonons. These interactions are always existent in the antiferromagnets, therefore, the natures of wave-corpuscle feature of the magnons cannot be changed, i.e., the wave-corpuscle feature of the magnons is its inherent nature, and then nonlinear quantum mechanics is a correct theory to describe the properties of the magnons in antiferromagnets. If theinteraction between the magnon and magnon and between themagnons and phonons are not considered, then the magnon has only a wave feature and is a spin wave; its properties were described in Section 8.4.

In practice, the above nonlinear interactions are always existent in all antiferromagnets, but their mechanism of generation of the magnon-phononcoupling and magnon-magnon interactionsaredifferent, although their effects for the generation of the nonlinear interaction and the formation of solitonsare the same [10,22-25]. In the first place, the mechanism of localized nonlinear collective excitation(magnon-soliton), caused by the first type, is the breaking of kinetic symmetry, that is, it is caused by the interaction between the magnon and lattice oscillation. In a steady state, the magnon-soliton and the localized deformation, which depend on the lattice oscillation, propagate together with the same speed along the antiferromagnetic chain. The mechanism of excitation caused by the second type is the spontaneous breaking of symmetry brought about by the magnon-magnon interactions in the single-axis anisotropic antiferromagnet. However, the collective excitations resulting from both mechanisms have the same characteristics, i.e., the structural anisotropy. As it was already mentioned earlier, once the two mechanisms cancel each other in isotropic antiferromagnetic chains, no soliton can exist, as in the case of the ferromagnetic chains. This also indicates that soliton excitation of magnons in such systems is determined by the anisotropy of the system. As long as the anisotropy exists in a given system, the magnon-phonon coupling and nonlinear interactions between the magnons will make the magnons "self-trapping" like a soliton in a range of dimension $2W_s$in the one-dimensional chains, and a stable magnon-soliton will propagate along the antiferrmagnetic chains. When the anisotropy changes, the amplitude, the momentum, and the number of the solitons all change accordingly.

On the other hand, the formation of magnon-solitons, due to nonlinear interactions in anisotropic antiferromagnetic chains, leads to many interesting physical phenomena. Indeed, many anomalies have been observed in experiments. Attempts have been made to explain them using the magnetic soliton model, even though analytic expressions in place of Eqs.(8.101)-(8.107) have not been obtained. (The readers are referred to work by Mikeska and Steiner [1,33]. For example, Boucher et al. [51,52], used the soliton concept to explain the phenomenon of nuclear spin-lattice relaxation (NSLR) in antiferromagnetic chain $(CH3)_4NMnCl_3$, even though theoretical expression for magnetic solitonshas not been obtained. Through measurement, Boucher et al. obtained the ratio T_1^{-1} of NSLR of N^{15} in the antiferromagnet, as a function of the external field $\overline{H}(2kAm^{-1} < \overline{H} < 80kAm^{-1})$ and temperature $T(2K \le T \le 4.2K)$, and observed that T_1^{-1} diverged exponentially with $\overline{H}/T$ at a certain temperature. With the analytic results for the soliton given above, we can explain the excitation and the behavior of the soliton in such systems, which in turn validate the correctness of the above theory.

We now find out the specific heat of the antiferromagnets using the above soliton theory, in which the temperature effect of the system should be considered. The inner energy of the magnon excited in the antiferromagnetic systems can be represented [8,37] by

$$U(T) = \sum_k < a_k^+ a_k >_T \hbar\omega_k + \sum_k < b_k^+ b_k >_T \hbar\omega_k \approx 2\sum_k \hbar\omega_k / [\exp(\hbar\omega_k / K_B T)] \quad (8.135)$$

In the cubic crystals we can represent approximately the frequency of the magnon, which is directly proportional to its wavevector k, thus its dispersion relation can be approximately denoted by $\hbar\omega_k \approx 8\hbar S(J_0^2 - \nu\xi_o^2)kr_0$ from Eq.(8.134) in long wave approximation of $kr_0 << 1$, which is the same as that of the acoustical phonon in the same systems. If we think the antiferromagnet is a bcc crystal, which is composed of mutually penetrated two sc sublattices, in which the side length of each sublattice is denoted by a,then $\hbar\omega_k >> K_B T$ the following relation [2, 8,37]can be obtained

$$U(T) = \frac{2Nr_0^3}{8\pi^3}.4\pi\int_0^\infty \frac{\hbar\omega_k k^2 dk}{[\exp(\hbar\omega_k / K_B T) - 1]} = \frac{2N(K_B T)^4}{2\pi^2[8S\hbar(J_0^2 - \nu\xi_0^2)]^3}\int_0^\infty x^3[e^{-x} + e^{-2x} + ...]dx \quad (8.136)$$

Thus we obtain the specific heat is the form of

$$C_m = \frac{\partial U(T)}{\partial T} = \frac{13.71K_B}{[48S\hbar(J_0^2 - \nu\xi_0^2)]^3}(K_B T)^3 \propto T^3 \quad (8.137)$$

This result is consistent with the experimental datum.

It should also be pointed out that effects of external fields are not included in the above discussion. If any such field is present and in the direction along the easy magnetic axis, then its effect, due to the opposite magnetization directions of the two sublattices, will be equivalent to a periodic external field of period $2r_0$, which will strengthen the discreteness of the lattice and thus invalidates the continuum approximation. However, if the direction of the external field is perpendicular to the antiferromagnetic spin direction, then the continuum approximation will still be valid. For this reason, earlier experimental and theoretical studies were concentrated mainly on transverse fields, rather than longitudinal fields.

8.6.3. The Nonlinear Properties of the Magnons in Different Antiferromagnetic Systems

Xu and Pang [53] further studied nonlinear excitations and properties of the magnons in antiferromagnetic molecular crystal, such as $Ni(C_2H_8N_2)_2{-}NO_2(ClO_4)$ (NENP) and $Ni(C_3H_{10}N_2)_2NO_2(ClO_4)$ (NINO), using the above model and method. Their molecular structure is shown in Figure 8.5, where NENP's space group is $Pn2_1$ a, but N INO's belongs to $Pbn2_1$.The crystal structureconstants are: a = 15. 223 nm, b = 8. 295 nm, c = 10. 300 nm for the NENP and a = 15. 384 nm, b = 8. 507 nm, c = 10. 590 nm for N INO, respectively. In the two molecular crystals, their molecular structures are all composed of Ni^{2+} ionic chains, which are arranged along the c axis, in which the distance between two Ni^{2+} ions is about 5.15nm in each chain, but between two Ni^{2+} ions the two chains it is about 8.295nm, as shown in Figure 8.5, where there are many nonmagnetic perchlorate anions between the two chains, the Ni^{2+} ions are related with both a nitrogen atom and an oxygen atom by means of covalent bonds formed by the nitrite base in the same chain.

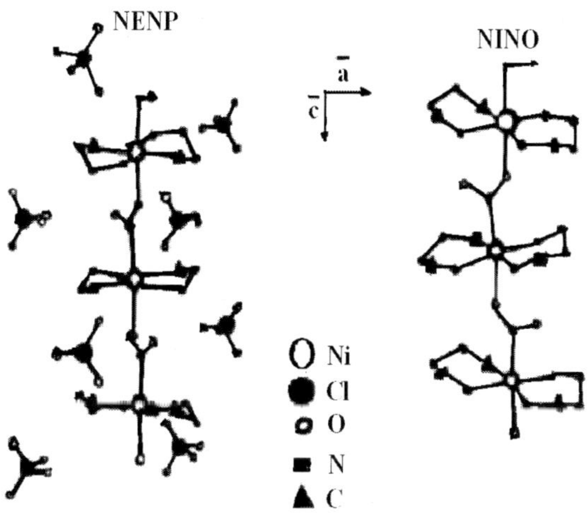

Figure 8.5. Molecular structures of NENP and NINO crystals [53].

The localized crystal structure of Ni^{2+} ions is a distorted octahedral, its fundamental surface is perpendicular to c axis and has the space group $Pn2_1a$ structure, there is also an inversion center in the point near Ni^{2+}. This structure makes Ni^{2+} have a high spin state with S = 1. Meanwhile, such octahedron configuration also makes Ni^{2+} ions form two magnetic tracks, one has d_{xy} symmetry in the fundamental surface, another has a d_Z^2 symmetry along the axis of the chain. There is also the strong superposition of the magnetic tracks of d_Z^2 due to the influence of nitrite bases along the chain axis. These properties of structures predicted in this structure: the Ni^{2+} ions in one chain forms approximate antiferromagnetic chains through antimagnetic interactions because of very weak interaction. In this case, Renard et al., [58-59] obtained the rate of the interactions between the chains relative to that in one chain is about 4×10^{-4}. Therefore, NENP and NIPO are a well anisotropic antiferromagnetic-chain with the Heisenberg model. Thus, we [53]gave its Hamiltonian, which is the form of

$$H = T + V + 2\sum_{nm} J_{nm} S_n^z S_m^z + 2\sum_{nm} \xi m (S_n^x S_m^x + S_n^y S_m^y) + 2D\sum_n (S_n^z)^2 + 2E\sum_n [(S_n^x)^2 - (S_n^y)^2]$$

where D and E are the anisotropic energies of single ions, and J_{nm} is the interaction energy of spin ions in one chain.

We also studied [54] the properties of the magnons in the antiferromagnetic molecular crystals with order parameter conservation, such as magnetic compound CeAs, which were first investigated by Bose [60-62].In our study the Hamiltonian of this system of interaction is represented by

$$H_{Bose} = \sum_{nm} \{ J_{nm} S_n^z S_m^z - \frac{1}{2}\sum_{nm} \Delta_{nm} (S_n^+ S_m^+ + S_n^- S_m^-) - h\sum_n S_n^z$$

For an antiferromagnetics with order-parameter conservation (OPCAFM), we gave its Hamiltonian of a nonlinear collective excitation, which contains the magnon-phonon interaction, by

$$H = T + V + \sum_{nm} J_{nm} S_n^z S_m^z - \frac{1}{2}\sum_{nm} (S_n^+ S_m^+ + S_n^- S_m^-) - h\sum_n S_n^z + h\sum_m S_m^z$$

On the other hand, we [55-57] also researched the properties of the magnons and its rules of motion in the one-dimensional Heisenberg antiferromagnet, such as $CsNiF_3$. CsNiF3 has a hexagonal crystal structure [1], its lattice constant is a = b= 62.1nm, and c =52nm, in this structure, the public face of the octahedron is composed of NF_6, which are homogeneously arranged along the c axis, and the chains are separated by the Cs ions. Thus, the coupling interaction between the Ni^{2+}s along the c axis possess the ferromagnetic feature, but there are antiferromagnetic features along the a and b chains [63-65]. Therefore, we may think that CsNiF3 is a ferromagnet with two chains and the chain-chain interaction. In such a case, its Hamiltonian is represented by

$$H = -J_1 \sum_{n\delta} S_{An}^z S_{An+\delta}^z - J_1 \sum_{n\delta} S_{Bn} S_{Bn+\delta} + J_2 \sum_n S_{An} S_{Bn} + D\sum_n (S_{An}^z)^2 + D\sum_n (B_n)^2 \ .$$

We used the above methods to obtain the dynamic equations of these magnons from the above Hamiltonians for these antiferromagnetic systems, respectively. These results indicated these dynamic equations are all similar to equation (8.119) or (8.132), in which their differences are only distinctions of the parameters and coefficients. This indicated clearly that the nonlinear interactions in these systems are still due to the interactions between the moved magnons and phonons, or between the moved magnon and mognon. Therefore, the properties and moved rules of the magnons also have the wave-corpuscle duality, thus they can still be described by the nonlinear quantum mechanics.

REFERENCES

[1] H. J. Mikeska and M. Steiner, *Advances in Physics*, 40(1991)191.
[2] F. Bloch, *Z .Phys*. 74(1932)295.
[3] U. Enz, *Helv.Phys.Acta*, 37(1964)245.
[4] W. Heisenberg, *Z. Phys*. 49(1928) 619.
[5] P. A. M. Dirac, *Proc. Roy.Soc*.A 123(1963)99.
[6] P. W. Anderson, *Solid state physics*, 14(1963099.
[7] C. Kittel, *Quantum theory of solid*, Chapter 4, John Wiley and Sons, Inc.,New York, 1963.
[8] Z. Z. Lee, *Solid state physics*, Advanced education Press (Chinese), Beijing, 1987.
[9] T. Holstein and H. Primakoff, *phys.Rev*.58(1940)1098.
[10] Pang Xiao feng, *quantum mechanics in nonlinear systems*, World Scientic Publishing Co. Singapore, 2005.
[11] N. N. Bogoliubov, *J.phys*.USSR, 11(1947)23.
[12] D. Pushkarov and Kh. Pushkarov, *Phys.Stat.So1*.(b)81(1979)703.
[13] D. Pushkarov and Kh. Pushkarov, *Phys.Stat.So1*.(b) 90(1978)36.
[14] D. Pushkarov and Kh. Pushkarov, *Phys.Stat.So1.(*b) 93(1979)735.
[15] D. Pushkarov and Kh. Pushkarov, *Phys.Stat.So1.(*b) 123(1984)573.
[16] D. Pushkarov and Kh. Pushkarov, *J.Phys. C* 10(1977)3711.
[17] A. S. Davydov and N. Kislukha, *Sov.phys.JETP* 44(1976)571.
[18] A. S. Davydov and N. Kislukha, *Phys.Sat.Sol*.(b)59(1973)465
[19] V. Makhankov, N.Makhaidiani and O.Pashaev, *Phys.Lett*.81A(198)167.
[20] V. G. Makhankov, *Phys.Lett*. A81(1981)158.
[21] V. G. Makhandov, and V.K.Fedyanin, *Phys.Rep*. 104(1984)1.
[22] Pang Xiao feng, *Nonlinear quantum mechanical theory*, Chongqing Press, Chongqing, 1994.
[23] Pang Xiao-feng, The behaviours and properties of microscopic particles in nonlinear systems, in *Handbook of solitons*, Research, technology and applications, eds S.P.Lang and S.H.Bedore, Nova Science Publishers, Inc. New York, 2000. Pp185-249.
[24] Pang Xiao feng, Establishment of nonlinear quantum mechanics and changes of property of microscopic particles as well as their experimental evidences in Contemporary Research in quantum systems, eds. Zoheir Wzziane, Nove Publishers, New York 2014, pp207-338.
[25] Pang Xiao Feng, *International Journal of New Innovation in Science and Technology*, 1(2013)25-78.
[26] V. Fedyanin, L. Yakushevich, *Phys.Lett*. A 85 (1981)100.
[27] V. K. Fedyanin and V. G. Makhaankov, *Phys.Scr*. 20(1979)552.
[28] V. I. Karpman and E. M. Maslov, *Sov.Phys.JETP*,46(1977)281.
[29] V. I. Karpman and E. M. Maslov, *Sov.Phys.JETP*, 48(1978)252.
[30] H. M. Ronnow, M. Euderle, D. F.Mcmorrow, et.al. *Phys.Rev.Lett*. 84(2000)4468.
[31] M. Enderle, H. M. Ronnow, D. F. Mcmerrow, et.al, *Phys.Rev.Lett*. 87(2001)177203.
[32] Y. F. Revarat, M. Horvatic, C. Berther, et.al, *Phys.Rev.Lett* 77(1996)1861.
[33] H. J. Mikeska *J.Phys.,* C11(1979)l28.
[34] D. I. Pushkarov and Kh. Pushkarov, *Phys. Lett*. A61(1977)339.

[35] R. J. Glauber, *Phys.Rev.* 131(1963)276.
[36] O. W. Dietrich et al.,*Phys. Rev*.B 14(1976)4932.
[37] F. Boch, *Z.Phys.* 61(1930)206.
[38] C. E. Cordeiro et al., *Phys.Lett.*A 66(1975)332.
[39] V. P. Yatsishin, *Theoretical and Mathematical Physics*,42(1980)187.
[40] Pang Xiao-feng, *J. Phys. Condensed matter* 2(1990) 9541.
[41] Pang Xiao feng, *Chin. J. Atom. Mol. Phys.* 5(1987)383.
[42] Pang Xiao feng, *Chin. Phys.* 22(1993)282.
[43] Pang Xiao feng, *J. Semicond. Sin.* 15(1992)917.
[44] Pang Xiao feng, *Chin. J. Light. Scattering*,4(1992)126.
[45] Pang Xiao-feng, *Phys.Stat.Sol.* (b) 189(1993)237.
[46] Pang Xiao-feng, *Acta.Math.phys.Sin.*10(1990)47.
[47] D. Pushkarov and Kh. Pushkarov, *Teor Mat. Fiz.* 30 (1977) 1.
[48] A. S. Davydov, *the soliton in molecular systems*, Dordrecht, Reidel, 1985.
[49] Guo, Bai-lin&Pang, Xiao-feng, solitons, Chinese Science Press, Beijing, 1987.
[50] Pang, Xiao-feng, Soliton physics, Sichuan Sci. and Tech. Press Sin., Chengdu, 2003.
[51] J. P. Boucher, R. Pynn, M. Remoissewet, L. P.Regnault, Y. Endo and J.P.Renard, Phys.Rev.Lett. 64(1990)1557.
[52] J. P. Boucher, R. Pynn, M. Remoissewet, L. P. Regnault, Y. Endo and J. P. Renard, *Phys. Rev.Lett.* 65(1980)488.
[53] Xu, Chang-tan and Pang Xiao-feng, *Chin.J.Atom.Mol.Phys*.13(1996)508.
[54] Xu, Chang-tan and Pang Xiao-feng, *Chin.J.Atom.Mol.Phys.* 13(1993)219.
[55] Xu, Chang-tan and Pang Xiao-feng, *Chin.J.Atom.Mol.Phys*. 15(1999)506.
[56] Xu, Chang-tan and Pang Xiao-feng, *Chia.J.Atom.Mol.Phys.* 16(2000)99.
[57] Xu, Chang-tan and Pang Xiao-feng, *Chia.J.Atom.Mol.Phys*. 16(2000)513.
[58] J. P. Renard, M. Verdaguer, L. P. Regnault, W. A. C. Erkelens, J. Rossat-Mignod, J. Ribas, W. G. Stirling, C. Vettier, *J.Appl. phys.* 63(1988)3538.
[59] M. Date and K. Kindo, *Phys.Rev. Lett*. 65(1990)1659.
[60] I. Bose, *Physics B* 174(1991)66.
[61] I. Bose and S.Chatterjee, *Phys.Rev*. B29(1984)2741.
[62] I. Bose, *J.Phys. C* 21(1988) L841.
[63] J. K. Ki ems and M. Steiner, *Phys. Rev. Lett*., 41(1978) 1137.
[64] J .M. Loveluk, T. Sc hneider, E. Stoll, et al., *Phys. Rev. Lett.*,54(1980)150 5.
[65] G. Reiter, *Phys. Rev. Lett*, 46 (1981)2 02.

INDEX

A

Abrams, 50
absorption spectra, 233, 261, 270, 271, 277, 280
absorption spectroscopy, 274
accelerator, 5, 9
Acetanilide, vii, 242, 245, 252, 255, 260, 264, 266, 270, 273, 274
acid, 218, 219
alanine, 394
algorithm, 132
amino, 218, 219, 280
amino acid, 218, 219, 280
anisotropy, x, 165, 214, 409, 425, 449
annealing, 234
annihilation, 66, 140, 141, 181, 189, 190, 219, 226, 238, 240, 248, 250, 251, 264, 292, 298, 299, 323, 345, 350, 351, 353, 381, 383, 385, 393, 394, 395, 415, 416, 420, 437
ansatz, 143, 306, 351, 352, 353, 354, 358, 360, 362, 365
antiferromagnet, 301, 419, 421, 436, 442, 443, 445, 449, 450, 452
antiferromagnetic systems, x, 407, 419, 422, 436, 444, 445, 449, 450, 453
approximate method, 23, 199
approximate theory, 17, 23, 24, 73, 211, 212, 433
argon, 270
ASI, 85
asymmetry, 52, 254
atomic theory, 303
atoms, ix, x, xi, 1, 7, 23, 43, 53, 64, 69, 75, 165, 198, 211, 218, 219, 225, 234, 284, 285, 287, 301, 302, 303, 304, 308, 309, 310, 313, 314, 316, 319, 323, 324, 329, 330, 331, 334, 341, 342, 343, 344, 345, 351, 398, 422
ATP, 218
authority, 19, 66

B

bandwidth, 270, 275, 277, 351, 353, 389
barriers, 380
base, 59, 273, 426, 451
BCS theory, 292, 299, 366, 381, 382
beams, 1, 38, 55, 63, 157, 273, 311, 313, 323
BEC states, 308
behaviors, 53, 123, 137, 148, 204, 324, 374
Beijing, 25, 27, 87, 160, 161, 229, 230, 281, 326, 328, 401, 405, 453, 454
Bell inequality, 19, 21, 46, 47
Biatynicki-Birula and Mycielski, 58
binding energy(s), 45, 51, 68, 98, 188, 223, 224, 249, 250, 310, 338, 340, 351, 353, 358, 360, 365, 393, 400
biological systems, ix
biomolecules, 69, 191, 211
biopolymers, 269
bleaching, 233, 273, 274
Bogoliubov, 287, 289, 290, 300, 306, 325, 328, 421, 453
Bogoliubov approximation, 306
Bogoliubov method, 289, 290
Bohm, 21, 24, 27, 30, 31, 58, 83, 84, 401
Bohr, ix, xi, 1, 9, 10, 13, 17, 18, 19, 20, 21, 22, 25, 333
Bollinger, 32, 45, 53, 68, 83, 84, 86
bonding, 383
bonds, 221, 224
Born, 1, 10, 14, 17, 19, 23, 25, 60, 63, 199, 257, 402
Born-Oppenheimer approximation, 23, 199
boson(s), vii, ix, x, 142, 143, 144, 146, 167, 190, 215, 283, 284, 285, 286, 287, 288, 289, 292, 293, 294, 297, 299, 300, 301, 302, 303, 304, 305, 306, 308, 309, 311, 313, 314, 318, 319, 323, 324, 325, 326, 328, 374, 383
Boson feature, 294, 304

Boson gas, 300, 301, 306, 325
Boson particle, ix, 283, 286, 289, 294, 305, 308, 383
Boson-Einstein condensation, 283, 288, 289, 294, 300, 304, 326
bounds, 31
branching, 50
breakdown, 283, 295, 297, 298, 299, 372
Brody, 86
Broglie, v, 1, 2, 8, 10, 18, 25, 29, 30, 31, 83, 93, 111, 136, 139, 183, 330, 337, 339, 341, 342
Broglie relationship, 2, 8, 10, 330
Bugajski, 58, 59, 85
Bullough, 288, 324, 326, 328

C

calculus, 68
capillary, 301
carbon, 223, 234
Castro, 68, 71
Cauchy problem, 61, 85
causality, 17, 19, 62, 65, 68
chain deformation, 242
chaos, 53
charge density, 344, 362
chemical, 218, 286, 301, 305, 306, 345
Chicago, 25, 401
China, xii, 87, 160, 229, 402, 403, 404
circulation, 310
classes, 287, 300
classical electrodynamics, 64
classical logic, 69
classical mechanics, 10, 66, 67, 110, 120, 174, 186
classification, 55
closure, 59, 61, 187
clusters, 288
coherence, 183, 306
collective dynamics, 320
collision(s), x, 4, 6, 8, 89, 100, 121, 123, 124, 125, 126, 128, 129, 130, 131, 132, 133, 134, 135, 136, 138, 154, 157, 186, 303
color, 5, 6, 155, 156, 262
combined effect, 311
communication, 48, 49, 50, 51, 70
compatibility, 54, 80, 81, 192
competition, 361, 409
compilation, 265
complementarity, 68
complexity, 22, 129, 204
composition, 59
compounds, 374
concordance, 308
condensation, x, 99, 165, 215, 283, 284, 287, 288, 289, 292, 294, 297, 299, 300, 301, 303, 304, 308, 309, 319, 326, 374
conduction, 390
conductivity, 370, 383, 392
conductor(s), 216, 375, 377, 380
conference, 85
configuration, 100, 285, 409, 452
confinement, 301
conflict, 62, 65
conjugation, 447
conservation, x, 8, 19, 50, 56, 66, 67, 70, 81, 89, 101, 115, 116, 118, 119, 120, 121, 130, 143, 148, 175, 452
construction, 51, 53, 55, 71, 144
contradiction, 13, 22, 62, 92, 330, 338, 342
controversial, xi
convergence, 60
cooling, 235, 273, 301, 303, 304
coordination, 399, 414
corpuscle feature, x, 1, 7, 8, 9, 10, 11, 13, 14, 15, 16, 30, 89, 91, 92, 97, 98, 105, 116, 121, 123, 124, 126, 131, 136, 147, 180, 185, 330, 339, 342, 431, 449
correlated ansatz, 352, 358
correlation(s), 21, 46, 47, 48, 49, 62, 67, 146, 184, 226, 314, 325, 352, 353, 354, 355, 356, 357, 360, 361, 365, 445, 447
correlation function, 67, 325
cosmic rays, 68
Coulomb interaction, 12, 330, 331, 332, 333, 334
Coulomb potential, 13, 330, 331, 332, 337, 338, 339, 341, 342, 343, 411
coupling constants, 71, 83, 226
covalent bond, 223, 224, 451
covering, 86
critical value, 151
criticism, 19, 22
Croatia, 230, 403
crystal structure, 234, 235, 451, 452
crystalline, 234, 266, 273, 276, 279, 300
crystallization, 300
crystals, ix, x, 1, 191, 218, 224, 231, 233, 236, 255, 261, 268, 279, 323, 329, 350, 382, 387, 394, 404, 422, 450, 451, 452
Czachor, 50, 67, 69, 70, 84, 86, 87

D

damping, 15, 51, 74
Davies, 58, 84, 85, 328, 404
Davisson and Gerner, 2
De Broglie wavelengths, 2

de Broglie's dual solution theory, 30
de Broglie's nonlinear wave theory, 30
decay, 51, 52, 57, 65, 196, 197, 208, 273, 302, 325
decoupling, 346
deduction, 157, 233
defects, 204, 224, 276, 392, 393
deficit, 306
deformation, 122, 159, 213, 219, 226, 236, 237, 238, 247, 249, 250, 254, 270, 344, 353, 359, 362, 365, 385, 388, 389, 391, 393, 399, 422, 425, 426, 449
degenerate, 195, 304, 377, 418, 419
Del Seta, 69, 86
density fluctuations, 355, 358
density functional theory, 60, 313, 315
density matrices, 55, 70
deposition, 273
depression, 223, 283, 294
depth, ix, 105, 131, 149, 151, 153, 154, 324, 344, 366, 368
derivatives, 101, 217, 218, 280, 387, 439, 447
detection, 303
detuning of resonant transition, 32, 44
deviation, 145, 315, 408, 422, 426
dielectric constant, 235
differential equations, 69, 77, 207, 307
diffraction, 1, 2, 3, 4, 19, 25, 29, 38, 63, 136, 186, 323, 375, 376, 393
diffraction of electrons, 1, 2, 3
diffusion, 98, 338, 392
dilute gas, 326
dimerization, 344, 352, 361, 365
diodes, 156
Dirac, 10, 24, 25, 31, 63, 65, 71, 86, 304, 401, 453
Dirac equation, 71
discontinuity, 129, 130
discrete variable, 32, 33
discreteness, 451
dispersion, 15, 92, 128, 158, 159, 189, 210, 213, 214, 216, 217, 236, 242, 246, 266, 287, 307, 309, 385, 391, 393, 394, 416, 422, 447, 450
displacement, 122, 129, 130, 221, 225, 237, 242, 261, 296, 331, 332, 338, 350, 351, 352, 353, 355, 365, 384, 398, 399, 434, 441
dissociation, 5
distortions, 236, 394
distribution, 3, 4, 7, 21, 30, 43, 64, 65, 66, 68, 72, 93, 146, 215, 216, 249, 286, 288, 302, 316, 324, 342, 407, 432, 435, 447
distribution function, 288
divergence, 211
diversification, 167
DNA, 223
dualism, 26
dynamical properties, 313

E

early universe, 68
editors, 85, 86
education, 453
eigenenergy, 188, 189, 191, 233, 264, 271, 330, 337, 339, 340, 341, 342, 343, 399, 414
eigenvalue problem, 43, 100, 187, 193
Einstein, vii, ix, x, xi, 7, 18, 19, 20, 21, 22, 26, 27, 30, 31, 48, 57, 65, 84, 99, 215, 226, 237, 283, 284, 285, 287, 288, 289, 294, 297, 300, 301, 302, 303, 304, 305, 306, 312, 313, 314, 318, 319, 324, 325, 326, 328, 370, 374, 398
election, 344
electric conductivity, 224
electric current, 378
electric field, 12, 42, 113, 114, 158, 159, 160, 323
electrical conductivity, 224
electromagnetic, 43, 52, 57, 60, 65, 215, 218, 324, 367, 371, 372, 373, 374, 375
electromagnetic fields, 375
electron diffraction, 93, 337, 341, 342
electron pairs, 375, 381
electron-phonon coupling, x, 350, 352, 356, 357
electrons, ix, 1, 2, 3, 4, 6, 8, 9, 22, 23, 25, 60, 75, 216, 218, 230, 324, 329, 336, 341, 342, 343, 344, 345, 353, 355, 358, 359, 362, 366, 367, 368, 370, 371, 372, 374, 375, 376, 377, 383, 392, 398, 400, 409, 411
elongation, 426
emission, 4, 43, 51, 52, 57, 65, 274, 277
endothermic, 300
Energy, vii, 68, 115, 161, 264, 339
energy conservation, 8, 119
energy density, 68, 305
energy transfer, 393
entropy, 47, 48, 66, 300
environment, 31, 214
EPR, 19, 48, 49, 50, 61, 62
EPR paradox, 19, 48
equilibrium, 30, 219, 224, 226, 257, 286, 293, 306, 311, 313, 319, 324, 331, 332, 345, 359, 385, 388, 391, 435
ESR, 32
Etchegoin, 69
Euler-Lagrange equations, 185
evidence, ix, 30, 46, 47, 62, 235, 272, 274, 277, 278, 279
evolution, 38, 46, 47, 48, 49, 50, 52, 60, 61, 65, 68, 69, 70, 80, 85, 100, 133, 137, 141, 143, 157, 193, 211, 288, 318, 346, 394, 395, 397

exciton, ix, x, 183, 218, 219, 220, 221, 223, 233, 236, 237, 238, 239, 240, 241, 242, 243, 244, 245, 247, 249, 250, 251, 254, 255, 260, 261, 262, 263, 267, 268, 269, 270, 271, 273, 276, 393
exclusion, 69, 304
experimental condition, 20
experimental evidence, ix, 30, 62, 147, 160, 164, 272, 274, 279, 404, 453
experimental results, x, 1, 4, 6, 7, 21, 51, 63, 89, 93, 132, 139, 150, 153, 155, 157, 160, 233, 262, 270, 278, 308
external field, 32, 41, 44, 46, 60, 204, 249, 314, 323, 367, 370, 371, 372, 373, 374, 375, 450

F

Fann, 273, 281
Fermi, 23, 51, 57, 64, 65, 84, 199, 215, 231, 235, 276, 288, 300, 306, 319, 321, 427
fermions, 304
ferrite, 407
ferromagnetic, x, 407, 409, 410, 413, 414, 418, 419, 422, 426, 429, 431, 432, 433, 434, 436, 443, 448, 449, 452
ferromagnets, 302, 407, 433, 445
fiber(s), x, 89, 128, 131, 155, 156, 157, 158, 159, 160, 163, 218, 231
fiber optics, 131
field theory, 22, 68, 71, 342
filters, 55
financial, xii
financial support, xii
finite element method, 208
flight, 132
fluctuations, 21, 144, 146, 350, 352, 354, 355, 356, 357, 362, 365, 396
fluid, 21, 148, 149, 152, 300, 301, 311, 312, 374
fluorescence, 51, 52
Fock space, 142
Fock-Planck equation, 21
force, 8, 12, 23, 30, 154, 156, 214, 300, 311, 320, 322, 330, 331, 333, 334, 345, 347, 350, 390, 441
formation, 288, 301, 302, 308, 314, 315, 350, 351, 352, 389, 393, 395, 445, 448, 449, 450
formula, 2, 7, 30, 52, 65, 102, 129, 171, 176, 204, 260, 266, 267, 286
foundations, x, 26, 29, 75, 76, 84
free energy, 306, 366, 367, 384
free will, 22
freedom, 6, 38, 53, 67, 81, 142, 191, 192, 234, 285
freezing, 23, 350
friction, 60, 389
functional analysis, 85
fundamental hypotheses, 10, 18, 22, 75, 177
fundamental principles, xi, 74

G

Gahler, 63
gauge invariant, 71
gauge theory, 67
generalized nonlinear Schrödinger equation, 68, 74
generalized Sine-Gordon equation, 74
geometry, 55, 71, 85
Germany, 86, 232, 404
Ginzburg-Landau(GL) equation, 75
Gisin, 45, 46, 47, 50, 54, 58, 61, 62, 65, 83, 84, 85
G-L equation, 76, 368, 370, 381
glutamate, 280
glycerin, 152
Goldin, 60, 61, 62, 65, 85, 86
GPA, 312
graduate students, ix
graph, 302, 376
grass, 159
gravitation, 311
gravitational constant, 313
gravitational field, 70, 312
gravity, 70, 71, 149, 313, 318
Grigorenko, 70, 87
Gross-Pitaerskii equation, 288, 311
growth, 365

H

Haas, 55, 58, 84
Hamiltonian density, 37, 76, 77, 78, 101, 109, 110, 215, 288, 289, 296
Hamiltonian equation, 89, 108, 247, 264, 337
Hamiltonian function, 35, 36, 38, 42, 53, 54, 70, 174
Hamiltonian operator, 22, 23, 70, 76, 78, 167, 174, 175, 187, 188, 189, 191, 199, 219, 289, 333, 337, 385
Hamm, 274, 278, 279, 280, 282
Hartree-Fock, 23, 199
Hartree-Fock approximation, 23, 199
He atom beam, 1, 3, 4
healing, 302
height, 153, 212, 225, 228
Heisenberg, ix, 1, 10, 17, 25, 79, 141, 142, 145, 174, 175, 227, 240, 245, 386, 401, 402, 410, 411, 413, 414, 415, 426, 427, 429, 434, 436, 437, 438, 439, 442, 446, 448, 452, 453
Heisenberg equations, 240, 439

helium, 5, 23, 75, 270, 283, 289, 300, 301, 308, 309, 310, 311, 312, 313
Hermitian operator, 33, 166, 169, 173, 174, 187
Hilbert space, 30, 55, 59, 60, 69, 75, 166
history, xi, 17, 18
homogeneity, 36, 37, 38, 39, 51, 53, 56, 65, 73, 76, 77, 166
homogeneous non-linear Schrödinger equations, 37
host, 394
Hubbard model, 324, 344, 346, 349
Hubsch, 69
hydrides, 394
hydrogen, xi, 9, 12, 13, 14, 23, 51, 54, 63, 183, 211, 218, 223, 224, 225, 227, 232, 234, 276, 278, 279, 288, 303, 317, 329, 330, 331, 332, 334, 337, 338, 339, 340, 341, 342, 394
hydrogen atom(s), 9, 12, 13, 14, 23, 54, 63, 211, 225, 329, 330, 331, 332, 334, 337, 338, 339, 340, 341, 342
hydrogen bonds, 218, 223
hydrogenation, 14
hydrolysis, 218
hydroxyl, 224
hyperfine interaction, 45, 303
hypothesis, 14, 19, 60, 76, 167, 173, 174
hysteresis, 384, 401

I

ID, 352
ideal, 19, 20, 152
idealization, 61
identity, 34, 35, 46, 191, 207
image(s), 3, 4, 13, 318, 319, 374, 408, 409, 418
improvements, 17
impurities, 204, 249, 261
independence, 24, 29, 76, 167, 174, 334
independent variable, 108, 206, 413
industry, 9
inequality, 19, 21, 46, 47
Infrared absorption spectra, 271
Infrared spectra, 275
ingredients, 60
initial state, 10, 41, 46, 69, 100, 127, 251, 257, 318
institutions, 18
insulation, 432
integration, 44, 75, 93, 94, 116, 117, 121, 139, 310, 314, 317, 318, 319, 324, 345, 372, 433, 442, 444
interface, x, 89, 136, 137, 138, 368
interface energy, 368
interference, 1, 4, 6, 7, 20, 29, 69, 155, 270, 301, 303, 311, 323
invariants, 55, 66
inventors, 301
inversion, 52, 69, 416, 451
involution, 81
ion crystal, 389
ionization, 4
ions, 4, 5, 32, 44, 45, 52, 223, 225, 226, 382, 410, 413, 419, 429, 432, 451, 452
IR spectra, 276
IR spectroscopy, 276
Islam, 163
isotope, 303
issues, xi, 18, 21, 22

J

Japan, 230, 231, 281, 402
Jaynes, 51, 57, 70, 84
Jordan, 50, 67, 70, 83, 84, 86, 87, 427
Josephson junction, 350

K

kinetics, 235
kinks, 322
Korea, 87, 163

L

Lagrange density function, 76, 77, 117
Lagrange equation, 109, 110, 118, 185
Lagrangian density, 108, 110
Landau-Lifshitz equation, 412
large polaron, 383, 384
lasers, 156, 302
lasing effect, 313
lattices, 23, 198, 218, 219, 324, 344, 382, 393, 401
laws, x, 21, 50, 67, 70, 81, 89, 110, 111, 112, 115, 116, 119, 121, 126, 175, 180, 183, 185, 186, 243, 349, 449
Lax form, 80, 168, 192
Lax method, 78, 79, 168
lens, 273
Lie algebra, 34, 70
LiF, 1, 2, 3, 4
lifetime, 51, 64, 72, 233, 270, 273, 274, 276, 278, 392
light, xi, 7, 8, 9, 14, 19, 27, 49, 52, 57, 62, 65, 70, 86, 133, 140, 148, 155, 157, 158, 159, 160, 213, 218, 247, 250, 251, 252, 253, 254, 255, 257, 270, 300, 311, 366, 367
light transmission, 155

linear Schrödinger equation, 10, 11, 14, 17, 22, 23, 37, 63, 67, 75, 109, 180, 184, 188, 198, 199, 200, 201, 203, 211, 213, 220, 287, 329, 340, 433
liquid phase, 300
liquids, 151, 152, 284
lithium, 288, 289, 301, 302
localization, ix, x, 20, 30, 60, 64, 69, 98, 99, 147, 148, 151, 165, 185, 188, 213, 214, 215, 223, 243, 244, 247, 250, 324, 338, 350, 390, 394, 426
Localization, viii, 221, 392, 404
Lochan, 64, 86
logarithmic Schrödinger equation, 37, 71
low temperatures, 261, 293, 303, 304, 319, 323
luminescence, 14, 342, 382
lying, 265, 345, 426, 427

M

macroscopic quantum effects, 230, 375
macroscopic wave function, 73, 308, 319, 366
magnetic field, 41, 42, 45, 46, 70, 301, 303, 345, 366, 373, 374, 407, 409, 413
magnetic moment, 344
magnetization, 418, 451
magnets, 350, 407, 411, 412
magnitude, 19, 44, 45, 54, 60, 63, 68, 312, 359
Magnon, 326, 413, 436, 445
majority, 21, 30, 78, 175, 293, 300
Makhankov, 81, 83, 88, 100, 161, 206, 230, 326, 343, 344, 345, 346, 347, 349, 386, 387, 388, 402, 428, 438, 446, 453
Manakov, 83, 88
manifolds, 54, 55, 265
mapping, 48
Mark free-electron laser, 273
Markov chain, 21
materials, 136, 215, 216, 234, 235, 300, 311, 342, 344, 374, 394, 419
mathematics, 207
matrix, 17, 33, 40, 46, 47, 49, 66, 79, 80, 82, 129, 169, 175, 189, 191, 199, 226, 237, 259, 263, 264, 267, 345, 386, 396, 398
matter, 1, 4, 7, 8, 13, 17, 21, 24, 32, 44, 71, 86, 93, 98, 115, 116, 128, 188, 196, 211, 232, 283, 284, 285, 286, 309, 313, 323, 330, 334, 338, 344, 351, 383, 403, 454
mean-field theory, 319, 350
measurement(s), 18, 20, 21, 30, 38, 41, 43, 45, 49, 50, 55, 58, 60, 61, 62, 64, 66, 69, 70, 156, 166, 167, 177, 180, 204, 235, 270, 273, 274, 450
mechanical properties, ix, x, 228, 233, 270, 308, 329, 383
media, 111, 136, 137, 161, 323, 394
median, 213, 384
medicine, 9
melt, 273
membranes, 224
methanol, 276, 277
microcavity, 319, 323
micrometer, 323
microscope, 6, 7
Mielnik, 55, 56, 58, 60, 61, 65, 84, 85, 86
migration, 224, 384
Minelli, 71, 87
mixing, 66, 323, 395
model system, 323, 409
modern science, ix, xi, 1, 10, 31
modifications, 53, 58, 68
molecular orientation, 4
molecular structure, 157, 232, 233, 274, 276, 279, 451
molecules, 23, 53, 151, 152, 183, 191, 198, 211, 218, 219, 234, 235, 236, 238, 273, 274, 276, 283, 303, 304, 342, 343, 384, 398, 405
Moscow, 87, 229, 281, 325, 403, 404
motivation, 32, 63
multiples, 369
multiplication, 33, 41, 67
Mycielski, 37, 60, 63, 85

N

NaCl, 4
naphthalene, 394
National Bureau of Standards, 44
Nattermann, 59, 60, 61, 62, 85, 86
Nd, 155
neglect, 23, 256, 260, 341, 389, 416, 418
neutrons, 1, 4, 7, 63
new nonlinear quantum mechanics, ix, x, 29, 89, 165
nickel, 1, 2
Nikolic, 66, 86
nitrite, 451, 452
nitrogen, 234, 266, 451
NMR, 32, 429
Nobel Prize, 301
nodes, 323
nonlinear dynamics, 66, 208
nonlinear equations, 30, 61, 67, 339, 346
nonlinear Hamiltonian systems, 35
nonlinear optics, 288, 324
nonlinear quantum mechanical, ix, x, xi, 29, 48, 54, 58, 72, 75, 77, 89, 157, 175, 206, 228, 233, 266, 270, 273, 283, 289, 295, 300, 308, 344, 370, 407, 445
nonlinear quantum mechanical features, xi

nonlinear quantum mechanical ideas, ix, x, xi, 29, 75
nonlinear quantum mechanical model, 48
nonlinear quantum mechanical theory, 54, 72, 77, 89, 157, 175, 233, 273, 283, 289, 300, 407
nonlinear quantum system, 61, 74, 81, 105, 112, 136, 141, 185, 391
nonlinear superposition principle, 172
nonlinear systems, ix, 26, 61, 73, 110, 111, 161, 163, 164, 191, 401, 405, 424, 453
nonlinear wave theory, 29
normalization constant, 123, 227, 286, 439
nuclear spins, 303
nuclei, 5, 23, 41, 43, 198, 303, 343
nucleons, x, 23, 165, 331, 342
nucleus, 44, 45, 51, 53, 303

O

obstacles, 61
obstruction, 62
OH, 224
one dimension, 12, 71, 126, 288, 295, 318, 335, 340
operations, 55, 58
opportunities, 313
optical fiber, 99, 131, 148, 155, 157, 159, 160
optical lattice, 313, 323, 324
optical solitons, 156, 157, 162
orbit, 9, 71, 393, 398
oscillation, 127, 151, 323, 385, 388, 389, 408, 439, 449
oscillators, 7, 32, 226, 255, 256
overlap, 133, 315, 385
oxygen, 224, 225, 234, 451

P

pairing, 365
parallel, 123, 419, 445
parallel algorithm, 123
Pardy, 63, 86
parity, 195
partial differential equations, 61, 77
particle collisions, 126
particle physics, 405
peptide(s), 218, 219, 221, 223, 233, 235, 236, 237, 238, 239, 241, 242, 244, 245, 263, 270, 279
perchlorate, 451
Peres, 47, 54, 66, 83
periodicity, 23, 35
perylene, 394, 399
phase shifts, 131, 133, 134, 135, 380
phase transitions, 297
phenomenology, 402
phonon(s), vii, 7, 219, 236, 238, 244, 247, 248, 249, 251, 255, 259, 260, 261, 262, 267, 270, 273, 277, 280, 329, 344, 350, 353, 355, 356, 357, 358, 359, 361, 363, 365, 383, 384, 385, 389, 392, 393, 395, 396, 397, 435, 436, 449, 453
photon(s), ix, x, 1, 4, 7, 8, 9, 19, 21, 43, 45, 50, 52, 62, 89, 148, 157, 159, 160, 218, 251, 270, 287, 323, 324, 392, 394
photon soliton, x, 89, 218
photon-electric effect, 1, 7
physical features, 53
physical mechanisms, 54, 77
physical phenomena, 10, 20, 50, 228, 247, 323, 329, 366, 450
physical properties, 234
physical theories, 17, 55
Planck constant, 30, 367
plane waves, 15, 183, 216, 220, 307, 323
playing, 19
point charges, 13, 330
polar, 384, 410, 412
polarization, 21, 46, 156, 251, 256, 329, 385
Polaron, 382, 392, 394
Polchinski, 48, 49, 50, 83
polymers, ix, 211, 218
polypeptide(s), 234, 235, 278
Popova, 70, 71, 87
population, 52, 324
potential operator, 10, 22
preparation, 47
prima facie, 62
principles, ix, xi, 24, 25, 55, 71, 74, 167, 401
probability, 11, 14, 17, 18, 19, 20, 21, 43, 54, 58, 59, 60, 64, 66, 69, 70, 73, 77, 116, 123, 148, 166, 167, 251, 252, 286, 308, 324, 365, 393, 396, 399, 400, 401, 439, 446
probability distribution, 66, 286
probe, 138, 233, 273, 274, 275, 276, 277, 278, 279, 280
project, 415
propagation, 15, 16, 30, 63, 91, 92, 100, 121, 124, 127, 129, 132, 136, 137, 139, 145, 155, 157, 159, 160, 212, 213, 253, 302, 322, 408
proteins, 69, 218, 223, 269, 274, 278, 279, 280
protons, 4, 6, 224, 225, 226
prototype, 276, 350
PTT, 46
publishing, 25, 86, 228, 401
Pushkarov, 422, 426, 429, 453, 454

Q

QCD, 67
quanta, 7, 181, 182, 183, 184, 191, 236, 262, 300, 407
quantization, 9, 57, 66, 67, 70, 71, 139, 189, 190, 238, 264, 308, 311, 314, 369, 370, 385
quantum computer, 50
quantum dot(s), 324
quantum dynamics, 69, 350
quantum electrodynamics, 31, 65
quantum field theory, 55, 59, 82, 289
quantum fluctuation(s), 144, 146, 147, 184, 186, 351, 352, 357, 361, 362, 363, 365
quantum gravity, 68
quantum jumps, 51, 52
quantum mechanical framework, 14, 330
quantum optics, 84, 350, 352
quantum phenomena, 319
quantum potential, 30, 31
quantum state, 17, 47, 48, 50, 55, 65, 142, 183, 191, 285, 305, 314
quantum well, 319
quasiparticles, 215, 216, 301, 308, 309, 383, 395

R

Rabi frequency, 51, 57, 58
radiation, 1, 7, 21, 41, 44, 51, 53, 57, 64, 65, 137, 215, 229, 257, 270, 323, 324
radio, 44
radius, 7, 13, 330, 333, 388, 389
Raman spectra, 250, 270, 272, 394
reality, 33, 50, 204
reasoning, 136
reconstruction, 287
recovery, 270, 273
red shift, 236, 250
reference frame, 197, 217, 247, 249
refractive index, 138, 157, 159, 257
Reinisch, 65, 86
rejection, 66
relativity, 8, 22, 55, 62
relaxation, 273, 278, 390, 450
relevance, 68, 69, 352
renormalization, 352, 387, 388, 389, 390, 392
repulsion, 156, 215, 303, 318, 321, 322, 324
requirements, 71, 444
researchers, ix, xi, 156, 224, 285, 393
resistance, 301, 370, 371
resolution, 20, 273
response, 274, 276, 277, 278, 279
restrictions, 58
restructuring, 128
rings, 2, 311
Roberts, 71, 87, 311
room temperature, 234, 266, 277, 300, 302, 303
root, 44, 90
rotations, 35, 224, 234, 416
Royal Society, 25, 86
rubidium, 284, 285, 288, 301, 302, 303
rules, ix, 11, 53, 59, 67, 77, 83, 105, 116, 124, 125, 128, 131, 136, 154, 167, 168, 186, 213, 220, 225, 228, 241, 247, 268, 308, 344, 433, 436, 452, 453

S

saltwater, 153
saturation, 273, 433
scalar field, 215, 298
scaling, 54, 150, 325
scattering, x, 1, 4, 8, 9, 43, 60, 91, 92, 100, 124, 125, 128, 129, 136, 138, 142, 143, 157, 161, 186, 193, 202, 203, 230, 231, 233, 234, 247, 250, 251, 252, 253, 254, 260, 266, 270, 288, 303, 304, 305, 320, 321, 322, 339, 384, 429
scattering intensity, 254
school, 18
Schuch, 25, 66, 67, 86
science, 9, 10, 11, 161, 186, 188, 341, 342
scope, 18
seed, 69
self-consistency, 69
self-focus mechanism, 148, 160
self-interaction(s), x, 60, 70, 71, 165, 214, 215, 216, 316, 317, 318, 342
self-localization, x, 69, 151, 165
self-trapping mechanism, 218, 223, 233, 235, 242, 243, 329, 334, 397, 424
semiconductor(s), x, 99, 323, 329, 382, 383, 392
set theory, 55
shape, 52, 56, 100, 131, 132, 155, 159, 178, 182, 212, 247, 254, 273, 302, 311, 320, 371, 373, 374
showing, 69, 138
Shull, 38, 63, 83
side chain, 280
signals, 48, 62, 65, 133, 138
signal-to-noise ratio, 273
signs, 215, 274
silicon, 270
simulation(s), 123, 132, 134, 138, 186, 208, 311, 316, 318, 344
sine-Gordon equation, 169, 409
Singapore, 27, 83, 84, 85, 86, 161, 228, 229, 232, 281, 326, 453

Slater, 66, 86
smoothness, 254
sodium, 288, 301
solid state, 262
solitary wave, x, 77, 89, 92, 124, 131, 136, 148, 149, 151, 153, 154, 163, 184, 186, 211, 214, 308, 321, 374, 375, 390, 401, 432, 433, 444, 449
sound speed, 222
space-time, 62, 68, 77, 85, 91, 92, 105, 112, 126, 167, 336
special relativity, 54
special theory of relativity, 20
specific heat, 7, 233, 235, 266, 267, 268, 269, 270, 297, 450
spectroscopy, 233, 274, 276, 278
speech, 21
speed of light, 20, 157, 257
Spin, viii, 407, 408, 425
spin dynamics, 70
spinning particle, 41
square lattice, 419
stability, 99, 100, 102, 185, 206, 280, 322, 324, 332, 337, 344, 358, 361, 428, 432
stabilization, 365
standard deviation, 179
Stark effect, 323
statistics, 301, 325
stretching, 235, 242, 270
strong interaction, 388
structure, x, 22, 54, 58, 65, 68, 69, 70, 85, 100, 123, 131, 136, 156, 160, 165, 172, 205, 206, 214, 215, 223, 225, 226, 228, 242, 249, 274, 276, 279, 280, 300, 305, 314, 318, 320, 322, 332, 374, 392, 407, 409, 451, 452
substitution, 147, 235, 371
Sun, 281
superconductivity, xi, 72, 75, 87, 297, 300, 344, 366, 368, 370, 374, 379, 381
superconductor, x, 76, 289, 292, 293, 299, 300, 329, 366, 367, 368, 369, 370, 372, 373, 374, 375, 377, 380, 381
superfluid, 75, 288, 289, 300, 301, 303, 307, 308, 309, 310, 311, 312, 313, 323
superfluidity, 75, 300, 301, 307, 308, 309, 311, 313, 324
superlattice, 323, 324
supersonic soliton, 247
suppression, 211
surface tension, 150, 151, 152, 153
Svelichny, 62, 85
Svetlichny, 62, 68, 85, 86
Switzerland, 85
symmetry, 34, 40, 42, 66, 67, 69, 70, 71, 75, 77, 194, 195, 198, 283, 294, 295, 297, 298, 299, 314, 315, 320, 349, 372, 392, 409, 413, 416, 437, 439, 441, 447, 449, 452
synthesis, 55, 185

T

target, 5, 7, 31
teachers, ix
techniques, 67, 302, 305, 313
technology, 27, 405, 453
telephone, 46
temperature dependence, 255, 261, 262, 270, 274, 277, 278, 287, 380, 394
tension, 152, 153
testing, 47
the free electron and average field approximations, 23, 199
the Heisenberg uncertainty principle, 17
thermodynamics, 47, 48, 54, 66
Thomson, 1, 2, 25
time resolution, 275
time-dependent nonlinear Schrödinger equation, 35, 340
topology, 60
torus, 35
total energy, 101, 123, 390, 425
TPA, 85
tracks, 130, 452
trajectory, 13, 126
transformations, 34, 39, 58, 61, 66, 68, 71, 77, 82, 85, 117, 170, 202, 208, 230, 231, 352, 353, 388, 415, 420, 421, 430, 438
transition temperature, 286, 301
translation, 35, 78, 119, 120, 126, 224, 234, 387, 389, 390, 392, 425
transmission, x, 48, 89, 136, 137, 138, 139, 155, 159, 186, 270, 273, 375, 376
transport, 235, 273, 323, 394, 405
transportation, 382
traveling waves, 136
treatment, 48, 53, 249, 266, 318, 351
trial, 207, 312, 386, 435
tunneling, 69, 71, 136, 323, 350, 351, 352, 361, 365, 396
tunneling effect, 71, 136, 361, 365
turbulence, 311
twins, 284

U

Ulmer, 68, 71, 86, 87
uncorrelated ansatz, 363
unhappiness, 19
unification, 66, 395
uniform, 12, 41, 76, 91, 112, 113, 114, 136, 137, 151, 219, 285, 302, 306, 309, 336, 368, 371, 373, 381
USA, 231
USSR, 231, 325, 453

V

vacuum, 68, 142, 144, 157, 248, 252, 360, 362, 435, 439
Valentini, 68
variables, 20, 21, 30, 31, 41, 47, 56, 67, 70, 107, 108, 121, 146, 158, 202, 208, 217, 287, 333, 396, 427, 428, 435
variations, 63, 175, 176, 184, 287, 354, 385
vector, 30, 51, 53, 56, 57, 59, 62, 79, 82, 83, 91, 157, 158, 159, 166, 189, 191, 213, 251, 253, 256, 264, 330, 331, 349, 367, 381, 393, 410, 413
vector nonlinear Schrödinger equations, 82
vibration, 122, 218, 221, 226, 235, 237, 238, 242, 244, 245, 270, 273, 331, 332, 333, 334, 385, 388, 392, 398
viscosity, 301, 311

W

Walsworth, 53, 54, 84
water, x, 89, 148, 149, 150, 151, 152, 153, 154, 160, 212, 218, 223, 224, 232, 283
water soliton, x, 89, 150, 151, 152, 153, 154
wave and corpuscle features, 1, 167
wave feature, x, xi, 1, 2, 4, 7, 8, 9, 11, 13, 14, 17, 23, 24, 89, 93, 97, 124, 136, 139, 177, 186, 188, 212, 213, 219, 220, 330, 332, 336, 339, 417, 418, 422, 449
wave number, 241, 309
wave packet, 16, 19, 50, 91, 220, 313, 370
wave propagation, 212, 213
wave vector, 111, 136, 157, 158, 159, 213, 216, 217, 238, 250, 251, 256, 394
wavelengths, 2, 318
weak interaction, 452
Weinberg, v, x, xi, 29, 31, 32, 34, 35, 37, 38, 39, 40, 41, 42, 43, 44, 45, 46, 47, 48, 49, 50, 51, 53, 54, 57, 58, 60, 65, 66, 67, 68, 69, 72, 73, 75, 83, 84
wells, 224, 323, 372, 380
Wigner, 53, 84
windows, 273
Wodkiewicz, 51, 57, 58, 84
Wodkiewicz and Scully, 57, 58
workers, 218, 235, 236, 237, 278, 301, 436
World War I, 20

Y

yield, 43, 68, 179, 207, 211

Z

Zakharov and Faddeev, 82
Zakharov and Shabat, 80, 82, 91, 92, 124, 125, 126, 127, 128, 129, 130, 168, 203